Theoretical and Practical Advancements for Fuzzy System Integration

Deng-Feng Li
Fuzhou University, China

A volume in the Advances in Computational Intelligence and Robotics (ACIR) Book Series

IGI GLOBAL
DISSEMINATOR OF KNOWLEDGE

www.igi-global.com

Published in the United States of America by
IGI Global
Information Science Reference (an imprint of IGI Global)
701 E. Chocolate Avenue
Hershey PA, USA 17033
Tel: 717-533-8845
Fax: 717-533-8661
E-mail: cust@igi-global.com
Web site: http://www.igi-global.com

Library of Congress Cataloging-in-Publication Data

Names: Li, Deng-Feng, 1965- editor.
Title: Theoretical and practical advancements for fuzzy system integration / Deng-Feng Li, editor.
Description: Hershey, PA : Information Science Reference, 2017.
Identifiers: LCCN 2016045999| ISBN 9781522518488 (hardcover) | ISBN 9781522518495 (ebook)
Subjects: LCSH: Fuzzy systems. | System analysis--Mathematical models. | Decision making--Mathematical models.
Classification: LCC QA402 .T469 2017 | DDC 006.3/3--dc23 LC record available at https://lccn.loc.gov/2016045999

This book is published in the IGI Global book series Advances in Computational Intelligence and Robotics (ACIR) (ISSN: 2327-0411; eISSN: 2327-042X)

British Cataloguing in Publication Data
A Cataloguing in Publication record for this book is available from the British Library.

Advances in Computational Intelligence and Robotics (ACIR) Book Series

Ivan Giannoccaro
University of Salento, Italy

ISSN:2327-0411
EISSN:2327-042X

Mission

While intelligence is traditionally a term applied to humans and human cognition, technology has progressed in such a way to allow for the development of intelligent systems able to simulate many human traits. With this new era of simulated and artificial intelligence, much research is needed in order to continue to advance the field and also to evaluate the ethical and societal concerns of the existence of artificial life and machine learning.

The **Advances in Computational Intelligence and Robotics (ACIR) Book Series** encourages scholarly discourse on all topics pertaining to evolutionary computing, artificial life, computational intelligence, machine learning, and robotics. ACIR presents the latest research being conducted on diverse topics in intelligence technologies with the goal of advancing knowledge and applications in this rapidly evolving field.

Coverage

- Agent technologies
- Algorithmic Learning
- Artificial Intelligence
- Cyborgs
- Automated Reasoning
- Brain Simulation
- Computer Vision
- Adaptive and Complex Systems
- Natural language processing
- Synthetic Emotions

IGI Global is currently accepting manuscripts for publication within this series. To submit a proposal for a volume in this series, please contact our Acquisition Editors at Acquisitions@igi-global.com or visit: http://www.igi-global.com/publish/.

The Advances in Computational Intelligence and Robotics (ACIR) Book Series (ISSN 2327-0411) is published by IGI Global, 701 E. Chocolate Avenue, Hershey, PA 17033-1240, USA, www.igi-global.com. This series is composed of titles available for purchase individually; each title is edited to be contextually exclusive from any other title within the series. For pricing and ordering information please visit http://www.igi-global.com/book-series/advances-computational-intelligence-robotics/73674. Postmaster: Send all address changes to above address.

Titles in this Series

For a list of additional titles in this series, please visit: www.igi-global.com

Multi-Agent-Based Simulations Applied to Biological and Environmental Systems
Diana Francisca Adamatti (Universidade Federal do Rio Grande, Brazil)
Information Science Reference • copyright 2017 • 406pp • H/C (ISBN: 9781522517566) • US $205.00 (our price)

Strategic Imperatives and Core Competencies in the Era of Robotics and Artificial Intelligence
Roman Batko (Jagiellonian University, Poland) and Anna Szopa (Jagiellonian University, Poland)
Business Science Reference • copyright 2017 • 302pp • H/C (ISBN: 9781522516569) • US $185.00 (our price)

Emerging Research on Applied Fuzzy Sets and Intuitionistic Fuzzy Matrices
Amal Kumar Adak (Jafuly Deshpran High School, India) Debashree Manna (Damda Jr. High School, India) and Monoranjan Bhowmik (Vidyasagar Teacher's Training College, India)
Information Science Reference • copyright 2017 • 375pp • H/C (ISBN: 9781522509141) • US $205.00 (our price)

Multi-Core Computer Vision and Image Processing for Intelligent Applications
Mohan S. (Al Yamamah University, Saudi Arabia) and Vani V. (Al Yamamah University, Saudi Arabia)
Information Science Reference • copyright 2017 • 292pp • H/C (ISBN: 9781522508892) • US $210.00 (our price)

Developing and Applying Optoelectronics in Machine Vision
Oleg Sergiyenko (Autonomous University of Baja California, Mexico) and Julio C. Rodriguez-Quiñonez (Autonomous University of Baja California, Mexico)
Information Science Reference • copyright 2017 • 341pp • H/C (ISBN: 9781522506324) • US $205.00 (our price)

Pattern Recognition and Classification in Time Series Data
Eva Volna (University of Ostrava, Czech Republic) Martin Kotyrba (University of Ostrava, Czech Republic) and Michal Janosek (University of Ostrava, Czech Republic)
Information Science Reference • copyright 2017 • 282pp • H/C (ISBN: 9781522505655) • US $185.00 (our price)

Integrating Cognitive Architectures into Virtual Character Design
Jeremy Owen Turner (Simon Fraser University, Canada) Michael Nixon (Simon Fraser University, Canada) Ulysses Bernardet (Simon Fraser University, Canada) and Steve DiPaola (Simon Fraser University, Canada)
Information Science Reference • copyright 2016 • 346pp • H/C (ISBN: 9781522504542) • US $185.00 (our price)

Handbook of Research on Natural Computing for Optimization Problems
Jyotsna Kumar Mandal (University of Kalyani, India) Somnath Mukhopadhyay (Calcutta Business School, India) and Tandra Pal (National Institute of Technology Durgapur, India)
Information Science Reference • copyright 2016 • 1015pp • H/C (ISBN: 9781522500582) • US $465.00 (our price)

Table of Contents

Detailed Table of Contents

The aim of this paper is to propose a satisfactory degree method by using nonlinear programming for solving multi-attribute decision making (MADM) problems in which ratings of alternatives on attributes is expressed via interval-valued intuitionistic fuzzy (IVIF) sets and preference information on attributes is incomplete. Concretely, a nonlinear programming model is firstly explored to determine the satisfactory degree which is the ratio of the square of the weight Euclidean distance between an alternative and the IVIF negative ideal solution (IVIFNIS) to the sum of the square of the weight Euclidean distance between the IVIF negative ideal solution (IVIFNIS) and the IVIF positive ideal solution (IVIFPIS). Another nonlinear programming model is also developed to obtain satisfactory intuitionistic fuzzy sets, and then the general satisfactory degrees of the satisfactory intuitionistic fuzzy sets are used to generate the ranking order of the alternatives. Finally, a real example is employed to verify the applicability of the proposed approach and illustrate its practicality and effectiveness.

In this paper, an intuitionistic fuzzy (IF) distance measure between two triangular intuitionistic fuzzy numbers (TIFNs) is developed. The metric properties of the proposed IF distance measures are also studied. Then, based on the IF distance, an extended TOPSIS is developed to solve multi-attribute decision making (MADM) problems with the ratings of alternatives on attributes of TIFNs. In this methodology, the IF distances between each alternative and the TIFN positive ideal-solution are calculated as well as the TIFN negative ideal-solution. Then the relative closeness degrees obtained of each alternative to the TIFN positive ideal solution are TIFNs. Based on the ranking methods of TIFNs the alternatives are ranked. A numerical example is examined to the validity and practicability of the method proposed in this paper.

Logistics activities are performed in order to balance the operational chains of firms. The selection of the Third Party Logistics (3PL) is a challenging task for each organization, which involves various factors and attributes. The presented methodology acts as a boon and aids the decision makers for effectively choosing the appropriate Third Party Logistics (3PL) network. In the revealed work, the authors explored fuzzy sets theory and presented a fuzzy AHP model to facilitate the managers of organizations to deal

This paper considers the problem of developing test procedures for testing credibility hypotheses about the variance of fuzzy normal distribution assuming the expected values of the distributions mentioned under null and alternative credibility hypotheses are known and equal. The cases where the underlying hypothesis is simple and composite (one sided) are considered. Tests have been derived with the help of the membership ratio criterion. Properties possessed by the developed tests, like best credibility rejection region and uniformly best rejection region have been studied. Examples are also given to illustrate the usage of the derived tests.

Based on studies, access to healthcare services and long waiting time is one of the main issues in many countries including Canada and United States. Healthcare organizations can't increase their limited resources nor treat all patients simultaneously. Then, patients' access to these services should be prioritized in a way that best uses the scarce resources and insures patients' safety. Prioritization is essential and inevitable not only because of resource shortage, which have not been improved during years, but also because it is a crucial issue that could contribute to the capability and stability of the healthcare systems, and most importantly to patients' safety. On the other hand, inappropriate prioritization of patients waiting for treatment, could affect directly on inefficiencies in healthcare delivery, quality of care, and most importantly on patients' safety and their quality of life and satisfaction. Inspired by these facts, in this chapter the importance of patients' prioritization and using fuzzy logic in this area will be discussed, and a novel hybrid framework using fuzzy soft sets for patients' prioritization will be proposed. The proposed framework may have a significant impact on patients' safety, and on both medical community and the public's faith in justice and equity.

Amongst various proposed SC philosophies, Lean, Green and Agile strategies of SC worked seminal to solve many problems of firms. Performance measurement acts as a tool to quantify overall efficiency via G-L-A SC activities of firm, where Green SC agenda is to control and diminish pollution and Lean SC helps to reduce waste of firm, while Agile SC target to manage SC more cum quick responding to clients. This chapter proposed a MCDM performance appraisement module (module constituted by mixing the segregated chain of green-lean-agile logistic activities and corresponding their interrelated metrics) conjunctive with Fuzzy Performance Index model in purpose to estimation the overall performance of individual firm. Furthermore, a centroid method coupled with fuzzy number set is proposed for classifying ill and strong G-L-A metrics of firm, so that managers could escalate their firms performance in case of non desirable performance. A hypothetical case research of two firms i.e. gear and shaft manufacturing firms are shown to measure their performance under G-LA supply chain.

Nowadays, the use of renewable energy has become increasingly significant, and cost-effective. Between all existing sources of energy, solar and wind energies are the highly exploited. However, solar and wind energies are not available all the time and their performance is affected by unpredictable weather changes, therefore, it is not always feasible to obtain an accurate mathematical model of the controlled system. Various mathematical modeling methods were used to predict wind and solar powers using natural parameters but considering multiple parameters in equations makes the solution more complex. In addition to complexity, some coefficients are uncertain and based on probability. Fuzzy logic is a perfect tool to model any kind of uncertainty related to vagueness. This chapter presents a computer algorithm based on fuzzy logic control (FLC) to estimate the wind and solar energies using natural factors. As input parameters, the wind speed was used to predict the wind power and the temperature and the lightening were used to estimate the solar power.

Preface

In most real situations, such as economic management and engineering management, fuzziness (or vagueness uncertainty) is inherited in decision making problems and decision process. Fuzziness seems to be suitably dealt with the fuzzy set introduced by L.A. Zadeh and the intuitionistic fuzzy set (IFS) introduced by K.T. Atanassov. The latter is an important and a successful extension of the former through adding a non-membership degree (or function). Researchers and practitioners in disciplines and fields have acknowledged the significance of the fuzzy set and the IFS. Therefore, it is very important to express and measure fuzziness in real decision making problems and hereby develop effective and practical models and methods. In fact, in recent years, the fuzzy set and the IFS have been extensively explored and achieved a great success in many fields such as business, management, economics, energy and natural resource, medical treatment, environment, safety, and transportation. In order to comprehensively reflect the advance in the fuzzy set and the IFS, the book titled "Theoretical and Practical Advancements for Fuzzy System Integration" has been edited to collect the original research in this area.

Sixteen proposals for chapters for the book titled "Theoretical and Practical Advancements for Fuzzy System Integration" have been received and carefully reviewed. Thirteen chapter submissions are finally accepted to publish in the aforementioned book. In other words, this edited book has 13 chapters.

Chapter 1 titled "Fuzziness: A Mathematical Tool" is written by A. K. Sahu, N. K. Sahu, and A. K. Sahu. They pointed out decision information usually is uncertain, vague, and inconsistent. In this case, the decision information seems to be easily collected in terms of the fuzzy set, which is observed as an important and a powerful tool that aids numerous researchers and learners in a variety of ways and in countless disciplines by transforming oral information into mathematical data. The majority of the conventional tools for computing, analysis, and modeling only deal with the defined, deterministic, certain, and precise information. Chapter 1 gave an idea to the learners and researcher about the basic concepts of the fuzzy sets and fuzzy numbers and their applications. The main purpose of Chapter 1 is to impart fuzzy knowledge to the readers and new learners so that they augment and disseminate the fuzzy set theory into new application fields to benefit the society and mankind.

Chapter 2 titled "Some Recent Defuzzification Methods" is written by H. Kumar. Defuzzification is a process that transforms a fuzzy number into a crisp value (i.e., real number). In fuzzy modeling, fuzzy optimization, and fuzzy control, defuzzification is employed to transform the fuzzy outputs from the systems into crisp values due to the fact that all fuzzy sets inferred by fuzzy inference in the fuzzy rules have to be aggregated into one single number as the output of the fuzzy model. Currently, there are numerous defuzzification methods. Chapter 2 discussed some recent defuzzification techniques such as the sums' center method, the height method, the magnitude method, the middle point method, and the Robust's ranking method as well as their applications.

Chapter 3 titled "Interval-Valued Intuitionistic Fuzzy Multi-Attribute Decision Making Based on Satisfactory Degree" is written by G. F. Yu, D.-F. Li, and J. M. Qiu. They focused on how to solve multi-attribute decision making problems in which ratings of alternatives on attributes are expressed with interval-valued IFSs and preference information on attributes is incomplete. Chapter 3 constructed a nonlinear programming model based on the idea of the determined intuitionistic fuzzy satisfactory degree which is the ratio of the square of the weighted Euclidean distance between an alternative and an interval-valued intuitionistic fuzzy negative ideal solution to the sum of the square of the weighted Euclidean distance between the interval-valued intuitionistic fuzzy negative ideal solution and the interval-valued intuitionistic fuzzy positive ideal solution. The general satisfactory degrees of alternatives are obtained. A new ranking method is developed to generate the order of the alternatives. The proposed model and method are illustrated with a real example.

Chapter 4 titled "Intuitionistic Fuzzy Distance-Based Intuitionistic Fuzzy TOPSIS Method and Application to MADM" is written by J. X. Nan, T. Wang, and J. J. An. In Chapter 4, the intuitionistic fuzzy distance is defined to measure difference between two triangular intuitionistic fuzzy numbers (TIFNs) and some important and useful metric properties are also discussed. Based on the concepts of the defined intuitionistic fuzzy distance and the Technique for order Preference by Similarity to Ideal Solution (TOPSIS), a new extended TOPSIS is developed to solve multi-attribute decision making (MADM) problems in which the ratings of alternatives on attributes are expressed with TIFNs. In this method, the intuitionistic fuzzy distances between alternatives and the TIFN positive ideal-solution as well as the TIFN negative ideal-solution are calculated. Then, the relative closeness degrees of alternatives to the TIFN positive ideal solution can be obtained. The relative closeness degrees are also TIFNs, which are used to generate the ranking order of alternatives according some ranking method of TIFNs. The proposed intuitionistic fuzzy distance and extended TOPSIS can be applicable to MADM problems in which the ratings of alternatives on attributes are expressed with intuitionistic fuzzy numbers, particularly trapezoidal intuitionistic fuzzy numbers.

Chapter 5 titled "Fuzzy-AHP: A Boon in 3PL Decision Making Process" is written by N. K. Sahu, A. K. Sahu, and A. K. Sahu, who discussed a kind of selection problems of the third party logistics for organizations (or firms). The third party logistics' selection which involves various factors and attributes is a challenging problem. Chapter 5 explored the fuzzy set theory and presented a fuzzy Analytical Hierarchy Process (AHP) model to facilitate the managers of organizations to deal with the third party logistics' decision making problems. The overall performances of the third party logistics' service providers are greatly influenced by many significant factors/attributes such as quality, reliability, service assurance, shipment cost, and customer relationship. In the constructed fuzzy AHP model, service level, financial security capabilities, location, global presence, relationship management, and client fulfillment are chosen to represent the first level's indices. The aforementioned factors have various sub-factors/attributes which are chosen to represent the second level's indices. Various researchers have constrained their works to the first level indices and have not considered the second level indices, which is a crucial part of today's practical decision making process for the third party logistics' selection. Hereby the AHP accompanied with the fuzzy set theory is used to solve industrial Logistics problems. The purpose of Chapter 5 is to develop a fuzzy based AHP method for solving benchmarking (preference ranking orders of alternatives under criteria) problems. The proposed method facilitates the managers of firms to make the verdict towards choosing the best third party logistics' service provider(s).

Chapter 6 titled "Intuitionistic Fuzzy Set Theory with Fair Share CPU Scheduler: A Dynamic Approach" is written by S. Raheja, who mainly discussed fair share CPU scheduler, which equally/statically distributes the processor time between groups, users and tasks, as opposed to the fixed distribution among tasks (round Robin schedulers). Fairness should be the major consideration for any scheduler. Traditional schedulers considered that every task belongs to a different user. Therefore, schedulers like round Robin logically distribute equal CPU services/processor time to all tasks. When several tasks need a processor, operating system often allots processor time slices on a round Robin fashion, i.e., scheduler distributes equal time to all tasks having same priority. However, the static approach may result in the distribution of small processor time to a busy group slows down the performance of a system. The main objective of Chapter 6 is to propose a dynamic approach to distributing the processor time of fair share schedulers based on the IFS theory. The fair share scheduler contains three modules. The first module is the intuitionistic fuzzy inference system IFIS-DCS, the second module is the intuitionistic fuzzy inference system IFIS-DP, and the third module is the scheduling algorithm. The proposed method can deal with the impreciseness and inexactness of tasks, dynamically shares processor time among groups, users as well as tasks, and provides dynamic priority to each task of active users. Moreover, the proposed method also improves the performance of system over the fair share scheduler in terms of average response time, average normalized turnaround time, and number of context switches.

Chapter 7 titled "MAGDM Problems with Correlation Coefficient of Triangular Fuzzy IFS" is written by J. P. Robinson and E. C. Henry Amirtharaj. The concepts of correlation coefficients of IFSs, interval-valued IFSs, triangular IFSs, and trapezoidal IFSs are very important and have been presented in many fields. Chapter 7 proposed the correlation coefficient of triangular fuzzy IFSs (TFIFSs) and developed a new method for solving uncertain multiple attribute group decision making (MAGDM) problems with TFIFSs. In this method, the triangular fuzzy intuitionistic fuzzy ordered weighted averaging (TFIFOWA) operator and the triangular fuzzy intuitionistic fuzzy ordered weighted geometric (TFIFOWG) operator are proposed. Hereby the new decision making models are constructed to deal with the decision situations where expert weights are completely unknown. Also some comparisons are made with existing ranking methods.

Chapter 8 titled "Credibility Hypothesis Testing of Variance of Fuzzy Normal Distribution" is written by S. Sampath and B. Ramya, who considered the problem of test procedures for testing credibility hypotheses about the variance of fuzzy normal distribution assuming that the expected values of the distributions mentioned under null and alternative credibility hypotheses are known and equal *a priori*. Chapter 8 discussed some cases where the underlying hypothesis is simple and composite (one sided). The tests have been derived with the help of the membership ratio criterion. Some important and useful properties possessed by the developed tests such as the best credibility rejection region and uniformly best rejection region have been studied. The validity and feasibility of the derived tests are illustrated with several examples.

Chapter 9 titled "Application of Fuzzy Soft Set in Patients' Prioritization" is written by S. A. Rahimi, who is concerned access to healthcare services and long waiting time which is one of the main issues in OECD countries including Canada and United States. Healthcare organizations (or hospitals) cannot increase their limited resources nor treat all patients simultaneously. As a result, patients' access to these services should be prioritized so that the scarce (or limited) resources are used in the best way. Hereby in patients' prioritization problems there appear some challenging aspects such as multiple conflicting criteria, incomplete and imprecise data, associated risks that threaten patients on waiting lists, uncertainties in clinicians' decisions, decision maker group's opinions, and system's dynamic behavior. Inappropriate prioritization of patient waiting for treatment may directly affect on inefficiencies in healthcare delivery, quality of care, and most importantly on patients' safety and their satisfaction. Inspired by these facts,

Chapter 9 discussed the importance of using fuzzy logic in medical area and proposed a novel hybrid framework to prioritize patients based on the fuzzy soft set.

Chapter 10 titled "Performance Estimation of Firms by G-L-A Supply Chain under Imperfect Data" is written by A. K. Sahu, N. K. Sahu, and A. K. Sahu. In recent decade, for supply chain management, many methods were proposed to solve various problems of industrial sectors, such as assessment of resiliency of supplier against disasters, waste management equipments availability in manufacturing firms, green performance, and responses of firms towards their clients. Performance measurement is utilized as a tool to quantify overall efficiency via the green, lean, and agile (G-L-A) supply chain activities of firms, where the green supply chain agenda is to control and diminish pollution and the lean supply chain helps to reduce the waste of firms, while the agile supply chain manages to respond to clients as quick as possible. Chapter 10 proposed a multi-criterion decision making performance appraisement module (constituted by mixing the segregated green, lean and agile supply chain measures and their corresponding interrelated metrics) conjunctive with fuzzy performance index model, which is used to evaluate the overall performance of firms. In addition, an incentre of the centroid method is proposed to classify ill and strong G-L-A metrics of firms so that managers could raise their firms' performance to standard levels.

Chapter 11 titled "Appraise the Economic Values of Logistic Handling System under Mixed Information" is written by A. K. Sahu, N. K. Sahu, and A. K. Sahu, who considered the problems of quantifying the economic values of material handling systems. In recent years, assessments of firm (or corporation) logistic activities are found as a historic topic in real industrial management. Each parameter (or factor, attribute) leads the trait role in relation to materialize the economic values of material handling systems, which may result in the so-called a kind of multi-criteria decision making problems with observed data rather than verbal data. Thus, it is observed that these problems are resolved with less computation task by exploration of new effective algorithmic techniques. Chapter 11 proposed a new multi-level multi-criterion decision making algorithmic technique for appraising the economic values of material handling systems (i.e., candidates or alternatives) under subjective (linguistic information provided by personnel) or objective or subjective cum objective information. The proposed technique can be implemented to resolve miscellaneous quandaries of industries, such as performance measurement of supply chain management, benchmarking of advanced manufacturing technologies, port selection, shipping logistic route evaluation, layout evaluation, and banking sector evaluation. Chapter 11 displayed a flexible decision support system for quantifying the economic values of material handling system alternatives.

Chapter 12 titled "Benchmarking of Advanced Manufacturing Machines Based on Fuzzy-TOPSIS Method" is written by A. K. Sahu, A. K. Sahu, and N. K. Sahu, who studied the evaluation and management of advanced manufacturing machines. It is observed that the evaluation of the most appropriate computer numerical control (CNC) machine tool is the foremost issue for sustaining the advanced manufacturing production units at the competitive global market place. In last years, many evaluation parameter/criteria/attribute such as the cost, tool length, tool diameter, metal removal rate, and X-Y-Z-axis movement of tool are taken into account except the productivity, precision, accuracy, and reliability prior to purchasing/electing CNC machine tools among the preferred alternatives. Chapter 12 developed a new extended TOPSIS algorithmic technique with trapezoidal fuzzy numbers for appraising and benchmarking the preferable CNC machine tools (i.e., options or alternatives). In this method, the qualitative and quantitative parameter/criteria/attributes are considered, where subjective assessments of experts' panel (the decision making group) on qualitative criteria are integrated into electing the best CNC machine tool. The proposed method in Chapter 12 may solve some real life decision making problems such as appraisement, evaluation, and benchmarking of cars, buses, trucks, plant layouts, aero planes, material handling equipments, and robots.

Chapter 13 titled "Prediction of Solar and Wind Energies by Fuzzy Logic Control" is written by S. Faquir, A. Yahyaouy, H. Tairi, and J. Sabor, who presented a study of photovoltaic panels and wind turbines and proposed a simulator based on fuzzy logic control (FLC) to estimate the wind and solar powers under any weather conditions. These two energy sources are very useful renewable energies that can be combined together in one hybrid system to increase system's efficiency. The choice of a mathematical model to determine the output power of each source depends on different parameters. Considering lots of parameters and different weather conditions makes the mathematical modeling too complex to be implemented. Chapter 13 proposed the fuzzy modeling to predict the photovoltaic and wind turbine powers from natural factors, temperature, lighting, and wind speed and considering any weather conditions. Results obtained from previous work were compiled and used to determine the values of the fuzzy variables' intervals. The photovoltaic and wind turbine power values predicted from the fuzzy modeling were compared to the data obtained from the simulation and the difference error was very small. The obtained fuzzy modeling values can be used as input values for a hybrid renewable system composed of photovoltaic panels, a wind turbine, and storage units such as batteries that can be implemented to supply a load demand for a site.

This edited book aims to provide relevant theoretical frameworks and the latest empirical research findings in the area of fuzzy sets, IFSs, and fuzzy systems. It is written for professionals who want to improve their understanding of fuzzy sets and IFSs which are applied to different fields under the background of the global economy and networks economy. Also this edited book aims to develop new theory and methods of fuzzy sets, IFSs, and their systems and extend the application ranges or fields.

It is addressed to people in theoretical researches and practical applications from different fields and disciplines such as:

- Decision science,
- Game theory,
- Management science,
- Operational research,
- Fuzzy sets or fuzzy mathematics,
- Applied mathematics,
- Industrial engineering,
- Finance,
- Applied economics,
- Expert system, and
- Social economy as well as artificial intelligence.

Moreover, it is also addressed to teachers, postgraduates, and doctors in colleges and universities in different disciplines or majors:

- Decision analysis,
- Management,
- Operation research,
- Fuzzy mathematics,

- Fuzzy system analysis,
- Applied mathematics,
- Systems engineering,
- Project management,
- Supply chain management,
- Industrial engineering,
- Applied economics, and
- Hydrology and water resources.

Finally, the editor would like to thank all the authors (contributors) of this edited book and all the reviewers who cost their valued times and energies to review this edited book and give excellent comments and constructive suggestions.

Deng-Feng Li
Fuzhou University, China
September 20, 2016

Chapter 1
Fuzziness:
A Mathematical Tool

Atul Kumar Sahu
Guru Ghasidas Central Vishwavidyalaya, India

Nitin Kumar Sahu
Guru Ghasidas Central Vishwavidyalaya, India

Anoop Kumar Sahu
J.K.I.E., India

ABSTRACT

The fuzzy sets theory is observed as an important and powerful tool, which aids numerous researchers and learners in a variety of ways and in countless disciplines by transforming oral information into mathematical data. The majority of our conventional tools for computing, analysis and modeling are based on clear cut, defined, deterministic and precise information (generally crisp in nature) and concentrate on true yes or true no type information rather than more or less type information. It is very easy to deal with the cases, where the statement refers true yes or true no type information, however it is critical to deal with more or less type of data information, which can be easily overcome by fuzzy sets theory. The presented chapter tries to give an idea to the learners and researcher about: the basic concepts of fuzzy, what is fuzzy and fuzzy number, what are the application fields of the developed fuzzy theory, how to import fuzzy data, what is linguistic and importance of linguistic in fuzzy theory, what are the existing developed techniques that integrate the fuzzy theory effectively, etc. The main objective of the presented chapter is to impart fuzzy knowledge to the new learners and readers, so that they augment and disseminate the fuzzy sets theory into new application fields to benefit the society and mankind.

AT A GLANCE

In today's era, majority of the work fields require decisions, which are pertaining to information which is unclear, vagueness, inconsistent. In this type of situation, information can be easily collect in terms of fuzzy scale. This chapter addresses the significance, necessity, use and applicability of fuzzy sets theory.

DOI: 10.4018/978-1-5225-1848-8.ch001

INTRODUCTION

The fuzzy sets theory was introduced by L. A. Zadeh (1965), to handle decision making problems, where a sense of vagueness, unclear, inaccurate and impreciseness information or data set is present. It has always been considered as a natural oral language of the decision maker to model the approximate situations or circumstances, because the fuzzy information can be easily generated by using linguistic scale. Fuzzy phenomena and indices exist in almost every work field's i.e.:

- Manufacturing,
- Decision theory,
- Quality,
- Computer science,
- Operations research information management,
- Robotics,
- Inspection,
- Artificial intelligence production control,
- Scheduling etc.

A small decision in the aforesaid fields can improve the performance and gain of such fields, but the basic problem to take small decision in these fields lies in collecting data information and data analysis, which can be easily diagnosed by collecting data information via exploring fuzzy sets theory. The presented work reveals the generation and applicability of fuzzy sets theory in different application fields. Li and Chang (2002) introduced the concept of fuzzy logic in game theory and proposed a computational method for auxiliary fuzzy linear programming for each player. Mula et al. (2015) proposed the concept of bi rough programming to measure the uncertainty of the variable. Fuzzy logic can effectively deals with the concept of completely true, completely false and partial information (where the gathered information may range between completely true and completely false). The information can be easily collected by fuzzy number. For example if someone wants to collect fuzzy information in case of a glass of water which is fully filled with water. The fuzzy number assigned should be 1, because it is completely filled with water, which represents the completely true information case and the assigned fuzzy number should be 0, to represents completely false information i.e. glass is empty, because the glass is completely filled with water.

Fuzzy logic also found their extended application where the captured information is neither completely true nor completely false or consists of partly true and partly false information. Suppose the water is filled more than half but less than completely full in the glass. In this case, the assigned fuzzy number can be 0.6, 0.7, 0.8, 0.9, which depends on the capability and analytical view of the decision maker. Fuzzy logic is an idea, which requires further extension, explanation and elaboration to understand the significance and applicability. Fuzzy sets theory can effectively handle the ambiguous and the vague property involved in the captured data via linguistic expression (Zadeh, 1975). As the complexity of a system increases, our ability to make precise and significant statements about its behavior diminishes. In this type of situations, the theories of probability are the principles theories to model the uncertainties of reality. Yu et al. (2015) highlights the role and the importance of fuzzy information aggregation operators in supplier selection application under group decision making (GDM). Li (2014) proposed a series of representation theorem, extension principle, ranking methods, linear and non-linear programming methods for fuzzy multi-attribute decision making and fuzzy matrix games. Sahu et al. (2016c) implemented a revised ranking approach accompanied with fuzzy performance important index to identify ill and strong measures in Agile Supply Chain Management (ASCM). Fuzzy numbers are fuzzy sets which can facilitate easy calculations and are employed as a tool and methodology on account of following several specific reasons:

1. It deals with the simplify information, knowledge and representation.
2. It can be easily applied to imprecise/ambiguous/unclear/contradictory inputs.
3. It is capable to handle the case of conflicting objectives,
4. It deals with linguistic terms which facilitated the individual thinking, feeling and idea to output in linguistic terms and make it easily understandable.

TYPES OF FUZZY NUMBERS

The following fuzzy numbers are generally used to characterize and to convert the captured data for analysis.

Interval Valued Fuzzy Numbers Sets (IVFNS$_S$)

According to Gorzalczany (1987), an IVFN is defined on $(-\infty, +\infty)$ and is given by:

$$A = \left\{[(x, \mu_A{}^L(x), \mu_A{}^U(x)])\right\} \tag{1}$$

where

$$\mu_A{}^L, \mu_A{}^U: X \rightarrow [0,1] \qquad \forall x \in X,\ \mu_A{}^L \leq \mu_A{}^U,$$

$\mu_A{}^L$ is the lower limit of degree of membership and $\mu_A{}^U$ is the upper limit of membership degree.

For the given two interval-valued fuzzy numbers $N_x = [N_x{}^-, N_x{}^+]$ and $M_x = [M_x{}^-, M_x{}^+]$, the following definitions are prominent:

Definition 1

The multiplication between two interval-valued fuzzy numbers $N_x = [N_x{}^-, N_x{}^+]$ and $M_x = [M_x{}^-, M_x{}^+]$

$$N_x \ominus M_x = [N_x{}^-. M_x{}^-, N_x{}^+. M_x{}^+]$$

Definition 2

The subtraction and division operations between two interval-valued fuzzy numbers $\tilde{N}$ and $\tilde{M}$ which are triangular in nature are as follows (Kuo, 2011):

$$\tilde{N} - \tilde{M} = [(N_1, N'_1); N_2; (N'_3, N_3)] - [(M_1, M'_1); M_2; (M'_3, M_3)]$$

$$[(N_1 - M_3, N'_1 - M'_3); N_2 - M_2; (N'_3 - M'_1, N_3 - M_1)]$$

$$\tilde{N} \div \tilde{M} = [(N_1, N'_1); N_2; (N'_3, N_3)] \div [(M_1, M'_1); M_2; (M'_3, M_3)]$$

$$[(N_1 \div M_3, N'_1 \div M'_3); N_2 \div M_2; (N'_3 \div M'_1, N_3 \div M_1)]$$

Definition 3

The intersection of two IVFNSs (Gorzalczany, 1987), is defined as the minimum of their respective lower and upper bounds of their membership intervals. For the given two intervals of $[0,1]$ and

$$N_x = [N_x^-; N_x^+] \subset [0,1], M_y = [M_y^-; M_y^+] \subset [0,1],$$

the minimum of both intervals is $K = Min(N_x, M_y) = [Min(N_x^-; M_y^-), Min(N_x^+; M_y^+)]$.

Definition 4

The union of two IVFNSs (Mousavi et al., 2012) is defined as the maximum of their respective lower and upper bounds of their membership intervals. For the given two intervals of $[0,1]$ and

$$N_x = [N_x^-; N_x^+] \subset [0,1], M_y = [M_y^-; M_y^+] \subset [0,1],$$

the maximum of both intervals is

$$K = Max(N_x, M_y) = [Max(N_x^-; M_y^-), Max(N_x^+; M_y^+)]$$

Definition 5

Absolute value: $\left|N_x\right| = Max\left\{\left|N_x^-\right|, \left|N_x^+\right|\right\}$.

Figure 1. Interval-valued Fuzzy number (Triangular in Nature)

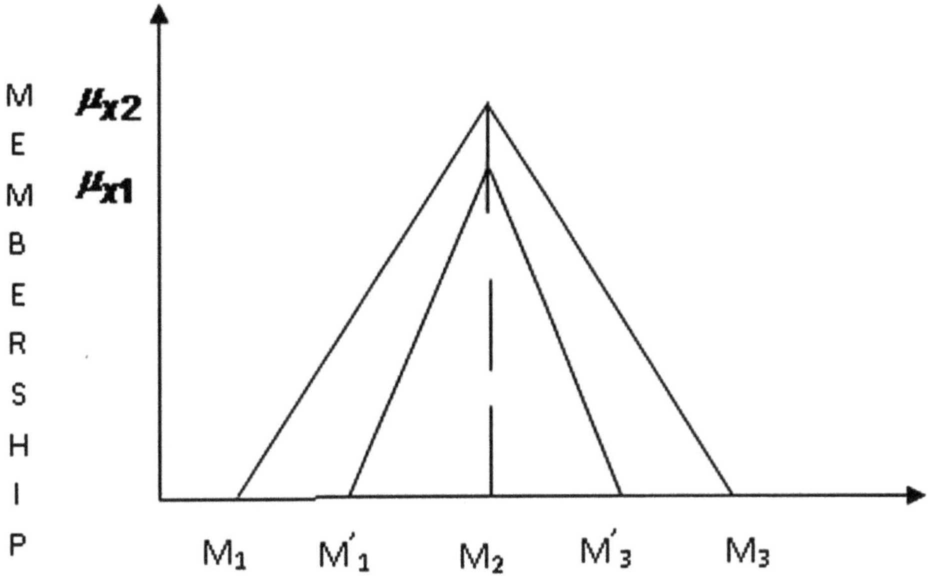

Definition 6

Let $\tilde{N}$ and $\tilde{M}$ be two interval-valued fuzzy numbers $\tilde{N}=[N_1,N'_1);N_2;(N'_3,N_3)]$ and $\tilde{M}=[M_1,M'_1);M_2;(M'_3,M_3)]$ as shown in Figure 1, which are triangular in nature can then be represented as follows:

$$h(\tilde{N})=\frac{N_1+N'_1+N_2+N'_3+N_3}{6}, \quad (2)$$

$$h(\tilde{M})=\frac{M_1+M'_1+M_2+M'_3+M_3}{6}, \quad (3)$$

Triangular Fuzzy Numbers

The concept of triangular fuzzy sets theory was introduced by Zadeh (1965), which was considered as a language of modeling to estimate the behavior of the situations, where lack or unclear information is available. In a universe X, a fuzzy subset A of universe X is defined by a membership function $f_A(x)$, representing a mapping between element in the closed interval [0, 1], where the value of $f_A(x)$ for the fuzzy set A is called a membership value, which represents the extent of X in the fuzzy set A (Figure 2).

Figure 2. Triangular Fuzzy number

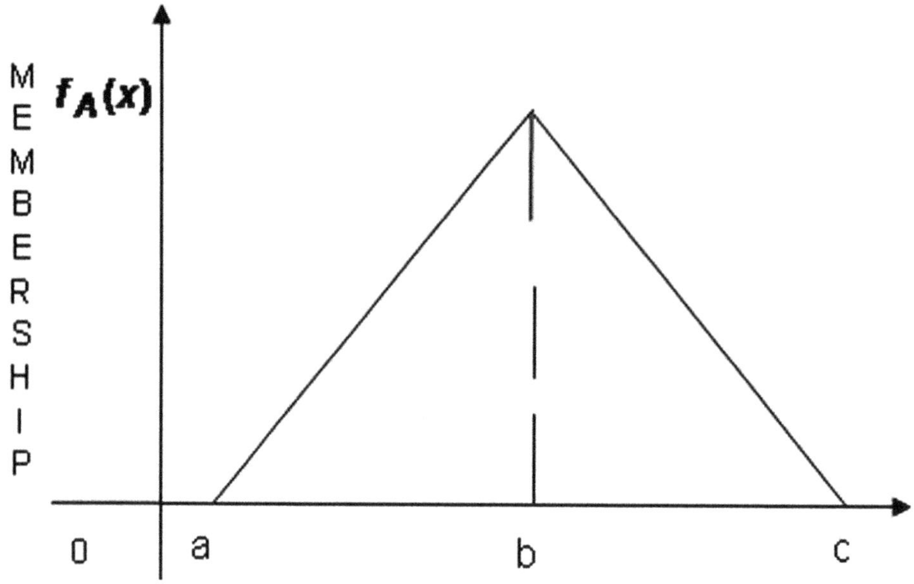

If a fuzzy set A on the universe R of real numbers satisfies the following conditions, we call it a fuzzy number (Zadeh, 1965).

1. A is a convex fuzzy set;
2. there is only one x_0 that satisfies $f_A(x_0) = 1$; and
3. $f_A(x)$ is continuous in an interval.

Let $\tilde{B} = (a, b, c)$ such that $a \le b \le c$ a, be a fuzzy set on $R = (-\infty, \infty)$ as shown in Figure 2. It is called a triangular fuzzy number, if its membership function is

$$\mu_{\tilde{B}}(x) = \begin{cases} \dfrac{x-a}{b-a}, \ if \ a \le x \le b \\ \dfrac{c-x}{c-b}, \ if \ b \le x \le c \\ 0, \quad otherwise \end{cases}$$

The triangular fuzzy number $\tilde{B} = (a, b, c)$ can be considered as trapezoid fuzzy number (a, b, b, c). Li (2012) utilized triangular fuzzy numbers and developed a valuable methodology for solving matrix games with payoffs expressed by triangular fuzzy numbers. As per the extension principle, we can derive the various operations of Triangular Fuzzy Number as (Zadeh, 1975; Tanaka, 1997):

$$\tilde{a} \oplus \tilde{b} = \left(a_1, a_2, a_3\right) \oplus \left(b_1, b_2, b_3\right)$$
$$= \left(a_1 + b_1, a_2 + b_2, a_3 + b_3\right) \tag{4}$$

$$\tilde{a} - \tilde{b} = \left(a_1, a_2, a_{3-}\right) - \left(b_1, b_2, b_3\right)$$
$$= \left(a_1 - b_1, a_2 - b_2, a_3 - b_3\right) \tag{5}$$

$$\tilde{a} \otimes \tilde{b} = \left(a_1, a_2, a_3\right) \otimes \left(b_1, b_2, b_3\right)$$
$$= \tilde{a} \otimes \tilde{b} = \left(a_1 \times b_1,\ a_2 \times b_2,\ a_3 \times b_3\right) \tag{6}$$

$$\frac{\tilde{a}}{\tilde{b}} = \frac{\left(a_1, a_2, a_3\right)}{\left(b_1, b_2, b_3\right)} = \left(\frac{a_1}{b_3}, \frac{a_2}{b_2}, \frac{a_3}{b_1}\right) \tag{7}$$

The Generalized Trapezoidal Fuzzy Numbers

A fuzzy set $\tilde{A}$ in a universe of discourse X is characterized by a membership function $\mu_{\tilde{A}}\left(x\right)$ which associates with each element x in X, a real number in the interval [0, 1]. The function value $\mu_{\tilde{A}}\left(x\right)$ is termed as the grade of membership of x in $\tilde{A}$. A trapezoidal fuzzy number can be defined as $\tilde{A} = \left(a_1, a_2, a_3, a_4; w_{\tilde{A}}\right)$ as shown in Figure 3 and the membership function $\mu_{\tilde{A}}\left(x\right) : R \rightarrow \left[0,1\right]$ is defined as follows:

$$\mu_{\tilde{A}}\left(x\right) = \begin{cases} \dfrac{x - a_1}{a_2 - a_1} \times w_{\tilde{A}}, & x \in \left(a_1, a_2\right) \\ w_{\tilde{A}}, & x \in \left(a_2, a_3\right) \\ \dfrac{x - a_4}{a_3 - a_4} \times w_{\tilde{A}}, & x \in \left(a_3, a_4\right) \\ 0, & x \in \left(-\infty,\ a_1\right) \cup \left(a_4,\ \infty\right) \end{cases} \tag{8}$$

Here, $a_1 \leq a_2 \leq a_3 \leq a_4$ and $w_{\tilde{A}} \in (0,1)$

Suppose $\tilde{a} = \left(a_1, a_2, a_3, a_4; w_{\tilde{A}}\right)$ and $\tilde{b} = \left(b_1, b_2, b_3, b_4; w_{\tilde{B}}\right)$ are two trapezoidal fuzzy numbers, then the operational rules of the trapezoidal fuzzy numbers $\tilde{a}$ and $\tilde{b}$ are shown as follows:

$$\tilde{a} \oplus \tilde{b} = \left(a_1, a_2, a_3, a_4; w_{\tilde{A}}\right) \oplus \left(b_1, b_2, b_3, b_4; w_{\tilde{B}}\right) =$$

$$\left(a_1 + b_1, a_2 + b_2, a_3 + b_3, a_4 + b_4; \min\left(w_{\tilde{A}}, w_{\tilde{B}}\right)\right) \tag{9}$$

$$\tilde{a} - \tilde{b} = \left(a_1, a_2, a_3, a_4; w_{\tilde{A}}\right) - \left(b_1, b_2, b_3, b_4; w_{\tilde{B}}\right) =$$

$$\left(a_1 - b_4, a_2 - b_3, a_3 - b_2, a_4 - b_1; \min\left(w_{\tilde{A}}, w_{\tilde{B}}\right)\right) \tag{10}$$

$$\tilde{a} \otimes \tilde{b} = \left(a_1, a_2, a_3, a_4; w_{\tilde{A}}\right) \otimes \left(b_1, b_2, b_3, b_4; w_{\tilde{B}}\right) =$$

$$\tilde{a} \otimes \tilde{b} = \left(a_1 \times b_1,\ a_2 \times b_2,\ a_3 \times b_3,\ a_4 \times b_4; \min\left(w_{\tilde{A}}, w_{\tilde{B}}\right)\right) \tag{11}$$

$$\tilde{a} / \tilde{b} = \left(a_1, a_2, a_3, a_4; w_{\tilde{A}}\right) \Big/ \left(b_1, b_2, b_3, b_4; w_{\tilde{B}}\right)$$

$$= \left(a_1 / b_4, a_2 / b_3, a_3 / b_2, a_4 / b_1; \min\left(w_{\tilde{A}}, w_{\tilde{B}}\right)\right) \tag{12}$$

Li (2013) utilized trapezoidal fuzzy numbers and developed an important methodology for solving matrix games with payoffs, expressed by trapezoidal fuzzy numbers. Liu and Yu (2014) highlight the concept of 2-dimension uncertain linguistic variables given by decision makers, to evaluate the reliability and to express fuzzy information. Sahu et al. (2016b) explored Ratio system, Reference point and full Multiplication form coupled with aggregated interval valued trapezoidal fuzzy number sets to evaluate and benchmark the CNC machine tool from preferred options.

Figure 3. Trapezoidal Fuzzy number

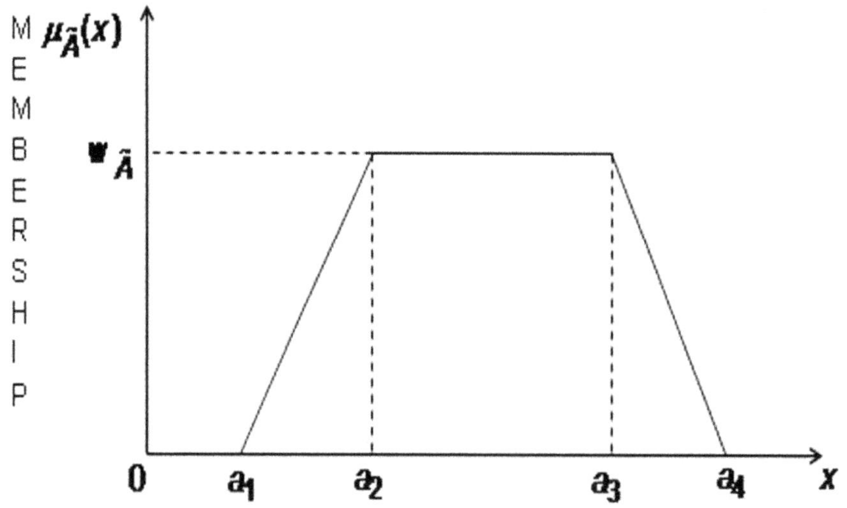

DEFUZZIFICATION

It is not possible by anyone to compare different fuzzy numbers directly. The fuzzy number has to be defuzzified to convert into the crisp value for the purpose of ranking and comparing the alternative. Fuzzy logic is capable of supporting to a reasonable extent, human type reasoning in natural form. Defuzzification is the practice of producing a quantifiable result into crisp (Leekwijck & Kerre 1999). It computes a crisp value for evaluation and comparison. It means mapping of the fuzzy numbers into real numbers, which can be then further assessed for magnitude. The assignment of a real value to a fuzzy number is called Defuzzification. It is the action to find the crisp solution from the process of interpolation or inference. Researchers have developed several methods for defuzzification, but there is no universal consensus. The following are the various techniques available for Defuzzification:

1. BOA (bisector of area)
2. Centroid method
3. FOM (first of maximum)
4. LOM (last of maximum)
5. MOM (middle of maximum)
6. Height Method
7. WFM (weighted fuzzy mean)
8. MeOM (mean of maxima)

BOA (Bisector of Area)

The BOA partitions the area into two regions with the same area (Lee, 1990; Timothy, 2010). Bisection of a shape or part may always be defined by the line that divides it into two equal parts as shown on Figure 4. Consider the trapezoid fuzzy number $\tilde{A} = (a, b, c, d)$, then by bisection of area

The defuzzified value $D_{\tilde{A}} = \dfrac{a+b+c+d}{4}$ (13)

Only if $b \le D_{\tilde{A}} \le c$ and for the triangular $\tilde{B} = (a, b, c)$, we can treat triangular fuzzy number set as trapezoidal fuzzy number set by converting it as (a, b, b, c) and by using the Equation 13, then

The defuzzified value $D_{\tilde{B}} = \dfrac{a+2b+c}{4}$ (14)

Centroid Method

It is the most common and physically appealing of all the defuzzification methods (Sugeno, 1985). A centroid of a triangle is the point where the three medians of the triangle meet as shown in Figure 5. The centroid is also called the center of gravity or centre of area. For example, if we try to balance the tri-

Figure 4. Illustration of bisection

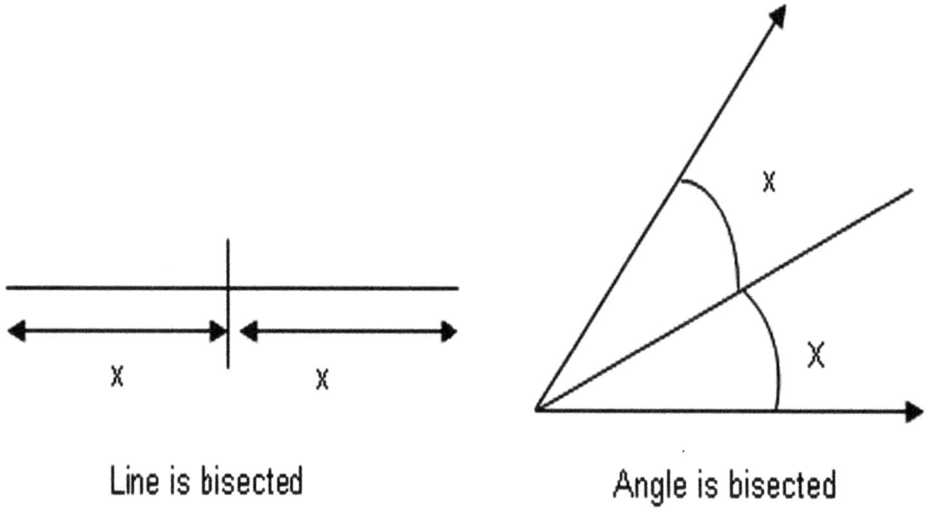

angular plate on our finger, then the point where it will balance the whole triangular body is known as centroid of that plate. Consider the trapezoid fuzzy number $\tilde{A} = (a,\ b,\ c,\ d)$, then by centroid method

The defuzzified value $D_{\tilde{A}} = \dfrac{c^2 + d^2 + cd - a^2 - b^2 - ab}{3(c + d - a - b)}$ (15)

and for the triangular $\tilde{B} = (a,\ b,\ c)$, by using the Equation 15 and treating triangular fuzzy number set as trapezoidal fuzzy number set by converting it as $(a,\ b, b,\ c)$

The Defuzzified value $D_{\tilde{B}} = \dfrac{a + b + c}{3}$ (16)

Allahviranloo and Saneifard (2012), proposed the centroid formula for defuzzification of the Generalized Trapezoidal Fuzzy Number set (GTFNs) $\tilde{A} = [a.b, c, d, \omega]$

Here, $[w_i = \omega\ ]$, represents the weight or value of membership function for Trapezoidal Fuzzy Number set

$$\overline{x}_0(A) = \frac{1}{3}[a + b + c + d - \frac{dc - ab}{(d + c) - (a + b)}]$$

$$\overline{y}_0(A) = \omega\frac{1}{3}[1 + \frac{c - b}{(d + c) - (a + b)}]$$

The Defuzzified value $R(A) = \sqrt{\overline{x}_0{}^2(A) + \overline{y}_0{}^2(A)}$ (17)

Figure 5. The centroid of triangle abc

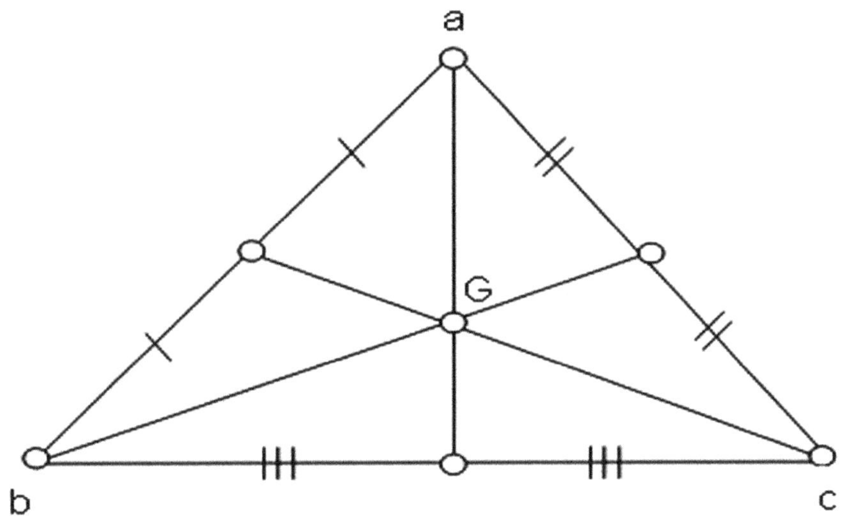

FOM (First of Maxima)

This method employs the principle of selecting first fuzzy number in a fuzzy set by considering maximum membership (Sadeghi, 2013; Timothy, 2010; Sugeno, 1985), as a plateau as shown in Figure 6. For example in case of trapezoidal fuzzy number the maximum membership, represents the two value of fuzzy number in a fuzzy set. Thus choosing the first fuzzy number as a defuzzified value representing the maximum membership is a key of this approach.

LOM (Last of Maximum)

This method employs the principle of selecting last fuzzy number in a fuzzy set by considering maximum membership (Sadeghi, 2013; Timothy, 2010; Sugeno, 1985), as a plateau as shown in Figure 6. For example in case of trapezoidal fuzzy number the maximum membership, represents the two value of fuzzy number in a fuzzy set. Thus choosing the last fuzzy number as a defuzzified value representing the maximum membership is a key of this approach.

MOM (Middle of Maximum)

This method employs the principle of selecting middle value amongst first and last fuzzy number in a fuzzy sets by considering maximum membership (Sadeghi, 2013; Timothy, 2010; Sugeno, 1985), as a plateau as shown in Figure 6. For example in case of trapezoidal fuzzy number the maximum membership represents the two value of fuzzy number in a fuzzy set. Thus choosing the middle value as a defuzzified value representing the maximum membership is a key of this approach.

For the trapezoidal fuzzy number as shown in Figure 6,

The Defuzzified value $D_{\tilde{B}} = \dfrac{a+c}{2}$ (18)

Figure 6. Illustration of defuzzification for trapezoidal fuzzy number

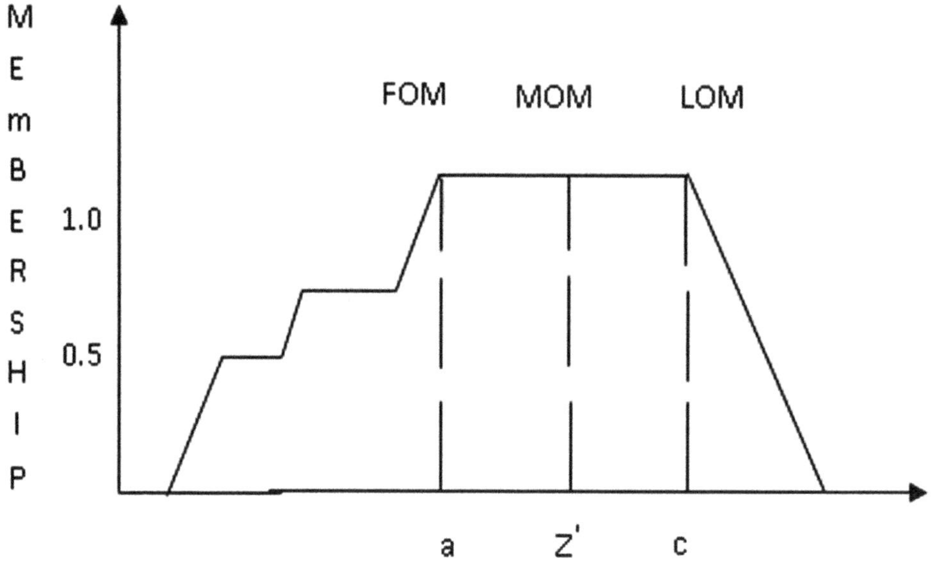

Height Method

This method employs the selection of defuzzified value from the fuzzy number set based on the concept of maximum membership principle (Timothy, 2010). The fuzzy number in a fuzzy set, which yields maximum height of the membership function, selected as a defuzzified value as shown in Figure 7. This method gives good result, in case of triangular fuzzy number sets.

Figure 7. Illustration of defuzzification for triangular fuzzy number

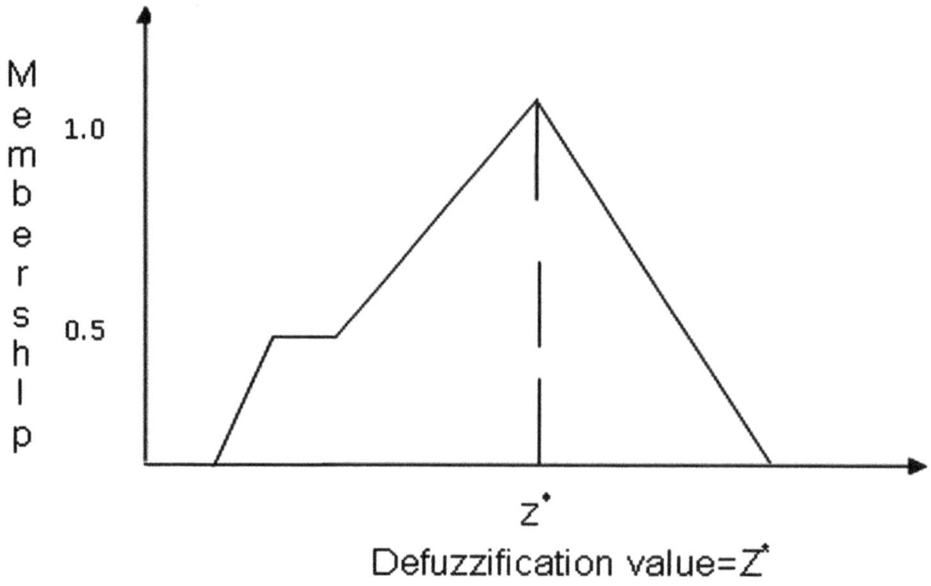

Weighted Average Method (Weighted Fuzzy Mean)

This method is employed to compute the composite score, in case of comparison between two fuzzy numbers in a fuzzy sets having symmetrical output membership functions as shown in Figure 8. The weighted average method is the more computationally efficient methods in fuzzy applications (Timothy, 2010). In this method the aggregated weighted value of first and second fuzzy number is considered as a defuzzified value, by considering membership function as weight of the corresponding fuzzy number. Let Z_1 is the middle value of first trapezoidal fuzzy number in a fuzzy set having a membership value μ_1 as 0.5 and Z_2 is the middle value of second trapezoidal fuzzy number in a fuzzy set having a membership value μ_2 as 1 as shown in Figure 8., then by Equation 19

The Defuzzified value $$Z^* = \frac{Z_1 \otimes \mu_1 + Z_2 \otimes \mu_2}{\mu_{1\oplus}\mu_2} \tag{19}$$

where, μ_1 *and* μ_2 are the corresponding values of the membership function for first and second fuzzy number respectively

MeOM (Mean of Maxima)

This method is employed to compute the composite score, to compare two fuzzy numbers in fuzzy sets having output membership function as shown in Figure 9. In this method, the aggregated value of first and second fuzzy number is considered, to compute defuzzified value Z* (Lee, 1990; Timothy, 2010). Let M_1 is the middle value of first trapezoidal fuzzy number in a fuzzy set having maximum member-

Figure 8. Illustration of defuzzification by weighted fuzzy mean

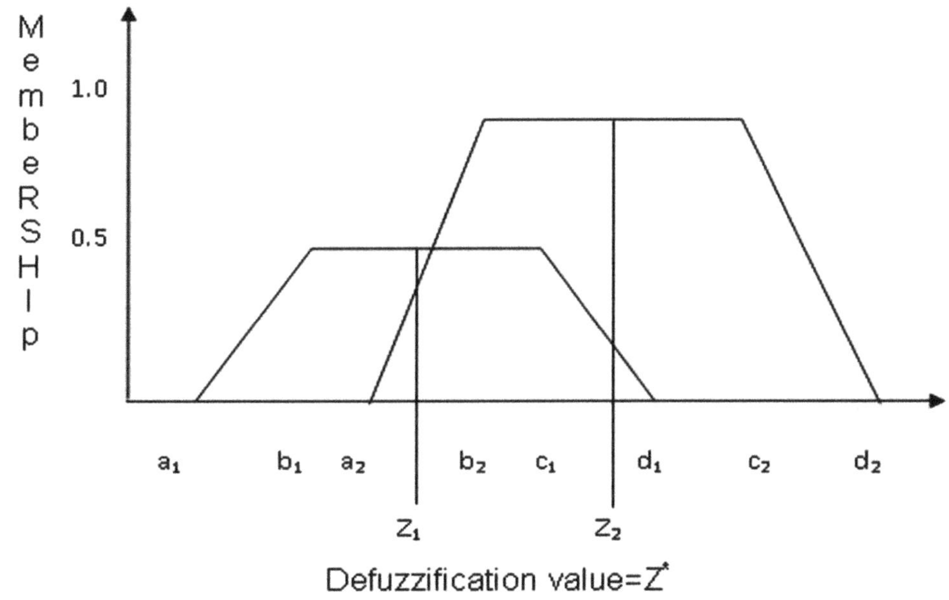

Figure 9. Illustration of defuzzification by means of maxima

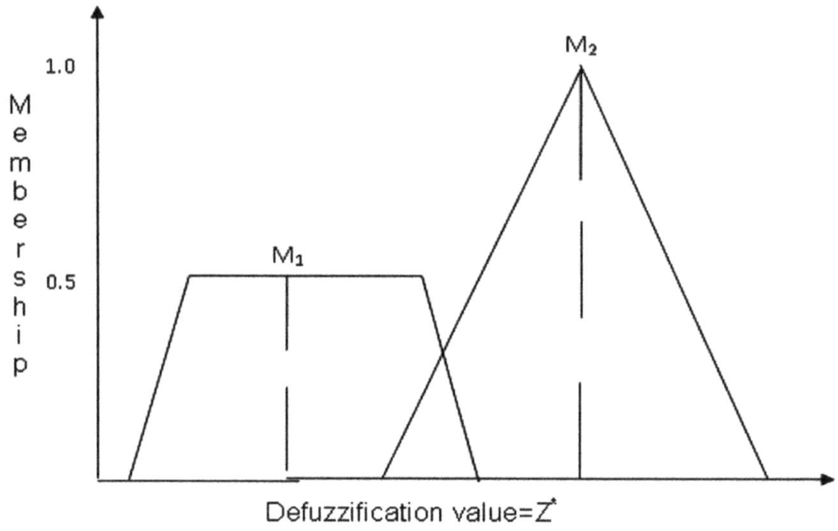

ship function value and M_2 is the middle value of second triangular fuzzy number in a fuzzy set having maximum membership function value, then by Equation 20

$$\text{The Defuzzified value } Z^* = \frac{M_1 + M_2}{2} \tag{20}$$

LINGUISTIC

The modeling of imprecise and qualitative knowledge as well as the transmission and handling of uncertainty at various stages are possible by use of fuzzy sets theory. In fuzzy sets theory, the non-numeric values are usually used to make under or over information to describe fact, situation or result. In many real life situations, it is not possible to collect information in quantitative term because of uncertainty and impreciseness. Linguistic variables are usually incorporated to deal with uncertain or imprecise information (Samantra et al., 2013; Sahu et al., 2013; Sahu et al., 2014; Sahu et al. 2015). Variables concept, i.e. Very low, Low, Average, High, Very high, are used in standard fuzzy sets theory, which replaces a numerical value in place of non numerical information (Zimmermann, 2010). These variables are referred to as the linguistic variables or linguistic terms in the fuzzy sets, to determine the correspondence with the numerical values.

Table 1 highlights the linguistic variables in terms of Triangular interval-valued fuzzy numbers. A linguistic variable is a linguistic expression in terms of one or more words which are used to roughly estimate the information. A membership function in fuzzy set is labeled by the terminology like:

- Hot coffee,
- Young age,
- Poor quality,
- High status etc.

Table 1. Linguistic Variables for Triangular interval-valued fuzzy numbers

Linguistic Variables	Triangular Interval-Valued Fuzzy Numbers
Very Poor (VP)	[(0,0); 0; (1,1.5)]
Poor (P)	[(0, 0.5); 1; (2.5, 3.5)]
Moderately Poor (MP)	[(0, 1.5); 3; (4.5, 5.5)]
Fair (F)	[(2.5, 3.5); 5; (6.5, 7.5)]
Moderately Good (MG)	[(4.5,5.5); 7; (8, 9.5)]
Good (G)	[(5.5, 7.5); 9; (9.5, 10)]
Very Good (VG)	[(8.5, 9.5); 10; (10, 10)]

Figure 10. Illustration of linguistic information

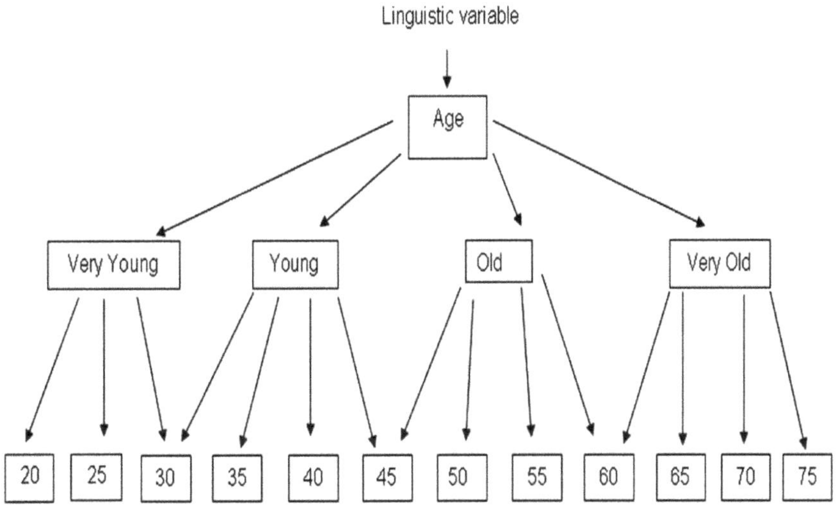

Figure 10 illustrates the case, where one can easily determine the age of a person followed by linguistic information, by selecting a particular domain in fuzzy sets theory.

APPLICATION FIELDS

The application of fuzzy numbers includes:

- Numerous distinctive fields of operation research,
- Optimization,
- Decision analysis,
- Data processing,
- Signal conditioning etc,

where the information is neither completely true nor completely false or which is partly true and partly false. Fuzziness may prevail in many other cases where impression, matching, similarity, implication, partial truth or the approximate information is present. In last few decades researchers have successfully applied fuzzy theory in the fields of:

- Supplier selection in supply chain management,
- 3PL service provider assessment and selection (Govindan & Murugesan 2011),
- Administration and accounting,
- Facility location and layout design,
- statistical research, material election in manufacturing,
- Shipping logistic management, linguistic analysis of human behavior,
- Crisis management,
- Quality management,
- Measurement and control,
- power relationship,
- translation work,
- Planning and control,
- information services,
- Appraisement of material handling equipments,
- meteorology,
- Industrial engineering,
- Hydrology and water resources etc. (Li, 2016b).

The crisp conversion and ranking of fuzzy number based on the appropriate value of crisp always seems to be the challenging work for various researchers, to apply the fuzzy concept in diverse application fields. Li (2010a) and Li et al. (2010b) developed the conception of arithmetical operations and Ranking of different fuzzy numbers.

INTEGRATED TECHNIQUES

Fuzzy sets theory can be effectively integrated with a variety of developed techniques. The main benefit of fuzzy sets theory is that, it can easily amalgamate with the quantitative data. Li (2011c) developed a methodology for MADM problems and used intuitionistic fuzzy sets to express, ratings of alternatives and weights of attributes under qualitative and quantitative information. It is well known fact, that the majority of the collected data is measurable in quantitative terms. While the information data which cannot be measurable in quantitative terms can be assessed by fuzzy linguistic scale. Sahu et al. (2016a), applied interval-valued fuzzy number set with modified technique for order of preference by similarity to ideal solution method to benchmark partner firms under similar agile supply chain platform. Liu and Shi (2015) proposed the multiple attribute group decision making with interval neutrosophic hesitant under fuzzy information. Figure 11 depicts some useful techniques, which can be integrated with the fuzzy sets theory:

Figure 11. Fuzzy techniques

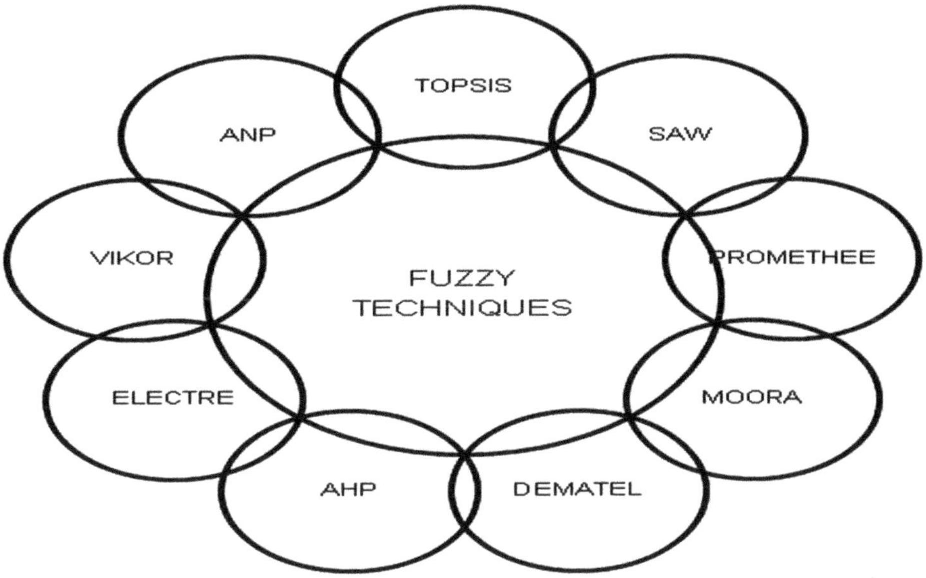

TOPSIS

The TOPSIS method was first proposed by Hwang and Yoon (1981). It is based on the concept of Positive Ideal Solution (PIS) as well as Negative Ideal Solution (NIS) known as Anti-Ideal Solution. The PIS minimizes the cost criteria and maximizes the benefit criteria, whereas the NIS maximizes the cost criteria and minimizes the benefit criteria. The so-called benefit criteria are those whose maximum values are favored; while the cost criteria are those whose minimum values are preferred. The best alternative solution is one, which is placed at closest to the PIS and at the farthest distance from the NIS. Sahu et al. (2015a) applied TOPSIS for the subjective information to benchmark and appraise the advanced manufacturing CNC machine tool. Chen (2000) proposed extensions of the TOPSIS method for group decision-making under fuzzy environment.

Suppose a MCDM problem has m alternatives $(A_1,................A_m)$ and n decision criterion $(C_1,................,C_n)$. all the alternatives are evaluated with respect to given n criterion. The ratings assigned to the alternatives with respect to each criterion form a decision-matrix denoted by $X = (x_{ij})_{m x n}$.

Let $W = (w_{1,}\ w_{2,............}, w_n)$ be the relative weight vector of the criterion, satisfying $\sum_{j=1}^{n} w_j = 1$. Then, the summarized TOPSIS method is as follows:

Step 1: Normalize the decision matrix $X = (x_{ij})_{m x n}$ using the following equation:

$$r_{ij} = \frac{x_{ij}}{\sqrt{\sum_{j=1}^{n} x^2_{ij}}}, \quad i = 1,2,3,.......m,, \quad j = 1,2,3,.......n \tag{21}$$

Here r_{ij} is the normalized criterion rating.

Step 2: Calculate the weighted normalized decision matrix $v = (v_{ij})_{mxn}$

Here w_j is the relative weight of the jth criterion or attribute, and $\sum_{j=1}^{n} w_j = 1$.

Step 3: Determine the PIS and NIS by:

$$A* = \{v_1{}^*,..........,v_n{}^*.\} = \left\{(\max_i v_{ij} (j \in \Omega_b], (\min_i v_{ij} (j \in \Omega_c]\right\}_j \tag{22}$$

$$A^- = \{v_1{}^-,..........,v_n{}^-.\} = \left\{(\min_i v_{ij} (j \in \Omega_b], (\max v_{ij} (j \in \Omega_c]\right\}_j \tag{23}$$

Here Ω_b and Ω_c are the sets of benefit criteria and cost criteria, respectively.

Step 4: Calculate the Euclidean distances of each alternative from the PIS and the NIS, respectively.

$$D^*_i = \sqrt{\sum_{j=1}^{n} (v_{ij} - v^*)^2_{ij}}, \quad i = 1,2,3,.......m \tag{24}$$

$$D^-_i = \sqrt{\sum_{j=1}^{n} (v_{ij} - v^*)^2_{ij}}, \quad i = 1,2,3,.......m \tag{25}$$

where, D^*_i and D^-_i are the PIS and NIS Euclidean distances of each alternative and v_{ij} is weighted normalized value of ith alternative and j^{th} decision criterion, whose value varies, such that $i = 1,2,3,.......m$ and $j = 1,2,3,.......n$.

Step 5: Calculate the relative closeness of each alternative with respect to the ideal solution. The relative closeness of the alternative A_i with respect to A^* is defined by:

$$RC_i = \frac{D_i^-}{D_i^* + D_i^-}, \quad i = 1,2,3,.......m \tag{26}$$

Step 6: Rank the alternatives according to their relative closeness to the ideal solution. The bigger the RC_i, the better the alternative A_i is. The best alternative is the one which is having the greatest relative closeness to the ideal solution. Sahu et al. (2012), proposed a well-organized Grey-TOPSIS method for appraisement of supplier platform, to promote environmentally concerned purchasing strategies of an organization.

VIKOR

The VIKOR was introduced in multi criteria decision making (MCDM) by Po-Lung Yu (1973) and is used to solve MCDM problem of complex systems. This method is employed for ranking and selecting best alternative from a set of available alternatives, in the presence of non commensurable and conflicting criteria. In this method each alternative is evaluated by decision maker against each criterion function, the compromise score for ranking is obtained by comparing the measure of closeness of each alternative by the ideal alternative. The following steps should be followed in successfully implementation of the said method:

Step 1: Formulation of the decision-making problem.
Step 2: Normalization of decision-making information.
Step 3: Construction of weighted decision-making matrix.
Step 4: Determine the positive ideal solution and the fuzzy negative ideal solution of the evaluation objects.

For benefit attributes:

$$\tilde{\tilde{V}}^+ = \left[\tilde{\tilde{v}}_j^+\right]_{1\times n}, \tilde{\tilde{V}}^- = \left[\tilde{\tilde{v}}_j^-\right]_{1\times n} \tag{27}$$

For cost attributes:

$$\tilde{\tilde{V}}^+ = \left[\tilde{\tilde{v}}_j^-\right]_{1\times n}, \tilde{\tilde{V}}^- = \left[\tilde{\tilde{v}}_j^+\right]_{1\times n} \tag{28}$$

Step 5: Compute the values $S_i, R_i \left(i = 1,2,...,m\right)$ by the relations,

$$S_i = \sum_{j=1}^{n} \frac{d\left(\tilde{\tilde{v}}_j^+, \tilde{\tilde{v}}_{ij}\right)}{d\left(\tilde{\tilde{v}}_j^+, \tilde{\tilde{v}}_j^-\right)} \tag{29}$$

$$R_i = \max_j \left[\frac{d\left(\tilde{\tilde{v}}_j^+, \tilde{\tilde{v}}_{ij}\right)}{d\left(\tilde{\tilde{v}}_j^+, \tilde{\tilde{v}}_j^-\right)} \right] \quad (30)$$

Step 6: Compute the values $Q_i, \left(i = 1, 2, ..., m\right)$ by the following relation:

$$Q_i = \nu \frac{\left(S_i - S^*\right)}{\left(S^- - S^*\right)} + \left(1 - \nu\right) \frac{\left(R_i - R^*\right)}{\left(R^- - R^*\right)} \quad (31)$$

where ν is introduced as weight of the strategy of the majority of criteria or the maximum group utility. Rank the alternatives by sorting the values of Q in ascending order

SAW

SAW stands for simple additive weighting also known as weighted sum model (WSM). SAW method is most likely the widely used and best known method for solving multiple criteria decision making problem due to its simplicity and ease implementation (Peter, 1967). It is based on the aggregation of weighted criteria. An appraisal score is computed for each available alternative by multiplying the objective value given to the attribute of that alternative with the corresponding weights assigned by the expert followed by addition of the products for all criteria. The following steps should be followed in successfully implementation of the said method:

Step 1: Construct a decision matrix $(m\,x\,n)$ that includes m personnel and n criterion.

Step 2: Calculate the normalized decision matrix:

For positive criteria:

$$r_{ij} = \frac{x_{ij}}{x_{ij}^{\max}}, \quad i = 1, 2, 3, ...m,, \quad j = 1, 2, 3, ...n \quad (32)$$

For negative criteria:

$$r_{ij} = \frac{x_{ij}^{\min}}{x_{ij}}, \quad i = 1, 2, 3, ...m,, \quad j = 1, 2, 3, ...n \quad (33)$$

where, r_{ij} is the normalized score of the ith alternative with respect to the jth criterion and $x_{ij}^{\max}$ is the maximum value of x_{ij} in the column of jth criterion and $x_{ij}^{\min}$ is the minimum value of x_{ij} in the column of jth criterion.

Step 3: Evaluate each alternative by the following formula:

$$A_{ij} = \sum_{j=1}^{n} w_{ij}.r_{ij} \tag{34}$$

where, A_{ij} is the composite score of ith alternative with respect to the jth criterion and w_{ij} is the weight of jth criterion.

DEMATEL

The Decision-Making Trial and Evaluation Laboratory (DEMATEL) method is used to build interrelations between criteria and factors (Govindan et al., 2016a; Detcharat & Pongpun, 2013). In DEMATEL, matrices and graphs of complicated causal relationships are constructed, to visualize the structure, nature of impact and interrelations on considered criteria's. The technique helps the decision makers in analyzing the role of involved independent criteria by separating the involved criteria of a system or subsystem into groups of cause and effect. It can be successfully applied with fuzzy sets theory to gather and assemble group ideas and to analyze the cause and effect relationship between criteria's of complex problems in fuzzy environments. The following steps should be followed in successfully implementation of the said method (Tzeng et al, 2007):

Step 1: Gather experts' opinion and calculate the average matrix Z.
Step 2: Calculate the normalized initial direct-relation matrix D.
Step 3: Derive the total relation matrix T.
Step 4: Calculate the arithmetic score of rows and columns of matrix T.
Step 5: Set the threshold value (α).
Step 6: Build a cause and effect relationship diagram.

MOORA METHOD

The Multi-Objective Optimization by Ratio Analysis (MOORA) method was introduced by (Brauers & Zavadakas, 2006). Brauers and Zavadakas (2010) extended the method to make it more robust via developing the concept of full multiplicative form. The method considers matrix X as input, where the matrix elements x_{ij} denote i^{th} alternative of jth objective $\left(i = 1,2,...,m; j = 1,2,...,n\right)$. The method includes attribute normalization of the corresponding alternatives and considers all the objectives as equally important. The method considers three approaches:

1. The Ratio System approach.
2. The Reference Point approach.
3. The Full Multiplicative Form approach.

The Ratio System implies Equation 35 to normalize data by comparing attributes of an alternative according to minimization and maximization of attribute

$$x_{ij}^{*} = \frac{x_{ij}}{\sqrt{\sum_{i=1}^{m} x_{ij}^{2}}} \tag{35}$$

Here x_{ij}^{*} denotes normalize value of ith alternative of jth objective. The summarizing index of each alternative is derived by using Equation 36 and 37:

$$y_{i}^{*} = \sum_{j=1}^{g} x_{ij}^{*} - \sum_{j=g+1}^{n} x_{ij}^{*}, \tag{36}$$

$$y_{i}^{*} = \sum_{j=1}^{g} w_{i} * x_{ij}^{*} - \sum_{j=g+1}^{n} w_{i} * x_{ij}^{*}, \tag{37}$$

Here w_{i}, represents the attribute/criterion weights and $g = 1,...,n$, denotes number of objectives to be maximized. Then computed ratio is used for ranking: the higher the index, the higher the rank. Sahu et al. (2014), applied ratio analysis methodology for multi objective optimization problem by utilizing grey number aligned with distinctive criterion to benchmark CNC machine tool via handling the vagueness and impreciseness information.

Analytic Hierarchy Process (AHP)

The analytic hierarchy process (AHP) was first introduced by Saaty (1980), is a powerful, flexible and systematic technique for organizing and analyzing decision problems with many criteria and alternatives. AHP can be successfully applied to handle qualitative and quantitative criteria. The subjective (qualitative) data are first converted into numerical values by Table 2.

Table 2. Scale for qualitative comparison of alternatives

Option	Numerical Scale(S)
Equal	1
Marginally Strong	3
Strong	5
Very Strong	7
Extremely Strong	9
Intermediate values to reflect fuzzy inputs	2,4,6,8
Reflecting dominance of second alternative compared with the first	Reciprocals

Figure 12. AHP hierarchy for choosing a student leader

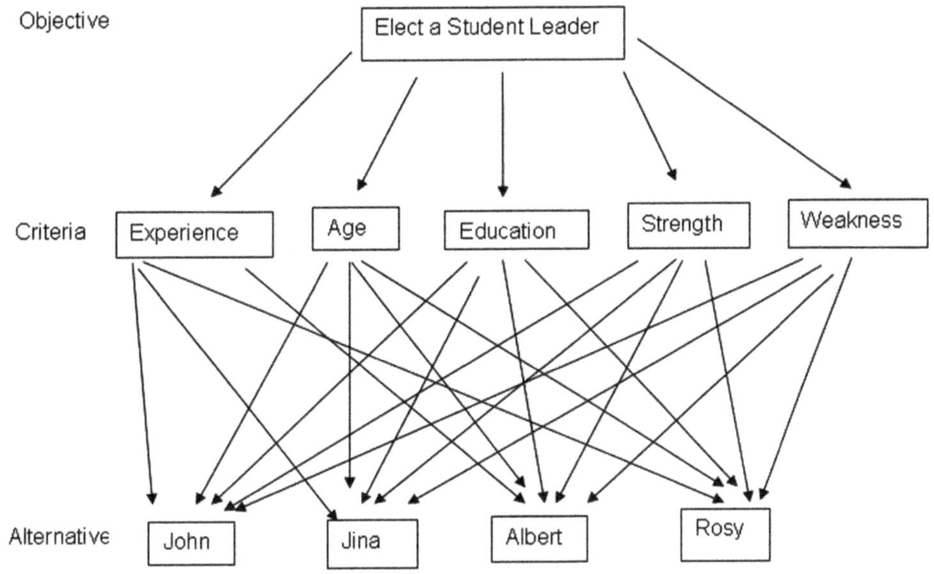

AHP aids the decision maker in setting the sequence of priorities and to make the best decision (Chang, 1996). The AHP technique can be explained with the help of following steps:

Step 1: The decision problem should be first arranged as a hierarchy structure in AHP, where the problem is structured into a hierarchy of goal, criteria, sub-criteria and alternatives as shown in Figure 12.

Step 2: Decision-makers are employed to collect and covert data qualitative data into quantitative information via following linguistic scale as shown in Table 2, for the constructed hierarchy structure. Experts can rate the comparison as equal, marginally strong, strong, very strong, and extremely strong. In AHP, we consider an $n \times n$ pair-wise comparison matrix A with positive elements such that

$$A = \begin{bmatrix} 1 & a_{12} & \cdots & a_{1n} \\ a_{21} & 1 & \cdots & a_{2n} \\ \vdots & \vdots & \ddots & \vdots \\ a_{n1} & a_{n2} & \cdots & 1 \end{bmatrix}_{n x n}$$

This matrix is reciprocal, where $a_{ij} = \frac{1}{a_{ji}}$ for each $1 \leq i,\ j \leq n$. We say that A is consistent if $a_{ij} * a_{jk} = a_{ik}$.

Step 3: The generated pair wise comparisons of various criteria are structured into a square matrix (Ramik & Korviny, 2010). The diagonal elements of the said matrix are 1. The geometric means method is explored to calculate the relative normalized weights of the criterion because of its easiness and simplicity to find out the maximum Eigen value and to reduce the inconsistency in judgment.

Step 4: The calculated principal Eigen value and the corresponding normalized eigenvector of the comparison matrix provide the relative importance of the various criteria. The elements of the normalized eigenvector are termed as weights with respect to the criteria or sub-criteria.

Step 5: Verify the consistency index (CI) and consistency ratio (CR) of the comparison matrix. This CI can be compared with that of a random matrix (RI). The consistency ratio (CR) is the ratio of CI/RI. The suggested of CR should be less than 0.1.

Step 6: The calculated weights of the sub-criteria is multiplied by the rating of each alternative and aggregated to get local ratings. The local ratings are again multiplied by the weights of the criteria and aggregated to get global ratings.

Due to vagueness, imprecision and uncertainty in the judgment of the decision makers, a conventional AHP may be unable to accurately capture the decision maker's judgment. Sahu et al. (2015 b) implemented AHP to third-party logistics (3PL) service provider's appraisement platform to umpire the best 3PL service provider.

PROMETHEE

PROMETHEE stands for Preference Ranking Organization method for Enrichment of Evaluations. This method has wide applications around the world in different fields of decision making,

- i.e. Logistic management,
- Supply chain management,
- Economic analysis in business,
- Warehouse selection,
- Location design etc.

It requires less information from the side of decision makers and facilitates in simulating the actual decision problem (Deshmukh, 2013). The methods can be effectively applied to decision situation dealing with:

- **Choice:** The selection of one alternative amongst distinctive given alternatives, carrying multiple decision criteria.
- **Prioritization:** Aids in shaping the relative merit between set of alternatives.
- **Resource Allocation:** Aids in the allocation of resources amongst available set of alternatives.
- **Ranking:** Ranking alternatives from most preferred/ relevant to least preferred/ relevant.

This method facilitates in solving multi criteria decision making problem which focused on building an outranking between different alternatives. The Promethee method aids the decision makers to locate the best alternative among various alternatives available that best suits their objective (Brans & Vincke, 1958). The two types of information are required in implementation of said method:

1. Information on the weights i.e. relative importance of the considered criteria.

2. Information on the preference function by the decision-makers, to compare the involvement of the available alternatives for each separate criterion.

The following steps should be followed in successfully implementation of the said method:

Step 1: Extension of the Notion of Criteria.
Step 2: Valued Outranking Relation.
Step 3: Exploitation of the Outranking Relation.

ELECTRE

It is generally classified as an "outranking method" for solving multi criteria decision making problems. The ELECTRE method is used to select the best act amongst given set of actions (Govindan & Jepsen, 2016b). It can be effectively applied to solve problem related to choosing, prioritizing, ranking and sorting (Gokhan & Gorener, 2016). This method aids the decision makers to locate the best alternative among various alternatives available that best suits their objective with maximum benefit and minimum conflict while dealing with various distinctive criteria. The following steps should be followed in successfully implementation of the said method:

Step 1: Preparation of Decision Matrix.
Step 2: Calculation of the normalized decision Matrix.
Step 3: Calculate the weighted normalized decision Matrix.
Step 4: Determine the concordance and discordance sets.
Step 5: Calculate the concordance matrix.
Step 6: Calculate the discordance matrix.
Step7: Calculate net concordance and discordance Matrix to make the ranking amongst alternatives.

ANP

The Analytic Network Process (ANP) is a generalization of the Analytic Hierarchy Process (AHP) used in multi-criteria decision analysis (Saaty, 2006). The method starts with the Preparation of Decision Matrix. The considered problem in ANP should be evidently acknowledged and decomposed into a rational system, constructed by a network relationship by considering the dependency between the elements of the constructed network. In this method the Decision Matrix can be prepared based on the opinion of decision maker by passing through brainstorming sessions or other appropriate methods (Hsu, 2012). ANP is represented by a network rather than a structured hierarchy because numerous decision problems cannot be arranged into hierarchy because of the involvement of the interaction and dependence of higher level elements on lower level elements. The following steps should be followed in successfully implementation of the said method:

Step 1: Model construction and problem structuring.
Step 2: Pair-wise comparisons matrices and priority vectors.

Step 3: Verify consistency index (CI) and consistency ratio (CR) of comparison matrix.
Step 4: Super matrix formation.
Step 5: Selection the best alternatives.
Step 6: The alternative with the largest overall priority should be selected.

CONCLUSION

In today's daily life humans are often operated under fuzzy environments. Fuzzy sets theory is effectively used to handle problems, where a sense of vagueness, unclear, inaccurate and impreciseness information or data set is present. It has always been measured as a natural oral language to model or deal with the approximate situations or circumstances, because the fuzzy information can be easily generated by using linguistic scale. In most of the cases the human being does not assess and compute the exact values of the situation, thus the individual impulsively applies quick estimates dealing with fuzziness, based upon his/her past experience to determine the output values. Li (2011a, 2016a) developed a group of effective and efficient linear programming models and methods to solve matrix games with payoffs in fuzzy numbers. Li (2011b) developed a closeness coefficient based on nonlinear programming method to solve multi attribute decision making problems. The presented chapter will be beneficial to the new readers and learners and help them to deal with the cases of acquiring fuzzy information under fuzzy circumstances. The presented work highlights the significant steps needed to apply fuzzy sets theory, in order that new readers and researchers can augment and disseminated the fuzzy theory into new application fields, which will be beneficial for the society and mankind. The chapter transfers the basic concept of fuzzy knowledge aligned with various integrated techniques under one roof.

ACKNOWLEDGMENT

The authors thankfully acknowledge the support provided by Prof. Deng-Feng Li, Editor of the Book of "Theoretical and Practical Advancements for Fuzzy System Integration". The authors also extended their gratitude towards the anonymous reviewer(s) to augment the quality of chapter. Once more the authors transfer their request to Prof. Deng-Feng Li for making his kind attention upon authors and providing them great occasion to be associated in archive of the Book entitled "Theoretical and Practical Advancements for Fuzzy System Integration".

REFERENCES

Allahviranloo, T., & Saneifard, R. (2012). Defuzzification method for ranking fuzzy numbers based on center of gravity. *Iranian Journal of Fuzzy Systems*, *9*(6), 57–67.

Brans, J. P., & Source, Ph. V. (1985). A Preference Ranking Organisation Method: The PROMETHEE Method for Multiple Criteria Decision-Making. *Management Science*, *31*(6), 647–656. doi:10.1287/mnsc.31.6.647

Brauers, W. K. M., & Zavadskas, E. K. (2006). The MOORA method and its application to privatization in transition economy. *Control and Cybernetics*, *35*(2), 443–468.

Brauers, W. K. M., & Zavadskas, E. K. (2010). Project management by MULTI-MOORA as an instrument for transition economies. *Technological and Economic Development of Economy*, *16*(1), 5–24. doi:10.3846/tede.2010.01

Chang, D. Y. (1996). Application of the extent analysis method on fuzzy AHP. *European Journal of Operational Research*, *65*(3), 649–655. doi:10.1016/0377-2217(95)00300-2

Chen, C. T. (2000). Extensions of the TOPSIS for group decision-making under fuzzy environment. *Fuzzy Sets and Systems*, *114*(1), 1–9. doi:10.1016/S0165-0114(97)00377-1

Deshmukh, S. C. (2013). Preference Ranking Organization Method of Enrichment Evaluation. *International Journal of Engineering Science Invention*, *2*(11), 28–34.

Detcharat, S., & Pongpun, A. (2013). Using DEMATEL Method to Analyze the Causal Relations on Technological Innovation Capability Evaluation Factors in Thai Technology-Based Firms. *International Transaction Journal of Engineering, Management, &. Applied Sciences & Technologies*, *4*(2), 81–103.

Gokhan, Y. M., & Gorener, A. (2016). Decision Making for Company Acquisition by ELECTRE Method. *Int. J Sup. Chain. Mgt. Exceling Tech Pub*, *5*(1), 75–83.

Gorzalczany, M. B. (1987). A method of inference in approximate reasoning based on Interval-valued fuzzy sets. *Fuzzy Sets and Systems*, *21*(1), 1–17. doi:10.1016/0165-0114(87)90148-5

Govindan, K., & Jepsen, M. B. (2016b). ELECTRE: A comprehensive literature review on methodologies and applications. *European Journal of Operational Research*, *250*(1), 1–29. doi:10.1016/j.ejor.2015.07.019

Govindan, K., Khodaverdi, R., & Vafadarnikjoo, A. (2016a). A grey DEMATEL approach to develop third-party logistics provider selection criteria. *Industrial Management & Data Systems*, *116*(4), 690–722. doi:10.1108/IMDS-05-2015-0180

Govindan, K., & Murugesan, P. (2011). Selection of third-party reverse logistics provider using fuzzy extent analysis. *Benchmarking: An International Journal*, *18*(1), 149–167. doi:10.1108/14635771111109869

Hsu, P. F. (2012). Selection model based on ANP and GRA for independent media agencies. *Quality & Quantity*, *46*(1), 1–17. doi:10.1007/s11135-010-9323-y

Hwang, C. L., & Yoon, K. (1981). *Multiple Attribute Decision Making Methods and Applications*. Berlin: Springer. doi:10.1007/978-3-642-48318-9

Kuo, M. S. (2011). A novel interval-valued fuzzy MCDM method for improving airlines service quality in Chinese cross-strait airlines. *Transportation Research Part E, Logistics and Transportation Review*, *47*(6), 1177–1193. doi:10.1016/j.tre.2011.05.007

Lee, C. C. (1990). Fuzzy Logic in Control Systems: Fuzzy Logic Controller. *Part II. IEEE Transactions on Systems, Man, and Cybernetics*, *20*(2), 419–435. doi:10.1109/21.52552

Leekwijck, V., & Kerre, E. E. (1999). Defuzzification: Criteria and classification. *Journal of Fuzzy Sets and Systems*, *108*(2), 159–178. doi:10.1016/S0165-0114(97)00337-0

Li, D.-, F. (2011a). Linear programming approach to solve interval-valued matrix games. OMEGA. *International Journal of Management Sciences*, *39*(6), 655–666.

Li, D.-, F. (2011b). Closeness coefficient based nonlinear programming method for interval-valued intuitionistic fuzzy multiattribute decision making with incomplete preference information. *Applied Soft Computing*, *11*(4), 3402–3418. doi:10.1016/j.asoc.2011.01.011

Li D.-, F. (2011c). The GOWA operator based approach to multi attribute decision making using intuitionistic fuzzy sets. *Mathematical and Computer Modelling*, *53*(5-6), 1182–1196. doi:10.1016/j.mcm.2010.11.088

Li, D.-, F. (2012). A fast approach to compute fuzzy values of matrix games with payoffs of triangular fuzzy numbers. *European Journal of Operational Research*, *223*(2), 421–429. doi:10.1016/j.ejor.2012.06.020

Li, D.-, F. (2013). An effective methodology for solving matrix games with fuzzy payoffs. *IEEE Transactions on Cybernetics*, *43*(2), 610–621. doi:10.1109/TSMCB.2012.2212885 PMID:22987531

Li, D.-, F. (2014). *Decision and Game Theory in Management with Intuitionistic Fuzzy Sets*. Heidelberg, Germany: Springer.

Li, D.-, F. (2016a). *Linear Programming Models and Methods of Matrix Games with Payoffs of Triangular Fuzzy Numbers*. Berlin: Springer.

Li, D.-, F. (2016b). *Models and Methods of Interval-Valued Cooperative Games in Economic Management*. Cham, Switzerland: Springer.

Li, D.-, F.,, & Cheng, C. T. (2002). Fuzzy multi objective programming methods for fuzzy constrained matrix games with fuzzy numbers. *International Journal of Uncertainty, Fuzziness and Knowledge-based Systems*, *10*(4), 385–400. doi:10.1142/S0218488502001545

Liu, P. D., & Shi, L. L. (2015). The Generalized Hybrid Weighted Average Operator Based on Interval Neutrosophic Hesitant Set and Its Application to Multiple Attribute Decision Making. *Neural Computing & Applications*, *26*(2), 457–471. doi:10.1007/s00521-014-1736-4

Liu, P. D., & Yu, X. C. (2014). 2-dimension uncertain linguistic power generalized weighted aggregation operator and its application for multiple attribute group decision making. *Knowledge-Based Systems*, *57*(1), 69–80. doi:10.1016/j.knosys.2013.12.009

Mousavi, S. M., Vahdani, B., Moghaddam, R. T., Ebrahimnejad, S., & Amiri, M. (2012). A multistage decision making process for multiple attributes analysis under an interval-valued fuzzy environment. *International Journal of Advanced Manufacturing Technology*, *64*(9-12), 1263–1273. doi:10.1007/s00170-012-4084-5

Mula, P., Roy, S. K., & Li, D.-, F. (2015). Birough programming approach for solving bi-matrix games with birough payoff elements. *Journal of Intelligent & Fuzzy Systems*, *29*(2), 863–875. doi:10.3233/IFS-151616

Peter, F. C. (1967). Letter to the Editor-Additive Utilities with Incomplete Product Sets: Application to Priorities and Assignments. *Operations Research*, *15*(3), 537–542. doi:10.1287/opre.15.3.537

Ramik, J., & Korviny, P. (2010). Inconsistency of pair wise comparison matrix with fuzzy elements based on geometric mean. *Fuzzy Sets and Systems*, *161*(11), 1604–1613. doi:10.1016/j.fss.2009.10.011

Saaty, T. L. (1980). *The Analytic Hierarchy Process*. New York: McGraw-Hill.

Saaty, T. L., & Vargas, L. G. (2006). *Decision Making with the Analytic Network Process: Economic, Political, Social and Technological Applications with Benefits, Opportunities, Costs and Risks*. New York: Springer.

Sadeghi, M., & Hosseini, H. M. (2013). Evaluation of Fuzzy Linear Programming Application in Energy Models. *International Journal of Energy Optimization and Engineering*, *2*(1), 50–59. doi:10.4018/ijeoe.2013010104

Sahu, A. K., Sahu, N. K., & Sahu, A. (2014). Appraisal of CNC machine tool by integrated MULTI MOORA-IGVN circumstances: An empirical study. *International Journal of Grey Systems: Theory and Application*, *4*(1), 104–123.

Sahu, A. K., Sahu, N. K., & Sahu, A. K. (2015a). Benchmarking CNC machine tool using hybrid fuzzy methodology a multi indices decision making approach. *International Journal of Fuzzy System Applications*, *4*(2), 28-46..

Sahu, A. K., Sahu, N. K., & Sahu, A. K. (2016a). Application of Integrated TOPSIS in ASC index: Partners Benchmarking perspective. *Benchmarking: An International Journal*, *23*(3), 540–563. doi:10.1108/BIJ-03-2014-0021

Sahu, A. K., Sahu, N. K., & Sahu, A. K. (2016b). Appraisal of Partner Enterprises under GTFNS Environment in Agile SC. *International Journal of Decision Support System Technology*, *8*(3), 1–19. doi:10.4018/IJDSST.2016070101

Sahu, A. K., Sahu, N. K., & Sahu, A. K. (2016c). Application of Modified MULTI-MOORA for CNC Machine Tool Evaluation in IVGTFNS Environment: An Empirical Study. *International Journal of Computer Aided Engineering and Technology*, *8*(3), 234–259. doi:10.1504/IJCAET.2016.077603

Sahu, N. K., Datta, S., & Mahapatra, S. S. (2012). Establishing green supplier appraisement platform using grey concepts. *Grey Systems: Theory and Application*, *2*(3), 395–418. doi:10.1108/20439371211273276

Sahu, N. K., Datta, S., & Mahapatra, S. S. (2013). Decision making for selecting 3PL service provider using three parameter interval grey numbers. *International Journal of Logistic Systems and Management*, *14*(3), 261–297. doi:10.1504/IJLSM.2013.052061

Sahu, N. K., Datta, S., & Mahapatra, S. S. (2014). Green Supplier Appraisement in Fuzzy Environment. *Benchmarking: An International Journal*, *21*(3), 412–429. doi:10.1108/BIJ-06-2012-0042

Sahu, N. K., Datta, S. & Mahapatra, S. S. (2015). Fuzzy Based Appraisement Module for 3PL Evaluation and Selection. *Benchmarking, an International Journal, 22*(3), 354-392.

Sahu, N. K., Sahu, A. K., & Sahu, A. K. (2015b). Appraisement and Benchmarking of Third Party Logistic Service Provider by Exploration of Risk Based Approach. Cogent business and management, 2, 1-21.

Samantra, C., Sahu, N. K., Datta, S., & Mahapatra, S. S. (2013). Decision-Making in Selecting Reverse Logistics Alternative using Interval-Valued Fuzzy Sets Combined with VIKOR Approach. *International Journal of Services and Operations Management*, *14*(2), 175–196. doi:10.1504/IJSOM.2013.051828

Sugeno, M. (1985). An introductory survey of fuzzy control. *Information Sciences*, *36*(1-2), 59–83. doi:10.1016/0020-0255(85)90026-X

Tanaka, K. (1997). *An Introduction to Fuzzy Logic for Practical Applications*. New York: Springer-Verlag.

Timothy, J. R. (2010). *Fuzzy Logic with Engineering Applications* (3rd ed.). John Wiley & Sons.

Tzeng, G. H., Chiang, C. H., & Li, C. W. (2007). Evaluating intertwined effects in e-learning programs: A novel hybrid MCDM model based on factor analysis and DEMATEL. *Expert Systems with Applications*, *32*(4), 1028–1044. doi:10.1016/j.eswa.2006.02.004

Yu, P. L. (1973). A Class of Solutions for Group Decision Problems. *Management Science*, *19*(8), 936–946. doi:10.1287/mnsc.19.8.936

Yu, D., Li, D.-F. , & Merigo, J. M. (2015). Dual hesitant fuzzy group decision making method and its application to supplier selection. *International Journal of Machine Learning and Cybernetics,* 1-13.

Zadeh, L. A. (1965). Fuzzy Sets. *Information and Control*, *8*(3), 338–353. doi:10.1016/S0019-9958(65)90241-X

Zadeh, L. A. (1975). The Concept of a Linguistic Variable and its Application to Approximate Reasoning. *Information Sciences*, *8*(3), 199–249. doi:10.1016/0020-0255(75)90036-5

Zimmermann, H. J. (2010). Fuzzy set theory. John Wiley & Sons Inc.

Chapter 2
Some Recent Defuzzification Methods

Harendra Kumar
Gurukula Kangari University, India

ABSTRACT

Defuzzification is a process that converts a fuzzy set or fuzzy number into a crisp value or number. Defuzzification is used in fuzzy modeling and in fuzzy control system to convert the fuzzy outputs from the systems to crisp values. This process is necessary because all fuzzy sets inferred by fuzzy inference in the fuzzy rules must be aggregated to produce one single number as the output of the fuzzy model. There are numerous techniques for defuzzifying a fuzzy set; some of the more popular techniques are included in fuzzy logic system. In the present chapter some recent defuzzification methods used in the literature are discussed with examples.

INTRODUCTION

Fuzzy set theory was introduced by Zadeh (1965). Fuzzy set is the extension of crisp set. In fuzzy, each and every element has the degrees of membership value which lies in [0,1]. Fuzzy set theory extends the classical set theory with memberships of its elements described by the classical characteristic function, to allow for partial membership described by a membership function. Thus, fuzzy set theory has great capabilities and flexibilities in solving many real-world problems which classical set theory does not intend or fails to handle. Fuzzy set theory was applied to control systems theory and engineering almost immediately after its birth. In many practical applications such as in fuzzy inference systems, the fuzzy results generated cannot be used as such to the applications, hence it is necessary to convert the fuzzy quantities into crisp quantities for further processing.

Defuzzification

Defuzzification is a process that converts a fuzzy set or fuzzy number into a crisp value or number. Defuzzification is used in fuzzy modeling and in fuzzy logic control to convert the fuzzy inputs or out-

DOI: 10.4018/978-1-5225-1848-8.ch002

puts (or both) from the systems to crisp values. In another word, defuzzification interprets the membership degrees of the fuzzy sets into a specific decision or real value. This process is necessary because all fuzzy sets inferred by fuzzy inference in the fuzzy rules must be aggregated to produce one single number as the output of the fuzzy model. There are numerous methods are available for defuzzifying crisp value or number in the literature. Unfortunately, we have no systematic procedure for choosing a good defuzzification method, and thus we have to select one in considering the properties of application case. Hellendoorn and Thomas (1995) describe some defuzzification method with their respective merits and shortcomings dependent on the rule and domain. Yager and D. P. Filev (1995) address the problem of defuzzification in situations where in addition to the usual fuzzy output of the controller there exists some ancillary restriction on the allowable defuzzified value. Yager and Filev (2006) discussed a problem of selecting a crisp element based on information provided by a fuzzy set, a problem which manifests itself in the defuzzification step in fuzzy logic controllers. The matrix games with fuzzy payoffs have been extensively discussed by Deng-Feng Li (2013). Two major kinds of solution methods have been devised. One is the defuzzification approach based on ranking functions. Another is the two-level linear programming method which can obtain membership functions of players' fuzzy values (or gain floor and loss ceiling). An effective methodology for solving constrained matrix games with payoffs of trapezoidal fuzzy numbers (TrFNs), which are a type of two-person non-cooperative games with payoffs expressed by TrFNs and players' strategies being constrained has been developed by Li and Hong (2013). Recently, a new approach for defuzzification of generalized fuzzy numbers is established by Rouhparvar and Panahi (2015). This method uses in centre point of a triangle where the three bisector lines of its angles meet. There are various methods available for defuzzification; some of them are defined in the following section.

In the following sections, the most often used defuzzification methods are defined.

- **Center of Area (COA) Method**

In the Center of Area (COA) defuzzification method, we first calculate the area under the scaled membership functions and within the range of the output variable, then we uses the following equation to calculate the geometric center of this area.

$$z^{*} = \frac{\int z.\mu_{\tilde{A}}\left(z^{*}\right)dz}{\int \mu_{\tilde{A}}\left(z^{*}\right)dz} \tag{1}$$

where COA is the center of area, x is the value of the linguistic variable, and x_{min} and x_{max} represent the range of the linguistic variable. The following figure illustrates the COA defuzzification method:

In figure 1, μ is the degree of membership, and the shaded portion of the graph represents the area under the scaled membership functions. This method is also known as center of gravity and centroid method. Center of area defuzzification is the most commonly used method, as it is very accurate method. It provides center of area under the curve of membership function. For complex membership functions, it puts high demands on computations. It is the most common used method for defuzzification.

Figure 1. Centre of Area method

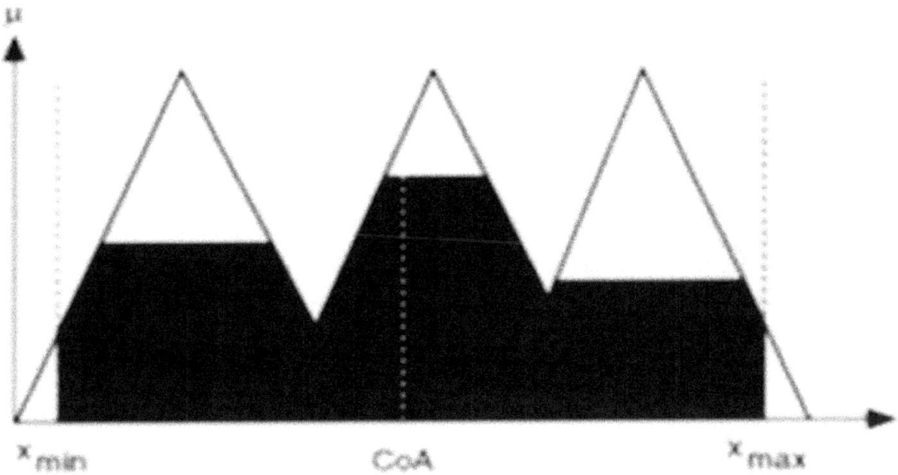

- **Center of Sums (COS) Method**

The defuzzified value z^* is given by the following expression:

$$z^* = \frac{\int_z \bar{z}.\sum_{k=1}^{n} \mu_{\tilde{A}_k}(z)\,dz}{\int_z \sum_{k=1}^{n} \mu_{\tilde{A}_k}(z)\,dz} \tag{2}$$

where $\bar{z}$ is the distance to the centroid of each of the respective membership functions.

This defuzzification method has been used by many researchers in different applications and researchers conclude that this method is faster than many defuzzification methods which are presently in use. This process involves the algebraic sum of individual output fuzzy sets, instead of their union. Two drawbacks to this method are that the intersecting areas are added twice, and the method also involves finding the centroid of the individual membership functions.

- **Center of Largest Area Method**

If the output fuzzy set has at least two convex sub-regions, then the center of gravity of the convex fuzzy sub-region with the largest area is used to obtain the defuzzified value z^* of the output and it is given by the following expression:

$$z^* = \frac{\int z.\mu_{\tilde{A}_m}(z)\,dz}{\int \mu_{\tilde{A}_m}(z)\,dz} \tag{3}$$

Where $\tilde{A}_m$ is the convex sub-region that has the largest area making up $\tilde{A}_k$. This condition applies in the case when the overall output $\tilde{A}_k$ is non convex and in the case when $\tilde{A}_k$ is convex (i.e. only one convex region), z^* is the same quantity as determined by the centroid method or the center of largest area method.

- **Weighted Average Method**

The weighted average method is the most frequently used in fuzzy applications since it is one of the more computationally efficient methods. Unfortunately it is restricted to symmetrical output membership functions. It is given by the following expression:

$$z^* = \frac{\sum \bar{z}.\mu_{\tilde{A}}(\bar{z})}{\sum \mu_{\tilde{A}}(\bar{z})} \tag{4}$$

Where $\bar{z}$ is the centroid of each symmetric membership function. This method is similar to the center of sum method, except in the weighted average method the weights are maximum membership values whereas in the center of sums method the weights are the areas of the respective membership functions.

- **Mean Max Membership Method**

This method is also known as middle of maxima method. It is commonly used when the maximum membership can be a plateau rather than a single point. It uses the average value of the aggregated membership functions of outputs. It is given by the following expression:

$$z^* = \frac{a+b}{2} \tag{5}$$

Where a and b are the points which have the maximum membership value. It can be defined by the following figure 2.

- **First of Maxima Method**

This method uses the overall output or union of all individual output fuzzy sets $\tilde{A}_k$ to determine the smallest value of the domain with maximized membership degree in $\tilde{A}_k$. It is given by the following expression:

$$z^* = \inf_{z\in Z}\left\{z \in Z : \mu_{\tilde{A}_k}(z) = ht(\tilde{A}_k)\right\} \tag{6}$$

Where $ht(\tilde{A}_k)$ is the height of fuzzy set $\tilde{A}_k$ and is defined as $ht(\tilde{A}_k) = \sup_{z\in Z}\mu_{\tilde{A}_k}(z)$.

Figure 2. Mean max membership defuzzification method

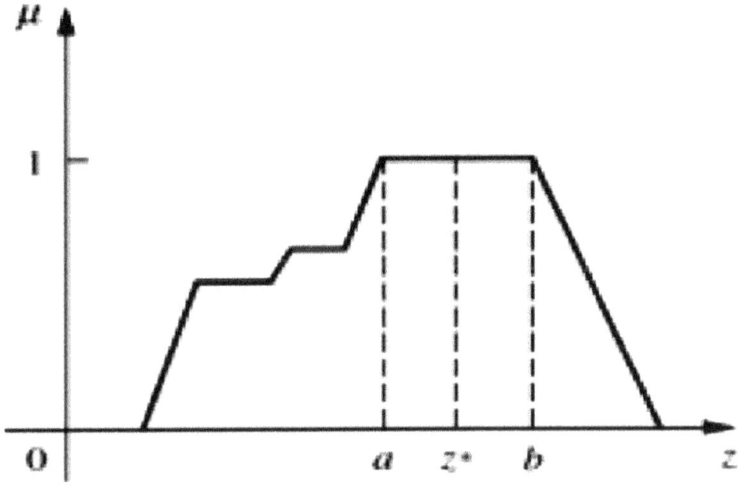

- **Last of Maxima Method**

This is an alternative method of first of maxima method. This method uses the overall output or union of all individual output fuzzy sets $\tilde{A}_k$ to determine the greatest value of the domain with maximized membership degree in $\tilde{A}_k$. It is given by the following expression:

$$z^* = \sup_{z\in Z}\left\{z \in Z : \mu_{\tilde{A}_k}(z) = ht(\tilde{A}_k)\right\} \tag{7}$$

- **Height Method**

It is also known as max membership principle; this method is limited to peaked output functions. It is closely related to mean max membership method. This method is given by the following expression:

$$\mu_{\tilde{A}}(z^*) \geq \mu_{\tilde{A}}(z)\ \forall z \in Z \tag{8}$$

Where z^* is the defuzzified value and it is shown graphically in the following figure 3.

Example 1: Let $\tilde{A}_1 = \begin{Bmatrix} 0.4(x-1); & 1 \leq x \leq 2 \\ 0.4; & 2 \leq x \leq 5 \\ 0.4(6-x); & 5 \leq x \leq 6 \\ 0; & otherwise \end{Bmatrix}$, $\tilde{A}_2 = \begin{Bmatrix} 0.8(x-4); & 4 \leq x \leq 5 \\ 0.8; & 5 \leq x \leq 7 \\ 0.8(8-x); & 7 \leq x \leq 8 \\ 0; & otherwise \end{Bmatrix}$ and

Figure 3. Height method

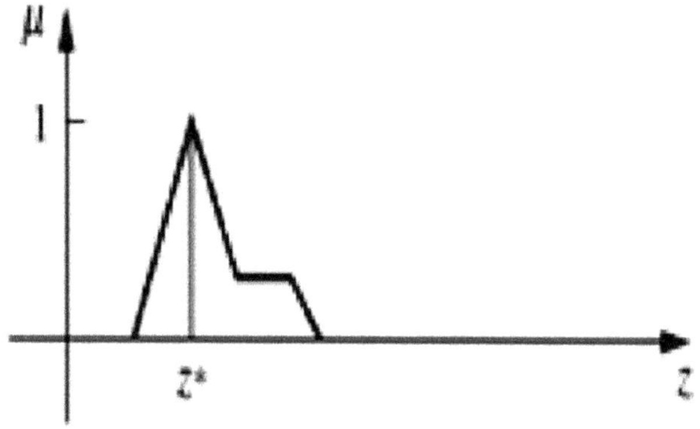

$$\tilde{A}_3 = \begin{cases} x-6; & 6 \le x \le 7 \\ 1.0; & 7 \le x \le 8 \\ (9-x); & 8 \le x \le 9 \\ 0; & otherwise \end{cases}$$ are three fuzzy numbers, which are also shown in figure 4, Figure 5, and Figure 6 respectively.

By center of area method (Figure 7), z^* can be calculated as

$$z^* = \frac{\int z.\mu_{\tilde{A}}(z^*)dz}{\int \mu_{\tilde{A}}(z^*)dz}$$

$$= \frac{\left[\int_1^2 0.4(z-1)z\,dz + \int_2^{4.5}(0.4z)\,dz + \int_{4.5}^{5} 0.8(z-4)z\,dz + \int_5^{6.8}(0.8z)\,dz + \int_{6.8}^{7}(z-6)z\,dz + \int_7^8 z\,dz + \int_8^9 (9-z)z\,dz\right]}{\left[\int_1^2 0.4(z-1)\,dz + \int_2^{4.5} 0.4\,dz + \int_{4.5}^{5} 0.8(z-4)\,dz + \int_5^{6.8} 0.8\,dz + \int_{6.8}^{7}(z-6)\,dz + \int_7^8 1\,dz + \int_8^9 (9-z)\,dz\right]}$$

$=7.312$

By center of sums method (Figure 8), z^* can be calculated as

$$z^* = \frac{\int_z \bar{z} \sum_{k=1}^{n} \mu_{\tilde{A}_k}(z)\,dz}{\int_z \sum_{k=1}^{n} \mu_{\tilde{A}_k}(z)\,dz}$$

Figure 4. Fuzzy set $\widetilde{A}_1$.

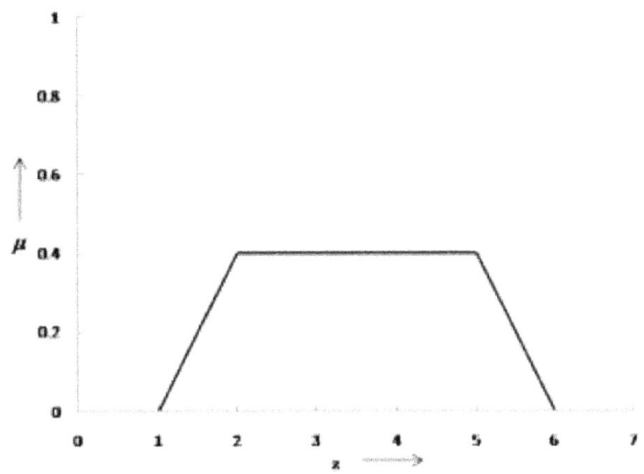

Figure 5: Fuzzy set $\widetilde{A}_2$.

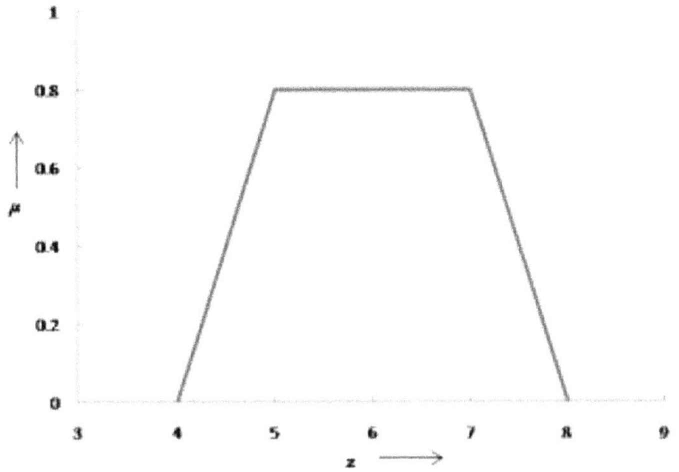

Figure 6. Fuzzy set $\widetilde{A}_3$

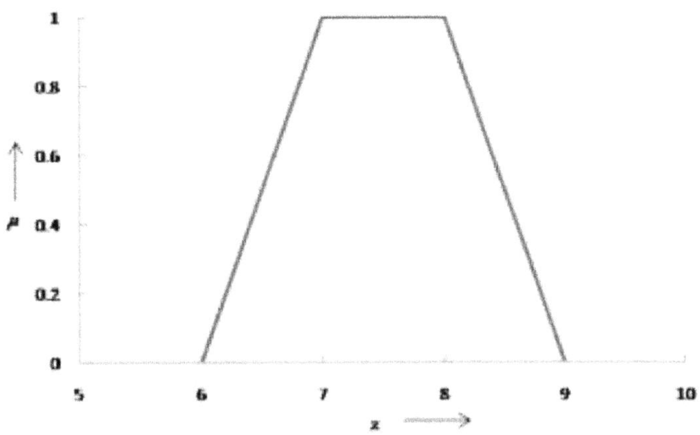

Figure 7. Center of area method for finding z^*

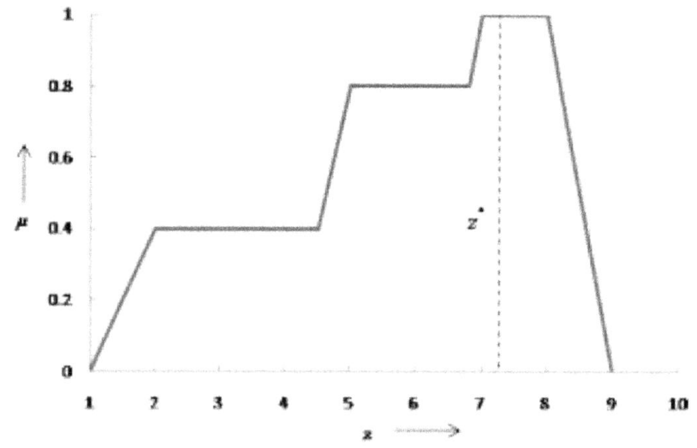

$$= \frac{[3.5\times 0.5\times 0.4(3+5)+6\times 0.5\times 0.8(2+4)+7.5\times 0.5\times 1(3+1)]}{[0.5\times 0.4(3+5)+0.5\times 0.8(2+4)+0.5\times 1(3+1)]}$$

$$= 5.83$$

The complete output fuzzy set is convex, as seen in the figure 9. Then z^* by center of largest area method provide the same results as the center of area method. Hence

$$z^* = \frac{\int z.\mu_{\tilde{A}_m}(z)dz}{\int \mu_{\tilde{A}_m}(z)dz} = 7.312$$

Figure 8. Center of sums method for finding z^*

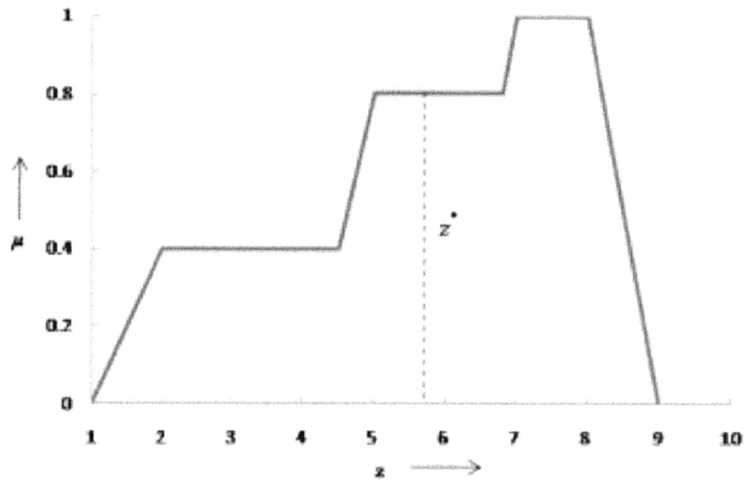

Figure 9. Center of largest area method

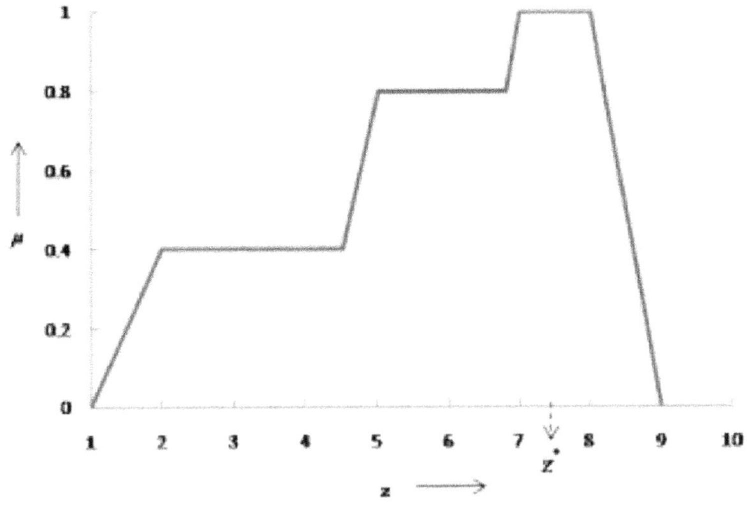

By weighted avearage method (Figure 10), z^* can be calculated as

$$z^* = \frac{\sum \bar{z}.\mu_{\tilde{A}_m}(\bar{z})}{\sum \mu_{\tilde{A}_m}(\bar{z})}$$

$$= \frac{[(0.4\times 3.5)+(0.8\times 6)+(1\times 7.5)]}{[0.4+0.8+1]}$$

$$= 6.227$$

Figure 10. Weighted average method for finding z^*

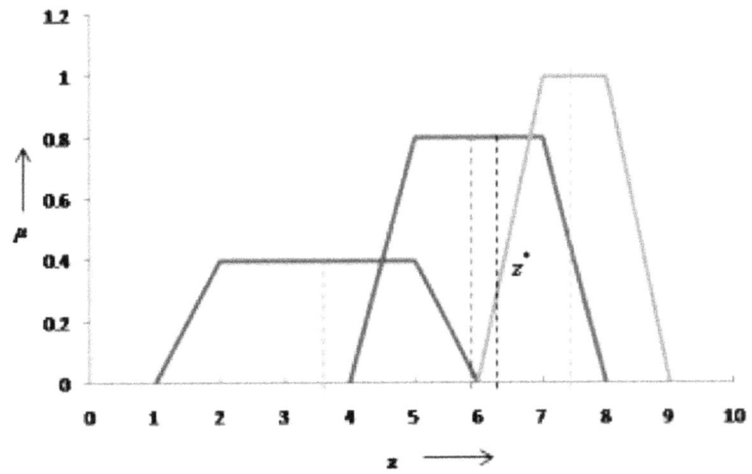

Figure 11. Mean max membership method for finding z^*

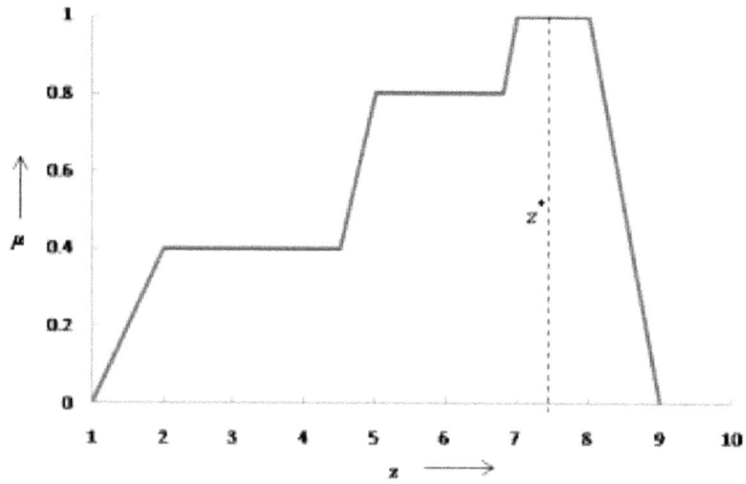

By mean max membership method (Figure 11), z^* can be calculated as

$$z^* = \frac{a+b}{2}$$
$$= \frac{7+8}{2}$$
$$= 7.5$$

Figure 12. First of maxima and last of maxima methods for finding the z^*

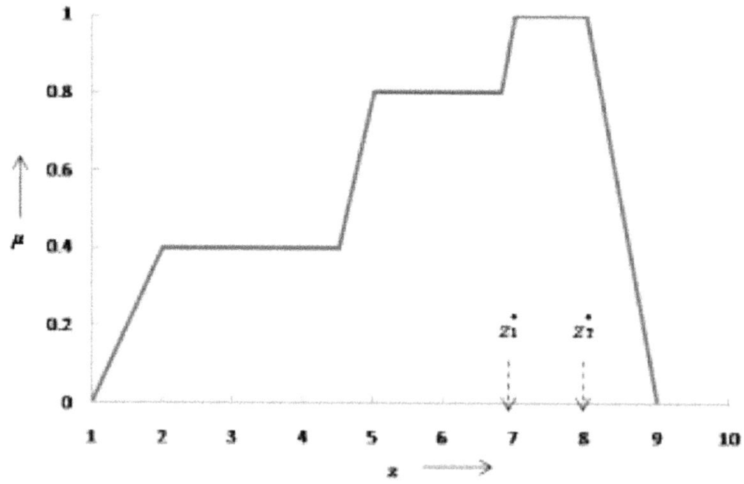

By first of maxima method (Figure 12), z_1^* can be calculated as

$$z_1^* = \inf_{z \in Z} \{ z \in Z : \mu_{\tilde{A}}(z) = ht(\tilde{A}_k) \} = 7$$

And by last of maxima method (Figure 12), z_2^* can be calculated as

$$z_2^* = \sup_{z \in Z} \{ z \in Z : \mu_{\tilde{A}}(z) = ht(\tilde{A}_k) \} = 8$$

Defuzzification Methods for Ranking the Fuzzy Numbers

Ranking fuzzy numbers plays a very important role in decision making and some other fuzzy application systems. In many fuzzy decision problems, the final score of alternatives are represented in terms of fuzzy numbers. In order to express a crisp preference of alternatives, we need a method for constructing a crisp total ordering from fuzzy numbers. Recently many different methods have been proposed to deal with ranking fuzzy numbers. Each method appears to have some advantages as well as disadvantages. In the context of each problem, some methods seem more useful than others. Here, we describe only some simple methods and illustrate them by examples.

- **Magnitude Method (Abbasbandy & Hajjari, 2009)**

The trapazoidal fuzzy number $u = (a, b, c, d)$, with two defuzzifier a,b, and left fuzziness $c > 0$ and right fuzziness $d > 0$ is a fuzzy set whose membership function is defined as following

$$u(z) = \begin{cases} \frac{1}{c}(z - a + c); & a - c \le z \le a \\ 1; & a \le z \le b \\ \frac{1}{d}(b - z + d); & b \le z \le b + d \\ 0; & otherwise \end{cases} \tag{9}$$

Its parametric form $u = (\underline{u}(r), \overline{u}(z))$, where $\underline{u}(z) = a - c + cr$, $\overline{u}(r) = b + d - dr$ and the magnitude of the trapazoidal fuzzy number is defined as the follwing expression:

$$mag(u) = \frac{1}{2}\left(\int_0^1 \left(\underline{u}(r) + \overline{u}(r) + a + b \right) f(r) dr \right) \tag{10}$$

Where $f(r)$ is a non-negative and increasing function on [0,1] with $f(0)=0, f(1)=1$ and $\int_0^1 f(r)dr=\frac{1}{2}$. Here $f(r)$ can be considered as the weighting function and in actual application $f(r)$ can be chosen according to the actual situation. It has the some properties such as

1. For all symmetric trapazoidal fuzzy number $u=(-a,a,c,c), mag(u)=0$.
2. Foranytwosyymetrictrapazoidalfuzzynumber $u=(a,b,c,c), v=(a,b,d,d), mag(u)=mag(v)$.
3. For two arbitrary trapazoidal fuzzy number u and v, $mag(u+v)=mag(u)+mag(v)$.

Provided that, $a=b$ then u is a triangular fuzzy number and we can write $u=(b,c,d)$ or $u=(a,c,d)$.

- **Middle Point Method (Asady & Zendehman, 2007)**

The trapazoidal fuzzy number $u=(a,b,c,d)$, with two defuzzifier a,b, and left fuzziness $c>0$ and right fuzziness $d>0$ is a fuzzy set whose membership function is defined as following

$$u(z)=\begin{cases}\frac{1}{c}(z-a+c); & a-c\le z\le a\\ 1; & a\le z\le b\\ \frac{1}{d}(b-z+d); & b\le z\le b+d\\ 0; & otherwise\end{cases}$$

Its parametric form $u=(\underline{u}(r),\bar{u}(z))$, where $\underline{u}(z)=a-c+cr$, $\bar{u}(r)=b+d-dr$ and the middle point is defined as the following expression:

$$M(u)=\frac{1}{2}\int_0^1(\underline{u}(r)+\bar{u}(z))dr \tag{11}$$

It has the following properties

1. If $u=(a,b,c,d)$ is a trapazoidal fuzzy number, then $M(u)=\frac{1}{2}(a+b)+\frac{1}{4}(d-c)$
2. If $u=(b,c,d)$ is a triangular fuzzy number, then $M(u)=b+\frac{1}{4}(d-c)$.
3. If $u=(b,c,d)$ is a symmetrical trapazoidal fuzzy number, then $M(u)=\frac{1}{2}(b+c)$.
4. If $u=(a,c)$ ia symmetrical triangular fuzzy number,then $M(u)=a$.
   - **Robust's Ranking Method (Yager, 1981)**

Defuzzification by Robust's ranking method is the oldest one. If $\left(a_\alpha^L, a_\alpha^U\right)$ is a α- cut for a fuzzy number u then its corresponding defuzzified crisp value is calculated by the following expression:

$$u^* = \frac{1}{2}\int_0^1 \left(a_\alpha^L + a_\alpha^U\right) d\alpha \tag{12}$$

Example 2: Let us consider four fuzzy numbers whose membership functions are defined as follows and these are also depicted in figure 13, figure 14, figure 15 and figure 16 respectively.

$$\mu_{\tilde{A}}(z) = \begin{cases} \frac{1}{6}(z+4); & -4 \le z \le 2 \\ 1; & z = 2 \\ \frac{1}{2}(4-z); & 2 \le z \le 4 \\ 0; & otherwise \end{cases}, \quad \mu_{\tilde{B}}(z) = \begin{cases} \frac{1}{3}(z+1.75); & -1.75 \le z \le 1.25 \\ 1; & z = 1.25 \\ \frac{1}{2}(3.25-z); & 1.25 \le z \le 3.25 \\ 0; & otherwise \end{cases}$$

$$\mu_{\tilde{C}}(z) = \begin{cases} \frac{1}{10}(z+7); & -7 \le z \le 3 \\ 1; & z = 3 \\ \frac{1}{2}(5-z); & 3 \le z \le 5 \\ 0; & otherwise \end{cases} \text{ and } \mu_{\tilde{D}}(z) = \begin{cases} \frac{1}{3}(z+2); & -2 \le z \le 1 \\ 1; & 1 \le z \le 2 \\ (3-z); & 2 \le z \le 3 \\ 0; & otherwise \end{cases}$$

Figure 13. Fuzzy set $\tilde{A} = (-4, 2, 4)$

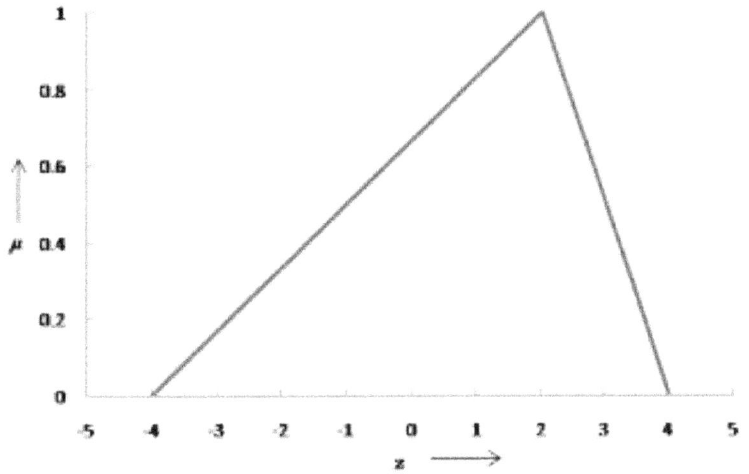

Figure 14: Fuzzy set $\widetilde{B} = (-1.75, 1.25, 3.25)$

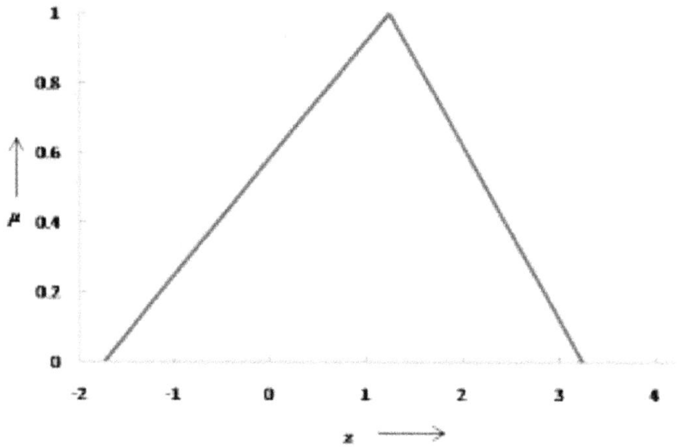

Figure 15. Fuzzy set $\widetilde{C} = (-7, 3, 5)$

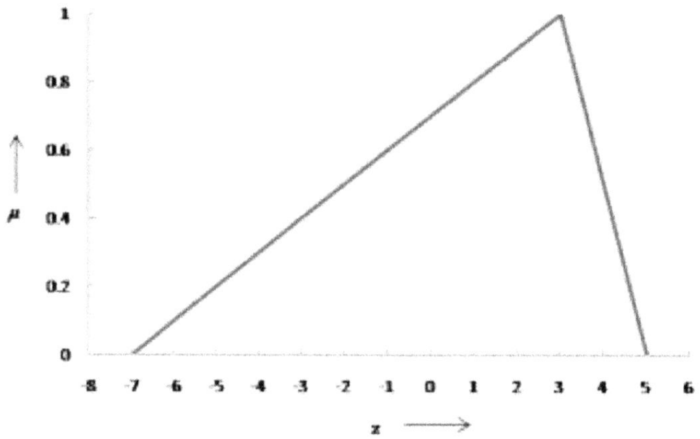

Figure 16: Fuzzy set $\widetilde{D} = (-2, 1, 3)$

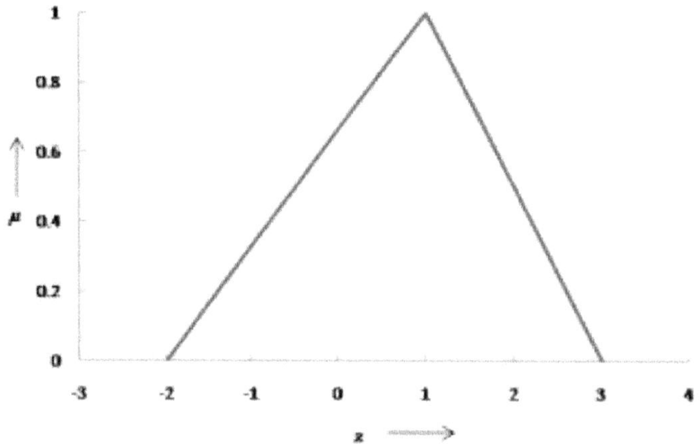

Using defuzzification by magnitude method, we have $mag(\widetilde{A}) = 1.667$, $mag(\widetilde{B}) = 1.167$, $mag(\widetilde{C}) = 2.334$, $mag(\widetilde{D}) = 1.334$ .Also using defuzzification by middle point method, we have $M(\widetilde{A}) = 1, M(\widetilde{B}) = 1, M(\widetilde{C}) = 1, M(\widetilde{D}) = 1$. Using defuzzification by Robust's ranking for the fuzzy numbers $\widetilde{A}, \widetilde{B}, \widetilde{C}$ and $\widetilde{D}$, we get $A = 1.0, B = 1.0, C = 1.0, D = 0.75$.

- **Area between the Centroid Point and Original Point Method (Chung & Chung, 2002)**

As the name suggested, it is the ranking method with an area between the centoid point and the original point of a fuzzy number. The centroid point of a fuzzy number corresponds to an $\bar{x}$ value on the horizontal axis and $\bar{y}$ value on the vertical axis and it can be defined as:

$$\bar{x}\left(\widetilde{A}\right) = \frac{\int_a^b x.\mu_{\widetilde{A}}^L dx + \int_b^c x dx + \int_c^d x.\mu_{\widetilde{A}}^R dx}{\int_a^b \mu_{\widetilde{A}}^L dx + \int_b^c dx + \int_c^d \mu_{\widetilde{A}}^R dx} \text{ and } \bar{y}\left(\widetilde{A}\right) = \frac{\int_0^w y.\eta_{\widetilde{A}}^L dy + + \int_0^w x.\eta_{\widetilde{A}}^R dy}{\int_0^w \eta_{\widetilde{A}}^L dy + + \int_0^w \eta_{\widetilde{A}}^R dy}$$

Where $\mu_{\widetilde{A}}^L$ and $\mu_{\widetilde{A}}^R$ are the left and right membership functions of fuzzy number $\widetilde{A}$ and $\eta_{\widetilde{A}}^L$ and $\eta_{\widetilde{A}}^R$ are the inverse function of $\mu_{\widetilde{A}}^L$ and $\mu_{\widetilde{A}}^R$. The area between the centroid point $\left(\bar{x}, \bar{y}\right)$ and the original point $\left(0, 0\right)$ of the fuzzy number $\widetilde{A}$ is defined as:

$$Z^*\left(\widetilde{A}\right) = \bar{x}.\bar{y} \tag{13}$$

Example 3: Let us consider three fuzzy sets $\widetilde{A}_1 = \left\{\begin{array}{ll} \frac{x-0.4}{0.1}; & 0.4 \le x \le 0.5 \\ 1; & x = 0.5 \\ \frac{1-x}{0.5}; & 0.5 \le x \le 1 \\ 0; & otherwise \end{array}\right\}$

$$\widetilde{A}_2 = \left\{\begin{array}{ll} \frac{x-0.4}{0.5}; & 0.4 \le x \le 0.9 \\ 1; & x = 0.9 \\ \frac{1-x}{0.1}; & 0.9 \le x \le 1 \\ 0; & otherwise \end{array}\right\} \text{ and } \widetilde{A}_3 = \left\{\begin{array}{ll} \frac{x-0.4}{0.3}; & 0.4 \le x \le 0.7 \\ 1; & x = 0.7 \\ \frac{1-x}{0.3}; & 0.3 \le x \le 1 \\ 0; & otherwise \end{array}\right\}$$

which are also depicted in figure 17.

Figure 17. Grapically representation of *fuzzy numbers* $\tilde{A}_1$, $\tilde{A}_2$ *and* $\tilde{A}_3$

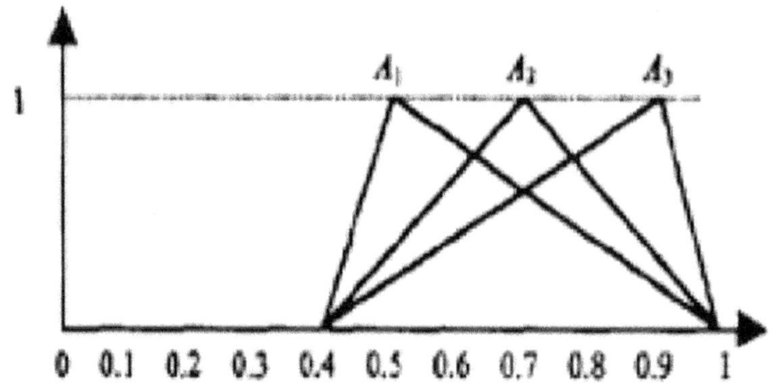

Defuzzified values of the fuzzy numbers are:

$$Z^*(\tilde{A}_1) = \bar{x}_1 \cdot \bar{y}_1 = 0.633 \times 0.472 = 0.299$$
$$Z^*(\tilde{A}_2) = \bar{x}_2 \cdot \bar{y}_2 = 0.7 \times 0.5 = 0.35$$
$$Z^*(\tilde{A}_3) = \bar{x}_3 \cdot \bar{y}_3 = 0.767 \times 0.521 = 0.4$$

Application of Defuzzification Methods

Defuzzification is used in fuzzy modeling and in fuzzy control system to convert the fuzzy outputs from the systems to crisp values. Defuzzification is a highly important part in the design of a fuzzy control system, since it will determine the action taken by the system. The general architecture of a fuzzy control system is shown in Figure 18.

Figure 18. Fuzzy Control System

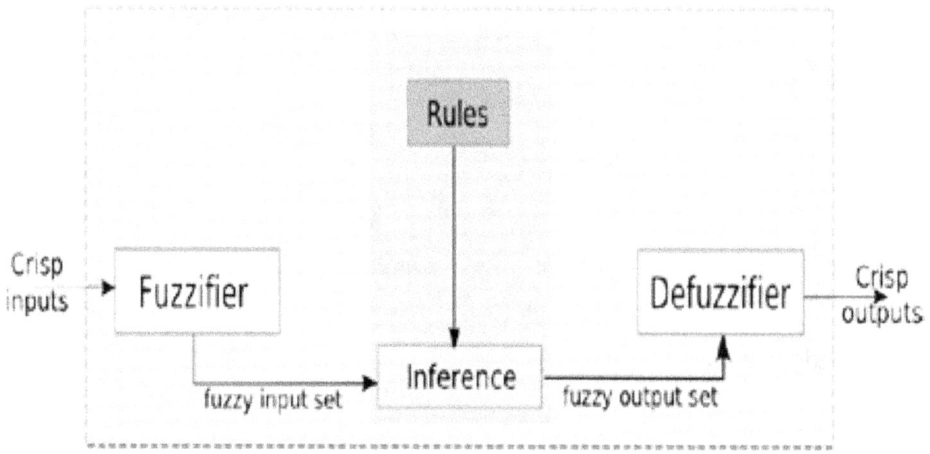

Generally a fuzzy control system consists of four main parts: fuzzifier, rules, inference engine and defuzzifier. Air conditioner system controlled by fuzzy logic system is one of the best examples of fuzzy control system. The Air conditioner system adjusts the temperature of the room according to the current temperature of the room and target value. The fuzzy engine periodically compares the room temperature and the target temperature and produces a command to cool the room.

Other applications which have benefited through the use of defuzzification processes are:

- Information retrieval systems
- A navigation system for automatic cars, air plane, automatic operation of trains
- Laboratory water level controllers
- Decision making.
- Financial planners
- Diagnostic systems
- Meteorological expert system

CONCLUSION

In recent years, numerous methods for fuzzy defuzzification have been suggested in literature. Each method appears to have some advantages as well as disadvantages. It is impossible to develop a general framework for defuzzification, including all cases. This is because in practical all models possess their characteristics. In the context of each application, some defuzzification methods have seen more appropriate than other. However, the issue of choosing a proper defuzzification method is still a subject of research.

REFERENCES

Abbasbandy, S., & Hajjari, T. (2009). A new approach for ranking of trapezoidal fuzzy numbers. *Computers & Mathematics with Applications (Oxford, England)*, *57*(3), 413–419. doi:10.1016/j.camwa.2008.10.090

Asady, B., & Zendehman, A. (2007). Ranking of fuzzy numbers by distance minimization. *Applied Mathematical Modelling*, *31*(11), 2589–2598. doi:10.1016/j.apm.2006.10.018

Bo, Y., & Klir, G. J. (1995). *Fuzzy Sets and Fuzzy Logic Theory and Applications*. PHI Learning Private Limited.

Chung, T. A., & Chung, T. (2002). Ranking fuzzy numbers with an area between the centroid point and the original point. *An International Journal of Computer and Mathematics with Application*, *43*(1-2), 111–117. doi:10.1016/S0898-1221(01)00277-2

Hellendoorn, H., & Thomas, C. (1995). On Quality Defuzzification. In Fuzzy Logic & its Application to Engineering. Information Sciences & Intelligent Systems. doi:10.1007/978-94-009-0125-4_16

Li, D.-F. (2013). An effective methodology for solving matrix games with fuzzy payoffs. *IEEE Transactions on Cybernetics*, *43*(2), 610–621. doi:10.1109/TSMCB.2012.2212885 PMID:22987531

Li, D.-F., & Hong, F.-X. (2013). Alfa-cut based linear programming methodology for constrained matrix games with payoffs of trapezoidal fuzzy numbers. *Fuzzy Optimization and Decision Making*, *12*(2), 191–213. doi:10.1007/s10700-012-9148-3

Rouhparvar, H., & Panahi, A. (2015). A new definition for defuzzification of generalized fuzzy numbers and its application. *Applied Soft Computing*, *30*, 577–584. doi:10.1016/j.asoc.2015.01.053

Yager, R. R. (1981). A procedure for ordering fuzzy subsets of the unit interval. *Information Sciences*, *24*(2), 143–161. doi:10.1016/0020-0255(81)90017-7

Yager, R. R., & Filev, D. P. (2006). On the issue of defuzzification and selection based on a fuzzy set. *Fuzzy Sets and Systems*, *55*(3), 255–272. doi:10.1016/0165-0114(93)90252-D

Yager, R. R., & Filev, D. P. (1995). Defuzzification with Constraints. In Fuzzy Logic & its Application to Engineering. Information Sciences & Intelligent Systems. doi:10.1007/978-94-009-0125-4_15

Chapter 3
Interval-Valued Intuitionistic Fuzzy Multi-Attribute Decision Making Based on Satisfactory Degree

Gao-Feng Yu
Sanming University, China

Deng-Feng Li
Fuzhou University, China

Jin-Ming Qiu
Sanming University, China

ABSTRACT

The aim of this paper is to propose a satisfactory degree method by using nonlinear programming for solving multi-attribute decision making (MADM) problems in which ratings of alternatives on attributes is expressed via interval-valued intuitionistic fuzzy (IVIF) sets and preference information on attributes is incomplete. Concretely, a nonlinear programming model is firstly explored to determine the satisfactory degree which is the ratio of the square of the weight Euclidean distance between an alternative and the IVIF negative ideal solution (IVIFNIS) to the sum of the square of the weight Euclidean distance between the IVIF negative ideal solution (IVIFNIS) and the IVIF positive ideal solution (IVIFPIS). Another nonlinear programming model is also developed to obtain satisfactory intuitionistic fuzzy sets, and then the general satisfactory degrees of the satisfactory intuitionistic fuzzy sets are used to generate the ranking order of the alternatives. Finally, a real example is employed to verify the applicability of the proposed approach and illustrate its practicality and effectiveness.

INTRODUCTION

Due to the high complexity of socioeconomic environments, there are some difficulties for a single decision maker (DM) to consider all important aspects of some decision problems in real decision

DOI: 10.4018/978-1-5225-1848-8.ch003

making. Therefore, group decision making under fuzzy environment problem is an interesting research topic and has received more and more attention from researchers during the last several years. A fuzzy multi-attribute decision making (MADM) problem is to choose the best solutions (i.e., alternatives) from all feasible alternatives on multiple attributes by using the fuzzy set introduced by Zadeh (1965). The fuzzy set has been successfully applied to many practical decision making problems. It only uses a membership function which assigns to each element of the universe of discourse a number from the interval [0, 1] in order to indicate the degree of membership to the fuzzy set under consideration. The degree of non-membership is just automatically the complement to 1 of the membership degree to the fuzzy set. However, a human being who expresses the degree of membership of a given element in a fuzzy set very often does not express the corresponding degree of non-membership as the complement to 1. Then Atanassov (1983) introduced the concept of an intuitionistic fuzzy (IF) set that is characterized by two functions which are used to express the degree of membership and the degree of nonmembership, respectively. The IF set seems to be more accurate for uncertainty quantification and provides an opportunity to precisely model the problem based on the existing knowledge and observations. It may not be easy to identify exact value for both the degree of membership and the degree of non-membership of an element to a given set. In this case, a range of values may be a more appropriate measurement to accommodate the uncertainty, imprecision or vagueness.

The interval-valued intuitionistic fuzzy (IVIF) set (Atanassov, 1989), characterized by a membership function and a non-membership function whose values are intervals rather than real numbers, seems to be more suitable for describing the MADM problems in which membership and non-membership degrees of alternatives on attributes cannot be expressed by exact numerical values. In recent years, many scholars discussed the IVIF set' basic operators and their:

- Properties (Chen et al., 2012; Lakshmana et al., 2011),
- Aggregating operators (Xu, 2007; Yu, 2013; Cai, 2014),
- Topological properties (Mondal, 2001,
- Information measures (Ye, 2009; Huang, 2013; Wei, 2011; Xu, 2010; Zhangetal., 2010) and
- Correlation coefcient(Bustince et al., 1995; Park et al., 2009; Wei et al., 2011; Li, 2011),
- Distances (Grzgorzewski, 2004).

These works have made great contribution to enrich IVIF set theory and have been applied to many fields, such as:

- Student answer scripts (Chen et al., 2013),
- Game theory (Li, 2010),
- MADM (Li, 2010) and
- Group decision making (Yue et al., 2013; zhang et al., 2015).

To the best of our knowledge, there is less investigation on how to solve MADM problems in which ratings of alternatives on attributes are expressed using IVIF sets and preference information on attributes is incomplete.

Although the aforementioned methods are very effective for solving many MADM problems with IVIF sets, they suffer for some limitations:

- The method (Ye, 2009; Wei et al., 2011; Li, 2014) ranked alternatives without considering DM's risk preference.

However, DM often has diverse risk attitudes. As DM often has different attitudes, decision results in the same MADM problems may vary.

To overcome the above limitations, we propose a new nonlinear programming method for solving the above MADM problems based on the concept of satisfactory degree. The satisfactory degree of alternatives to the IVIFPIS may be regarded as continuous and monotonic functions with respect to the membership and non-membership degrees of alternatives on attributes. As a result, the satisfactory degrees of alternatives to the IVIFPIS are IF sets, called the satisfactory IF sets, which are used to rank the alternatives according to the general satisfactory degree of the alternatives.

The rest of the paper is organized as follows. In Section 2, we briefly review the notions of an IF set and an IVIF set. In Section 3, we discuss a MADM problem in which ratings of alternatives on attributes are expressed with IVIF sets and preference information on the attributes is incomplete. In Section 4, we establish an auxiliary nonlinear programming model based on the concept of satisfactory degrees and then develop a new methodology for solving such a MADM problem. Also simpler nonlinear programming models are used to obtain satisfactory IF sets, and the general satisfactory degrees of IF sets are used to generate the ranking order of the alternatives, where the general satisfactory degree can reflect the preference of decision makings. In Section 5, a real example is analyzed in detail. In Section 6, we make a comparison of the proposed model with Li's method (Li, 2014) and Wei's method (Wei, et al., 2011) to show its advantages and effectiveness as well as the difference among these three kinds of methods. The paper is concluded with a brief summary in Section 7.

Basic Definitions

In this section, we introduce some basis concepts related to the IF and IVIF sets.

Definition 1 (Atanassov, 1983)

Let $X = \left\{x_1, x_2, \ldots x_m\right\}$ be a finite universal set. An IF set A on X is defined as:

$$\mathrm{A} = \left\{\left\langle x_j, \mu_A(x_j), \nu_A(x_j)\right\rangle \middle| x_j \in X\right\},$$

where

$$\mu_A : X \to \left[0,1\right], \quad x_j \to \mu_A\left(\mathrm{x}_j\right) \in \left[0,1\right]$$

and

$$v_A : X \to \left[0,1\right], \quad x_j \to \mu_A\left(\mathrm{x}_j\right) \in \left[0,1\right]$$

are called the degrees of membership and non-membership of the element x_j to the set A, respectively, and they satisfy $0 \le \mu_A(x_j) + \nu_A(x_j) \le 1$.

Let

$$\tau_A(x_j) = 1 - \mu_A(x_j) - \nu_A(x_j)$$

be called the IF index of an element x_j in the IF set A. It is the degree of indeterminacy membership of the element x_j to the IF set A. Obviously,

$$0 \le \tau_A(x_j) \le 1.$$

Definition 2 (Atanassov et al., 1989)

Let $X = \{x_1, x_2, \ldots x_m\}$ be a finite universal set and I be the set of all closed sub intervals of the interval [0, 1]. An IVIF set D on X is defined as:

$$D = \{\langle x_j, u_D(x_j), \nu_A(x_j)\rangle | x_j \in X\},$$

where

$$\mu_D : X \to [0,1], \quad x_j \to \mu_D(\mathrm{x}_j) \in [0,1]$$

and

$$\upsilon_D : X \to [0,1], \quad x_j \to \mu_D(\mathrm{x}_j) \in [0,1]$$

are called the interval-valued degrees of membership and non-membership of the element x_j to the set D, respectively, and they satisfy

$$0 \le \sup\{\mu_D(\mathrm{x}_j)\} + \sup\{\upsilon_D(\mathrm{x}_j)\} \le 1.$$

The lower and upper bounds $\mu_D(x_j)$ of are $\mu_D^l(\mathrm{x}_j)$ and $\mu_D^u(\mathrm{x}_j)$, respectively, and those of $\nu_D(x_j)$ are $\nu_D^l(\mathrm{x}_j)$ and $\nu_D^u(\mathrm{x}_j)$, respectively. Then, the IVIF set D can be expressed as:

$$D = \{\langle x_j, [\mu_D^l(x_j), \mu_D^u(x_j)], [\nu_D^l(x_j), \nu_D^u(x_j)]\rangle \mid x_j \in X\}$$

MADM Problems with IVIF Sets and Incomplete Preference Information

The discussed problems in this paper mainly are a type of MADM problems which have a finite numbers of feasible alternatives evaluated on multi-attributes . Mathematically assume that a MADM problems has m alternatives $x_i (i = 1, 2, ..., m)$ evaluated on n attitudes a_j (j $= 1, 2, ..., n)$ .Denote the alternative set and the attitude set by X $= \{x_1, x_2, ..., x_m\}$ and A$= \{a_1, a_2, ..., a_n\}$.

Assume that $[\mu_j^l(\mathrm{x}_i), \mu_j^u(\mathrm{x}_i)]$ and $[\nu_j^l(\mathrm{x}_i), \nu_j^u(\mathrm{x}_i)]$ be intervals of degrees of membership and non-membership of alternatives $x_i \in X$ on attributes $\mathrm{a}_j \in A$ with respect to the concert "excellence", respectively, where $0 \le \mu_{ij}^l \le \mu_{ij}^u \le 1, 0 \le \nu_{ij}^l \le \nu_{ij}^u \le 1$ and $\mu_{ij}^u + \nu_{ij}^u \le 1$.In other words, ratings of the alternatives $x_i \in X$ on attributes $\mathrm{a}_j \in A$ are IVIF sets. Denoted by $X_{ij} = \{\langle x_i, [\mu_{ij}^l, \mu_{ij}^u], [\nu_{ij}^l, \nu_{ij}^u]\rangle\}$.

For short, denote $X_{ij} = \langle x_i, [\mu_{ij}^l, \mu_{ij}^u], [\nu_{ij}^l, \nu_{ij}^u]\rangle$.Thus, a MADM problem with IVIF sets can be expressed concisely in the interval-valued matrix format as follows:

$$\mathrm{M} = \left(\langle[\mu_{ij}^l, \mu_{ij}^u], [\nu_{ij}^l, \nu_{ij}^u]\rangle\right)_{m \times n} =$$

$$\begin{pmatrix} \langle[\mu_{11}^l, \mu_{11}^u], [\nu_{11}^l, \nu_{11}^u]\rangle & \langle[\mu_{12}^l, \mu_{12}^u], [\nu_{12}^l, \nu_{12}^u]\rangle & \cdots \langle[\mu_{1n}^l, \mu_{1n}^u], [\nu_{1n}^l, \nu_{1n}^u]\rangle \\ \langle[\mu_{21}^l, \mu_{21}^u], [\nu_{21}^l, \nu_{21}^u]\rangle & \langle[\mu_{22}^l, \mu_{22}^u], [\nu_{22}^l, \nu_{22}^u]\rangle & \cdots \langle[\mu_{2n}^l, \mu_{2n}^u], [\nu_{2n}^l, \nu_{2n}^u]\rangle \\ \cdots & \cdots & \cdots \quad \cdots \\ \langle[\mu_{m1}^l, \mu_{m1}^u], [\nu_{m1}^l, \nu_{m1}^u]\rangle & \langle[\mu_{m2}^l, \mu_{m2}^u], [\nu_{m2}^l, \nu_{m2}^u]\rangle & \cdots \langle[\mu_{mn}^l, \mu_{mn}^u], [\nu_{mn}^l, \nu_{mn}^u]\rangle \end{pmatrix}$$

which is referred to as an IVIF decision matrix used to represent the MADM problem with IVIF sets.

Let $M_i = \left(\langle[\mu_{ij}^l, \mu_{ij}^u], [\nu_{ij}^l, \nu_{ij}^u]\rangle\right)_{1 \times n}$ represent an IVIF set row vector of n-dimension, often x_i may be interchangeably used with M_i.

In real decision process, the DM may pay more attention to importance of some attributes, i.e., specify some preference relations on weights of attributes according to his/her knowledge, experience and judgment. Such information of attribute weights is incomplete. Usually incomplete information of attribute weights can be obtained according to partial preference relations on weights given by the DM and has several different forms of structures. Summarizing earlier research (Li, 2011) these incomplete weight information structures may be expressed in the following five basic relations among attribute weights, which are denoted by subsets Λ_s (s $= 1, 2, 3, 4, 5)$ of weight vectors in Λ_0, respectively.

A Week Ranking:

$$\Lambda_1 = \{\omega \in \Lambda_0 \,|\, \omega_i \ge \omega_j \text{ for all } i \in T_1 \text{ and } \mathrm{j} \in \mathrm{J}_1\},$$

where T1 and J1 are two disjoint subsets of the subscript index set $N = \{1, 2, ..., n\}$ of all attributes. Thus, Λ_1 is a set of all weight vectors in Λ_0 with the property that the weight of an attribute in the set T_1 is greater than or Equationual to that of an attribute in the set J_1.

A Strick Ranking:

$$\Lambda_2 = \left\{ \omega \in \Lambda_0 \left| \alpha_{ij} \leq \omega_i - \omega_j \leq \beta_{ij} \text{ for all } i \in T_2 \text{ and } j \in J_2 \right. \right\},$$

where $\alpha_{ij} > 0$ and $\beta_{ij} > 0$ are constants, satisfying β_{ij} and α_{ij}; T_2 and J_2 are two disjoint subsets of N. Thus, Λ_2 is a set of all weight vectors in Λ_0 with the property that the weight of an attribute in the set T_2 is greater than or Equationual to that of an attribute in the set J_2 but their difference does not exceed some range, i.e., a closed interval $\left[\alpha_{ij}, \beta_{ij}\right]$.

A Ranking with Multiple:

$$\Lambda_3 = \left\{ \omega \in \Lambda_0 \left| \omega_{ij} \geq \xi_{ij} \omega_j \text{ for all } i \in T_3 \text{ and } j \in J_3 \right. \right\},$$

where $\xi_{ij} > 0$ is a constant; T_3 and J_3 are two disjoint subsets of N. Thus, Λ_3 is a set of all weight vectors in Λ_0 with the property that the weight of an attribute in the set T_3 is greater than or Equationual to ξ_{ij} multiple of that of an attribute in the set J_3.

An Interval Form:

$$\Lambda_4 = \left\{ \omega \in \Lambda_0 \left| \eta_j \leq \omega_j \leq \gamma_j \text{ for all } j \in T_4 \right. \right\},$$

where $\eta_j > 0$ and $\gamma_j > 0$ are constants, satisfying $\eta_j < \gamma_j$; J_4 is a subset of N. Thus, Λ_4 is a set of all weight vectors in Λ_0 with property that the weight of an attribute in the set J_4 does not exceed some range, i.e., a closed interval $\left[\eta_j, \gamma_j\right]$.

A Ranking of Difference:

$$\Lambda_5 = \left\{ \omega \in \Lambda_0 \left| \omega_i - \omega_j \geq \omega_k - \omega_s \text{ for all } i \in T_5, j \in J_5, k \in K_5 \text{ and } \mathrm{s} \in \mathrm{S}_5 \right. \right\},$$

where T_5, J_5, K_5 and S_5 are four disjoint subsets of N. Thus, Λ_5 is a set of all weight vectors in Λ_0 with the property that the difference between weights of attributes in the sets T_5 and J_5 is greater than or Equationual to that of attributes in the sets K_5 and S_5.

In reality, usually the preference information structure Λ of attribute importance may consist of several sets of the above basic sets Λ_s $(s = 1,2,3,4,5)$. For example, the DM may provide a preference information structure expressed as follows (Li, 2014):

$$\widehat{\Lambda} = \{\omega \in \Lambda_0 \mid 0.15 \le \omega_1 \le 0.55, 0.2 \le \omega_2 \le 0.65, 0.1 \le \omega_3 \le 0.35, \omega_2 \ge 1.2\omega_1, 0.02 \le \omega_2 - \omega_3 \le 0.45\},$$

which may be decomposed into the following three basic subsets:

$$\widehat{\Lambda}_2 = \{\omega \in \Lambda_0 \mid 0.02 \le \omega_2 - \omega_3 \le 0.45\},\ \widehat{\Lambda}_3 = \{\omega \in \Lambda_0 \mid \omega_2 \ge 1.2\omega_1\}$$

and

$$\widehat{\Lambda}_4 = \{\omega \in \widehat{\Lambda}_0 \mid 0.15 \le \omega_1 \le 0.55, 0.2 \le \omega_2 \le 0.65, 0.1 \le \omega_3 \le 0.35\},$$

where

$$\Lambda_0 = \{\omega = (\omega_1, \omega_2, \omega_3)^{\mathrm{T}} \mid \omega_1 + \omega_2 + \omega_3 = 1,\ \omega_i \ge \varepsilon\ (j = 1,2,3)\}.$$

In other words, the information structure $\widehat{\Lambda}$ consists of the above three sets $\widehat{\Lambda}_2$, $\widehat{\Lambda}_3$ and $\widehat{\Lambda}_4$.

A Principle and Method for IVIF MADM with Incomplete Preference Information

An IVIFPIS and an IVIFNIS are defined as x^+ and x^-, respectively. Each IVIF set of x^+ on attributes $\mathrm{a}_j \in A$ is chosen as $\{\langle \mathrm{a}_j, [1,1], [0,0] \rangle\}$. Namely, the degrees of membership and the non-membership of x^+ on $\mathrm{a}_j \in A$ are 1 and 0, respectively. They are also denoted by $\langle [1,1], [0,0] \rangle$ for short. Thus, the IVIF set vector of the IVIFPIS x^+ on all attributes is expressed concisely in the vector format as follows:

$$\left(\left\langle [\mu_j^{l+}, \mu_j^{u+}], [\nu_j^{l+}, \nu_j^{u+}] \right\rangle\right)_{1\times n} = \left(\left\langle [1,1], [0,0] \right\rangle\right)_{1\times n}$$

In practice, an IVIF set $x_i = \left(\left\langle [\mu_{ij}^{l+}, \mu_{ij}^{u+}], [\nu_{ij}^{l+}, \nu_{ij}^{u+}] \right\rangle\right)_{1\times n}$ can be interpreted mathematically as follows: the interval $[\mu_{ij}^l, \mu_{ij}^u]$ means that the membership degree μ_{ij} of the alternative $x_i \in X$ on the attribute $\mathrm{a}_j \in A$ takes any value between μ_{ij}^l and μ_{ij}^u, i.e., $\mu_{ij} \in [\mu_{ij}^l, \mu_{ij}^u]$.Likewise, $[\nu_{ij}^l, \nu_{ij}^u]$ means that the non-membership degree of the alternative $x_i \in X$ on the attribute $\mathrm{a}_j \in A$ takes any value between ν_{ij}^l

and ν_{ij}^{u}, i.e. $\nu_{ij} \in \left[\nu_{ij}^{l}, \nu_{ij}^{u}\right]$. Thus, in order to compare alternatives $x_i (i = 1, 2, ..., m)$, some methods should be used to measure the distance between alternatives x_i and x^- as well as that between x^+ and x^-. Here the weighted Euclidean distance between alternatives x_i and x^- and that between x^+ and x^- are respectively defined as follows (Atanassov, 1999; Lakshmana, 2011):

$$d\left(x_i, x^-\right) = \sqrt{\sum_{j=1}^{n} \left\{\left(\omega_j \mu_{ij}\right)^2 + \left[\omega_j (1 - \nu_{ij})\right]^2\right\}} \tag{1}$$

and

$$d\left(x^+, x^-\right) = \sqrt{\sum_{j=1}^{n} (\omega_j^2 + \omega_j^2)} = \sqrt{2}\sqrt{\sum_{j=1}^{n} \omega_j^2} \tag{2}$$

It is worthwhile to notice that the greater the value of $d\left(x_i, x^-\right)$ the better the alternative x_i. The satisfactory degree of the alternative x_i is defined as follows:

$$\rho_i\left(\left(\mu_{ij}\right)_{m\times n}, \left(\nu_{ij}\right)_{m\times n}, \left(\omega_j\right)_{n\times 1}\right) = \frac{d\left(x_i, x^-\right)^2}{d\left(x^+, x^-\right)^2}$$

which is ratio of the square of the weight Euclidean distance between x_i and x^- to the square of the weight Euclidean distance between x^+ and x^-, where $\left(\mu_{ij}\right)_{m\times n}$ and $\left(\nu_{ij}\right)_{m\times n}$ represent matrices of $m \times n$ with $\mu_{ij} \in \left[\mu_{ij}^{l}, \mu_{ij}^{u}\right]$ and $\nu_{ij} \in \left[\nu_{ij}^{l}, \nu_{ij}^{u}\right]$, and $\omega = \left(\omega_j\right)_{n\times 1}$ is the weight vector in the preference information structure Λ defined above. Obviously,

$$0 \le d\left(x_i, x^-\right) \le d\left(x^+, x^-\right);$$

yielding $0 \le \rho_i\left(\left(\mu_{ij}\right)_{m\times n}, \left(\nu_{ij}\right)_{m\times n}, \left(\omega_j\right)_{n\times 1}\right) \le 1$.

According to Eqs. (1) and (2), the satisfactory degree of the alternative x_i can explicitly be rewritten out as follows:

$$\rho_i\left(\left(\mu_{ij}\right)_{m\times n}, \left(\nu_{ij}\right)_{m\times n}, \left(\omega_j\right)_{n\times 1}\right) = \frac{\sum_{j=1}^{n} \left\{\left(\omega_j \mu_{ij}\right)^2 + \left[\omega_j (1 - \nu_{ij})\right]^2\right\}}{2\sum_{j=1}^{n} \omega_j^2} \tag{3}$$

It is noticing that the satisfactory degree defined by Eq. (3) is different from that in the TOPSIS (Hwang, 1981). The reason is that the defined satisfactory degree ensures that the constructed nonlinear programming model in this paper is easier solved and costs less computation time than the nonlinear programming models derived from the concept of the TOPSIS in which the objective functions are ratios of the square roots for the sums of the squares that are very difficult to be computed.

Obviously, $\rho_i\left(\left(\mu_{ij}\right)_{m\times n},\left(\nu_{ij}\right)_{m\times n},\left(\omega_j\right)_{n\times 1}\right)$ is a continuous function of $\left(2m+1\right)n$ variables with $\mu_{ij}\in\left[\mu_{ij}^l,\mu_{ij}^u\right]$, $\nu_{ij}\in\left[\nu_{ij}^l,\nu_{ij}^u\right]$ and $\left(\omega_j\right)_{n\times 1}\in\Lambda$. The bigger

$\rho_i\left(\left(\mu_{ij}\right)_{m\times n},\left(\nu_{ij}\right)_{m\times n},\left(\omega_j\right)_{n\times 1}\right)$

the higher the satisfactory degree of the alternative xi . Now we prove the monotonicity of the aforementioned function $\rho_i\left(\left(\mu_{ij}\right)_{m\times n},\left(\nu_{ij}\right)_{m\times n},\left(\omega_j\right)_{n\times 1}\right)$.

Partial differential of $\rho_i\left(\left(\mu_{ij}\right)_{m\times n},\left(\nu_{ij}\right)_{m\times n},\left(\omega_j\right)_{n\times 1}\right)$ $(i=1,2,...,\mathrm{m})$ with respect to $\mu_{ij}(j=1,2,...,\mathrm{n})$ are respectively computed as follows:

$$\frac{\partial\rho_i((\mu_{ij})_{m\times n},(\nu_{ij})_{m\times n},(\omega_j)_{m\times n})}{\partial\mu_{ij}}=\frac{\omega_j^2\mu_{ij}}{\sum_{j=1}^{n}\omega_j^2} \tag{4}$$

Since $\mu_{ij}\geq 0$ it is easily seen from Eq. (4) that

$$\frac{\partial\rho_i((\mu_{ij})_{m\times n},(\nu_{ij})_{m\times n},(\omega_j)_{m\times n})}{\partial\mu_{ij}}\geq 0$$

Therefore, $\rho_i((\mu_{ij})_{m\times n},(\nu_{ij})_{m\times n},(\omega_j)_{m\times n})$ $(i=1,2,\ldots,m)$ are monotonic and non-decreasing functions of the variables $\mu_{ij}\in\left[\mu_{ij}^l,\mu_{ij}^u\right]$ $(j=1,2,...,\mathrm{n})$.

Partial differential of $\rho_i((\mu_{ij})_{m\times n},(\nu_{ij})_{m\times n},(\omega_j)_{m\times n})$ $(i=1,2,\ldots,m)$ with respect to $\nu_{ij}(j=1,2,...,\mathrm{n})$ are respectively computed as follows:

$$\frac{\partial\rho_i((\mu_{ij})_{m\times n},(\nu_{ij})_{m\times n},(\omega_j)_{m\times n})}{\partial\nu_{ij}}=\frac{\omega_j^2(1-\nu_{ij})}{\sum_{j=1}^{n}\omega_j^2} \tag{5}$$

Since $1-\nu_{ij}\geq 0$, it is easily seen from Eq. (5) that

$$\frac{\partial \rho_i((\mu_{ij})_{m\times n},(\nu_{ij})_{m\times n},(\omega_j)_{m\times n})}{\partial \nu_{ij}} \le 0$$

Therefore, $\rho_i((\mu_{ij})_{m\times n},(\nu_{ij})_{m\times n},(\omega_j)_{n\times 1})$ $(i=1,2,...,m)$ are monotonic and non-increasing functions of the variables $\nu_{ij} \in \left[\mu_{ij}^l,\mu_{ij}^u\right]$ $(j=1,2,...,\mathrm{n})$.

Thus, $\rho_i((\mu_{ij})_{m\times n},(\nu_{ij})_{m\times n},(\omega_j)_{n\times 1})$ reaches its maximum at the upper bound μ_{ij}^u of the interval $\left[\mu_{ij}^l,\mu_{ij}^u\right]$ and at the lower bound ν_{ij}^l of the interval $\left[\nu_{ij}^l,\nu_{ij}^u\right]$.As a result, the nonlinear optimization model can be established as follows:

$$\rho_i^u = \max\left\{\rho_i((\mu_{ij})_{m\times n},(\nu_{ij})_{m\times n},(\omega_j)_{n\times 1})_{n\times 1}\right\}$$
$$s.t(\omega_j)_{n\times 1} \in \Lambda$$

Similarly, $\rho_i((\mu_{ij})_{m\times n},(\nu_{ij})_{m\times n},(\omega_j)_{n\times 1})$ reaches its minimum at the lower bound μ_{ij}^l of the interval $\left[\mu_{ij}^l,\mu_{ij}^u\right]$ and at the upper bound ν_{ij}^u of the interval $\left[\nu_{ij}^l,\nu_{ij}^u\right]$ As a result, the nonlinear optimization model can be established as follows:

$$\rho_i^l = \min\left\{\rho_i((\mu_{ij})_{m\times n},(\nu_{ij})_{m\times n},(\omega_j)_{n\times 1})_{n\times 1}\right\}$$
$$s.t(\omega_j)_{n\times 1} \in \Lambda$$

It is easily known that

$$\rho_i^l + (1-\rho_i^u) = 1-(\rho_i^l - \rho_i^u) \le 1.$$

Hence, ρ_i^l and ρ_i^u satisfy the following conditions: $0 \le \rho_i^l \le 1, 0 \le 1-\rho_i^u \le 1$ and $1-(\rho_i^l-\rho_i^u) \le 1$ .According to the definition of the IF set (Zaheh, 1975; Atanassov, 1983), the satisfactory degree $\left[\rho_i^l,\rho_i^u\right]$ of x_i can Equationuivalently be expressed as an IF set $\rho_i = \left\langle \rho_i^l, 1-\rho_i^u \right\rangle$, which means that satisfactory and non-satisfactory degrees of the alternative $x_i \in X$ to x^+ are ρ_i^l and $1-\rho_i^u$,respectively. $\tau_{\rho i} = \rho_i^u - \rho_i^l$ reflects that there is an uncertainty on satisfactory degree of the alternative $x_i \in X$ to x^+.

Let $\rho_k = \left\langle \rho_k^l, 1-\rho_k^u \right\rangle$ and $\rho_p = \left\langle \rho_p^l, 1-\rho_p^u \right\rangle$ be two satisfactory IF sets.

$$LG(\rho k) = \rho_k^l + \lambda(\rho_k^u - \rho_k^l)$$

and

$$LG(\rho p) = \rho_p^l + \lambda(\rho_p^u - \rho_p^l)$$

are the general satisfactory degrees of x_k and x_p, respectively, where $\lambda \in [0,1]$ is a weight which represents the decision maker's preference information.

$\lambda \in [0.5,1]$ shows that the DM prefers uncertainty or negative feeling;
$\lambda \in [0,0.5]$ shows that the DM prefers certainty or positive feeling;
$\lambda = 0.5$ shows that the DM is indifferent between positive feeling and negative feeling.

Let x_k and x_p be two alternatives. A rule based on the general satisfactory degrees can be summarized as follows:

Rule 1: If $LG(\rho k) > LG(\rho p)$, then the alternative x_k is better than the alternative x_p;
Rule 2: If $LG(\rho k) = LG(\rho p)$, then the alternative x_k is as good as the alternative x_p;
Rule 3: If $LG(\rho k) < LG(\rho p)$, the alternative x_k is better than the alternative x_p.

Based on the above discussions, we put forward a practical algorithm for solving MADM problems with IVIF sets and incomplete preference information. The procedure is described as follows:

Step 1: Identify the evaluation attitudes and alternatives.
Step 2: Pool the DM's opinion to get the ratings of alternatives on attributes, i.e., the IVIF decision matrix $\mathrm{M} = (\langle [\mu_{ij}^l, \mu_{ij}^u], [\nu_{ij}^l, \nu_{ij}^u] \rangle)_{m\times n}$.
Step 3: Pool the decision maker's opinion to get a specific preference information structure on attributes, i.e, Λ.
Step 4: Construct auxiliary nonlinear programming models for alternatives $x_i (i = 1,2,...,m)$.
Step 5: Solve the above constructed auxiliary nonlinear programming models using the nonlinear programming method and obtain satisfactory IF sets $\rho_i = \langle \rho_i^l, 1 - \rho_i^u \rangle (i = 1,2,...,m)$.
Step 6: Compute the general satisfactory degrees of $\mathrm{LG}(\rho_i)(i = 1,2,...,m)$.
Step 7: Utilize the general satisfactory degrees of $\mathrm{LG}(\rho_i)(i = 1,2,...,m)$ to rank the alternatives $x_i (i = 1,2,...,m)$, and select the most desirable one (s).
Step 8: End.

A Real Example Analysis

Green supplier management as one of important part of supply chain is a critical issue for effective green supply chain management. supplier selection is the direct outcome of supply chain. The characteristics of green supplier selection are friendly to environment, focusing the long-time cooperation, ettering the system of analyzing the suppliers. The green purchase is the most important linkage in green supply chain management. The success of purchase activity is highly dependent on successful selection of green suppliers.

Fujian Science and Technology Company Limited is the national high-tech enterprise of China. Currently, the company has a staff of 1000. After preliminary scanning, five candidate green suppliers

x_1, x_2, x_3, x_4 and x_5 remain for further evaluation. Through discussion with other researchers and practitioners, four independent attributes are chosen for this study, involving achievement index a_1, productive capacity index a_2, economic index a_3 and service index a_4.

We invite some related experts in the field of wafer testing (including the president, vice presidents, managers of testing departments, staffs, and many stakeholders) to participate in the interview and filling up questionnaires. After data eliciting and statistical treatment, the ratings of the five suppliers on the four attributes are obtained as follows:

$$M = \begin{pmatrix} \langle[0.4,0.5],[0.3,0.4]\rangle & \langle[0.4,0.6],[0.2,0.4]\rangle & \langle[0.3,0.4],[0.4,0.5]\rangle & \langle[0.5,0.6],[0.1,0.3]\rangle \\ \langle[0.5,0.6],[0.2,0.3]\rangle & \langle[0.6,0.7],[0.2,0.3]\rangle & \langle[0.5,0.6],[0.3,0.4]\rangle & \langle[0.6,0.8],[0.1,0.2]\rangle \\ \langle[0.3,0.5],[0.3,0.4]\rangle & \langle[0.1,0.3],[0.5,0.6]\rangle & \langle[0.2,0.5],[0.4,0.5]\rangle & \langle[0.2,0.3],[0.4,0.6]\rangle \\ \langle[0.2,0.5],[0.3,0.4]\rangle & \langle[0.4,0.7],[0.1,0.2]\rangle & \langle[0.4,0.5],[0.3,0.5]\rangle & \langle[0.5,0.8],[0.1,0.2]\rangle \\ \langle[0.3,0.4],[0.1,0.3]\rangle & \langle[0.7,0.8],[0.1,0.2]\rangle & \langle[0.5,0.6],[0.2,0.4]\rangle & \langle[0.6,0.7],[0.1,0.2]\rangle \end{pmatrix}$$

Assume that the DM provides a preference information structure on the attributes $a_i (i = 1,2,3,4)$, which is expressed with the set as follows:

$$\Lambda = \left\{W \in \Lambda_0 \middle| 0.15 \le \omega_1 \le 0.25, 0.2 \le \omega_2 \le 0.35, 0.35 \le \omega_3 \le 0.40, \omega_4 \le 0.3\omega_2\right\}.$$

Using the developed method in this paper, the aforementioned specific green suppliers selection problem can be solved:

Step 1: As stated earlier, it is easy to see that there are four evaluation attributes and five evaluation alternatives.

Step 2: Using the statistic and knowledge engineering, ratings of alternatives on the four attributes are given by DM. The IVIF decision matrix

$$\mathrm{M} = (\langle[\mu_{ij}^l, \mu_{ij}^u], [\nu_{ij}^l, \nu_{ij}^u]\rangle)_{5\times4}.$$

Step 3: According to the experience, knowledge and preference, the decision maker may give a specific preference information structure Λ.

Step 4: The optimization models for alternative x_1 are established as follows:

$$\rho_i^u = \max\left\{\frac{0.74\omega_1^2 + \omega_2^2 + 0.52\omega_3^2 + 1.17\omega_4^2}{2\left(\omega_1^2 + \omega_2^2 + \omega_3^2 + \omega_4^2\right)}\right\}$$

$$s.t\begin{cases} 0.15 \le \omega_1 \le 0.25 \\ 0.2 \le \omega_2 \le 0.35 \\ 0.35 \le \omega_3 \le 0.40 \\ \omega_4 \ge 0.3\omega_2 \\ \omega_1 + \omega_2 + \omega_3 + \omega_4 = 1 \\ \omega_1 \ge 0, \omega_2 \ge 0, \omega_3 \ge 0, \omega_4 \ge 0 \end{cases}$$

and

$$\rho_i^l = \max\left\{\frac{0.52\omega_1^2 + 0.52\omega_2^2 + 0.34\omega_3^2 + 0.74\omega_4^2}{2\left(\omega_1^2 + \omega_2^2 + \omega_3^2 + \omega_4^2\right)}\right\}$$

$$s.t\begin{cases} 0.15 \le \omega_1 \le 0.25 \\ 0.2 \le \omega_2 \le 0.35 \\ 0.35 \le \omega_3 \le 0.40 \\ \omega_4 \ge 0.3\omega_2 \\ \omega_1 + \omega_2 + \omega_3 + \omega_4 = 1 \\ \omega_1 \ge 0, \omega_2 \ge 0, \omega_3 \ge 0, \omega_4 \ge 0 \end{cases}$$

Step 5: Optimal objective function values of the above nonlinear programming models are obtained as follows:

$\rho_1^l = 0.2162, \quad \rho_1^u = 0.5897.$

Thus the satisfactory IF set of the alternative x_1 is $\rho_1 = \langle 0.2162, 0.5897 \rangle$.

In a similar manner, i.e., repeating steps 4 and 5 for the alternatives $x_i (i = 1,2,3,4,5)$, we can construct the auxiliary nonlinear programming models for the alternatives x_i. By solving these nonlinear programming models, the satisfactory degree IF sets of the alternatives $x_i (i = 1,2,3,4,5)$ are obtained as follows:

$\rho_2 = \langle 0.3393, 0.4503 \rangle, \rho_3 = \langle 0.2447, 0.5056 \rangle,$
$\rho_4 = \langle 0.2231, 0.4731 \rangle, \rho_5 = \langle 0.3412, 0.3945 \rangle,$

Step 6: The general satisfaction degrees $\mathrm{LG}(\rho_i)(i = 1,2,3,4,5)$ are computed as follows:

$$\mathrm{LG}(\rho_1) = 0.2162 + 0.1940\lambda$$
$$\mathrm{LG}(\rho_2) = 0.3393 + 0.2104\lambda$$
$$\mathrm{LG}(\rho_3) = 0.2247 + 0.2497\lambda$$
$$\mathrm{LG}(\rho_4) = 0.2231 + 0.3038\lambda$$
$$\mathrm{LG}(\rho_5) = 0.3412 + 0.2643\lambda$$

Step 7: It is easy to know that

$$\mathrm{LG}(\rho_5) > \mathrm{LG}(\rho_4) > \mathrm{LG}(\rho_3) > \mathrm{LG}(\rho_2) > \mathrm{LG}(\rho_1)$$

holds for any given weight $\lambda \in [0,0.4)$.Hence, the ranking order of the five companies is $x_5 > x_2 > x_3 > x_4 > x_1$.In this case, the best selection is the company x_5.

However, if $\lambda \in (0.4,1]$, then

$$\mathrm{LG}(\rho_5) > \mathrm{LG}(\rho_2) > \mathrm{LG}(\rho_4) > \mathrm{LG}(\rho_3) > \mathrm{LG}(\rho_1),$$

the ranking order of the five companies is $x_5 > x_2 > x_4 > x_3 > x_1$, and the best selection is the company x_5.

Especially, if $\lambda = 0.4$, then $\mathrm{LG}(\rho_5) > \mathrm{LG}(\rho_2) > \mathrm{LG}(\rho_4) = \mathrm{LG}(\rho_3) > \mathrm{LG}(\rho_1)$, the ranking order of the five companies is $x_5 > x_2 > x_4 = x_3 > x_1$, and the best selection is the company x_5. Obviously, the ranking order of the five alternatives is related to the attitude parameter $\lambda \in [0,1]$.

Discussion

We compare the results obtained by Li's method, Wei's method and the method proposed in this paper.

Li's Method (Li, 2011)

Firstly, We shall use the closeness coefficient based nonlinear programming method to solve the above IVIFMADM for short (Li, 2011)

Step 1: See Step 1 in section5.
Step 2: See Step 2 in section5.
Step 3: See Step 3 in section5.
Step 4: According to Eqs. (11) and (12) in (Li, 2011). Two auxiliary nonlinear programming models can be constructed for 1 for alternative x_1 as follows:

$$\begin{cases} c_i^u = \max \dfrac{0.74w_1^2 + w_2^2 + 0.52w_3^2 + 1.17w_4^2}{0.74w_1^2 + w_2^2 + 0.52w_3^2 + 1.17w_4^2 + 0.34w_1^2 + 0.2 * w_2^2 + 0.52w_3^2 + 0.17w_4^2} \\ s.t \\ (w_j)_{n\times 1} \in \Lambda. \end{cases}$$

and

$$\begin{cases} c_i^l = \min \dfrac{0.52w_1^2 + 0.52w_2^2 + 0.34w_3^2 + 0.74w_4^2}{0.52w_1^2 + 0.52w_2^2 + 0.34w_3^2 + 0.74w_4^2 + 0.52w_1^2 + 0.52w_2^2 + 0.74w_3^2 + 0.34w_4^2} \\ s.t \\ (w_j)_{n\times 1} \in \Lambda. \end{cases}$$

respectively.

Step 5: Using existing nonlinear programming methods (6.1) and (6.2) in (Li, 2011) optimal objective function values of above nonlinear programming model are obtained as follows:

$\rho_1^l = 0.4041, \quad \rho_1^u = 0.6773.$

respectively, ie., the satisfactory IF set of the alternative x_1 is $\rho_1 = (0.4041, 0.3227)$.

In the same way, i. e., repeating step 4 and step 5 for the alternatives $x_1 (i = 2, 3, 4)$, we can construct the auxiliary nonlinear programming models for the alternatives x_1, which are similar to Eq. (6.1) and (6.2). Solving these nonlinear programming models, the satisfactory degree IF sets of the alternatives $x_1 (i = 2, 3, 4)$ are obtained as follows:

$\rho_2 = (0.6468, 0.1459), \quad \rho_3 = (0.1969, 0.4389),$ respectively.
$\rho_4 = (0.4560, 16630), \rho_5 = (0.6437, 0.1050)$

Step 6: Use Eq. (13) in (Li, 2011) the inclusion comparison probability of the IF sets c_2 and c_3 can be calculated as follows:

$$p_{12} = p(c_1 \supseteq c_2) =$$
$$\max\{1 - \max\{\frac{0.851 - 0.4041}{(0.6773 - 0.4041) + (0.8541 - 0.6468)}\}\} = 0.0635$$

In the same way, we obtain:

$$p_{13} = 0.7253, p_{14} = 0.4209, p_{15} = 0.2196,$$
$$p_{23} = 1.000, p_{24} = 0.7481, p_{25} = 0.6054,$$
$$p_{33} = 0.1417, p_{34} = 0.6054, p_{45} = 0.3021$$

Likewise, the inclusion comparison probabilities of all other IF sets can be obtained and expressed in the matrix format as follows:

$$p = \begin{pmatrix} 0.5000 & 0.9365 & 0.2747 & 0.5791 & 0.7804 \\ 0.0635 & 0.5000 & 0.0000 & 0.2519 & 0.3946 \\ 0.7253 & 1.0000 & 0.5000 & 0.8583 & 0.3946 \\ 0.4209 & 0.7481 & 0.1417 & 0.5000 & 0.6979 \\ 0.2196 & 0.6054 & 0.6054 & 0.3021 & 0.5000 \end{pmatrix}$$

Step 7: Use Eq. (15) in (Li, 2011), and $m = 5$,the optimal membership degrees of the alternatives $x_i (i = 1, 2, 3, 4, 5)$ are computed as follow:

$$\theta_1 = \frac{1}{m(m-1)}(\prod_{k=1}^{m} p_{1k} + \frac{m}{2} - 1) =$$
$$\frac{1}{5(5-1)}(0.5000 + 0.0635 + 0.7537 + 0.3400 + 0.0641 + 1.5) = 0.1611$$
$$\theta_2 = \frac{1}{m(m-1)}(\prod_{k=1}^{m} p_{2k} + \frac{m}{2} - 1) =$$
$$\frac{1}{5(5-1)}(0.9365 + 0.5000 + 1.0000 + 0.6805 + 0.4588 + 1.5) = 0.2538$$
$$\theta_3 = \frac{1}{m(m-1)}(\prod_{k=1}^{m} p_{3k} + \frac{m}{2} - 1) =$$
$$\frac{1}{5(5-1)}(0.2463 + 0.0000 + 0.5000 + 0.1417 + 0.5968 + 1.5 = 0.1492$$
$$\theta_4 = \frac{1}{m(m-1)}(\prod_{k=1}^{m} p_{4k} + \frac{m}{2} - 1) =$$
$$\frac{1}{5(5-1)}(0.6600 + 0.3195 + 0.8583 + 0.5000 + 0.3021 + 1.5) = 0.2070$$
$$\theta_5 = \frac{1}{m(m-1)}(\prod_{k=1}^{m} p_{5k} + \frac{m}{2} - 1) =$$
$$\frac{1}{5(5-1)}(0.9359 + 0.5412 + 0.4032 + 0.6979 + 0.5000 + 1.5 = 0.2289$$

Step 8: It is easy to see that the decreasing order of the optimal satisfactory scores of the alternatives $x_i (i = 1,2,3,4,5)$ is obtained as follow:

$$\theta_5 > \theta_2 > \theta_4 > \theta_1 > \theta_3$$

Hence, the ranking order of the four alternatives is generated as follows:

$$x_5 > x_2 > x_4 > x_1 > x_3$$

and the best alternative is x_5, i.e, the TV company.

Wei's Method (Wei et al, 2011)

In the following, we shall use the correlation coefficient to solve the above IVIFMADM problems for short (Wei et al, 2011)

Step 1: Determine the negative ideal alternative

$$\tilde{r}^- = [([0,0],[1,1]),([0,0],[1,1]),([0,0],[1,1]),([0,0],[1,1])]$$

Step 2: Utilize the model to establish the following single-objective programming model:

$$\begin{cases} \max \lambda_1 \\ s.t \\ 0.550w_1 + 0.600w_2 + 0.450w_3 + 0.675w_4 \geq \lambda_1 \\ 0.650w_1 + 0.700w_2 + 0.600w_3 + 0.700w_4 \geq \lambda_1 \\ 0.525w_1 + 0.325w_2 + 0.450w_3 + 0.375w_4 \geq \lambda_1 \\ 0.500w_1 + 0.700w_2 + 0.525w_3 + 0.750w_4 \geq \lambda_1 \\ 0.575w_1 + 0.800w_2 + 0.625w_3 + 0.750w_4 \geq \lambda_1 \\ (w_j)_{n\times 1} \in \Lambda \end{cases} \tag{5.1}$$

Solving this model, we get the weight vector of attributes: $w = (0.15, 0.20, 0.35, 0.30)^T$.

Step 3: Utilize the weight vector w^*, we

$$\Delta(\tilde{r}_i) = ([a_i, b_i],[c_i, d_i]) = \\ ([1 - \prod_{j=1}^{n}(1 - a_i j)^{w_j}, 1 - \prod_{j=1}^{n}(1 - b_i j)^{w_j}],[\prod_{j=1}^{n} c_i j^{w_j}, \prod_{j=1}^{n} d_i j^{w_j}])$$

We obtain the overall value $\Delta(\widetilde{r}_i)$ of the alternatives $x_i (i = 1,2,3,4,5)$ and the overall value $\Delta(\widetilde{r}_i)$ of the negative ideal alternatives x^*.

$$\Delta(\widetilde{r}_1) = ([0.4345, 0.4969], [0.2536, 0.3669])$$
$$\Delta(\widetilde{r}_2) = ([0.4940, 0.6255], [0.2071, 0.2656])$$
$$\Delta(\widetilde{r}_3) = ([0.2163, 0.3026], [0.4331, 0.4899])$$
$$\Delta(\widetilde{r}_4) = ([0.4860, 0.6345], [0.2071, 0.2558])$$
$$\Delta(\widetilde{r}_5) = ([0.5998, 0.6330], [0.1625, 0.2125])$$

Step 4: Calculate the information intuitionistic energy of an alternative $x_i (i = 1,2,3,4,5)$ and the negative ideal alternative x^*.

$$E_1(x_1) = 0.3173, E_2(x_2) = 0.3743$$
$$E_3(x_3) = 0.2830, E_4(x_4) = 0.3735$$
$$E_5(x_5) = 0.4100, E^*(x^*) = 1.0000$$

Step 5: Calculate the correlation between an alternative $x_i (i = 1,2,3,4,5)$ and the negative ideal alternative x^*.

$$C_1(x_1, x^*) = 0.5508, C_2(x_2, x^*) = 0.3863$$
$$C_3(x_3, x^*) = 0.8676, C_4(x_4, x^*) = 0.3787$$
$$C_5(x_5, x^*) = 0.2928$$

Step 6: Rank all the alternatives $x_i (i = 1,2,3,4,5)$ and select the best one in accordance with $K_i (i = 1,2,3,4,5)$:

$$x_5 > x_2 > x_4 > x_1 > x_3.$$

and thus the best desirable alternatives is x_5.

As we can see, depending on the different approaches, the ordering of the alternatives is slightly different. Therefore, depending on the different approaches, the result may be lead to different decisions. However, the best alternative is x_5.

The example was solved by Li's method, the ranking order of the five alternatives is as follows: $x_5 > x_2 > x_4 > x_1 > x_3$, and the best alternative is x_5. However, the example was solved by Wei's method, rank all the alternatives $x_i (i = 1,2,3,4,5)$ as follows:

$$x_5 > x_4 > x_2 > x_1 > x_3.$$

But the best alternative is x_5 as well.

As we can see, the ranking order of the alternatives obtained by the proposed method is remarkably different from those obtained by Li's method and Wei's method. Furthermore, DMs with different preference attitudes have different ranking orders of the alternatives. Namely, a risk-taking DM may prefer $x_5 > x_2 > x_3 > x_4 > x_1$, where as a risk-averse decision maker may prefer

$$x_5 > x_2 > x_4 > x_3 > x_1.$$

These factors cannot be reflected in both Li's and Wei's methods, and the ranking order of the alternatives obtained by Li's and Wei's methods are a part of results in this paper.

CONCLUSION

In this paper, we study MADM problems in which ratings of alternatives on attributes are expressed by IVIFs and preference information on attributes is incomplete. A new method is developed for solving such problems. The main contributes of this study are summarized as follows:

1. We put forward the notion of satisfactory degree which is defined as a ratio of the square of the weight Euclidean distance between an alternative and the IVIF negative ideal solution to the square of the weight Euclidean distance between the IVIFNIS and the IVIFPIS. Nonlinear programming models also are developed to determine the satisfactory degrees. Moreover, simpler nonlinear programming models are deduced to obtain the satisfactory IF sets, and the general satisfactory degrees of the alternatives are used to generate the ranking order of the alternatives.
2. The IF and IVF sets are two special cases of the IVIF sets. Therefore, the proposed method in this paper can be applied to MADM problems with either IF sets or IVF sets, and different preference information structures between completely known and completely unknown ones.
3. We consider the DM's preference during the decision-making process. The DM's with different preference attitudes have different ranking orders of the alternatives, which can provide more choices for the DM's and greatly enhance the decision-making flexibility.
4. We have studied a type of heterogeneous decision making problems with some structure of incomplete information, which is expressed by different preference information structures.

ACKNOWLEDGMENT

This research was supported by the Key Program of National Natural Science Foundation of China (No. 71231003), the National Natural Science Foundation of China (Nos. 71171055, 71101033, 71001015 and 11401341), the National Natural Science Foundation of Fujian (No.2015J01287) and The Social science planning project of Fujian Province (FJ2016C029).

REFERENCES

Atanassov, K. T. (1983). *Intuitionistic fuzzy sets*. In Seventh Scientific Session of ITKR, Sofia.

Atanassov, K. T. (1999). *Intuitionistic fuzzy sets*. Heidelberg, Germany: Springer-Verlag. doi:10.1007/978-3-7908-1870-3

Atanassov, K. T., & Gargov, G. (1989). Interval valued intuitionistic fuzzy sets. *Fuzzy Sets and Systems*, *31*(3), 343–349. doi:10.1016/0165-0114(89)90205-4

Bustince, H., & Burillo, P. (1995). Correlation of interval-valued intuitionistic fuzzy sets. *Fuzzy Sets and Systems*, *74*(2), 237–244. doi:10.1016/0165-0114(94)00343-6

Cabrerizo, F. J., Perez, I. J., & Herrera Viedma, E. (2010). Managing the consensus in group decision making in an unbalanced fuzzy linguistic context with incomplete information. *Knowledge-Based Systems*, *23*(2), 169–181. doi:10.1016/j.knosys.2009.11.019

Cai, X. S., & Han, L. G. (2014). Some induced Einstein aggregation operators based on the data mining with interval-valued intuitionistic fuzzy information and their application to multiple attribute decision making. *Journal of Intelligent & Fuzzy Systems*, *27*, 331–338.

Chen, S. M., & Chiou, C. H. (2015). Multiattribute decision making based on interval-valued intuitionistic fuzzy sets, PSO techniques and evidential reasoning methodology. *IEEE Transactions on Fuzzy Systems*, *23*(6), 1905–1912. doi:10.1109/TFUZZ.2014.2370675

Chen, S. M., & Horng, Y. J. (1996). Finding inheritance hierarchies in interval-valued fuzzy concept networks. *Fuzzy Sets and Systems*, *84*(1), 75–83. doi:10.1016/0165-0114(95)00315-0

Chen, S. M., Lee, L. W., Liu, H. C., & Yang, S. W. (2012). Multi-attribute decision making based on interval valued intuitionistic fuzzy values. *Expert Systems with Applications*, *39*(12), 10343–10351. doi:10.1016/j.eswa.2012.01.027

Chen, S. M., & Li, T. S. (2013). Evaluating students answer scripts based on interval-valued intuitionistic fuzzy sets. *Information Sciences*, *235*, 308–322. doi:10.1016/j.ins.2012.12.031

Grzgorzewski, P. (2004). Distance between intuitionistic fuzzy sets and/or interval fuzzy sets based on Hausdorff metric. *Fuzzy Sets and Systems*, *148*(2), 319–328. doi:10.1016/j.fss.2003.08.005

Herrera, F., & Herrera, V. E. (2000). Linguistic decision analysis: Steps for solving decision problems under linguistic information. *Fuzzy Sets and Systems*, *115*(1), 67–82. doi:10.1016/S0165-0114(99)00024-X

Huang, B., Zhuang, Y. L., & Li, H. X. (2013). Information granulation and uncertainty measures in interval-valued intuitionistic fuzzy information systems. *European Journal of Operational Research*, *23*(1), 162–170. doi:10.1016/j.ejor.2013.05.006

Hung, K. C., & Wang, P. K. (2014). An integrated intuitionistic fuzzy similarity measures for medical problems. *International Journal of Computational Intelligent Systems*, *7*(2), 327–343. doi:10.1080/18756891.2013.856259

Hwang, C. L., & Yoon, K. (1981). *Multiple Attribute Decision Making: Methods and Applications. A State of the Art Survey*. Springer- Verlag. doi:10.1007/978-3-642-48318-9

Lakshmana, V., Nayagam, G., Muralikrishnan, G., & Sivaraman, G. (2011). Multi-criteria decision-making method based on intervalvalued intuitionistic fuzzy sets. *Expert Systems with Applications*, *38*(3), 1464–1467. doi:10.1016/j.eswa.2010.07.055

Li, D.-F. (2010). Mathematical-programming approach to matrix games with payoffs represented by Atanassovs interval-valued intuitionistic fuzzy sets. *IEEE Transactions on Fuzzy Systems*, *18*(6), 1112–1128. doi:10.1109/TFUZZ.2010.2065812

Li, D.-F. (2010). Linear programming method for MADM with interval valued intuitionistic fuzzy sets. *Expert Systems with Applications*, *37*(8), 5939–5945. doi:10.1016/j.eswa.2010.02.011

Li, D.-F. (2010). TOPSIS-based nonlinear-programming methodology for multiattribute decision making with interval-valued intuitionistic fuzzy sets. *IEEE Transactions on Fuzzy Systems*, *18*, 299–311.

Li, D.-F. (2011). Closeness coefficient based nonlinear programming method for interval-valued intuitionistic fuzzy multiattribute decision making with incomplete preference information. *Applied Soft Computing*, *11*(4), 3402–3418. doi:10.1016/j.asoc.2011.01.011

Li, D.-F. (2014). Decision and Game Theory in Management with Intuitionistic fuzzy sets. Springer. doi:10.1007/978-3-642-40712-3

Liu, P. D. (2013). Some generalized dependent aggregation operators with intuitionistic linguistic numbers and their application to group decision making. *Journal of Computer and System Sciences*, *79*(1), 131–143. doi:10.1016/j.jcss.2012.07.001

Liu, P. D. (2013). Some geometric aggregation operators based on interval intuitionistic uncertain linguistic variables and their application to group decision making. *Applied Mathematical Modelling*, *37*(4), 2430–2444. doi:10.1016/j.apm.2012.05.032

Liu, P. D. (2014). Some hamacher aggregation operators based on the interval valued intuitionistic fuzzy numbers and their application to group decision making. *IEEE Transactions on Fuzzy Systems*, *22*(1), 83–97. doi:10.1109/TFUZZ.2013.2248736

Liu, P. D., & Jin, F. (2012). A multi-attribute group decision-making method based on weighted geometric aggregation operators of interval-valued trapezoidal fuzzy numbers. *Applied Mathematical Modelling*, *36*(6), 2498–2509. doi:10.1016/j.apm.2011.09.006

Liu, P. D., Liu, Z. M., & Zhang, X. (2014). Some intuitionistic uncertain linguistic heronian mean operators and their application to group decision making. *Applied Mathematics and Computation*, *230*, 570–586. doi:10.1016/j.amc.2013.12.133

Liu, P. D., & Wang, Y. W. (2014). Multiple attribute group decision making methods based on intuitionistic linguistic power generalized aggregation operators. *Applied Soft Computing*, *7*, 90–104. doi:10.1016/j.asoc.2013.12.010

Liu, P. D., Zhang, X., & Jin, F. (2012). A multi-attribute group decisionmaking method based on interval valued trapezoidal fuzzy numbers hybrid harmonic averaging operators. *Journal of Intelligent & Fuzzy Systems*, *23*, 159–168.

Mondal, K. T., & Samanta, S. K. (2001). Topology of interval-valued intuitionistic fuzzy sets. *Fuzzy Sets and Systems*, *119*(3), 483–494. doi:10.1016/S0165-0114(98)00436-9

Park, D. G., Kwun, Y. C., Park, J. H., & Park, I. Y. (2009). Correlation coefficient of interval-valued intuitionistic fuzzy sets and its application to multiple attribute group decision making problems. *Mathematical and Computer Modelling*, *50*(9-10), 1279–1293. doi:10.1016/j.mcm.2009.06.010

Pradhan, R., & Madhumangal, P. (2014). The generalized inverse of Atanassovs intuitionistic fuzzy matrices. *International Journal of Computational Intelligence Systems*, *7*(6), 1083–1095. doi:10.1080/18756891.2014.963976

Szmidt, E., & Kacprzyk, J. (2001). Entropy for intuitionistic fuzzy sets. *Fuzzy Sets and Systems*, *118*(3), 467–477. doi:10.1016/S0165-0114(98)00402-3

Urea, M. R., Chiclana, F., Morente Molinera, J. A., & Herrera Viedma, E. (2015). Managing incomplete preference relations in decision making: A review and future trends. *Information Sciences*, *302*, 14–32. doi:10.1016/j.ins.2014.12.061

Wei, C. P., Wang, P., & Zhang, Y. Z. (2011). Entropy, similarity measure of interval-valued intuitionistic fuzzy sets and their applications. *Information Sciences*, *181*(19), 4273–4286. doi:10.1016/j.ins.2011.06.001

Wei, G. W., Wang, H. J., & Lin, R. (2011). Application of correlation coefficient to interval-valued intuitionistic fuzzy multiple attribute decision-making with incomplete weight information. *Knowledge and Information Systems*, *26*(2), 337–349. doi:10.1007/s10115-009-0276-1

Xu, Z. S. (2007). Methods for aggregating interval-valued intuitionistic fuzzy information and their application to decision making. *Control and Decision*, *2*, 215–219.

Xu, Z. S. (2010). A method based on distance measure for intervalvalued intuitionistic fuzzy group decision making. *Information Sciences*, *180*(1), 181–190. doi:10.1016/j.ins.2009.09.005

Ye, J. (2009). Multicriteria fuzzy decision-making method based on a novel accuracy function under interval-valued intuitionistic fuzzy environment. *Expert Systems with Application, 36*, 6899-6902.

Yu, D. J. (2013). Decision making based on generalized geometric operator under interval-valued intuitionistic fuzzy environment. *Journal of Intelligent & Fuzzy Systems*, *25*, 471–480.

Yue, Z. L., & Jia, Y. Y. (2013). An application of soft computing technique in group decision making under interval-valued intuitionistic fuzzy environment. *Applied Soft Computing*, *13*, 308–322.

Zadeh, L. A. (1965). Fuzzy sets. *Information and Control*, *8*(3), 338–356. doi:10.1016/S0019-9958(65)90241-X

Zaheh, L. A. (1975). The concept of a linguistic variable and its application to approximate reasoning. *Information Sciences*, *8*(3), 199–249. doi:10.1016/0020-0255(75)90036-5

Zhang, Q. S., Jiang, S. Y., Jia, B. G., & Luo, S. H. (2010). Some information measures for interval-valued intuitionistic fuzzy sets. *Information Sciences*, *180*(24), 5130–5145. doi:10.1016/j.ins.2010.08.038

Zhang, X. L., & Xu, Z. S. (2015). Soft computing based on maximizing consensus and fuzzy TOPSIS approach to interval-valued intuitionistic fuzzy group decision making. *Applied Soft Computing*, *26*, 42–56. doi:10.1016/j.asoc.2014.08.073

Chapter 4
Intuitionistic Fuzzy Distance-Based Intuitionistic Fuzzy TOPSIS Method and Application to MADM

Jiangxia Nan
Guilin University of Electronic Technology, China

Ting Wang
Guilin University of Electronic Technology, China

Jingjing An
Fuzhou University, China

ABSTRACT

In this paper, an intuitionistic fuzzy (IF) distance measure between two triangular intuitionistic fuzzy numbers (TIFNs) is developed. The metric properties of the proposed IF distance measures are also studied. Then, based on the IF distance, an extended TOPSIS is developed to solve multi-attribute decision making (MADM) problems with the ratings of alternatives on attributes of TIFNs. In this methodology, the IF distances between each alternative and the TIFN positive ideal-solution are calculated as well as the TIFN negative ideal-solution. Then the relative closeness degrees obtained of each alternative to the TIFN positive ideal solution are TIFNs. Based on the ranking methods of TIFNs the alternatives are ranked. A numerical example is examined to the validity and practicability of the method proposed in this paper.

INTRODUCTION

In the application of fuzzy numbers, distance measures between fuzzy numbers play a very important role. Thus some researchers of Cheng (1998), Chakraborty (2006), Guha and Chakraborty (2010), Tran and Duckstein (2002), Voxman (1998), Xu and Li (2001) have focused their attention to compute the distances between fuzzy numbers (Cheng, 1998; Chakraborty, 2006; Guha & Chakraborty, 2010; Tran

DOI: 10.4018/978-1-5225-1848-8.ch004

& Duckstein, 2002; Voxman, 1998; Xu & Li, 2001). Atanassov (1999) proposed the intuitionistic fuzzy (IF) sets, which is an extension of the fuzzy set and it is better than fuzzy set in dealing with vagueness and uncertainty of more flexible and practical. Based on the study of distance measures between fuzzy numbers, some researchers studied the distance between IF sets (Grzegorzewski, 2004; Szmidt & Kacprzyk, 2000; Wang & Xin, 2005). As a specialization of IF-set, the concept of intuitionistic fuzzy numbers (IFNs) can express information more abundant and flexible than the fuzzy number. However, there is very little investigation on distance measure between IFNs. Grzegorzewski (2003) studied the distance of IFNs, which is expressed as crisp values. It is unreasonable, the reason is that from a viewpoint of logic, a distance measure between IFNs should be an IFN. In other word, the uncertainty or hesitation or lack of knowledge presented in defining IFN should inherently be involved in their corresponding distance measures. Guha and Chakraborty (2010) studied the distance measure between two IFNs, which is IFN. In this paper, extending the fuzzy distance given by Tran and Duckstein (2002), a new IF distance measure between two triangular intuitionistic fuzzy numbers (TIFNs) defined by Li (2010) is proposed based on interval difference. Here the distance between two TIFNs is also TIFN.

The TOPSIS method is an important method for MADM problems. Some researchers studied classical TOPSIS method (Hwang, 1981; Lai, 1994), and some researchers studied TOPSIS with IF information (Zhang & Xu, 2015; Chen, 2015; Li & Nan, 2011; Li, 2010, 2012). The basic premise of the TOPSIS is that the chosen alternative should be the shortest distance to the positive ideal solution, and the longest distance from the negative ideal solution simultaneously. Thus, the distance measure plays an important role in TOPSIS method. Yu (2014) studied the Hamming distance of TrIFN, and give the TOPSIS method based on TrIFN. However, these distance measures existed TOPSIS were expressed as crisp values, which is unreasonable. So far as we know, there is no investigation on using the IF distance measure in TOPSIS to solve MADM with TIFNs. Thus, based on the new IF distance measure defined in this paper, the TOPSIS is extended to solve MADM problems in which the ratings of alternatives on attributes of TIFNs. The distances of each alternative from the TIFNs positive ideal solution, the TIFNs negative ideal solution and the relative closeness degrees of each alternative to the TIFNs positive ideal solution are calculated and they are still TIFNs. Then according to the ranking method of the weighted value-indexes and weighted ambiguity-indexes of TIFNs given by Li (2015), the alternatives are ranked.

The rest of the paper is organized as follows. Section 2 describes the basic definition and the operations of TrIFNs and TIFNs. The ranking order relation of TIFNs based on weighted value-indexes and weighted ambiguity-indexes is introduced in this section. In section 3, Some exist distances of IFNs are introduced. A new intuitionistic fuzzy distance between TIFNs is defined, respectively. The metric properties are studied. Section 4 introduce the principle of TOPSIS method for MADM with TIFNs and TrIFNs. An improved TOPSIS method be proposed. In section 5, a numerical example of personnel recruitment is used to illustrate the feasibility and practicability of the proposed extended TOPSIS. A short conclusion is given in section 6.

PRELIMINARIES

The Definition of IFNs

How to express an ill-known quantity using the intuitionistic fuzzy set is very important in game modeling. Inspired by the concept of the fuzzy number (Dubois, 1980), an IFN $\tilde{a}$ which is a special intuitionistic fuzzy set on the real number set R is defined as follow.

Definition 1 (Li, 2015)

An IFN $\tilde{a}$ is defined as a special intuitionistic fuzzy set on the real number set R, whose membership function $\mu_{\tilde{a}} : R \to [0, w_{\tilde{a}}]$ and non-membership function $\upsilon_{\tilde{a}} : R \to [u_{\tilde{a}}, 1]$ should satisfy the four conditions (1)-(4) as follows:

1. There exist at least two real numbers $x_0' \in R$ and $x_0'' \in R$ such that $\mu_{\tilde{a}}(x_0') = w_{\tilde{a}}$ and $\upsilon_{\tilde{a}}(x_0'') = u_{\tilde{a}}$;
2. $\mu_{\tilde{a}}$ is quasi concave and upper semi-continuous on R;
3. $\nu_{\tilde{a}}$ is quasi convex and lower semi-continuous on R;
4. The support sets $\{x | \mu_{\tilde{a}}(x) > 0, x \in R\}$ and $\{x | \upsilon_{\tilde{a}}(x) < 1, x \in R\}$ are bounded.

According to definition 1, we can easily define the follow general IFN.

Definition 2 (Li, 2015)

A general IFN $\tilde{a} = \langle (\underline{a}, a_1, a_2, \overline{a}), f_l, f_r, w_{\tilde{a}}; g_l, g_r, u_{\tilde{a}} \rangle$ is a special intuitionistic fuzzy set on the real number set R, whose membership and non-membership function are defined as follows:

$$\mu_{\tilde{a}}(x) = \begin{cases} 0 & (x < \underline{a}) \\ f_l(x) & (\underline{a} \le x < a_1) \\ w_{\tilde{a}} & (a_1 \le x \le a_2) \\ f_r(x) & (a_2 < x \le \overline{a}) \\ 0 & (x > \overline{a}) \end{cases} \tag{1}$$

and

$$\upsilon_{\tilde{a}}(x) = \begin{cases} 1 & (x < \underline{a}) \\ g_l(x) & (\underline{a} \le x < a_1) \\ u_{\tilde{a}} & (a_1 \le x \le a_2), \\ g_r(x) & (a_2 < x \le \overline{a}) \\ 1 & (x > \overline{a}) \end{cases} \tag{2}$$

respectively, depicted as in Figure 1, where $w_{\tilde{a}}$ and $u_{\tilde{a}}$ represent the maximum membership degree and the minimum non-membership degree such that they satisfy the following conditions: $0 \le w_{\tilde{a}} \le 1$, $0 \le u_{\tilde{a}} \le 1$ and $0 \le w_{\tilde{a}} + u_{\tilde{a}} \le 1$. Furthermore, f_l and g_r are non-decreasing and piecewise upper semi-continuous functions; f_r and g_l are non-increasing and piecewise lower semi-continuous functions.

Figure 1.

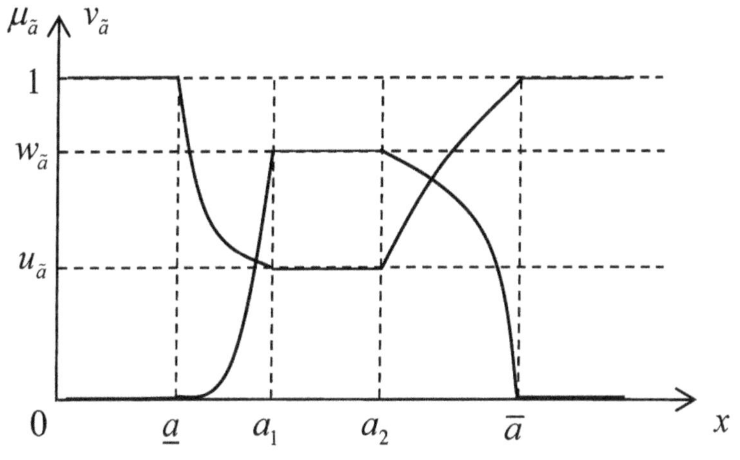

An IFN $\tilde{a}=\langle(\underline{a},a_1,a_2,\overline{a}),f_l,f_r,w_{\tilde{a}};g_l,g_r,u_{\tilde{a}}\rangle$

Let $\pi_{\tilde{a}}(x)=1-\mu_{\tilde{a}}(x)-v_{\tilde{a}}(x)$, which is called the intuitionistic fuzzy index of an element x in the IFN $\tilde{a}$. It is the degree of indeterminacy membership of the element x to IFN $\tilde{a}$.

Particularly, when $f_l(x)$, $f_r(x)$, $g_l(x)$ and $g_r(x)$ are linear functions, we have the notation of TrIFN as follow.

Definition 3 (Li, 2010)

An TrIFN $\tilde{a}=<(\underline{a},a_1,a_2,\overline{a});w_{\tilde{a}},u_{\tilde{a}}>$ is a special IF-set on the real number set R, whose membership function and non-membership function are defined as follows:

$$\mu_{\tilde{a}}(x)=\begin{cases} w_{\tilde{a}}(x-\underline{a})/(a_1-\underline{a}) & \text{if} \quad \underline{a}\le x<a_1 \\ w_{\tilde{a}} & \text{if} \quad a_1\le x\le a_2 \\ w_{\tilde{a}}(\overline{a}-x)/(\overline{a}-a_2) & \text{if} \quad a_2<x\le\overline{a} \\ 0 & \text{if} \quad x<\underline{a} \text{ or } x>\overline{a} \end{cases} \tag{3}$$

and

$$v_{\tilde{a}}(x)=\begin{cases} [a_1-x+u_{\tilde{a}}(x-\underline{a})]/(a_1-\underline{a}) & \text{if} \quad \underline{a}\le x<a_1 \\ u_{\tilde{a}} & \text{if} \quad a_1\le x\le a_2 \\ [x-a_2+u_{\tilde{a}}(\overline{a}-x)]/(\overline{a}-a_2) & \text{if} \quad a_2<x\le\overline{a} \\ 1 & \text{if} \quad x<\underline{a} \text{ or } x>\overline{a} \end{cases}, \tag{4}$$

respectively, depicted as in Figure 2.

Figure 2.

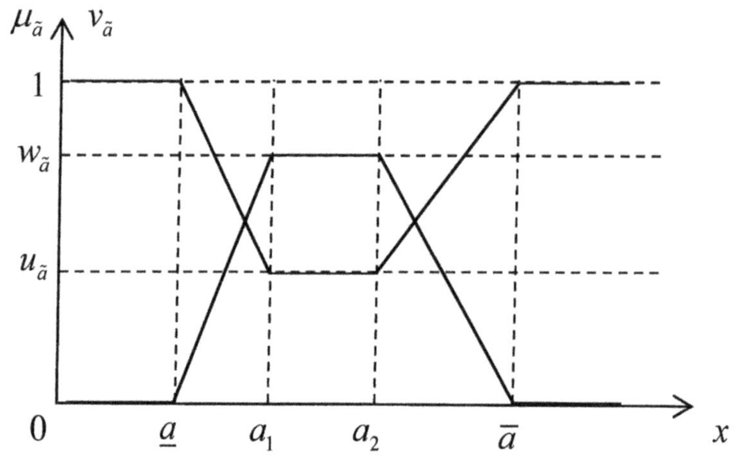

An TrIFN $\tilde{a} = \langle (\underline{a}, a_1, a_2, \overline{a}) w_{\tilde{a}}; u_{\tilde{a}} \rangle$

Where the values $w_{\tilde{a}}$ and $u_{\tilde{a}}$ represent the maximum degree of membership, and the minimum degree of non-membership such that they satisfy the following conditions:

$0 \le w_{\tilde{a}} \le 1$, $0 \le u_{\tilde{a}} \le 1$, $0 \le w_{\tilde{a}} + u_{\tilde{a}} \le 1$.

Let

$$\pi_{\tilde{a}}(x) = 1 - \mu_{\tilde{a}}(x) - v_{\tilde{a}}(x), \tag{5}$$

which is called the intuitionistic fuzzy index of an element x in $\tilde{a}$. It is the degree of indeterminacy membership of the element x to TrIFN $\tilde{a}$.

Definition 4 (Li, 2010)

Let $\tilde{a} =< (\underline{a}, a_1, a_2, \overline{a}); w_{\tilde{a}}, u_{\tilde{a}} >$ and $\tilde{b} =< (\underline{b}, b_1, b_2, \overline{b}); w_{\tilde{b}}, u_{\tilde{b}} >$ be two TrIFNs with $w_{\tilde{a}} \ne w_{\tilde{b}}$ and $u_{\tilde{a}} \ne u_{\tilde{b}}$. λ is a real number. The arithmetic operations of TrIFNs are defined as follows:

$$\tilde{a} + \tilde{b} =< (\underline{a} + \underline{b}, a_1 + b_1, a_2 + b_2, \overline{a} + \overline{b}); w_{\tilde{a}} \wedge w_{\tilde{b}}, u_{\tilde{a}} \vee u_{\tilde{b}} >, \tag{6}$$

$$\lambda\tilde{a} = \begin{cases} < (\lambda\underline{a}, \lambda a_1, \lambda a_2, \lambda\overline{a}); w_{\tilde{a}}, u_{\tilde{a}} > & \text{if} \quad \lambda > 0 \\ < (\lambda\overline{a}, \lambda a_2, \lambda a_1, \lambda\underline{a}); w_{\tilde{a}}, u_{\tilde{a}} > & \text{if} \quad \lambda > 0 \end{cases}, \tag{7}$$

$$\tilde{a} / \tilde{b} = \begin{cases} < (\underline{a} / \overline{b}, a_1 / b_2, a_2 / b_1, \overline{a} / \underline{b}); w_{\tilde{a}} \wedge w_{\tilde{b}}, u_{\tilde{a}} \vee u_{\tilde{b}} > & \text{if} \quad \tilde{a} > 0, \tilde{b} > 0 \\ < (\overline{a} / \overline{b}, a_2 / b_2, a_1 / b_1, \underline{a} / \underline{b}); w_{\tilde{a}} \wedge w_{\tilde{b}}, u_{\tilde{a}} \vee u_{\tilde{b}} > & \text{if} \quad \tilde{a} < 0, \tilde{b} > 0, \\ < (\overline{a} / \underline{b}, a_2 / b_1, a_1 / b_2, \underline{a} / \overline{b}); w_{\tilde{a}} \wedge w_{\tilde{b}}, u_{\tilde{a}} \vee u_{\tilde{b}} > & \text{if} \quad \tilde{a} < 0, \tilde{b} < 0 \end{cases} \tag{8}$$

where the symbols "$\wedge$" and "$\vee$" are the min and max operators, respectively.

Easy to see, if the closed interval $[a_1, a_2]$ degrade into a bit, as $a_1 = a_2$, then TrIFN is reduced into TIFN.

Definition 5 (Li, 2010)

An TIFN $\tilde{a} =< (\underline{a}, a, \overline{a}); w_{\tilde{a}}, u_{\tilde{a}} >$ is a special IF-set on the real number set R, whose membership function and non-membership function are defined as follows:

$$\mu_{\tilde{a}}(x) = \begin{cases} w_{\tilde{a}}(x-\underline{a}) \,/\, (a-\underline{a}) & \text{if} \quad \underline{a} \le x < a \\ w_{\tilde{a}} & \text{if} \quad x = a \\ w_{\tilde{a}}(\overline{a}-x) \,/\, (\overline{a}-a) & \text{if} \quad a < x \le \overline{a} \\ 0 & \text{if} \quad x < \underline{a} \text{ or } x > \overline{a} \end{cases} \tag{9}$$

and

$$v_{\tilde{a}}(x) = \begin{cases} [a - x + u_{\tilde{a}}(x-\underline{a})] \,/\, (a-\underline{a}) & \text{if} \quad \underline{a} \le x < a \\ u_{\tilde{a}} & \text{if} \quad x = a \\ [x - a + u_{\tilde{a}}(\overline{a}-x)] \,/\, (\overline{a}-a) & \text{if} \quad a < x \le \overline{a} \\ 1 & \text{if} \quad x < \underline{a} \text{ or } x > \overline{a} \end{cases}, \tag{10}$$

respectively, depicted as in Figure 3.

Figure 3.

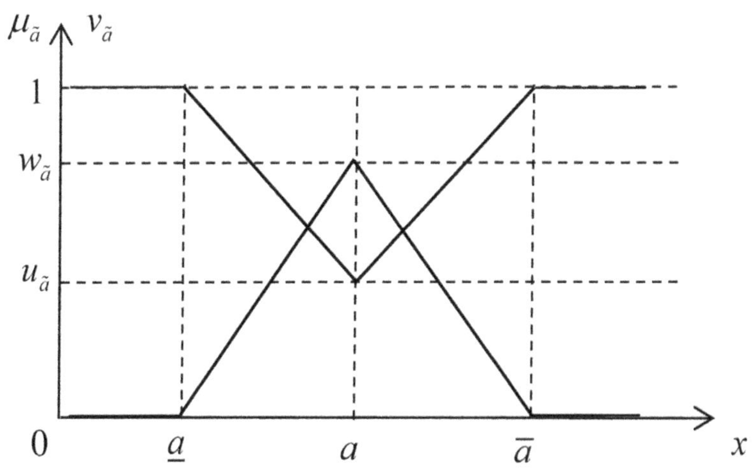

An TIFN $\tilde{a} = \langle (\underline{a}, a, \overline{a}) w_{\tilde{a}}; u_{\tilde{a}} \rangle$

Where the values $w_{\tilde{a}}$ and $u_{\tilde{a}}$ represent the maximum degree of membership, and the minimum degree of non-membership such that they satisfy the following conditions:

$0 \leq w_{\tilde{a}} \leq 1$, $0 \leq u_{\tilde{a}} \leq 1$, $0 \leq w_{\tilde{a}} + u_{\tilde{a}} \leq 1$.

Let

$$\pi_{\tilde{a}}(x) = 1 - \mu_{\tilde{a}}(x) - v_{\tilde{a}}(x), \tag{11}$$

which is called the intuitionistic fuzzy index of an element x in $\tilde{a}$. It is the degree of indeterminacy membership of the element x to TIFN $\tilde{a}$.

If $\underline{a} \geq 0$ and one of the three values $\underline{a}$, a and $\overline{a}$ is not equal to 0, then the TIFN $\tilde{a} = \left\langle \left(\underline{a}, a, \overline{a}\right); w_{\tilde{a}}, u_{\tilde{a}} \right\rangle$ is called as a positive TIFN, denoted by $\tilde{a} \geq 0$. Likewise, if $\overline{a} \leq 0$ and one of the three values $\underline{a}$, a and $\overline{a}$ is not equal to 0, then the TIFN $\tilde{a} = \left\langle \left(\underline{a}, a, \overline{a}\right); w_{\tilde{a}}, u_{\tilde{a}} \right\rangle$ is called as a negative TIFN, denoted by $\tilde{a} < 0$.

The arithmetic operations of TIFNs may be defined as follows.

Definition 6 (Li,2010)

Let $\tilde{a} =< (\underline{a}, a, \overline{a}); w_{\tilde{a}}, u_{\tilde{a}} >$ and $\tilde{b} =< (\underline{b}, b, \overline{b}); w_{\tilde{b}}, u_{\tilde{b}} >$ be two TIFNs with $w_{\tilde{a}} \neq w_{\tilde{b}}$ and $u_{\tilde{a}} \neq u_{\tilde{b}}$. λ is a real number. The arithmetic operations of TIFNs are defined as follows:

$$\tilde{a} + \tilde{b} =< (\underline{a} + \underline{b}, a + b, \overline{a} + \overline{b}); w_{\tilde{a}} \wedge w_{\tilde{b}}, u_{\tilde{a}} \vee u_{\tilde{b}} >, \tag{12}$$

$$\lambda\tilde{a} = \begin{cases} < (\lambda\underline{a}, \lambda a, \lambda\overline{a}); w_{\tilde{a}}, u_{\tilde{a}} > & \text{if} \quad \lambda > 0 \\ < (\lambda\overline{a}, \lambda a, \lambda\underline{a}); w_{\tilde{a}}, u_{\tilde{a}} > & \text{if} \quad \lambda > 0 \end{cases}, \tag{13}$$

$$\tilde{a} / \tilde{b} = \begin{cases} < (\underline{a} / \overline{b}, a / b, \overline{a} / \underline{b}); w_{\tilde{a}} \wedge w_{\tilde{b}}, u_{\tilde{a}} \vee u_{\tilde{b}} > & \text{if} \quad \tilde{a} > 0, \tilde{b} > 0 \\ < (\overline{a} / \overline{b}, a / b, \underline{a} / \underline{b}); w_{\tilde{a}} \wedge w_{\tilde{b}}, u_{\tilde{a}} \vee u_{\tilde{b}} > & \text{if} \quad \tilde{a} < 0, \tilde{b} > 0 \\ < (\overline{a} / \underline{b}, a / b, \underline{a} / \overline{b}); w_{\tilde{a}} \wedge w_{\tilde{b}}, u_{\tilde{a}} \vee u_{\tilde{b}} > & \text{if} \quad \tilde{a} < 0, \tilde{b} < 0 \end{cases}, \tag{14}$$

where the symbols "$\wedge$" and "$\vee$" are the min and max operators, respectively.

For the convenience of calculation, we only introduce relevant theories about TIFN.

Cut Sets of an TIFN

Definition 7 (Li, 2010)

A $\alpha - cut$ set of $\tilde{a} =< (\underline{a}, a, \overline{a}); w_{\tilde{a}}, u_{\tilde{a}} >$ is defined for $\tilde{a}^{\alpha} = \{x \mid \mu_{\tilde{a}}(x) \geq \alpha, 0 \leq \alpha \leq w_{\tilde{a}}\}$, which is a closed interval of R, denoted by $\tilde{a}^{\alpha} = [L^{\alpha}(\tilde{a}), R^{\alpha}(\tilde{a})]$, which can be calculated as follows:

$$[L^{\alpha}(\tilde{a}), R^{\alpha}(\tilde{a})] = [\underline{a} + \alpha(a - \underline{a}) / w_{\tilde{a}}, \overline{a} - \alpha(\overline{a} - a) / w_{\tilde{a}}].$$

Definition 8 (Li, 2010)

A $\beta - cut$ set of $\tilde{a} =< (\underline{a}, a, \overline{a}); w_{\tilde{a}}, u_{\tilde{a}} >$ is defined for $\tilde{a}_{\beta} = \{x \mid v_{\tilde{a}}(x) \leq \beta, u_{\tilde{a}} \leq \beta \leq 1\}$, which is a closed interval of R, denoted by $\tilde{a}_{\beta} = [L_{\beta}(\tilde{a}), R_{\beta}(\tilde{a})]$, which can be calculated as follows

$$[L_{\beta}(\tilde{a}), R_{\beta}(\tilde{a})] = [[(1-\beta)a + (\beta - u_{\tilde{a}})\underline{a}] / (1 - u_{\tilde{a}}), [(1-\beta)a + (\beta - u_{\tilde{a}})\overline{a}] / (1 - u_{\tilde{a}})].$$

The Ranking Order Relation of TIFNs

In the sequence, we review the ranking method of TIFNs based on weighted value-indexes and weighted ambiguity-indexes, which is defined by Li et al. (2010).

Convenient to discuss, we choose $f(\alpha) = \alpha / w_{\tilde{a}} (\alpha \in [0, w_{\tilde{a}}])$, $g(\beta) = (1-\beta) / (1 - u_{\tilde{a}})$ $(\beta \in [u_{\tilde{a}}, 1])$ in the article. Then the value-index about the membership function of TIFN $\tilde{a}$ is obtained as follow:

$$V_{\mu}(\alpha) = \frac{(\underline{a} + 4a + \overline{a}) w_{\tilde{a}}}{12}.$$

In the same way, the value-index about the non-membership function of TIFN $\tilde{a}$ can be also obtained as follow:

$$V_{v}(\alpha) = \frac{(\underline{a} + 4a + \overline{a})(1 - u_{\tilde{a}})}{12}.$$

Similarly, assume that a $\alpha -$ cut set of $\tilde{a} =< (\underline{a}, a, \overline{a}); w_{\tilde{a}}, u_{\tilde{a}} >$ is defined for $\tilde{a}_{\alpha} = [L_{\tilde{a}}(\alpha), R_{\tilde{a}}(\alpha)]$ and a $\beta -$ cut set of $\tilde{a} =< (\underline{a}, a, \overline{a}); w_{\tilde{a}}, u_{\tilde{a}} >$ is defined for $\tilde{a}_{\beta} = [L_{\tilde{a}}(\beta), R_{\tilde{a}}(\beta)]$, respectively. Denoted by

$$A_{\mu}(\tilde{a}) = \int_{0}^{w_{\tilde{a}}} [(R_{\tilde{a}}(\alpha) - L_{\tilde{a}}(\alpha))] f(\alpha) d\alpha$$

and

$$A_{v}(\tilde{a}) = \int_{u_{\tilde{a}}}^{1} [(R_{\tilde{a}}(\beta) - L_{\tilde{a}}(\beta))] g(\beta) d\beta.$$

We say that $A_\mu(\tilde{a})$ and $A_v(\tilde{a})$ are ambiguity-index about the membership function and non-membership function of TIFN $\tilde{a}$, respectively.

Then, we can obtain

$$A_\mu(\tilde{a}) = \frac{(\overline{a} - \underline{a})w_{\tilde{a}}}{6}$$

and

$$A_v(\tilde{a}) = \frac{(\overline{a} - \underline{a})(1 - u_{\tilde{a}})}{6}.$$

Definition 9 (Li, 2010)

Assume that $\tilde{a} =< (\underline{a}, a, \overline{a}); w_{\tilde{a}}, u_{\tilde{a}} >$ be a TIFNs. $\lambda \in [0,1]$ is a real number. Then we denote:

$$V_\lambda(\tilde{a}) = \lambda V_\mu(\tilde{a}) + (1-\lambda)V_v(\tilde{a}),$$

$$A_\lambda(\tilde{a}) = \lambda A_\mu(\tilde{a}) + (1-\lambda)A_v(\tilde{a}).$$

Definition 10 (Li, 2010)

Assume that $\tilde{a} =< (\underline{a}, a, \overline{a}); w_{\tilde{a}}, u_{\tilde{a}} >$ and $\tilde{b} =< (\underline{b}, b, \overline{b}); w_{\tilde{b}}, u_{\tilde{b}} >$ are two TIFNs. $\lambda \in [0,1], V_\lambda(\tilde{a}) = \lambda V_\mu(\tilde{a}) + (1-\lambda)V_v(\tilde{a}), A_\lambda(\tilde{a}) = \lambda A_\mu(\tilde{a}) + (1-\lambda)A_v(\tilde{a})$ are the weighted value-indexes and weighted ambiguity-indexes of TIFNs, respectively. Let

$H_\lambda(\tilde{b}) = V_\lambda(\tilde{b}) - A_\lambda(\tilde{b})$ and $H_\lambda(\tilde{a}) = V_\lambda(\tilde{a}) - A_\lambda(\tilde{a})$,

we have:

1. $H_\lambda(\tilde{a}) > H_\lambda(\tilde{b})$ if and only if $\tilde{a}$ is bigger than $\tilde{b}$,denoted by $\tilde{a} > \tilde{b}$;
2. $H_\lambda(\tilde{a}) = H_\lambda(\tilde{b})$ if and only if $\tilde{a}$ is equal to $\tilde{b}$,denoted by $\tilde{a} = \tilde{b}$;
3. $H_\lambda(\tilde{a}) \geq H_\lambda(\tilde{b})$ if and only if $\tilde{a} > \tilde{b}$ or $\tilde{a} = \tilde{b}$.

DISTANCE OF IFNS

As the distance of fuzzy numbers is important in fuzzy MADM problems, the distance of TIFNs is also important in MADM problems in which the ratings of alternatives on attributes are expressed TIFNs. In this subsection, the exist distances of TIFNs are defined.

The Definition About the Distance Between TIFNs

Definition 11 (Guha, 2010)

Let $\tilde{a}$, $\tilde{b}$ and $\tilde{c}$ be IFNs. A real function $D: IFNs \times IFNs \to IFNs$ is called the intuitionistic fuzzy distance measure of two IFNs, if D satisfies the following properties:

1. $D\left(\tilde{a},\tilde{b}\right) \geq 0$;
2. $D\left(\tilde{a},\tilde{b}\right) = D\left(\tilde{b},\tilde{a}\right)$;
3. $D\left(\tilde{a},\tilde{c}\right) \leq D\left(\tilde{a},\tilde{b}\right) \oplus D\left(\tilde{b},\tilde{c}\right)$.

Where "$\oplus$" denotes the extended sum operator.

Definition 12

Let $\tilde{a} =< (\underline{a}, a_1, a_2, \overline{a}); w_{\tilde{a}}, u_{\tilde{a}} >$ and $\tilde{b} =< (\underline{b}, b_1, b_2, \overline{b}); w_{\tilde{b}}, u_{\tilde{b}} >$ be two TrIFNs with $w_{\tilde{a}} \neq w_{\tilde{b}}$ and $u_{\tilde{a}} \neq u_{\tilde{b}}$. Then the Hamming distance between $\tilde{a}$ and $\tilde{b}$ can be defined as follows:

$$\begin{aligned} d(\tilde{a},\tilde{b}) = \frac{1}{8}\{&\left|(1+w_{\tilde{a}}-u_{\tilde{a}})\underline{a}-(1+w_{\tilde{b}}-u_{\tilde{b}})\underline{b}\right| + \left|(1+w_{\tilde{a}}-u_{\tilde{a}})a_1-(1+w_{\tilde{b}}-u_{\tilde{b}})b_1\right| \\ &+\left|(1+w_{\tilde{a}}-u_{\tilde{a}})a_2-(1+w_{\tilde{b}}-u_{\tilde{b}})b_2\right| + \left|(1+w_{\tilde{a}}-u_{\tilde{a}})\overline{a}-(1+w_{\tilde{b}}-u_{\tilde{b}})\overline{b}\right|\}. \end{aligned} \tag{15}$$

Theorem 1: $d(\tilde{a},\tilde{b})$ given in Definition 12 satisfies the properties in Definition 11.

Proof: Obviously, $d(\tilde{a},\tilde{b})$ satisfies the properties (1) and (2) in Definition 11. In the following, $d(\tilde{a},\tilde{b})$ will be proved to satisfy property (3).

$$\begin{aligned} d(\tilde{a},\tilde{b}) = \frac{1}{8}\{&|(1+w_{\tilde{a}}-u_{\tilde{a}})\underline{a}-(1+w_{\tilde{b}}-u_{\tilde{b}})\underline{c}+(1+w_{\tilde{b}}-u_{\tilde{b}})\underline{c}-(1+w_{\tilde{b}}-u_{\tilde{b}})\underline{b}| \\ &+|(1+w_{\tilde{a}}-u_{\tilde{a}})a_1-(1+w_{\tilde{b}}-u_{\tilde{b}})c_1+(1+w_{\tilde{b}}-u_{\tilde{b}})c_1-(1+w_{\tilde{b}}-u_{\tilde{b}})b_1| \\ &+|(1+w_{\tilde{a}}-u_{\tilde{a}})a_2-(1+w_{\tilde{b}}-u_{\tilde{b}})c_2+(1+w_{\tilde{b}}-u_{\tilde{b}})c_2-(1+w_{\tilde{b}}-u_{\tilde{b}})b_2| \\ &+|(1+w_{\tilde{a}}-u_{\tilde{a}})\overline{a}-(1+w_{\tilde{b}}-u_{\tilde{b}})\overline{c}+(1+w_{\tilde{b}}-u_{\tilde{b}})\overline{c}-(1+w_{\tilde{b}}-u_{\tilde{b}})\overline{b}|\} \\ \leq \frac{1}{8}\{&|(1+w_{\tilde{a}}-u_{\tilde{a}})\underline{a}-(1+w_{\tilde{b}}-u_{\tilde{b}})\underline{c}|+|(1+w_{\tilde{b}}-u_{\tilde{b}})\underline{c}-(1+w_{\tilde{b}}-u_{\tilde{b}})\underline{b}| \\ &+|(1+w_{\tilde{a}}-u_{\tilde{a}})a_1-(1+w_{\tilde{b}}-u_{\tilde{b}})c_1|+|(1+w_{\tilde{b}}-u_{\tilde{b}})c_1-(1+w_{\tilde{b}}-u_{\tilde{b}})b_1| \\ &+|(1+w_{\tilde{a}}-u_{\tilde{a}})a_2-(1+w_{\tilde{b}}-u_{\tilde{b}})c_2|+|(1+w_{\tilde{b}}-u_{\tilde{b}})c_2-(1+w_{\tilde{b}}-u_{\tilde{b}})b_2| \\ &+|(1+w_{\tilde{a}}-u_{\tilde{a}})\overline{a}-(1+w_{\tilde{b}}-u_{\tilde{b}})\overline{c}|+|(1+w_{\tilde{b}}-u_{\tilde{b}})\overline{c}-(1+w_{\tilde{b}}-u_{\tilde{b}})\overline{b}|\} \\ = d(\tilde{a}&,\tilde{c})+d(\tilde{c},\tilde{b}). \end{aligned}$$

Hence the proof of Theorem 1 is completed.

Distance Measure for TrIFNs

Definition 13 (Allahviranloo, 2010)

For any two TrIFNs $\tilde{a} =< (a_1, a_2, a_3, a_4); w_{\tilde{a}}, u_{\tilde{a}} >$ and $\tilde{b} =< (b_1, b_2, b_3, b_4); w_{\tilde{b}}, u_{\tilde{b}} >$, we can evaluate value and ambiguity indexes $V(\tilde{a})$, $V(\tilde{b})$, $A(\tilde{a})$, $A(\tilde{b})$, respectively. Based on the proposed distance measure, the distance measure for two TrIFNs have been defined as follows:

$$D(\tilde{a},\tilde{b}) = \begin{cases} d^{(p)}([A(\tilde{a}),V(\tilde{a})],[A(\tilde{b}),V(\tilde{b})]) & V(\tilde{a}) \geq 0 \\ d^{(p)}([V(\tilde{a}),A(\tilde{a})],[V(\tilde{b}),A(\tilde{b})]) & V(\tilde{a}) < 0 \end{cases}$$
$$= \begin{cases} (\int_0^1 | A(\tilde{a})x + (1-x)V(\tilde{a}) - A(\tilde{b})x - (1-x)V(\tilde{b}) |^P \,\mathrm{d}x)^{1/P} & V(\tilde{a}) \geq 0 \\ (\int_0^1 | V(\tilde{a})x + (1-x)A(\tilde{a}) - V(\tilde{b})x - (1-x)A(\tilde{b}) |^P \,\mathrm{d}x)^{1/P} & V(\tilde{a}) < 0 \end{cases}.$$

Metric Properties of Distance Measure for IFNs

The proposed distance measure is a metric distance and satisfies the metric properties:

1. $D(\tilde{a},\tilde{b}) \geq 0$.

   Proof: According to definition

$$D(\tilde{a},\tilde{b}) = (\int_0^1 \left|A(\tilde{a})x + (1-x)V(\tilde{a}) - A(\tilde{b})x - (1-x)V(\tilde{b})\right|^p dx)^{1/p}.$$

The proof follows from the fact that

$$\left|A(\tilde{a})x + (1-x)V(\tilde{a}) - A(\tilde{b})x - (1-x)V(\tilde{b})\right| \geq 0.$$

2. $D(\tilde{a},\tilde{b}) = 0$ if $\tilde{a} = \tilde{b}$.

   Proof: The proof follows from the fact that $\tilde{a} = \tilde{b}$ if $\tilde{a}_\alpha = \tilde{b}_\alpha$ and $\tilde{a}_\beta = \tilde{b}_\beta$. Consequently, $\tilde{a} = \tilde{b}$ if and only if $V(\tilde{a}) = V(\tilde{b})$ and $A(\tilde{a}) = A(\tilde{b})$.
3. $D(\tilde{a},\tilde{b}) = D(\tilde{b},\tilde{a})$.

   Proof:

$$\begin{aligned} D(\tilde{a},\tilde{b}) &= (\int_0^1 | A(\tilde{a})x + (1-x)V(\tilde{a}) - A(\tilde{b})x - (1-x)V(\tilde{b}) |^P \, dx)^{1/p} \\ &= (\int_0^1 | -(A(\tilde{b})x + (1-x)V(\tilde{b}) - A(\tilde{a})x - (1-x)V(\tilde{a})) |^P \, dx)^{1/p} \\ &= (\int_0^1 | A(\tilde{b})x + (1-x)V(\tilde{b}) - A(\tilde{a})x - (1-x)V(\tilde{a}) |^P \, dx)^{1/p} \\ &= D(\tilde{b},\tilde{a}). \end{aligned}$$

4. $D(\tilde{a},\tilde{c}) + D(\tilde{c},\tilde{b}) \geq D(\tilde{a},\tilde{b})$.
   Proof:

$$\begin{aligned} D(\tilde{a},\tilde{c}) + D(\tilde{c},\tilde{b}) &= (\int_0^1 | A(\tilde{a})x + (1-x)V(\tilde{a}) - A(\tilde{c})x - (1-x)V(\tilde{c}) |^P \, dx)^{1/P} \\ &\quad + (\int_0^1 | A(\tilde{c})x + (1-x)V(\tilde{c}) - A(\tilde{b})x - (1-x)V(\tilde{b}) |^P \, dx)^{1/P} \\ &\geq (\int_0^1 |A(\tilde{a})x + (1-x)V(\tilde{a}) - A(\tilde{c})x - (1-x)V(\tilde{c}) \\ &\quad + A(\tilde{c})x + (1-x)V(\tilde{c}) - A(\tilde{b})x - (1-x)V(\tilde{b}) |^P \, dx)^{1/P} \\ &= (\int_0^1 | A(\tilde{a})x + (1-x)V(\tilde{a}) - A(\tilde{b})x - (1-x)V(\tilde{b}) |^P \, dx)^{1/P} \\ &= D(\tilde{a},\tilde{b}). \end{aligned}$$

The above arguments show the metric properties for the proposed distance measure when $V(\tilde{a}) \geq 0$. Similarly, it can be proved for the case when $V(\tilde{a}) < 0$.

New Distance of TIFNs

Definition 14

Let $\tilde{a}$ and $\tilde{b}$ be two TIFNs. For any given $\alpha \in [0,w]$, $\tilde{a}_\mu^\alpha = [\tilde{a}_\mu^L(\alpha), \tilde{a}_\mu^R(\alpha)]$ and $\tilde{b}_\mu^\alpha = [\tilde{b}_\mu^L(\alpha), \tilde{b}_\mu^R(\alpha)]$ are $\alpha - cut$ sets of $\tilde{a}$ and $\tilde{b}$, while $\tilde{a}_v^\beta = [\tilde{a}_v^L(\beta), \tilde{a}_v^R(\beta)]$ and $\tilde{b}_v^\beta = [\tilde{b}_v^L(\beta), \tilde{b}_v^R(\beta)]$ are $\beta - cut$ sets of $\tilde{a}$ and $\tilde{b}$ for any given $\beta \in [u,1]$, respectively.

Now, the distance between $\tilde{a}_\mu^\alpha = [\tilde{a}_\mu^L(\alpha), \tilde{a}_\mu^R(\alpha)]$ and $\tilde{b}_\mu^\alpha = [\tilde{b}_\mu^L(\alpha), \tilde{b}_\mu^R(\alpha)]$ for all $\alpha \in [0,w]$ is one of the following:

1. $[\tilde{a}_\mu^L(\alpha), \tilde{a}_\mu^R(\alpha)] - [\tilde{b}_\mu^L(\alpha), \tilde{b}_\mu^R(\alpha)]$ if $a \geq b$ or
2. $[\tilde{b}_\mu^L(\alpha), \tilde{b}_\mu^R(\alpha)] - [\tilde{a}_\mu^L(\alpha), \tilde{a}_\mu^R(\alpha)]$ if $a < b$

In order to consider both the notations together, an indicator variable η is introduced such that

$$\begin{aligned} [d_\mu^L, d_\mu^R] = \eta([\tilde{a}_\mu^L(\alpha), \tilde{a}_\mu^R(\alpha)] - [\tilde{b}_\mu^L(\alpha), \tilde{b}_\mu^R(\alpha)]) + \\ (1-\eta)([\tilde{b}_\mu^L(\alpha), \tilde{b}_\mu^R(\alpha)] - [\tilde{a}_\mu^L(\alpha), \tilde{a}_\mu^R(\alpha)]) \end{aligned}$$

for $\eta = \begin{cases} 1 & \text{if } a \geq b \\ 0 & \text{if } a < b \end{cases}$.

Where

$$d_\mu^L = \eta(\tilde{a}_\mu^L(\alpha) - \tilde{b}_\mu^L(\alpha) + \tilde{a}_\mu^R(\alpha) - \tilde{b}_\mu^R(\alpha)) + (\tilde{b}_\mu^L(\alpha) - \tilde{a}_\mu^R(\alpha)),$$

$$d_\mu^R = \eta(\tilde{a}_\mu^L(\alpha) - \tilde{b}_\mu^L(\alpha) + \tilde{a}_\mu^R(\alpha) - \tilde{b}_\mu^R(\alpha)) + (\tilde{b}_\mu^R(\alpha) - \tilde{a}_\mu^L(\alpha)).$$

Similarly, the distance between $\tilde{a}_v^\beta = [\tilde{a}_v^L(\beta), \tilde{a}_v^R(\beta)]$ and $\tilde{b}_v^\beta = [\tilde{b}_v^L(\beta), \tilde{b}_v^R(\beta)]$ for all $D(\tilde{a}, \tilde{b})$ is one of the following:

1. $[\tilde{a}_v^L(\beta), \tilde{a}_v^R(\beta)] - [\tilde{b}_v^L(\beta), \tilde{b}_v^R(\beta)]$ if $a \geq b$ or
2. $[\tilde{b}_v^L(\beta), \tilde{b}_v^R(\beta)] - [\tilde{a}_v^L(\beta), \tilde{a}_v^R(\beta)]$ if $a < b$

In order to consider both the notations together, an indicator variable η is introduced such that

$$[d_v^L(\beta), d_v^R(\beta)] = \eta([\tilde{a}_v^L(\beta), \tilde{a}_v^R(\beta)] - [\tilde{b}_v^L(\beta), \tilde{b}_v^R(\beta)]) + (1-\eta)([\tilde{b}_v^L(\beta), \tilde{b}_v^R(\beta)] - [\tilde{a}_v^L(\beta), \tilde{a}_v^R(\beta)])$$

for $\eta = \begin{cases} 1 & \text{if } a \geq b \\ 0 & \text{if } a < b \end{cases}$.

Where

$$d_v^L(\beta) = \eta(\tilde{a}_v^L(\beta) - \tilde{b}_v^L(\beta) + \tilde{a}_v^R(\beta) - \tilde{b}_v^R(\beta)) + (\tilde{b}_v^L(\beta) - \tilde{a}_v^R(\beta)),$$

$$d_v^R(\beta) = \eta(\tilde{a}_v^L(\beta) - \tilde{b}_v^L(\beta) + \tilde{a}_v^R(\beta) - \tilde{b}_v^R(\beta)) + (\tilde{b}_v^R(\beta) - \tilde{a}_v^L(\beta)).$$

Then the IF distance between two TIFNs $\tilde{a}$ and $\tilde{b}$ is defined as follows:

$$\tilde{D}(\tilde{a}, \tilde{b}) = < (\underline{d}, d, \overline{d}); w, u >,$$

where, $\underline{d} = \lambda \max\{\int_0^w d_\mu^L(\alpha) d\alpha, 0\} + (1-\lambda)\max\{\int_u^1 d_v^L(\beta) d\beta, 0\}$,

$$d = \frac{\lambda}{2}(d_\mu^L(w) + d_\mu^R(w)) + \frac{1-\lambda}{2}(d_v^L(u) + d_v^R(u)),$$

$$\overline{d} = \lambda \int_0^w d_\mu^R(\alpha) d\alpha + (1-\lambda)\int_u^1 d_v^R(\beta) d\beta,\ w = w_{\tilde{a}} \wedge w_{\tilde{b}},\ u = u_{\tilde{a}} \vee u_{\tilde{b}}. \quad (16)$$

Obviously, $\tilde{D}(\tilde{a},\tilde{b})$ is a positive TIFN. The new intuitionistic fuzzy distance between TIFNs defined above is extension of fuzzy distance (Tran & Duckstein, 2002). Namely, fuzzy distance is a special case of intuitionistic fuzzy distance.

Theorem 2: Let $\tilde{a}$, $\tilde{b}$ and $\tilde{c}$ be TIFNs. $D(\tilde{a},\tilde{b})$ defined in Equation (16) satisfies the following properties:

1. $D(\tilde{a},\tilde{b}) \geq 0$;
2. $D(\tilde{a},\tilde{b}) = D(\tilde{b},\tilde{a})$;
3. $D(\tilde{a},\tilde{c}) \leq D(\tilde{a},\tilde{b}) \oplus D(\tilde{b},\tilde{c})$.

Proof: Obviously, $D(\tilde{a},\tilde{b})$ satisfies properties (1) and (2). In the following, $D(\tilde{a},\tilde{b})$ will be proved to satisfy property (3).

Let $\tilde{a}$, $\tilde{b}$ and $\tilde{c}$ be three TIFNs. Their $\alpha - cut$ sets are expressed as $\tilde{a}_\mu^\alpha = [\tilde{a}_\mu^L(\alpha), \tilde{a}_\mu^R(\alpha)]$, $\tilde{b}_\mu^\alpha = [\tilde{b}_\mu^L(\alpha), \tilde{b}_\mu^R(\alpha)]$ and $\tilde{c}_\mu^\alpha = [\tilde{c}_\mu^L(\alpha), \tilde{c}_\mu^R(\alpha)]$, respectively and their $\beta - cut$ sets are expressed as $\tilde{a}_v^\beta = [\tilde{a}_v^L(\beta), \tilde{a}_v^R(\beta)]$, $\tilde{b}_v^\beta = [\tilde{b}_v^L(\beta), \tilde{b}_v^R(\beta)]$ and $\tilde{c}_v^\beta = [\tilde{c}_v^L(\beta), \tilde{c}_v^R(\beta)]$, respectively.

Now, if $a < b < c$, we have the following fuzzy distance for

1. $\tilde{a}$ and $\tilde{b}$: $\begin{cases} d_\mu^L = \tilde{b}_\mu^L(\alpha) - \tilde{a}_\mu^R(\alpha) \\ d_\mu^R = \tilde{b}_\mu^R(\alpha) - \tilde{a}_\mu^L(\alpha) \end{cases}, \begin{cases} d_v^L = \tilde{b}_v^L(\beta) - \tilde{a}_v^R(\beta) \\ d_v^R = \tilde{b}_v^R(\beta) - \tilde{a}_v^L(\beta) \end{cases};$
2. $\tilde{b}$ and $\tilde{c}$: $\begin{cases} \hat{d}_\mu^L = \tilde{c}_\mu^L(\alpha) - \tilde{b}_\mu^R(\alpha) \\ \hat{d}_\mu^R = \tilde{c}_\mu^R(\alpha) - \tilde{b}_\mu^L(\alpha) \end{cases}, \begin{cases} \hat{d}_v^L = \tilde{c}_v^L(\beta) - \tilde{b}_v^R(\beta) \\ \hat{d}_v^R = \tilde{c}_v^R(\beta) - \tilde{b}_v^L(\beta) \end{cases};$
3. $\tilde{a}$ and $\tilde{c}$: $\begin{cases} \breve{d}_\mu^L = \tilde{c}_\mu^L(\alpha) - \tilde{a}_\mu^R(\alpha) \\ \breve{d}_\mu^R = \tilde{c}_\mu^R(\alpha) - \tilde{a}_\mu^L(\alpha) \end{cases}, \begin{cases} \breve{d}_v^L = \tilde{c}_v^L(\beta) - \tilde{a}_v^R(\beta) \\ \breve{d}_v^R = \tilde{c}_v^R(\beta) - \tilde{a}_v^L(\beta) \end{cases}.$

Let the fuzzy distance of $\tilde{a}$ and $\tilde{b}$ be $D(\tilde{a},\tilde{b}) = \tilde{d} = \langle (\underline{d}, d, \overline{d}); w, u \rangle$, the fuzzy distance of $\tilde{b}$ and $\tilde{c}$ be $D(\tilde{b},\tilde{c}) = \tilde{d}' = \langle (\underline{d}', d', \overline{d}'); w, u \rangle$ and the fuzzy distance of $\tilde{a}$ and $\tilde{c}$ is $D(\tilde{a},\dot{c}) = \tilde{d}'' = \langle (\underline{d}'', d'', \overline{d}''); w, u \rangle$, where $w = w_{\tilde{a}} \wedge w_{\tilde{b}} \wedge w_{\tilde{c}}$, $u = u_{\tilde{a}} \vee u_{\tilde{b}} \vee u_{\tilde{c}}$.

Hence, $D(\tilde{a},\tilde{c}) \leq D(\tilde{a},\tilde{b}) + D(\tilde{b},\tilde{c})$ is equivalent to the following equation

$$\underline{d} + \overline{d} + \underline{d}' + \overline{d}' \geq \underline{d}'' + \overline{d}''. \tag{17}$$

Therefore the L.H.S. of Equation (17) can be rewritten as

$$\begin{aligned} \underline{d} + \overline{d} + \underline{d}' + \overline{d}' = & \lambda \max\left\{\int_0^w \left(\tilde{b}_\mu^L(\alpha) - \tilde{a}_\mu^R(\alpha)\right) d\alpha, 0\right\} + (1-\lambda) \max\left\{\int_u^1 \left(\tilde{b}_v^L(\beta) - \tilde{a}_v^R(\beta)\right) d\beta, 0\right\} \\ & + \lambda \int_0^w \left(\tilde{b}_\mu^R(\alpha) - \tilde{a}_\mu^L(\alpha)\right) d\alpha + (1-\lambda) \int_u^1 \left(\tilde{b}_v^R(\beta) - \tilde{a}_v^L(\beta)\right) d\beta \\ & + \lambda \max\left\{\int_0^w \left(\tilde{c}_\mu^L(\alpha) - \tilde{b}_\mu^R(\alpha)\right) d\alpha, 0\right\} + (1-\lambda) \max\left\{\int_u^1 \left(\tilde{c}_v^L(\beta) - \tilde{b}_v^R(\beta)\right) d\beta, 0\right\} \\ & + \lambda \int_0^w \left(\tilde{c}_\mu^R(\alpha) - \tilde{b}_\mu^L(\alpha)\right) d\alpha + (1-\lambda) \int_u^1 \left(\tilde{c}_v^R(\beta) - \tilde{b}_v^L(\beta)\right) d\beta. \end{aligned}$$

Due to the $\max\left\{a,b\right\} = \dfrac{a+b+\left|a+b\right|}{2}$, we have

$$D(A_i, A^+) = \sum_{j=1}^{n} D(\tilde{g}_{ij}, \tilde{g}_j^+) \tag{18}$$

Equation (18) is equivalent to the following Equation (19), since $\left|a\right| + \left|b\right| \geq \left|a+b\right|$.

$$\rho_i = \frac{D(A_i, A^+)}{D(A_i, A^+) + D(A_i, A^-)} \tag{19}$$

According to $\left|a-b\right| \geq \left|a\right| - \left|b\right|$, the Equation 19is equivalent to the following Equation 20, i.e.

$$\rho_i \tag{20}$$

It is true that $\tilde{b}_\mu^L(\alpha) \leq \tilde{b}_\mu^R(\alpha)$ and $\tilde{b}_v^L(\beta) \leq \tilde{b}_v^R(\beta)$, thus Equation 20 is equivalent to the following Equation (21).

$$\begin{aligned} \underline{d} + \bar{d} + \underline{d}' + \underline{d}' \geq \frac{\lambda}{2}\left[\int_0^w \left(\tilde{c}_\mu^L(\alpha) - \tilde{a}_\mu^R(\alpha)\right) d\alpha + \left|\int_0^w \left(\tilde{c}_\mu^L(\alpha) - \tilde{a}_\mu^R(\alpha)\right) d\alpha\right|\right] \\ + \lambda \int_0^w \left(\tilde{c}_\mu^R(\alpha) - \tilde{a}_\mu^L(\alpha)\right) d\alpha \\ + \frac{1\text{-}\lambda}{2}\left[\int_u^1 \left(\tilde{c}_v^L(\beta) - \tilde{a}_v^R(\beta)\right) d\beta + \left|\int_u^1 \left(\tilde{c}_v^L(\beta) - \tilde{a}_v^R(\beta)\right) d\beta\right|\right] \\ + (1-\lambda) \int_u^1 \left(\tilde{c}_v^R(\beta) - \tilde{a}_v^L(\beta)\right) d\beta. \end{aligned} \tag{21}$$

Equation (21) is equivalent to the following equation. i.e.

$$\begin{aligned} \underline{d} + \bar{d} + \underline{d}' + \underline{d}' \geq \lambda \max\left\{\int_0^w \left(\tilde{c}_\mu^L(\alpha) - \tilde{a}_\mu^R(\alpha)\right) d\alpha, 0\right\} \\ + (1-\lambda) \max\left\{\int_u^1 \left(\tilde{c}_v^L(\beta) - \tilde{a}_v^R(\beta)\right) d\beta, 0\right\} \\ + \lambda \int_0^w \left(\tilde{c}_\mu^R(\alpha) - \tilde{a}_\mu^L(\alpha)\right) d\alpha \\ + (1-\lambda) \int_u^1 \left(\tilde{c}_v^R(\beta) - \tilde{a}_v^L(\beta)\right) d\beta \\ = \underline{d}'' + d''. \end{aligned}$$

Hence the proof of Theorem 2 is completed.

TOPSIS METHOD

In order to solve the problem of multi-attribute decision making problems, TOPSIS method is given by Hwang Chinglai in 1981. The method is that the chosen alternative should be the shortest distance to the positive ideal solution, and the longest distance from the negative ideal solution simultaneously. Thus, the distance measure plays an important role in TOPSIS method.

The Principle of Classical TOPSIS Method

Suppose that there exists an alternative set $A = \{A_1, A_2, \cdots, A_m\}$. Each solution is evaluated by n properties. Let $X = \left\{x_1, x_2, \cdots, x_n\right\}$ is attribute set, the each solution $A_i (i = 1, \cdots, m)$ of alternative set $A = \{A_1, A_2, \cdots, A_m\}$ is composed by $X = \left\{x_1, x_2, \cdots, x_n\right\}$, it as a point of dimensional space can only represent scheme A_i. Due to the importance of each attribute for makers is different, so we need to give different weights for each attribute. Assumes that the weights of attributes are $\omega_j (j = 1, 2, \cdots, n)$, and they satisfy the conditions: $\omega_j \in \left[0,1\right]$, $\sum_{j=1}^{n} \omega_j = 1$. All the weights of attributes available represented as a vector form $\omega = (\omega_1, \omega_2, \cdots, \omega_n)^T$.

Ideal solution A^* is imaginary of the optimal scheme. In the same way, the negative ideal solution is not solution concentration of a scheme, but the worst scheme by our fiction. In n dimensional space, to compare the distance between A_i and A^*, A_i and A^0, which close to the ideal solution but far away from the negative ideal solution scheme is the optimum scheme. Then according to the distance between each scheme and the positive and negative ideal solution to determine the pros and cons of each solution sequence (Figure 4).

Figure 4.

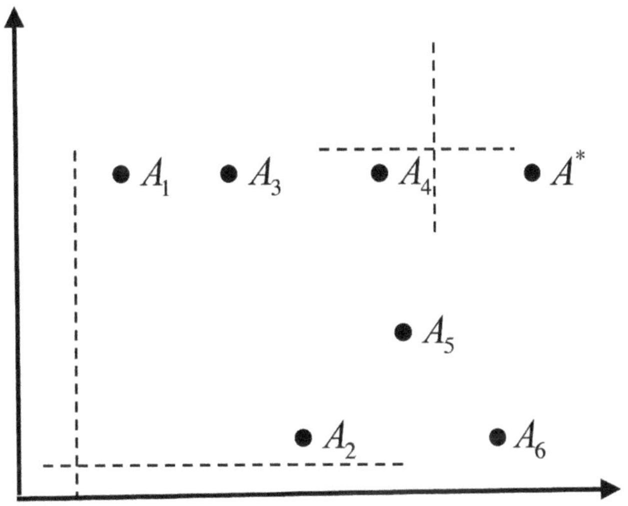

The positive and negative ideal solution

The general steps for TOPSIS can be summarized as follows:

Normalize the decision matrix, which each of the elements of decision matrix be normalized into [0,1]. For different types of attribute value, the standard method is also different. The type of cost and efficiency attribute values as follows:

$$r_{ij} = \begin{cases} x_{ij} / x_j^+ & \forall j \in N, j \in B \\ x_j^- / x_{ij} & \forall j \in N, j \in C \end{cases}, \tag{22}$$

where B and C are the subscript sets of benefit attributes and cost attributes respectively. Using the above transformation, decision matrix D can be converted to

$$D = \begin{bmatrix} x_{11} & x_{12} & \cdots & x_{1n} \\ x_{21} & x_{22} & \cdots & x_{2n} \\ \vdots & \vdots & & \vdots \\ x_{m1} & x_{m2} & \cdots & x_{mn} \end{bmatrix} \Rightarrow D' \begin{bmatrix} r_{11} & r_{12} & \cdots & r_{1n} \\ r_{21} & r_{22} & \cdots & r_{2n} \\ \vdots & \vdots & & \vdots \\ r_{m1} & r_{m2} & \cdots & r_{mn} \end{bmatrix}$$

As x_{ij} is real number, also r_{ij} is the precise value; When x_{ij} is fuzzy number, r_{ij} is fuzzy number. $\tilde{x}_j^+$ and $\tilde{x}_j^-$ can be obtained. Denote $\tilde{x}_j^+ = (a_j^+, b_j^+, c_j^+, d_j^+)$, $\tilde{x}_j^- = (a_j^-, b_j^-, c_j^-, d_j^-)$, then Equation (22) transform into

$$\tilde{r}_{ij} = \begin{cases} \dfrac{\tilde{x}_{ij}}{\tilde{x}_j^+} = (\dfrac{a_{ij}}{d_j^+}, \dfrac{b_{ij}}{c_j^+}, \dfrac{c_{ij}}{b_j^+}, \dfrac{d_{ij}}{a_j^+}) & j \in B \\ \dfrac{\tilde{x}_j^-}{\tilde{x}_{ij}} = (\dfrac{a_j^-}{d_{ij}}, \dfrac{b_j^-}{c_{ij}}, \dfrac{c_j^-}{b_{ij}}, \dfrac{d_j^-}{a_{ij}}) & j \in C \end{cases} \tag{23}$$

Constructed the weighted decision matrix V, where v_{ij} can be denoted as

$$v_{ij} = r_{ij}\omega_{ij}, \forall i \in M; j \in N. \tag{24}$$

Determine the positive ideal solution and negative ideal solution. The positive ideal solution and negative ideal solution, denote $\tilde{x}^+$ and $\tilde{x}^-$, respectively, are defined as follows:

$$\tilde{x}^+ = \left\{\tilde{v}_1^+, \tilde{v}_2^+, \cdots, \tilde{v}_n^+\right\}$$

and

$$\tilde{x}^- = \left\{\tilde{v}_1^-, \tilde{v}_2^-, \cdots, \tilde{v}_n^-\right\},$$

where, $\tilde{v}_j^+ = \max \tilde{v}_{ij}$, $\tilde{v}_j^- = \min \tilde{v}_{ij}$.

Calculate the distances of each choice from the positive ideal solution $\tilde{x}^+$ and negative ideal solution $\tilde{x}^-$.

The distance denote S_i^+ and S_i^-, respectively, are defined as follows:

$$S_i^+ = \sum_{j=1}^{m} z_{ij}^+, \forall i \in M$$

and

$$S_i^- = \sum_{j=1}^{m} z_{ij}^-, \forall i \in M,$$

where $z_{ij}^+ = \left|v_{ij} - v_j^+\right|$, $z_{ij}^- = \left|v_{ij} - v_j^-\right|$, respectively.

5. Calculate the relative closeness degree of each choice to the ideal solution as follows:

$$C_i = S_i^- / (S_i^- + S_i^+), \forall i \in M.$$

The alternatives are ranked according to the relative closeness degree to positive ideal solution. Obviously, the larger C_i the better A_i.

The Extended TOPSIS Method Based on TrIFNs

Extending the classical TOPSIS, the principle of TOPSIS method with TrIFNs can be summarized as follows:

Suppose that there exists an alternative set $A = \{A_1, A_2, \cdots, A_m\}$, which is consisted of m alternatives and each alternatives is assessed on n attributes. Denote the set of all attributes by $X = \{X_1, X_2, \cdots, X_n\}$. Assume that ratings of alternatives on attributes are given using TrIFNs. Namely, the rating of any alternatives $A_i \in A$ $(i = 1, 2, \cdots, m)$ on each attribute $X_j \in X$ $(j = 1, 2, \cdots, n)$ is an TIFN $\tilde{a}_{ij} = < (\underline{a}_{ij}, a_{1ij}, a_{2ij}, \overline{a}_{ij}); w_{\tilde{a}_{ij}}, u_{\tilde{a}_{ij}} >$. Thus an MADM problem with TrIFNs can be expressed concisely in the matrix format as follows:

Normalize the TrIFN decision matrix. In order to eliminate the effect of different physical dimensions on the final decision making results, the normalized TrIFN decision matrix can be calculated using the following formulate:

$$\tilde{r}_{ij} = < (\frac{\underline{a}_{ij}}{\overline{a}_j^+}, \frac{a_{1ij}}{\overline{a}_j^+}, \frac{a_{2ij}}{\overline{a}_j^+}, \frac{\overline{a}_{ij}}{\overline{a}_j^+}); w_{\tilde{a}_{ij}}, u_{\tilde{a}_{ij}} > (i = 1, 2, \cdots, m; j \in B) \tag{25}$$

and

$$\tilde{r}_{ij} = <(\frac{\underline{a}_j^-}{\overline{a}_{ij}}, \frac{\underline{a}_j^-}{a_{1ij}}, \frac{\underline{a}_j^-}{a_{2ij}}, \frac{\underline{a}_j^-}{\underline{a}_{ij}}); w_{\tilde{a}_{ij}}, u_{\tilde{a}_{ij}} > (i = 1, 2, \cdots, m; j \in C), \quad (26)$$

where B and C are the subscript sets of benefit attributes and cost attributes, respectively,

$$\overline{a}_j^+ = \max\{\overline{a}_{ij} \mid i = 1, 2, \cdots, m\} (j \in B)$$

and

$$\underline{a}_j^- = \min\{\underline{a}_{ij} \mid i = 1, 2, \cdots, m\} (j \in C).$$

Calculate the weighted normalized TrIFN decision matrix. Based on Equation(5) of definition 2, the weighted normalized TrIFN decision matrix can be calculated as $(\tilde{g}_{ij})_{m \times n}$, where

$$\tilde{g}_{ij} = \omega_j \tilde{r}_{ij}. \quad (27)$$

Determine TrIFNs positive ideal solution and TrIFNs negative ideal solution. The TrIFNs positive ideal solution and TrIFNs negative ideal solution, denote A^+ and A^-, respectively, are defined as follows:

$$A^+ = (\tilde{g}_1^+, \tilde{g}_2^+, \cdots, \tilde{g}_n^+) \quad (28)$$

and

$$A^- = (\tilde{g}_1^-, \tilde{g}_2^-, \cdots, \tilde{g}_n^-), \quad (29)$$

respectively, where

$$\tilde{g}_j^+ = <(\max_{1 \le i \le m}\{\underline{g}_{ij}\}, \max_{1 \le i \le m}\{g_{ij}\}, \max_{1 \le i \le m}\{\overline{g}_{ij}\}); \max_{1 \le i \le m}\{w_{\tilde{g}_{ij}}\}, \min_{1 \le i \le m}\{u_{\tilde{g}_{ij}}\} > \text{for } j \in B \quad (30)$$

and

$$\tilde{g}_j^- = <(\min_{1 \le i \le m}\{\underline{g}_{ij}\}, \min_{1 \le i \le m}\{g_{ij}\}, \min_{1 \le i \le m}\{\overline{g}_{ij}\}); \min_{1 \le i \le m}\{w_{\tilde{g}_{ij}}\}, \max_{1 \le i \le m}\{u_{\tilde{g}_{ij}}\} > \text{for } j \in B \quad (31)$$

or

$$\tilde{g}_j^+ =< (\min_{1\le i\le m}\{\underline{g}_{ij}\}, \min_{1\le i\le m}\{g_{ij}\}, \min_{1\le i\le m}\{\overline{g}_{ij}\}); \min_{1\le i\le m}\{w_{\tilde{g}_{ij}}\}, \max_{1\le i\le m}\{u_{\tilde{g}_{ij}}\} > \text{for } j\in C \quad (32)$$

and

$$\tilde{g}_j^- =< (\max_{1\le i\le m}\{\underline{g}_{ij}\}, \max_{1\le i\le m}\{g_{ij}\}, \max_{1\le i\le m}\{\overline{g}_{ij}\}); \max_{1\le i\le m}\{w_{\tilde{g}_{ij}}\}, \min_{1\le i\le m}\{u_{\tilde{g}_{ij}}\} > \text{for } j\in C\,. \quad (33)$$

Calculate the distances of each choice from the TrIFNs positive ideal solution A^+ and TIFNs negative ideal solution A^-.Using Equation (16),the distance of alternative $A_i \in A$ from A^+ and A^- can be obtained as follows:

$$d(A_i, A^+) = \sum_{j=1}^{n} d(\tilde{g}_{ij}, \tilde{g}_j^+) \quad (34)$$

and

$$d(A_i, A^-) = \sum_{j=1}^{m} d(\tilde{g}_{ij}, \tilde{g}_j^-)\,. \quad (35)$$

5. Calculate the relative closeness degree of each choice to the TrIFNs positive ideal solution as follows:

$$\rho_i = \frac{d(A_i, A^+)}{d(A_i, A^+) + d(A_i, A^-)} (i = 1, 2, \cdots, m). \quad (36)$$

6. Based on the ranking order relation of TrIFNs defined in definition 6, the alternatives are ranked according to the relative closeness degree to the TrIFNs positive ideal solution. Obviously, the larger ρ_i the better A_i.

A NUMERICAL EXAMPLE

We analyze a personnel recruitment problem. Suppose that a software company desires to hire a system analyst. After the primaries, three candidates A_1, A_2 and A_3 remain for further evaluation. The decision making committee assesses the three candidates based on five attributes, including emotional steadiness (X_1), oral communication skill (X_2), personality (X_3), past experience (X_4) and self-confidence ability (X_5). The ratings of the candidates with respect to the attributes are given as in Table 1. Assume that five weights of attributes are $w_1 = 0.20$, $w_2 = 0.22$, $w_3 = 0.18$, $w_4 = 0.24$ and $w_5 = 0.16$, respectively. Since the five attributes are all benefit attributes, according to Equation (37), the normalized TIFN decision matrix is obtained as in Table 2. Using Equation (39), the weighted normalized TIFN decision matrix is obtained as in Table 3.

Table 1. The TIFN decision matrix

Alternatives	Attributes				
	X_1	X_2	X_3	X_4	X_5
A_1	<(5.7,7.7,9.3);0.7,0.2>	<(5,7,9);0.6,0.3>	<(5.7,7.7,9);0.8,0.1>	<(8.3,9.6,10);0.6,0.4>	<(3,5,7);0.6,0.3>
A_2	<(6.5,8.6,10);0.4,0.5>	<(8,9,10);0.6,0.3>	<(8.3,9.7,10);0.7,0.2>	<(8,9,10);0.6,0.3>	<(7,9,10);0.6,0.2>
A_3	<(6.5,8.2,9.3);0.8,0.1>	<(7,9,10);0.7,0.2>	<(6,9,10);0.5,0.2>	<(6,8,9);0.6,0.2>	<(6.3,8.3,9.7);0.7,0.2>

Table 2. The normalized TIFN decision matrix

Alternatives	Attributes				
	X_1	X_2	X_3	X_4	X_5
A_1	<(0.57,0.77,0.93);0.7,0.2>	<(0.5,0.7,0.9);0.6,0.3>	<(0.57,0.77,0.9);0.8,0.1>	<(0.83,0.96,1);0.6,0.4>	<(0.3,0.5,0.7);0.6,0.3>
A_2	<(0.65,0.86,1);0.4,0.5>	<(0.8,0.9,1);0.6,0.3>	<(0.83,0.97,1);0.7,0.2>	<(0.8,0.9,1);0.6,0.3>	<(0.7,0.9,1);0.6,0.2>
A_3	<(0.65,0.82,0.93);0.8,0.1>	<(0.7,0.9,1);0.7,0.2>	<(0.6,0.9,1);0.5,0.2>	<(0.6,0.8,0.9);0.6,0.2>	<(0.63,0.83,0.97);0.7,0.2>

Table 3. The weighted normalized TIFN decision matrix

Alternatives	Attributes				
	X_1	X_2	X_3	X_4	X_5
A_1	<(0.11,0.15,0.19);0.7,0.2>	<(0.11,0.15,0.2);0.6,0.3>	<(0.1,0.14,0.16);0.8,0.1>	<(0.20,0.23,0.24);0.6,0.4>	<(0.05,0.08,0.11);0.6,0.3>
A_2	<(0.13,0.17,0.2);0.4,0.5>	<(0.18,0.19,0.2);0.6,0.3>	<(0.15,0.17,0.18);0.7,0.2>	<(0.19,0.22,0.24);0.6,0.3>	<(0.11,0.14,0.16);0.6,0.2>
A_3	<(0.13,0.16,0.19);0.8,0.1>	<(0.15,0.19,0.2);0.7,0.2>	<(0.11,0.16,0.18);0.5,0.2>	<(0.14,0.18,0.22);0.6,0.2>	<(0.10,0.13,0.16);0.7,0.2>

Where <(5.7,7.7,9.3);0.7,0.2> in the Table 1 is an TIFN which indicates that the mark of the candidate A_1 with respect to the attribute X_1 is about 7.7 with the maximum satisfaction degree is 0.7, while the minimum non-satisfaction degree is 0.2. In other words, the hesitation degree is 0.1. Other TIFNs in Table 1 are explained similarly.

According to Equations (40-45),TIFNs positive ideal solution and TIFNs negative ideal solution are:

$$A^{+}=\{<(0.13,0.17,0.2);0.8,0.1>,\\ <(0.18,0.19,0.2);0.7,0.2>,\\ <(0.15,0.17,0.18);0.8,0.1>,$$

$$<(0.2,0.23,0.24);0.6,0.2>,\\ <(0.11,0.14,0.16);0.7,0.2>\}$$

$$A^{-}=\{<(0.11,0.15,0.19);0.4,0.5>,\\ <(0.11,0.15,0.2);0.6,0.3>,\\ <(0.1,0.14,0.16);0.5,0.2>,$$

$$<(0.14,0.18,0.22);0.6,0.4>,\\ <(0.05,0.08,0.11);0.6,0.3>\}$$

According to Equation (9), Equation (46) and Equation (47). let parameter $\lambda = 1/2$,we have

$$D(A_1, A^{+}) =<(0.04,0.11,0.22);0.6,0.4>,$$

$$D(A_2, A^{+}) =<(0,0.01,0.16);0.4,0.5>$$

$$D(A_3, A^{+}) =<(0.01,0.08,0.16);0.5,0.2>$$

$$D(A_1, A^{-}) =<(0.01,0.04,0.15);0.4,0.5>,$$

$$D(A_2, A^{-}) =<(0.07,0.22,0.24);0.4,0.5>$$

$$D(A_3, A^{-}) =<(0.01,0.12,0.21);0.4,0.5>$$

Using Equations(10-12) in definition 4 and Equation (48), the relative closeness degrees $\rho_i (i = 1,2,3)$ are calculated as follows:

$$\rho_1 =<(0.03,0.27,3);0.4,0.5>,$$

$$\rho_2 =<(0.18,0.96,3.43);0.4,0.5>,$$

$$\rho_3 =<(0.03,0.6,10.5);0.4,0.5>$$

Then according to the ranking methods of TIFNs given in definition 8, $H_\lambda(\rho_2) > H_\lambda(\rho_1) > H_\lambda(\rho_3)$ is obtained, so the best candidate is A_2 and the optimal ranking order of the three candidates is $A_2 > A_1 > A_3$.

CONCLUSION

In this paper, an IF distance measure of TIFNs and TrIFNs are defined, respectively. Based on the IF distance measure, this paper extends TOPSIS method to solve MADM problems with TIFNs. The method proposed in this paper is easily implemented. Using the proposed method, the distance of each alternative from the TIFNs positive ideal solution, TIFNs negative ideal solution and the relative closeness degree of each alternative to the TIFNs positive ideal solution are obtained and they are still TIFNs. The proposed method presented in this paper is illustrated by a personal selection problem, however, it can also be applied to problems such as information project selection, material selection and many other areas of decision problems.

ACKNOWLEDGMENT

The authors would like to thank the valuable comments and constructive suggestions from the anonymous referees. This research was supported by Program for the Key Program of National Natural Science Foundation of China (No. 71231003), the National Natural Science Foundation of China (Nos. 71171055,71461005 and 71462004), the New Century Excellent Talents in University (the Ministry of Education of China, NCET-10-0020), the Science Foundation of Guangxi Province (2014GXNSFFA1 18001,2014GXNSFFAA118010) and the Graduate Student Innovation Project (YJCXS201553).

REFERENCES

Allahviranloo, T., & Firozja, M. A. (2010). Ranking of fuzzy numbers by a new metric. *Soft Computing*, *14*(7), 773–782. doi:10.1007/s00500-009-0464-7

Atanassov, K. T. (1999). *Intuitionistic fuzzy sets*. Heidelberg, Germany: Springer. doi:10.1007/978-3-7908-1870-3

Chakraborty, C., & Chakraborty, D. (2006). A theoretical development on a fuzzy distance measure for fuzzy numbers. *Mathematical and Computer Modelling*, *43*(3-4), 254–261. doi:10.1016/j.mcm.2005.09.025

Chen, T. Y. (2015). The inclusion-based TOPSIS method with interval-valued intuitionistic fuzzy sets for multiple criteria group decision making. *Applied Soft Computing*, *26*, 57–73. doi:10.1016/j.asoc.2014.09.015

Cheng, C. H. (1998). A new approach for ranking fuzzy numbers by distance method. *Fuzzy Sets and Systems*, *95*(3), 307–317. doi:10.1016/S0165-0114(96)00272-2

Dubois, D., & Prade, H. (1980). *Fuzzy Sets and Systems Theory and Applications*. New York, NY, USA.

Grzegorzewski, P. (2003). Distances and orderings in a family of intuitionistic fuzzy numbers. Conference of the European Society for Fuzzy Logic and Technology3rd (EUSFLAT'03), Zittau, Germany.

Grzegorzewski, P. (2004). Distance between intuitionistic fuzzy sets and/or interval-valued fuzzy sets based on the Hausdorff metric. *Fuzzy Sets and Systems*, *148*(2), 319–328. doi:10.1016/j.fss.2003.08.005

Guha, D., & Chakraborty, D. (2010). A new approach to fuzzy distance measure and similarity measure between two generalized fuzzy numbers. *Applied Soft Computing*, *10*(1), 90–99. doi:10.1016/j.asoc.2009.06.009

Guha, D., & Chakraborty, D. (2010). A Theoretical development of distance measure for intuitionistic fuzzy numbers. *International Journal of Mathematics and Mathematical Sciences*, *2010*, 1–25. doi:10.1155/2010/949143

Hwang, C. L., & Yoon, K. (1981). *Multiple Attribute Decision Making-Methods and Applications*. Berlin: Springer. doi:10.1007/978-3-642-48318-9

Lai, Y. J., Liu, T. Y., & Hwang, C. L. (1994). TOPSIS for MODM. *European Journal of Operational Research*, *76*(3), 486–500. doi:10.1016/0377-2217(94)90282-8

Li, D.-F. (2010a). TOPSIS-based nonlinear-programming methodology for multi-attribute decision making with interval-valued intuitionistic fuzzy sets. *IEEE Transactions on Fuzzy Systems*, *18*(2), 299–311.

Li, D.-F. (2010b). A ratio ranking method of triangular intuitionistic fuzzy numbers and its application to MADM problems. *Computers & Mathematics with Applications (Oxford, England)*, *60*(6), 1557–1570. doi:10.1016/j.camwa.2010.06.039

Li, D.-F. (2012). TOPSIS for multi-attribute group decision making using IFSs. *Advances in Fuzzy Sets and Systems*, *11*(2), 105–129.

Li, D.-F. (2014). *Decision and Game Theory in Management with Intuitionistic Fuzzy Sets*. Heidelberg, Germany: Springer. doi:10.1007/978-3-642-40712-3

Li, D.-F., & Liu, J. C. (2015). A parameterized non-linear programming approach to solve matrix games with payoffs of I-fuzzy numbers. *IEEE Transactions on Fuzzy Systems*, *23*(4), 885–896. doi:10.1109/TFUZZ.2014.2333065

Li, D.-F., & Nan, J. X. (2011). Extension of the TOPSIS for multi-attribute group decision making under Atanassov IFS environments. *International Journal of Fuzzy System Applications*, *1*(4), 47–61. doi:10.4018/ijfsa.2011100104

Li, D.-F., Nan, J. X., & Zhang, M. J. (2010). A ranking method of triangular intuitionistic fuzzy numbers and application to decision making. *International Journal of Computational Intelligence Systems*, *3*(5), 522–530. doi:10.1080/18756891.2010.9727719

Li, D.-F., & Yang, J. (2015). A difference-index based ranking method of trapezoidal intuitionistic fuzzy numbers and application to multiattribute decision making. *Mathematical and Computational Applications*, *20*(1), 25–38. doi:10.19029/mca-2015-003

Szmidt, E., & Kacprzyk, J. (2000). Distance between intuitionistic fuzzy sets. *Fuzzy Sets and Systems, 114*(3), 505–518. doi:10.1016/S0165-0114(98)00244-9

Tran, L., & Duckstein, L. (2002). Comparison of fuzzy numbers using a fuzzy distance measure. *Fuzzy Sets and Systems, 130*(3), 331–341. doi:10.1016/S0165-0114(01)00195-6

Vahdani, B., Mousavi, S. M., & Tavakkoli, M. R. (2011). Group decision making based on novel fuzzy modified TOPSIS method. *Applied Mathematical Modelling, 35*(9), 4257–4269. doi:10.1016/j.apm.2011.02.040

Voxman, W. (1998). Some remarks on distances between fuzzy numbers. *Fuzzy Sets and Systems, 100*(1-3), 353–365. doi:10.1016/S0165-0114(97)00090-0

Wang, J. Q., Nie, R. R., Zhang, H. Y., & Chen, X. (2013). Intuitionistic fuzzy multi-criteria decision-making method based on evidential reasoning. *Applied Soft Computing, 13*(4), 1823–1831. doi:10.1016/j.asoc.2012.12.019

Wang, W., & Xin, X. (2005). Distance measure between intuitionistic fuzzy sets. *Pattern Recognition Letters, 26*(13), 2063–2069. doi:10.1016/j.patrec.2005.03.018

Xu, R., & Li, C. (2001). Multidimensional least-aquares fitting with a fuzzy model. *Fuzzy Sets and Systems, 119*(2), 215–223. doi:10.1016/S0165-0114(98)00350-9

Yu, P. J., & Zhang, J. (2014). An algorithmic method to extend TOPSIS for multiple attribute decision making under intuitionistic fuzzy setting. *Journal of Intelligent and Fuzzy Systems, 26*(5), 2315–2322.

Zhang, X. L., & Xu, Z. S. (2015). Soft computing based on maximizing consensus and fuzzy TOPSIS approach to interval-valued intuitionistic fuzzy group decision making. *Applied Soft Computing, 26*, 42–56. doi:10.1016/j.asoc.2014.08.073

Chapter 5
Fuzzy-AHP: A Boon in 3PL Decision Making Process

Nitin Kumar Sahu
Guru Ghasidas Central Vishwavidyalaya, India

Atul Kumar Sahu
Guru Ghasidas Central Vishwavidyalaya, India

Anoop Kumar Sahu
J.K.I.E., India

ABSTRACT

Logistics activities are performed in order to balance the operational chains of firms. The selection of the Third Party Logistics (3PL) is a challenging task for each organization, which involves various factors and attributes. The presented methodology acts as a boon and aids the decision makers for effectively choosing the appropriate Third Party Logistics (3PL) network. In the revealed work, the authors explored fuzzy sets theory and presented a fuzzy AHP model to facilitate the managers of organizations to deal with the Third Party Logistics (3PL) decision making problems. The overall performance of defined Third Party Logistics (3PL) Service Providers are greatly influenced by many significant parameters: quality, reliability, service assurance, shipment cost, customer relationship, etc. The authors have considered various significant parameters: service level, financial security capabilities, location, global presence, relationship management, and client fulfillment representing first level indices. These parameters have chain of various sub-parameters, represented as second level indices, whose importance is affecting the judgment of the decision makers. Various researches have constraint their work up to first level indices and have not considered the second level indices, which is a crucial part of today's practical decision making process. The authors have considered this issue as research gap and transformed this research gap into research agenda. The authors applied an AHP (Analytical Hierarchy Process) accompanied with fuzzy set theory in order to solve industrial Logistics problems. The objective of chapter is to propose a fuzzy based AHP method towards solve benchmarking (preference orders of defined alternatives under criteria) problems. The presented method facilitates the managers of firms to make the verdict towards choosing the best Third Party Logistics (3PL) service provider. A numerical illustration is provided to validate the method application upon module.

DOI: 10.4018/978-1-5225-1848-8.ch005

AT A GLANCE

The fuzzy set theory is ever considered as an effectual mathematical tool for modeling non crisp, vagueness and unclear information into mathematical data. The proposed chapter deals with the application of AHP (Analytical Hierarchy Process) accompanied with fuzzy set theory in order to diagnose industrial Logistics problems. The aim of chapter is to display effectiveness of a fuzzy based AHP method towards solve benchmarking (preference orders of defined alternatives under criteria) problems.

INTRODUCTION

Thousands of decisions are made every day in extent of Logistics activities. All decisions are neither participating in building the reputation nor rupturing the organization profit, but every decision measures the success or failure of organization. Hence decision making is an important aspect, which employs choosing a particular course of action, after considering the possible options. In today's competitive globalization world, every organization has its prime focus compensate customer's demands and also provide high quality product and low cost under less lead time. Logistics processes help the enterprises to achieve their competitiveness and enhance their commercial status by providing systems and processes to advance their customer service quality.

According to Lambert (2008), "Logistics is the process of planning, implementing, and controlling the efficient, cost effective flow of raw materials, in-process inventory, finished goods and related information from the point of origin/consumption to the point of consumption/ origin for the purpose of capturing value or proper disposal" (p. 163). According to Folinas (2012), "Logistics is the process of planning, implementing and controlling procedures for the efficient and effective transportation and storage of goods including services and related information from the point of origin to the point of consumption for the purpose of conforming to customer requirements and includes inbound, outbound, internal and external movements" (p. 389). To satisfy their customer's requirement, the organizations are outsourcing logistic services from separate enterprises, which are not the core part of the organization known as Third Party Logistics (3PL) Service Provider (Figure 1). Hence Logistics outsourcing or Third Party Logistics (3PL) service provider is an emerging trend in the global market.

Figure 1. A 3PL Supply Chain Management Network

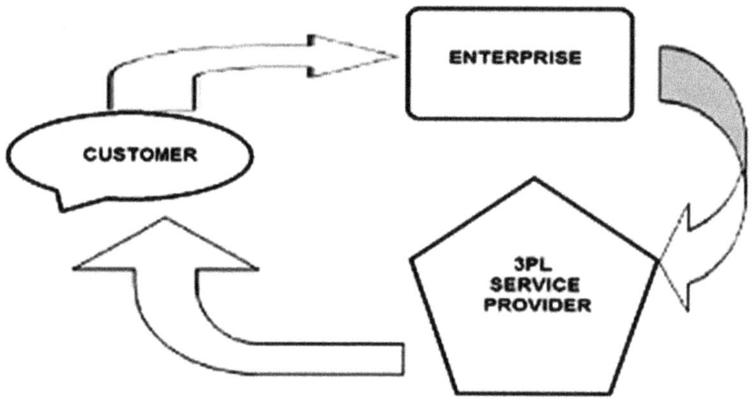

A Third Party Logistics (3PL) Service Provider is a medium that can be an organization, firm, company, enterprise etc., which provides Logistics services in the network of supply chain management. Logistics activities are carried out to transfer inputs from organization into outputs to customer or some other organization. According to Akkucuk (2015), "Third Party Logistics (3PL) is a firm that provides multiple Logistics services for use by customers. Preferably, these services are integrated, or bundled together, by the provider. Among the services Third Party Logistics (3PLs) provide:

- Transportation,
- Warehousing,
- cross-docking,
- Inventory management,
- Packaging, and
- Freight forwarding" (p. 137).

Logistics is presently considered as the area of study, where organizations have to impinge in order to reduce costs and advancement in customer service. Logistics activities assist organizations in competitiveness with their rival's by building their commercial reputation and by providing systems and processes to help customers in taking their return products and components for repair, reuse or disposal. In this chapter, the authors suggested a merged methodology of AHP with fuzzy set theory to make the verdict towards the best inclusive evaluation process of Third Party Logistics (3PL) performance appraisement. In every business, there will always be an enterprise and customers; however the existence of Third Party Logistics (3PL) Service Provider is not mandatory. But, the enterprises are outsourcing the Third Party Logistics (3PL) Service Provider in order to help their customers in taking their return products and components for repair, reuse or disposal as shown in Figure 1, as customer satisfaction is prime importance in every business organization.

Each major action from procurement to distribution generates waste and decline of these wastes is a chief target of eco friendly environment business practices. The massive developments of different varieties of products, have twisted a speedily growing volume of scrap material together with their undersized life cycles. Increasing environmental concern has forced the manufacturers to point the Logistics activities in their decision making. In today's scenario every country has started to focus on green issues via considering their Logistics supply chain network, in order to increased interest in environmentalism and recycling due to various regulations imposed on them by local bodies, Regional bodies, National and international organizations.

Logistics alliances allow the outsourcing company to concentrate on the core competence, reduce the transportation cost, improve the service, increase the efficiency and restructure the supply chains. Figure 2 depicts the important aspects of Third Party Logistics (3PL) Service Provider selection. The important job of the Logistics departments is to bring noteworthy savings for the organization. While assessing and selecting the appropriate Logistics service provider there will be always conflict to the Logistics managers that whether the selection will satisfy whole needs of the organization. Hence, the proper selection of Third Party Logistics (3PL) Service Provider is an important and extremely crucial aspect for the development and enhancing proficiency of an enterprise.

Figure 2. Important Aspects of 3PL Service Provider

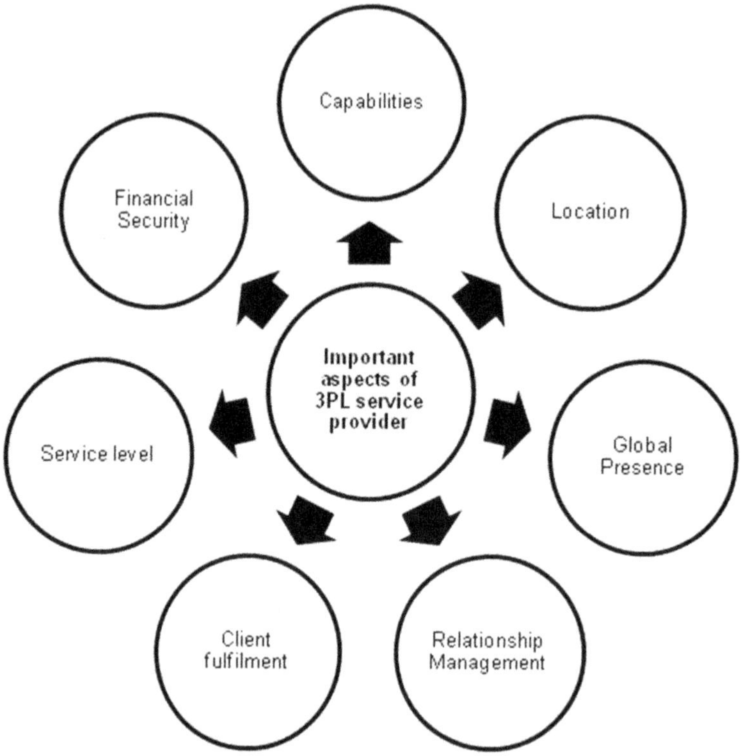

FUZZY LOGIC

Fuzzy logic is a way of representing the views, opinions, thinking of the subjective information of the society in objective information. the Fuzzy logic usually employed to hold the concept of partial information, where the data value may range between completely true and completely false called crisp number in fuzzy logic, Furthermore, when the degrees of information lies or managed between specific domain then linguistic variables are used and by specific membership functions the unspecified information is converted into crisp. It is a form in which the truth values of variables may be any real number between 0 and 1.

Let a cup be filled with coffee. Then to describe the temperature of coffee generally there are two ways either hot or cold as depicted in Figure 3. The meaning can be represented by a certain fuzzy set where cold represents real number 0 and hot represents real number 1 known as crisp information in fuzzy logic describing the nature of the coffee.

The other case to describe the partial hot or cold nature of the coffee can be represented by 70% hot or 30% cold. Thus this subjective information in fuzzy logic can be represented by below 0.7 or can be (0.5, 0.6, 0.7), known as membership function in triangular fuzzy number which can be further transform into crisp. Thus fuzzy logic uses degrees of truth for making mathematical model of vagueness or inconsistent information.

Figure 3. Triangular Fuzzy Number

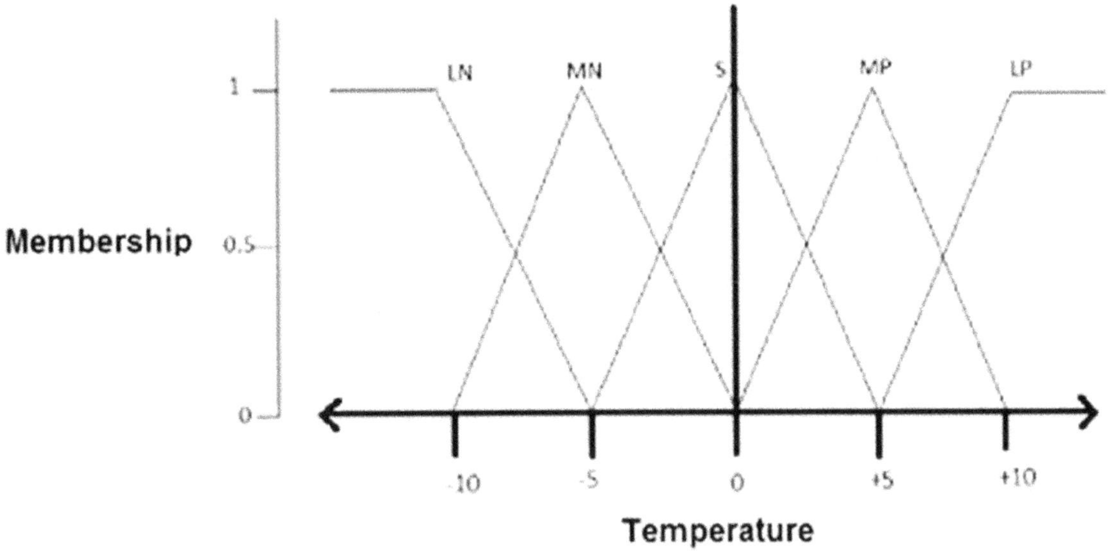

STATE OF ART AND PROBLEM FORMULATION

To evaluate the best alternative amongst preferred alternatives, a Multi criteria decision making (MCDM) appraisement module is formed by incorporating chain of multiple criteria, and is set up with defined alternatives (Li & Ren, 2015; Wan & Li, 2013, 2014, 2015, Li & Wan, 2014; Li, 2010, 2011, 2016;Samantra, Sahu, Datta & Mahapatra, 2013; Sahu, Datta & Mahapatra, 2013, 2014, 2015; Nan & Zhang, 2010). During the past few years the outsourcing of Logistics activities has become a universal practice by numerous organizations, as the Third Party Logistics (3PL) Service Provider have become imperative players in several supply chains and business, as they take part in the profit augmentation, the enhancement of the service quality of their customers and cost reduction. The demand of Third Party Logistics (3PL) Service Provider becomes an increasingly important issue for corporations seeking customer satisfaction, which is directly proportional for the profit maximization and organizational development. Third Party Logistics (3PL) Service Provider selection has gained great attention in Logistics management literature as they play a key role in the Logistic activities between the outsourcing company, the marketplace, and the customers. Rajesh, Pugazhendhi and Ganesh (2010), proposed a classification framework and suggested some new research settings based on a comprehensive literature review of 46 refereed academic journal articles of Third Party Logistics (3PL) Service Provider. Yan, Chaudhry & Chaudhry (2003), postulated a case based reasoning model framework for Third Party Logistics (3PL) evaluation and selection system. Liu and Wang (2009), presented an integrated fuzzy approach for the evaluation and selection of Third Party Logistics (3PL) Service Provider, consists of three different techniques like use of fuzzy Delphi method to identify important evaluation criteria, apply fuzzy inference method to eliminate unsuitable Third Party Logistics (3PL) Service Provider and develop a fuzzy linear assignment approach for the final selection.

Zhang, Li, Liu, Li and Zhang (2004), formulated an AHP-based model for examining the function in Third Party Logistics (3PL) vendor selection. Karagul and Albayrakoglu (2007), presented a multi-criterion decision making (MCDM) model for outsourcing Logistics services in the Turkish automotive industry. An analytic hierarchy process (AHP) model was created to select the best Third Party Logistics (3PL) provider among multiple candidates. Koh and Tan (2005), investigated how the selected foreign and domestic Third Party Logistics (3PL) enterprises in China using e-commerce to gain competitive advantages. Bayazit (2012) also applied AHP in selection of a Third Party Logistics (3PL) Service Provider.

Isiklar, Alptekin and Buyukozkan (2007), proposed an intelligent decision support framework for effective Third Party Logistics (3PL) evaluation and selection. The proposed framework integrated case-based reasoning, rule-based reasoning and compromise programming techniques in fuzzy environment. This real-time decision making approach dealt with uncertain and imprecise decision situations. Gol and Catay (2007), presented hand-on experiences of a pilot project conducted at an automotive company to redesign its Logistics operations and to select a global Logistics Service Provider, using an analytic hierarchy process for multi-criteria decision-making. This paper provided some practical insights from an automotive company that attempts to streamline and efficiently run its supply chain through Logistics outsourcing.

So, Kim, Cheong and Cho (2006) applied the analytic hierarchy process (AHP) to evaluate the service quality of Third Party Logistics (3PL) Service Provider. The authors first conceptualized five dimensions of Third Party Logistics (3PL) service quality (i.e. tangibles, reliability, responsiveness, assurance and empathy); then AHP method was adapted to determine the relative weights of the five service quality dimensions and eventually select the best Third Party Logistics (3PL) Service Provider. Zhou, Min, Xu and Cao (2008) identified factors that significantly affected the operational efficiency of Chinese Third Party Logistics (3PLs) and proposed ways to improve the competitiveness of Third Party Logistics (3PLs). Wang, Zantow, Lai and Wang (2006), investigated the impact of IT on the financial performance of 3PL in China and found that greater IT uptake could offer 3PL companies better financial performance. Green, Turner, Roberts, Nagendra and Wininger (2008), reported on choosing a 3PL partner and examines the perspective of one major company heavily involved as a broker and provider of Logistics services.

Cheng, Chen and Chuang (2008), studied to figure out the evaluation factors and their weights to aid the selection of 4PL for businesses. The primary criteria to evaluate 4PL were established with Fuzzy Delphi Method (FDM), and then Fuzzy Analytic Hierarchy Process (FAHP) was employed to calculate the weights of these criteria, so as to build the Fuzzy Multi-criteria model of 4PL. Hamdan and Rogers (2008), introduced data envelopment analysis (DEA) as a tool to evaluate the efficiency of a group of Third Party Logistics (3PL) warehouse Logistics operations. Cakir, Tozan and Vayvay (2009), also proposed Logistics Service Provider selection decision support system based on the fuzzy analytic hierarchy process (FAHP) method.

Liu and Fang (2009), examined the relationship between customer satisfaction and Logistics costs, and used the grey correlation analysis method to study the relationship between customer satisfaction and various activity centers. The research result could offer decision-making references to save costs and enhance customer satisfaction for 3PL enterprises. Qureshi, Kumar and Kumar (2009a), proposed a methodology based on combined approach of Analytic Hierarchy Process (AHP) and Graph theory and derived a LSP selection index which evaluates and ranks the Third Party Logistics (3PL) Service Provider, to help shippers and also, proposed coefficients of similarity, coefficients of dissimilarity and the identification sets for further comparison of Third Party Logistics (3PL) Service Provider. In another paper, Qureshi, Kumar and Kumar (2009b), used the fuzzy analytic hierarchy process (FAHP) approach to support a generic Logistics benchmarking process and benchmarked performance levels of the LOGINET, a Third Party Logistics (3PL) Service Provider.

Percin (2009), recommended the use of a-two-phase analytical hierarchy process (AHP) and *technique for order preference by similarity to ideal solution* (TOPSIS) approach in the evaluation of Third Party Logistics (3PL) Service Provider. Govindan (2009), proposed a structured model for evaluating and selecting the best Third Party Reverse Logistics Provider (3PRLP) under fuzzy environment for the battery industry. In this paper, the multi-criteria decision-making tools such as analytic hierarchy process (AHP) and fuzzy analytic hierarchy process (FAHP) were adopted to solve the problem of selection of 3PRLP. Sahu (2016a) developed interval-valued fuzzy number set conjunction with modified Technique for order preference by similarity to ideal solution methodology for benchmarking supplier/partner alternative firms under similar agile supply chain architectures. Sahu (2016b) implemented a revised ranking approach accompanied with fuzzy performance important index to identify the barriers/ill-strong measures in Agile Supply Chain Management. Sahu (2016c) explored Ratio system, Reference point, and full Mutification form coupled with aggregated interval valued trapezoidal fuzzy number set in perceptive to evaluating and benchmarking the CNC machine tool from preferred options.

Vinay, Govindan, Sasikumar (2009), explained the concept of supply chain, Logistics, the need for the novel trends for Logistics outsourcing, the newer terms like Fifth Party Logistics Providers (5PLP) and Seventh Party Logistics (7PL), benefits and also highlighted the barriers of Third Party Logistics (3PL) Service Provider implementation. Rajesh, Sachdeva and Bhardwaj (2010), proposed a classification framework and suggested some new research settings are suggested in selection of the Third Party Logistics (3PL) Service Provider as Third Party Logistics (3PL) plays a key role in the Logistics activities between the outsourcing company, the marketplace, and the customers.

Hsiao, Kemp, Vorst and Omta, (2010), investigated the impact of outsourced Logistics functions on service performance within the food processing industry in The Netherlands and Taiwan, and concluded that outsourcing had no direct impact on service performance. Saen (2010), introduced a model to rank Third Party Reverse Logistics Providers (3PRLP) with regard to various criteria, including dual-role factors, which was based on Data Envelopment Analysis (DEA). Soh (2010), proposed an evaluation framework and methodology based on fuzzy analytic hierarchy process for selecting a suitable Third Party Logistics (3PL) Service Provider and illustrated the process of evaluation and selection through a case study. Bhatti, Kumar and Kumar (2010), examined to model the choice parameters for selection of Third Party Logistics (3PL) Service Provider in global lead Logistics provider (LLP) environments. Analytical hierarchy process (AHP) modeling was carried out after questionnaire-based survey, results of which were moderated with inputs from experts from industry and academics. The findings of this study revealed the under lying sub-parameters which come into play while rating/choosing or evaluating Service Provider in the global LLP situations of today.

Lin, Li, Jin, Wanga, Wange and Yang (2012), proposed an indicator system for data integration towards Third Party Logistics (3PL) Service Provider selection through analyzing the features and role of Third Party Logistics (3PL). The authors also established a comprehensive evaluation model for Third Party Logistics (3PL) suppliers based on fuzzy sets. Sahu (2015a), integrated a hairy-based approach and implemented novel developed multi hierarchical Third Party Logistics (3PL) Service Provider appraisement index in purpose to umpire the Third Party Logistics (3PL) provider for their strong and ill's core indices. Sahu (2015b) applied TOPSIS to emphasize the decision making scenario at the subjective information evaluation index platform to benchmark the advanced manufacturing CNC machine tool.

Chen and Wu (2011), developed a decision-making method combining the Delphi method and analytical network process (ANP), to help the electronic companies that need to evaluate and select the Logistics Service Provider type. Results of this research showed the best Logistics service type for electronic companies was the international Logistics Service Provider type. This research proposed not only a solution for solving the interdependent Logistics service provider selection problem, but also a useful reference in Logistics Service Provider selection for electronic companies which plan to invest in Southeast Asia. Rajesh, Sachdeva and Bhardwaj (2010), proposed a framework to select the Third Party Logistics (3PL) Service Provider using fuzzy Delphi method to shortlist the most important criteria and most probable Service Provider and fuzzy TOPSIS to choose the best Service Provider by finding the closeness to the Positive Ideal Solution (PIS).

Olugu and Wong (2011) aimed at evaluating the performance of the reverse Logistics process in the automotive industry. Fuzzy logic was applied in assessing the performance of the reverse Logistics process using a case study. Ding and Chou (2011), aimed to develop a fuzzy multiple criteria decision-making (MCDM) model, to select middle managers for Third Party Logistics (3PL). Sahu (2012), proposed an efficient grey TOPSIS and Grey COPRAS methodology for supplier appraisement platform in order to foster environmentally concerned purchasing strategies in an enterprise with enhanced degree of awareness and frequent applications of "green" purchasing. Sahu (2014) applied multi-objective optimization by ratio analysis methodology conjunction with grey number against criterion from the prospectus of handling the vagueness, inconsistency, impreciseness in order to appraise and benchmark the candidate CNC machine tool alternatives.

Wong (2012), proposed a decision support system (DSS), based on fuzzy analytic network process (FANP) and preemptive fuzzy integer goal programming (PFIGP). The proposed DSS for Third Party Logistics (3PL) Service Provider selection was assumed to take into consideration flexible resource and interactions among providers. Shan (2012) proposed the criteria for evaluating providers of green Logistics service by combining the traditional evaluation index system, and the criteria were determined as:

- Compatibility with the users,
- Cost of service,
- Quality of service,
- Service ability and
- Adaptation with environment.

In this paper, method of MCDM based on intuitionistic language fuzzy entropy, was adopted to evaluate and select providers of green Logistics service, and the scientific and feasible nature of this decision-making process was then verified through empirical studies. Third Party Logistics (3PL), which is growing around the world is drawing the due attention at government, industrial, academicians and practitioner's levels. The worldwide trend in globalization has led many companies to outsource so as to focus on their core competencies (Gupta, Sachdeva & Bhardwaj, 2010 & 2011).

In the field of non-deterministic decision-making theory and analytical technique, Fuzzy decision-making theory and technique, especially when triangular and trapezoidal fuzzy numbers acted as attribute values, received more attentions along with research results in recent years (Kahraman, Cebeci & Ulukan, 2003; Soh, 2010; Govindan & Murugesan, 2011). Triangular fuzzy numbers, instead of fixed values, which allow variation between certain limits, be treated as attribute values in comprehensive decision making based on fuzzy information. As stated by Govindan and Murugesan (2011), in practice, While fuzzy AHP requires weighty computations, it is a more systematic method than the others, and it is more capable of capturing a human's appraisal of ambiguity when complex multi-attribute decision-making problems are considered, so it has theoretical significance and practical value to introduce triangular fuzzy numbers combined with AHP in fuzzy system theory.

Motivated by the growing significance of reverse logistics activities in an increasingly competitive global market, this study has aimed to establish a decision support system (DSS) for performance appraisement as well as selection of Third Party Logistics (3PL) Service Provider by exploring the concept of triangular fuzzy numbers combined with AHP in fuzzy system theory. Numerical illustration has also been provided through an empirical case study.

The Third Party Logistics (3PL) performance appraisement plays a major role and contributes directly or indirectly in organization wealth and development. From the past literature review, it is found that evaluation and selection of Third Party Logistics (3PL) Service Provider is an important and crucial aspect in multi-criteria decision making (MCDM) problem. This study formulated a robust decision support system (DSS) for performance appraisement as well as selection of Third Party Logistics (3PL) Service Provider by exploring the conception of triangular fuzzy numbers combined with AHP in fuzzy system theory.

THEORY OF FUZZY ANALYTIC HIERARCHY PROCESS (CHANG, 1996, GOVINDAN ET AL., 2011)

In fuzzy AHP, the triangular fuzzy numbers are used in the pair-wise comparison process to express subjective judgments. The triangular fuzzy numbers are defined by three real numbers, expressed as $\tilde{M} = \left(l,\ m,\ u \right)$. The parameters l, m, and u indicate the smallest possible value, the most promising value, and the largest possible value that describe a fuzzy event, respectively. Their membership functions are described as equation (1). The relations between the linguistic scales and their corresponding triangular fuzzy numbers, used by experts in this study, are given in Table 2 (See Appendix).

Definition 1

A fuzzy number $\tilde{M}$ in R is a Triangular Fuzzy Number if its membership function $\mu_{\tilde{M}}\left(x\right) : R \to \left[0,1\right]$ satisfies Equation (1):

$$\mu_{\tilde{M}}\left(x\right) = \begin{cases} \left(x-l\right)/\left(m-l\right), & l \le x \le m \\ \left(u-x\right)/\left(u-m\right), & m \le x \le u \\ 0, & otherwise \end{cases} \tag{1}$$

From Equation(1), l and u mean the lower and upper bounds of the fuzzy number $\tilde{M}$, and m is the modal value for $\tilde{M}$ (Figure 1).

Definition 2

The algebraic operational laws of Triangular Fuzzy Number $M_1 = (l_1, m_1, u_1)$ and $M_2 = (l_2, m_2, u_2)$ are performed as follows Equations (2)–(7).

- **Addition of the fuzzy number** $\oplus$

$$M_1 \oplus M_2 = (l_1, m_1, u_1) \oplus (l_2, m_2, u_2) = (l_1 + l_2,\ m_1 + m_2,\ u_1 + u_2) \tag{2}$$

- **Multiplication of the fuzzy number** $\otimes$

$$M_1 \otimes M_2 = (l_1, m_1, u_1) \otimes (l_2, m_2, u_2) = (l_1 l_2,\ m_1 m_2,\ u_1 u_2) \tag{3}$$

- **Subtraction of the fuzzy number** $(-)$

$$M_1 - M_2 = (l_1, m_1, u_1) - (l_2, m_2, u_2) = (l_1 - u_2,\ m_1 - m_2,\ u_1 - l_2) \tag{4}$$

- **Division of a fuzzy number** $(/)$

$$M_1 / M_2 = (l_1, m_1, u_1) / (l_2, m_2, u_2)$$

$$= \left({l_1}/{u_2},\ {m_1}/{m_2},\ {u_1}/{l_2} \right) \text{ for } l_1, l_2 > 0,\ m_1, m_2 > 0,\ u_1, u_2 > 0 \tag{5}$$

- **Reciprocal of the fuzzy number**

$$M_1^{-1} \approx (l_1, m_1, u_1)^{-1} \approx \left({1}/{u_1},\ {1}/{m_1},\ {1}/{l_1} \right) \tag{6}$$

- **Multiplication of constant**

$$(\lambda, \lambda, \lambda) \otimes (l_1, m_1, u_1) = (\lambda l_1, \lambda m_1, \lambda u_1),\ \lambda \succ 0,\ \lambda \in R \tag{7}$$

Definition 3: Value of Fuzzy Synthetic Extent

Let $X = \{ x_1, x_2, x_3, ...x_n \}$ be an object set, and $U = \{ u_1, u_2, u_3, ...u_m \}$ be a goal set. According to the method of extent analysis, we now take each object and perform extent analysis for each goal respectively. Therefore, we can get m extent analysis values for each object, with the following signs:

$$M_{g_i}^1, M_{g_i}^2, ...M_{g_i}^m \qquad i = 1, 2, ..., n,$$

where all the $M_{g_i}^j \left(j = 1, 2, ..., m \right)$ are triangular fuzzy numbers.

The steps of Chang's extent analysis can be given as follows:

Step 1: The value of fuzzy synthetic extent with respect to the ith object is defined as:

$$S_i = \sum_{j=1}^{m} M_{g_i}^j \otimes \left[\sum_{i=1}^{n} \sum_{j=1}^{m} M_{g_i}^j \right]^{-1} \tag{8}$$

Step 2: Since $\tilde{M}_1 = \left(l_1, m_1, u_1 \right)$ and $\tilde{M}_2 = \left(l_2, m_2, u_2 \right)$ are two triangular fuzzy numbers, the degree of possibility $M_1 \geq M_2$ defined as:

$$V\left(\tilde{M}_1 \geq \tilde{M}_2 \right) = \sup_{x \geq y} \left[\min \left(\mu_{M_1}(x), \mu_{M_2}(y) \right) \right]$$

$$V\left(\tilde{M}_2 \geq \tilde{M}_1 \right) = hgt\left(\tilde{M}_1 \cap \tilde{M}_2 \right) = \mu_{M_2}(d)$$

$$= \begin{cases} 1, & if\ m_2 \geq m_1, \\ 0, & if\ l_1 \geq u_2, \\ \dfrac{l_1 - u_2}{\left(m_2 - u_2 \right) - \left(m_1 - l_1 \right)} & otherwise, \end{cases} \tag{9}$$

where; d is the ordinate of the highest intersection point D between μ_{M_1} and μ_{M_2} (Figure 4). To compare M_1 and M_2 both value of $V\left(M_1 \geq M_2 \right)$ and $V\left(M_2 \geq M_1 \right)$ are required.

Figure 4. The Intersection between M_1 and M_2

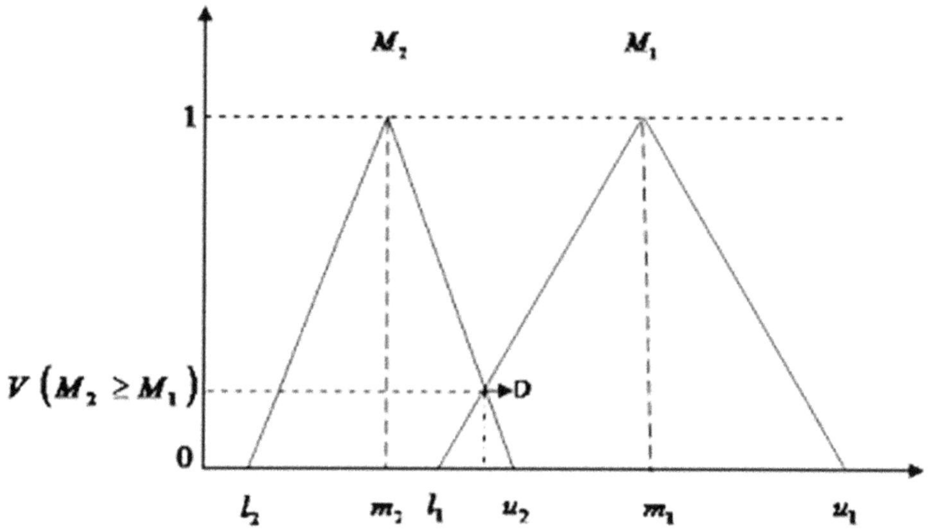

Step 3: The degree possibility for a convex fuzzy number to be greater than k convex fuzzy $M_i\left(i=1,2,\ldots\ldots,k\right)$ numbers can be defined by;

$$V\left(M \geq M_1, M_2, \ldots\ldots, M_k\right) = V\left[\left(M \geq M_1\right) and \left(M \geq M_2\right)\ and \ldots\ldots\ldots\ldots and\ \left(M \geq M_k\right)\right]$$

$$= \min V\ \left(M \geq M_i\right),\quad i = 1\ ,\ 2, \ldots\ldots, k. \tag{10}$$

Step 4: Assume that $d\left(X_i\right) = \min\ V\ \left(S_i \geq S_k\right)$ for $k = 1\ ,\ 2\ ,\ldots\ldots, n;\ \ k \neq i$. Then, the weight vector is given by

$$W' = \left(d'\left(X_1\right),\ d'\left(X_2\right), \ldots\ldots., d'\left(X_n\right)\right)^T \tag{11}$$

where; $X_i = \left(1\ ,\ 2\ ,\ldots\ldots\ldots, n\right)$ are n elements.

Step 5: Via normalization, the normalized weight vectors are:

$$W = \left(d\left(X_1\right),\ d\left(X_2\right), \ldots\ldots., d\left(X_n\right)\right)^T \tag{12}$$

where; W is a non-fuzzy number that gives the priority weights of one criterion over another.

The normalized weight vectors are calculated as:

$$NW_i = \frac{W_i}{\sum W_i} \tag{13}$$

Definition 4: *Consistency of Pair-Wise Comparison Matrix* (Saaty, 1980, Alonso & Lamata, 2006, Ramik & Korviny, 2010):

In classical AHP we consider an $n \times n$ pair-wise comparison matrix A with positive elements such that

$$A = \begin{bmatrix} 1 & a_{12} & \cdots & a_{1n} \\ a_{21} & 1 & \cdots & a_{2n} \\ \vdots & \vdots & \ddots & \vdots \\ a_{n1} & a_{n2} & \cdots & 1 \end{bmatrix} \tag{14}$$

This matrix is reciprocal, if $a_{ij} = \frac{1}{a_{ji}}$ for each $1 \leq \mathrm{i},\ \mathrm{j} \leq \mathrm{n}$. We say that A is *consistent* if

$$\text{for each } 1 \leq \mathrm{i}, \mathrm{j},\ \mathrm{k} \leq \mathrm{n};\ a_{ij} * a_{jk} = a_{ik} \tag{15}$$

If for some $\mathrm{i},\ \mathrm{j},\ \mathrm{k}$, Equation (15) does not hold, than A is said to be *inconsistent*. In analytic hierarchy process (AHP), it is assumed that $\frac{1}{9} \leq a_{ij} \leq 9$, for all $1 \leq \mathrm{i},\ \mathrm{j} \leq \mathrm{n}$, (Saaty, 1991). The inconsistency of A is measured by the *consistency index* CI_n as

$$CI_n = \frac{\lambda_{\max} - n}{n - 1} \tag{16}$$

where $\lambda_{\max}$ is the *principle eigen value* of A. It holds $CI_n \geq 0$.

If A is an $n \times n$ positive reciprocal (PR) matrix, then A is consistent if $CI_n = 0$. To provide a measure independent of the order of the matrix, n, Saaty proposed the *consistency ratio* (CR). This is obtained by taking the ratio between CI_n to its expected value over a large number of positive reciprocal matrices of order n, whose entries are randomly chosen in the set of values $S = \left\{\frac{1}{9}, \ldots\ldots\ldots, 9\right\}$. For this consistency measure, Saaty proposed a 10% threshold for the (CR) to accept the estimation. In practical decision situations inconsistency is "acceptable" if $CR < 0.1$.

To calculate the Consistency Ratio for the set of judgements using the CI for the corresponding value from large samples of matrices of purely random judgments using the table below, derived from Saaty's book, in which the upper row is the order of the random matrix, and the lower is the corresponding index of consistency for random judgements. Saaty defined the consistency ratio $\left(CR\right)$ as:

$$CR = \frac{CI}{RI} \tag{17}$$

where, RI is the average value of CI for random matrices using the (Saaty, 1980,) given scale.

NUMERICAL EXAMPLE

The present chapter deal to select the best Third Party Logistics (3PL) Service Provider amongst various Third Party Logistics (3PL) Service Provider under various important sizzling issues. The presented chapter exhibits the DSS to facilitate the organizations to make managerial decision under vagueness and uncertainty. The primary task in Third Party Logistics (3PL) Service Provider selection decision making is to structure a decision-making group having adequate knowledge and proficiency in logistics actions. The decision makers are selected from the various cross functional departments of organizations such as:

- Sales,
- Promotion,
- Production,
- Accounts, and
- Transportation.

Later, the decision makers team scrutinized and shortlisted important possible criteria amongst several criterions before decision-making. Here, Seven significant criteria followed the chain of forty one sub criteria are considered (Kannan, G., 2009; Soh, 2010; Govindan & Murugesan, 2011) as a hypothetical case research. Four Third Party Logistics (3PL) Service Provider have been considered for the decision alternatives, and these are alternative S_1, S_2, S_3 and S_4. Table 1 (See Appendix) displayed the Third Party Logistics (3PL) Service Provider's appraisement module. Now each decision maker (DM) is requested to choose suitable linguistic terms to allocate the importance weight against each of the evaluation criterion. The ground behind that the verdict made by individual is often vague, incomplete and imprecise in nature.

Therefore, the priory weight is evaluated in terms of linguistic variables. For each criterion, a linguistic variables i.e. Just equal (JE), Equally important (EI), Weakly important (WI), Strongly more important (SMI), Very strongly more important (VSMI) and Absolutely more important (AMI) is elected to assign criteria weights as shown in Table 2 (See Appendix). Next, the pair wise comparison matrix is formed as per opinion of decision makes as shown in Table 3 (See Appendix). The value of fuzzy synthetic extent with respect to the objective is computed by employing Equation 8 and then compared by using Equation 9. Equations 10, 11 and 12 are employed to compute the minimum degree of possibility of superiority of each criterion vs. another is obtained which further decides the weight vectors of the criteria. The normalized value of this vector decides the priority weights of each criterion over another. The normalized weight vectors are calculated using the Equation 13 and shown in Table 11 (See Appendix). By following the similar process, the pair-wise comparison matrix for sub criteria is built as per expert team and shown in Table 4–10 (See Appendix), and the priority weights of metrics were calculated as given in Table 12 (See Appendix).

The pair-wise comparison judgment matrix of the nine point linguistic scale is used. The idealized weights of:

- Outstanding,
- Excellent,
- Very Good,
- Above Average,
- Average,
- Below Average,
- Fair,
- Poor,
- Very Poor,

were determined as 1.0000, 0.5616, 0.5095, 0.3519, 0.2411, 0.1652, 0.1141, 0.0802, 0.0581, respectively, as shown in Table 13 (See Appendix). The Consistency Index and Consistency Ratio are computed by using Equation 16 and Equation 17 respectively. The overall score against four Third Party Logistics (3PL) Service Provider alternatives is computed and is shown in Table 14 (See Appendix). S_1 is found as a most appropriate alternative amongst four Third Party Logistics (3PL) Service Provider as it has the apex overall score (**0.2625**) than score obtained by S_2, S_3, S_4. Hence, Third Party Logistics (3PL) Service Provider preferences order is shown below.

$S_1 > S_4 > S_2 > S_3$

MANAGERIAL IMPLICATION

In this global scenario, the customers' demands are more challenging as well as the requirement for timely delivery of the product is also becomes indeed essential. It became imperative to focus on the entire upstream and downstream supply chain, so that the products shall reach to the consumer under lead time, which enables the organization to maximize its profit. In today's competitive market, Third Party Logistics (3PL) Service Provider have become a central part of a company's supply chain. Decision making to subcontract a Third Party Logistics (3PL) Service Provider to perform various supply chain functions is a crucial task which requires intellectual capabilities, because it ultimately results in the organizational development. The proposed FAHP based appraisement platform enables the managers of the organizations in making planned or operational decisions properly and supports in the organizational competitiveness. The proposed chapter enables them in selecting the best Third Party Logistics (3PL) Service Provider amongst the various under the different critical aspects, which plays a key role for the organizational competitive growth. The proposed decision support system (DSS) does not require sophisticated software's as the computation is carried out using simple techniques and algorithmic methods. The module also provides flexibility to the managers to incorporate other more important criteria and alternative subjected to the availability.

CONCLUSION

In present world, Third Party Logistics (3PL) Service Provider selection is found imperative to manage the supply chain actions of organizations. Choosing the best alternative plays a vital role in the organizational development and also provides valuable insights to their customer's satisfaction, which ultimately results in the growth of organization. The present chapter provides decision support system (DSS) which consists of AHP accompanied with fuzzy set theory to determine the preferences order against the defined Third Party Logistics (3PL) Service Provider selection alternatives. The main purpose of using triangular fuzzy number is to deal with the uncertainty associated with criteria, which is undertaken by the subjective assessment, while AHP is utilized to evaluate the performance scores against defined Third Party Logistics (3PL) due to its versatility for making decision under short computation. Apart from that, proposed DSS has more utility for wide range of organizations, as it does not requires the highly skill employee, computation can be done by any normal skill employee. DSS found its applications to diagnostic real life dilemmas, i.e. appraise the economic worth of:

- Cars,
- Scooters,
- Bikes,
- Buses,
- Aero planes,
- Helicopters etc. under MCDM hierarchical module.

The transaction of presented DSS can be done with international organization, because transaction can be brought by email or electronic media. Besides that, it has the great access for whole seller and retailer towards solving alternative election problems such as election of the best transporter for transferring local stuffs up to retailer, election of best flight Service Provider for transferring parcel etc. Moreover, the proposed module is flexible towards incorporating the broad parameters / criterions or more interrelated metrics with respect to the wide ranges of alternatives. The proposed module can also be totally replaced by other developing module remaining with Fuzzy-AHP for solving others industrial problems such as:

- Election of the feasible robot,
- Conveyor,
- Machine tool,
- Mono truck,
- JCB,
- Trolleys,
- Mono rail,
- Green supplier etc.

REFERENCES

Akkucuk, U. (2015). *Handbook of Research on Waste Management Techniques for Sustainability*. Hershey, PA: IGI Global.

Alonso, J. A., & Lamata, M. T. (2006). Consistency in the Analytic Hierarchy Process: A new approach. *International Journal of Uncertainty, Fuzziness and Knowledge-based Systems*, *14*(4), 445–459. doi:10.1142/S0218488506004114

Bayazit, O. (2012). An Application of The AHP In Selection of a Third Party Logistics Provider.*Orlando International Academic Conference*, (pp. 315-317).

Bhatti, R. S., Kumar, P., & Kumar, D. (2010). Analytical modeling of third party service provider selection in lead Logistics provider environments. *Journal of Modelling in Management*, *5*(3), 275–286. doi:10.1108/17465661011092641

Cakir, E., Tozan, H., & Vayvay, O. (2009). A Method For Selecting Third Party Logistic Service Provider Using Fuzzy AHP. *Journal of Naval Science and Engineering*, *5*(3), 38–54.

Chang, D.-Y. (1996). Application of the extent analysis method on fuzzy AHP. *European Journal of Operational Research*, *65*(3), 649–655. doi:10.1016/0377-2217(95)00300-2

Chen, K. Y., & Wu, W. T. (2011). Applying Analytic Network Process in Logistics Service Provider selection– A Case Study of the industry investing in Southeast Asia. *International Journal of Electronic Business Management*, *9*(1), 24–36.

Cheng, J. H., Chen, S. S., & Chuang, Y. W. (2008). Using Fuzzy Multiple Criterion Methods for Fourth Party Logistics Criteria Selection, *7th WSEAS Int. Conf. on Artificial Intelligence, Knowledge Engineering and Data Bases (AIKED'08)*, (pp. 282-288).

Ding, J. F., & Chou, C. C. (2011). Middle managers selection for third-party Logistics service providers. *International Journal of the Physical Sciences*, *6*(3), 610–619.

Folinas, D. (2012). *Outsourcing Management for Supply Chain Operations and Logistics Service*. Hershey, PA: IGI Global.

Gol, H., & Catay, B. (2007). Third-party Logistics provider selection: Insights from a Turkish automotive company. *Supply Chain Management: An International Journal*, *12*(6), 379–384. doi:10.1108/13598540710826290

Govindan, K. (2009). Fuzzy approach for the selection of third party reverse Logistics provider. *Asia Pacific Journal of Marketing and Logistics*, *21*(3), 397–416. doi:10.1108/13555850910973865

Govindan, K., & Murugesan, P. (2011). Selection of third-party reverse Logistics provider using fuzzy extent analysis. *Benchmarking: An International Journal*, *18*(1), 149–167. doi:10.1108/14635771111109869

Green, F. B., Turner, W., Roberts, S., Nagendra, A., & Wininger, E. (2008). A Practitioner's Perspective on the role of a Third-Party Logistics Provider. *Journal of Business & Economics Research*, *6*(6), 9–14.

Gupta, R., Sachdeva, A., & Bhardwaj, A. (2010). Selection of 3pl Service Provider using Integrated Fuzzy Delphi and Fuzzy TOPSIS.*Proceedings of the World Congress on Engineering and Computer Science 2010 (WCECS 2010)*, (pp. 1092-1097).

Gupta, R., Sachdeva, A., & Bhardwaj, A. (2011). Criteria of Selecting 3pl Provider: A Literature Review. *World Academy of Science. Engineering and Technology*, *59*, 2020–2024.

Hamdan, A., & Rogers, K. J. (2008). Evaluating the efficiency of 3PL Logistics operations. *International Journal of Production Economics*, *113*, 235–244. doi:10.1016/j.ijpe.2007.05.019

Hsiao, H. I., Kemp, R. G. M., Vorst, V. J. G. A. J., & Omta, S. W. F. (2010). A classification of Logistics outsourcing levels and their impact on service performance: Evidence from the food processing industry. *International Journal of Production Economics*, *124*, 75–86. doi:10.1016/j.ijpe.2009.09.010

Isiklar, G., Alptekin, E., & Buyukozkan, G. (2007). Application of a hybrid intelligent decision support model in Logistics outsourcing. *Computers & Operations Research*, *34*(12), 3701–3714. doi:10.1016/j.cor.2006.01.011

Kahraman, C., Cebeci, U., & Ulukan, Z. (2003). Multi-criteria supplier selection using fuzzy AHP. *Logistics Information Management*, *16*(6), 382–394. doi:10.1108/09576050310503367

Karagul, H., & Albayrakoglu, M. M. (2007). *Selecting a Third-Party Logistics Provider for an Automotive Company: An Analytic Hieararchy Process Model*. Vina Del Mar, Chile: ISAHP.

Koh, S. C. L., & Tan, Z. (2005). Using e-commerce to gain a competitive advantage in 3PL enterprises in China. *International Journal of Logistics Systems and Management*, *1*(2/3), 87–210. doi:10.1504/IJLSM.2005.005970

Lambert, D. M. (2008). *Supply Chain Management: Processes, Partnerships, Performance*. Sarasota, FL: Supply Chain Management Institute.

Li, D.-F. (2011). Closeness coefficient based nonlinear programming method for interval-valued intuitionistic fuzzy multiattribute decision making with incomplete preference information. *Applied Soft Computing*, *11*(4), 3402–3418. doi:10.1016/j.asoc.2011.01.011

Li, D.-F. (2016). *Linear Programming Models and Methods of Matrix Games with Payoffs of Triangular Fuzzy Numbers*. Berlin, Heidelberg: Springer. doi:10.1007/978-3-662-48476-0

Li, D.-F., & Ren, H. P. (2015). Multi-attribute decision making method considering the amount and reliability of intuitionistic fuzzy information. *Journal of Intelligent & Fuzzy Systems*, *28*, 1877–1883.

Li, D.-F., & Wan, S. P. (2014). Fuzzy heterogeneous multiattribute decision making method for outsourcing provider selection. *Expert Systems with Applications*, *41*(6), 3047–3059. doi:10.1016/j.eswa.2013.10.036

Li, D.-F., & Wan, S. P. (2014). A fuzzy inhomogenous multiattribute group decision making approach to solve outsourcing provider selection problems. *Knowledge-Based Systems*, *67*, 71–89. doi:10.1016/j.knosys.2014.06.006

Lin, F., Li, L., Jin, C., Wanga, R., Wange, H., & Yang, L. (2012). A 3PL supplier selection model based on fuzzy sets. *Computers & Operations Research*, *39*(8), 1879–1884. doi:10.1016/j.cor.2011.06.022

Liu, H. T., & Wang, W. K. (2009). An integrated fuzzy approach for provider evaluation and selection in third-party Logistics. *Expert Systems with Applications*, *36*(3), 4387–4398. doi:10.1016/j.eswa.2008.05.030

Liu, S., & Fang, Z. (2009). Study on the Relationship between Customer Satisfaction and 3PL Costs. *International Journal of Business and Management*, *4*(6), 23–28. doi:10.5539/ijbm.v4n6p23

Noci, G., & Toletti, G. (2000). Selecting quality-based programmes in small firms: A comparison between the fuzzy linguistic approach and the analytic hierarchy process. *International Journal of Production Economics*, *67*(2), 113–133. doi:10.1016/S0925-5273(99)00131-0

Olugu, E. U., & Wong, K. Y. (2011). Fuzzy logic evaluation of reverse Logistics performance in the automotive industry. *Scientific Research and Essays*, *6*(7), 1639–1649.

Percin, S. (2009). 'Evaluation of third-party Logistics (3PL) providers by using a two-phase AHP and TOPSIS methodology. *Benchmarking: An International Journal*, *16*(5), 588–604. doi:10.1108/14635770910987823

Qureshi, M. N., Kumar, P., & Kumar, D. (2009a). Selection of 3PL service providers: A combined approach of AHP and Graph theory. *International Journal of Services Technology and Management*, *12*(1), 35–60. doi:10.1504/IJSTM.2009.025035

Qureshi, M. N., Kumar, P., & Kumar, D. (2009b). Framework for benchmarking Logistics performance using fuzzy AHP.[IJBPSCM]. *International Journal of Business Performance and Supply Chain Modelling*, *1*(1), 82–98. doi:10.1504/IJBPSCM.2009.026267

Rajesh, R., Pugazhendhi, S., & Ganesh, K. (2010). A survey of literature on selection of third party Logistics service provider.[IJBPSCM]. *International Journal of Business Performance and Supply Chain Modelling*, *2*(2), 146–170. doi:10.1504/IJBPSCM.2010.036167

Ramik, J., & Korviny, P. (2010). Inconsistency of pair wise comparison matrix with fuzzy elements based on geometric mean. *Fuzzy Sets and Systems*, *161*(11), 1604–1613. doi:10.1016/j.fss.2009.10.011

Saen, R. F. (2010a). A New Model for Ranking 3PL Providers. *Australian Journal of Basic and Applied Sciences*, *4*(8), 3762–3769.

Saen, R. F. (2010b). A new model for selecting third-party reverse Logistics providers in the presence of multiple dual-role factors. *International Journal of Advanced Manufacturing Technology*, *46*(1-4), 405–410. doi:10.1007/s00170-009-2092-x

Sahu, A. K., Sahu, N. K., & Sahu, A. (2014). Appraisal of CNC machine tool by integrated MULTI MOORA-IGVN circumstances: An empirical study. *International Journal of Grey Systems: Theory and Application*, *4*(1), 104–123.

Sahu, A. K., Sahu, N. K., & Sahu, A. K. (2015b). Benchmarking CNC machine tool using hybrid fuzzy methodology a multi indices decision making approach. *International Journal of Fuzzy System Applications*, *4*(2), 28–46.

Sahu, A. K., Sahu, N. K., & Sahu, A. K. (2016a). Application of Integrated TOPSIS in ASC index: Partners Benchmarking perspective. *Benchmarking: An International Journal*, *23*(3), 540–563. doi:10.1108/BIJ-03-2014-0021

Sahu, A. K., Sahu, N. K., & Sahu, A. K. (2016b). Appraisal of Partner Enterprises under GTFNS Environment in Agile SC.[IJDSST]. *International Journal of Decision Support System Technology*, *8*(3), 1–19. doi:10.4018/IJDSST.2016070101

Sahu, A. K., Sahu, N. K., & Sahu, A. K. (2016c). Application of Modified MULTI-MOORA for CNC Machine Tool Evaluation in IVGTFNS Environment: An Empirical Study. *International Journal of Computer Aided Engineering and Technology*, *8*(3), 234–259. doi:10.1504/IJCAET.2016.077603

Sahu, N. K., Datta, S., & Mahapatra, S. S. (2012). Establishing green supplier appraisement platform using grey concepts. *Grey Systems: Theory and Application*, *2*(3), 395–418. doi:10.1108/20439371211273276

Sahu, N. K., Datta, S., & Mahapatra, S. S. (2013). Decision making for selecting 3PL service provider using three parameter interval grey numbers. *International Journal of Logistic Systems and Management*, *14*(3), 261–297. doi:10.1504/IJLSM.2013.052061

Sahu, N. K., Datta, S., & Mahapatra, S. S. (2014). Green Supplier Appraisement in Fuzzy Environment. *Benchmarking: An International Journal*, *21*(3), 412–429. doi:10.1108/BIJ-06-2012-0042

Sahu, N. K., Datta, S. & Mahapatra, S. S. (2015). Fuzzy Based Appraisement Module for 3PL Evaluation and Selection. *Benchmarking, an International Journal, 22*(3), 354-392.

Sahu, N. K., Sahu A. K., & Sahu, A. K (2015a). Appraisement and Benchmarking of Third Party Logistic Service Provider by Exploration of Risk Based Approach. *Cogent Business and Management, 2*, 1-21.

Samantra, C., Sahu, N. K., Datta, S., & Mahapatra, S. S. (2013). Decision-Making in Selecting Reverse Logistics Alternative using Interval-Valued Fuzzy Sets Combined with VIKOR Approach. *International Journal of Services and Operations Management*, *14*(2), 175–196. doi:10.1504/IJSOM.2013.051828

Selviaridis, K., & Spring, M. (2007). Third party Logistics: A literature review and research agenda. *International Journal of Logistics Management*, *18*(1), 125–150. doi:10.1108/09574090710748207

Shan, L. (2012). Research on Green Logistics Service Providers Selection based on Intuitionistic Language Fuzzy Entropy. *Journal of Computers*, *7*(2), 540–546. doi:10.4304/jcp.7.2.540-546

So, S., Kim, J., Cheong, K., & Cho, G. (2006). Evaluating The Service Quality of Third party Logistics Service Providers using the Analytic Hierarchy Process. *Journal of Information Systems and Technology Management*, *3*(3), 261–270. doi:10.4301/S1807-17752006000300001

Soh, S. (2010). A decision model for evaluating third-party Logistics providers using fuzzy analytic hierarchy process. *African Journal of Business Management*, *4*(3), 339–349.

Vinay, V. P., Govindan, K., & Sasikumar, P. (2009). Conceptual study on 3PL/4PL/ new trends for service industry. *International Journal of Services Technology and Management*, *12*(1), 3–22. doi:10.1504/IJSTM.2009.025033

Wan, S. P., & Li, D.-F. (2013). Fuzzy LINMAP approach to heterogeneous MADM considering comparisons of alternatives with hesitation degrees. *Omega: The International Journal of management. Science*, *41*(6), 925–940.

Wan, S. P., & Li, D.-F. (2014). Atanassovs intuitionistic fuzzy programming method for heterogeneous multiattribute group decision making with Atanassovs intuitionistic fuzzy truth degrees. *IEEE Transactions on Fuzzy Systems*, *22*(2), 300–312. doi:10.1109/TFUZZ.2013.2253107

Wan, S. P., & Li, D.-F. (2015). Fuzzy mathematical programming approach to heterogeneous multiattribute decision-making with interval-valued intuitionistic fuzzy truth degrees. *Information Sciences*, *325*, 484–503. doi:10.1016/j.ins.2015.07.014

Wang, Q., Zantow, K., Lai, F., & Wang, X. (2006). Strategic postures of third party Logistics providers in Mainland China. *International Journal of Physical Distribution & Logistics Management*, *36*(10), 793–819. doi:10.1108/09600030610714607

Wong, J. T. (2012). DSS for 3PL provider selection in global supply chain: Combining the multi-objective optimization model with experts opinions. *Journal of Intelligent Manufacturing*, *23*(3), 599–614. doi:10.1007/s10845-010-0398-z

Yan, J., Chaudhry, P. E., & Chaudhry, S. S. (2003). A model of a decision support system based on case-based reasoning for third-party Logistics evaluation. *Expert Systems: International Journal of Knowledge Engineering and Neural Networks*, *20*(4), 196–207. doi:10.1111/1468-0394.00244

Zaim, S., Sevkli, M., & Tarim, M. (2003). Fuzzy Analytic Hierarchy Based Approach for Supplier Selection. *Journal of Euromarketing*, *12*(3/4), 147–176. doi:10.1300/J037v12n03_09

Zhang, H., Li, X., Liu, W., Li, B., & Zhang, Z. (2004). An Application of the AHP in 3PL Vendor Selection of a 4PL System.*IEEE International Conference on Systems, Man and Cybernetics*, (pp. 1255-1260). doi:10.1109/ICSMC.2004.1399797

Zhou, G., Min, H., Xu, C., & Cao, Z. (2008). Evaluating the comparative efficiency of Chinese third-party Logistics providers using data envelopment analysis. *International Journal of Physical Distribution & Logistics Management*, *38*(4), 262–279. doi:10.1108/09600030810875373

APPENDIX

Table 1. 3PL supplier appraisement cum evaluation criterions

Goal	Attributes	Sub-Attributes
Evaluation and Selection of 3 PL provider	Service level	Reliability and timeliness
		Quality of service
		Flexibility and responsiveness
		Employee satisfaction level
		Agile service
	Financial Security	Financial stability
		Logistics costs
		Market share
		Total asset turnover
		Profit growth rate
	Capabilities	IT capability
		Warehouse management capability
		Optimization capability
		Inventory replenishment
		Logistics manpower
		Enterprise ability
	Location	Infrastructure
		Control
		Information level
		Execution
		Reporting
	Global Presence	Geographic spread and access to retailers
		Flexibility in operations and delivery
		Response time against complaint
		Environmental Awareness
		Quality of management
	Relationship Management	Trust and fairness
		Benefit and risk sharing
		Quality of management
		Compatibility
		Market knowledge
		Cost of relationship
		Resiliency against Disaster

continued on following page

Table 1. Continued

Goal	Attributes	Sub-Attributes
	Client fulfilment	Effective communication
		Service improvement
		Cost saving
		Overall working relations
		Payment conditions
		On-time delivery rate
		Contract fulfilment on scheduled time
		Information sharing

Table 2. Fuzzy Importance scale with Triangular Fuzzy Number for Criteria and Sub-Criteria

Verbal Judgment	Triangular Fuzzy Number	Reverse Triangular Fuzzy Number
Just equal (JE)	(1, 1, 1)	(1, 1, 1)
Equally important (EI)	(1/2, 1, 3/2)	(2/3, 1, 2)
Weakly important (WI)	(1, 3/2, 2)	(1/2, 2/3, 1)
Strongly more important (SMI)	(3/2, 2, 5/2)	(2/5, 1/2, 2/3)
Very strongly more important (VSMI)	(2, 5/2, 3)	(1/3, 2/5, 1/2)
Absolutely more important (AMI)	(5/2, 3, 7/2)	(2/7, 1/3, 2/5)

Table 3. . Fuzzy Pair-wise Comparison matrix at Criteria level

Criterions	C_1	C_2	C_3	C_4	C_5	C_6	C_7
C_1	(1,1,1)	(2/3, 1, 2)	(3/2, 2, 5/2)	(2/3, 1, 2)	(5/2, 3, 7/2)	(2/3, 1, 2)	(3/2, 2, 5/2)
C_2	(1/2, 1, 3/2)	(1,1,1)	(2/3, 1, 2)	(1/3, 2/5, 1/2)	(2/3, 1, 2)	(2/5, 1/2, 2/3)	(1/3, 2/5, 1/2)
C_3	(2/5, 1/2, 2/3)	(1/2, 1, 3/2)	(1,1,1)	(1/2, 2/3, 1)	(2/7, 1/3, 2/5)	(1/2, 2/3, 1)	(2/7, 1/3, 2/5)
C_4	(1/2, 1, 3/2)	(2, 5/2, 3)	(1, 3/2, 2)	(1,1,1)	(2/3, 1, 2)	(2, 5/2, 3)	(2/3, 1, 2)
C_5	(2/7, 1/3, 2/5)	(1/2, 1, 3/2)	(5/2, 3, 7/2)	(1/2, 1, 3/2)	(1,1,1)	(1/2, 1, 3/2)	(2, 5/2, 3)
C_6	(1/2, 1, 3/2)	(3/2, 2, 5/2)	(1, 3/2, 2)	(1/3, 2/5, 1/2)	(2/3, 1, 2)	(1,1,1)	(1/2, 1, 3/2)
C_7	(2/5, 1/2, 2/3)	(2, 5/2, 3)	(5/2, 3, 7/2)	(1/2, 1, 3/2)	(1/3, 2/5, 1/2)	(2/3, 1, 2)	(1,1,1)

Table 4. Fuzzy Pair-wise Comparison matrix at Sub-criteria level for (C_1)

Sub-criteria	C_{11}	C_{12}	C_{13}	C_{14}	C_{15}
C_{11}	(1, 1, 1)	(2, 5/2, 3)	(1/3, 2/5, 1/2)	(5/2, 3, 7/2)	(1, 3/2, 2)
C_{12}	(1/3, 2/5, 1/2)	(1, 1, 1)	(3/2, 2, 5/2)	(2/3, 1, 2)	(1, 3/2, 2)
C_{13}	(2, 5/2, 3)	(2/5, 1/2, 2/3)	(1, 1, 1)	(2, 5/2, 3)	(3/2, 2, 5/2)
C_{14}	(2/7, 1/3, 2/5)	(1/2, 1, 3/2)	(1/3, 2/5, 1/2)	(1, 1, 1)	(1/3, 2/5, 1/2)
C_{15}	(1/2, 2/3, 1)	(1/2, 2/3, 1)	(2/5, 1/2, 2/3)	(2, 5/2, 3)	(1, 1, 1)

Table 5. Fuzzy Pair-wise Comparison matrix at Sub-criteria level for (C_2)

Sub-criteria	C_{21}	C_{22}	C_{23}	C_{24}	C_{25}
C_{21}	(1, 1, 1)	(1/2, 2/3, 1)	(3/2, 2, 5/2)	(2, 5/2, 3)	(2/7, 1/3, 2/5)
C_{22}	(1, 3/2, 2)	(1, 1, 1)	(1, 1, 1)	(2/5, 1/2, 2/3)	(2/7, 1/3, 2/5)
C_{23}	(2/5, 1/2, 2/3)	(1, 1, 1)	(1, 1, 1)	(2/3, 1, 2)	(2/5, 1/2, 2/3)
C_{24}	(1/3, 2/5, 1/2)	(3/2, 2, 5/2)	(1/2, 1, 3/2)	(1, 1, 1)	(1, 3/2, 2)
C_{25}	(5/2, 3, 7/2)	(5/2, 3, 7/2)	(3/2, 2, 5/2)	(1/2, 2/3, 1)	(1, 1, 1)

Table 6. Fuzzy Pair-wise Comparison matrix at Sub-criteria level for (C_3)

Sub-Criteria	C_{31}	C_{32}	C_{33}	C_{34}	C_{35}	C_{36}
C_{31}	(1, 1, 1)	(2/3, 1, 2)	(2, 5/2, 3)	(1/3, 2/5, 1/2)	(1/2, 2/3, 1)	(2/3, 1, 2)
C_{32}	(1/2, 1, 3/2)	(1, 1, 1)	(3/2, 2, 5/2)	(5/2, 3, 7/2)	(2, 5/2, 3)	(1, 3/2, 2)
C_{33}	(1/3, 2/5, 1/2)	(2/5, 1/2, 2/3)	(1, 1, 1)	(3/2, 2, 5/2)	(2, 5/2, 3)	(1/2, 2/3, 1)
C_{34}	(2, 5/2, 3)	(2/7, 1/3, 2/5)	(2/5, 1/2, 2/3)	(1, 1, 1)	(1/2, 1, 3/2)	(3/2, 2, 5/2)
C_{35}	(1, 3/2, 2)	(1/3, 2/5, 1/2)	(1/3, 2/5, 1/2)	(2/3, 1, 2)	(1, 1, 1)	(1/2, 2/3, 1)
C_{36}	(1/2, 1, 3/2)	(1/2, 2/3, 1)	(1, 3/2, 2)	(2/5, 1/2, 2/3)	(1, 3/2, 2)	(1, 1, 1)

Table 7. Fuzzy Pair-wise Comparison matrix at Sub-criteria level for (C_4)

Sub-Criteria	C_{41}	C_{42}	C_{43}	C_{44}	C_{45}
C_{41}	(1, 1, 1)	(1, 3/2, 2)	(3/2, 2, 5/2)	(2, 5/2, 3)	(2/7, 1/3, 2/5)
C_{42}	(1/2, 2/3, 1)	(1, 1, 1)	(1/2, 2/3, 1)	(2/5, 1/2, 2/3)	(1, 3/2, 2)
C_{43}	(2/5, 1/2, 2/3)	(1, 3/2, 2)	(1, 1, 1)	(2/3, 1, 2)	(2/5, 1/2, 2/3)
C_{44}	(1/3, 2/5, 1/2)	(3/2, 2, 5/2)	(1/2, 1, 3/2)	(1, 1, 1)	(1/2, 1, 3/2)
C_{45}	(5/2, 3, 7/2)	(1/2, 2/3, 1)	(3/2, 2, 5/2)	(2/3, 1, 2)	(1, 1, 1)

Table 8. Fuzzy Pair-wise Comparison matrix at Sub-criteria level for (C_5)

Sub-Criteria	C_{51}	C_{52}	C_{53}	C_{54}	C_{55}
C_{51}	(1, 1, 1)	(2/3, 1, 2)	(1/3, 2/5, 1/2)	(1/2, 2/3, 1)	(2/5, 1/2, 2/3)
C_{52}	(1/2, 1, 3/2)	(1, 1, 1)	(2/5, 1/2, 2/3)	(1, 1, 1)	(1/2, 2/3, 1)
C_{53}	(2, 5/2, 3)	(3/2, 2, 5/2)	(1, 1, 1)	(1/3, 2/5, 1/2)	(2/3, 1, 2)
C_{54}	(1, 3/2, 2)	(1, 1, 1)	(2, 5/2, 3)	(1, 1, 1)	(1/2, 2/3, 1)
C_{55}	(3/2, 2, 5/2)	(1, 3/2, 2)	(1/2, 1, 3/2)	(1, 3/2, 2)	(1, 1, 1)

Table 9. Fuzzy Pair-wise Comparison matrix at Sub-criteria level for (C_6)

Sub-Criteria	C_{61}	C_{62}	C_{63}	C_{64}	C_{65}	C_{66}	C_{67}
C_{61}	(1, 1, 1)	(2/5, 1/2, 2/3)	(1, 3/2, 2)	(2/7, 1/3, 2/5)	(2/3, 1, 2)	(2, 5/2, 3)	(2/5, 1/2, 2/3)
C_{62}	(3/2, 2, 5/2)	(1, 1, 1)	(1/2, 1, 3/2)	(1/3, 2/5, 1/2)	(3/2, 2, 5/2)	(1, 1, 1)	(2/3, 1, 2)
C_{63}	(1/2, 2/3, 1)	(2/3, 1, 2)	(1, 1, 1)	(2/5, 1/2, 2/3)	(1/2, 2/3, 1)	(2/3, 1, 2)	(1/3, 2/5, 1/2)
C_{64}	(5/2, 3, 7/2)	(2, 5/2, 3)	(3/2, 2, 5/2)	(1, 1, 1)	(3/2, 2, 5/2)	(3/2, 2, 5/2)	(5/2, 3, 7/2)
C_{65}	(1/2, 1, 3/2)	(2/5, 1/2, 2/3)	(1, 3/2, 2)	(2/5, 1/2, 2/3)	(1, 1, 1)	(1/2, 1, 3/2)	(2/3, 1, 2)
C_{66}	(1/3, 2/5, 1/2)	(1, 1, 1)	(1/2, 1, 3/2)	(2/5, 1/2, 2/3)	(2/3, 1, 2)	(1, 1, 1)	(2/7, 1/3, 2/5)
C_{67}	(3/2, 2, 5/2)	(1/2, 1, 3/2)	(2, 5/2, 3)	(2/7, 1/3, 2/5)	(1/2, 1, 3/2)	(5/2, 3, 7/2)	(1, 1, 1)

Table 10. Fuzzy Pair-wise Comparison matrix at Sub-criteria level for (C_7)

Sub-Criteria	C_{71}	C_{72}	C_{73}	C_{74}	C_{75}	C_{76}	C_{77}	C_{78}
C_{71}	(1, 1, 1)	(2/3, 1, 2)	(2/3, 1, 2)	(3/2, 2, 5/2)	(2/3, 1, 2)	(3/2, 2, 5/2)	(2/3, 1, 2)	(2/3, 1, 2)
C_{72}	(1/2, 1, 3/2)	(1, 1, 1)	(2, 5/2, 3)	(1/2, 1, 3/2)	(1/3, 2/5, 1/2)	(1/2, 1, 3/2)	(2, 5/2, 3)	(1/2, 1, 3/2)
C_{73}	(1/2, 1, 3/2)	(1/3, 2/5, 1/2)	(1, 1, 1)	(3/2, 2, 5/2)	(2/3, 1, 2)	(3/2, 2, 5/2)	(2/3, 1, 2)	(3/2, 2, 5/2)
C_{74}	(2/5, 1/2, 2/3)	(2/3, 1, 2)	(2/5, 1/2, 2/3)	(1, 1, 1)	(1/2, 2/3, 1)	(2/7, 1/3, 2/5)	(2/7, 1/3, 2/5)	(1/2, 2/3, 1)
C_{75}	(1/2, 1, 3/2)	(2, 5/2, 3)	(1/2, 1, 3/2)	(1, 3/2, 2)	(1, 1, 1)	(2/3, 1, 2)	(1, 1, 1)	(2/3, 1, 2)
C_{76}	(2/5, 1/2, 2/3)	(2/3, 1, 2)	(2/5, 1/2, 2/3)	(5/2, 3, 7/2)	(1/2, 1, 3/2)	(1, 1, 1)	(1/2, 2/3, 1)	(1, 1, 1)
C_{77}	(1/2, 1, 3/2)	(1/3, 2/5, 1/2)	(1/2, 1, 3/2)	(5/2, 3, 7/2)	(1, 1, 1)	(1, 3/2, 2)	(1, 1, 1)	(2/3, 1, 2)
C_{78}	(1/2, 1, 3/2)	(2/3, 1, 2)	(2/5, 1/2, 2/3)	(1, 3/2, 2)	(1/2, 1, 3/2)	(1, 1, 1)	(1/2, 1, 3/2)	(1, 1, 1)

Table 11. The normalized priority weight vectors of the main criteria

Main Criterions	Priority Weight
Service level	0.1117
Financial Security	0.0501
Capabilities	0.0233
Location	0.1075
Global Presence	0.1003
Relationship Management	0.0816
Client fulfilment	0.5256

Table 12. The normalized priority weight vectors of the Sub criterions

Sub Criterions	Priority Weight
Reliability and timeliness	0.3202
Quality of service	0.2042
Flexibility and responsiveness	0.3249
Employee satisfaction level	0.0000
Agile service	0.1507
Financial stability	0.2095
Logistics costs	0.1961
Market share	0.0018
Total asset turnover	0.1741
Profit growth rate	0.4185
IT capability	0.1641
Warehouse management capability	0.2909
Optimization capability	0.1577
Inventory replenishment	0.1702
Logistics manpower	0.0852
Enterprise ability	0.1319
Infrastructure	0.2677
Control	0.1219
Information level	0.1458
Execution	0.1836
Reporting	0.2809
Geographic spread and access to retailers	0.1068
Flexibility in operations and delivery	0.1196
Response time against complaint	0.2600
Environmental Awareness	0.2501
Quality of management	0.2635

continued on following page

Table 12. Continued

Sub Criterions	Priority Weight
Trust and fairness	0.0578
Benefit and risk sharing	0.5275
Quality of management	0.0187
Compatibility	0.2153
Market knowledge	0.0465
Cost of relationship	0.0000
Resiliency against Disaster	0.1342
Effective communication	0.1397
Service improvement	0.1432
Cost saving	0.1432
Overall working relations	0.0617
Payment conditions	0.1390
On-time delivery rate	0.1215
Contract fulfilment on scheduled time	0.1374
Information sharing	0.1143

Table 13. Pairwise comparison judgment matrix for Nine-point rating scale

Sub Criteria	Outstanding	Excellent	Very Good	Above Average	Average	Below Average	Fair	Poor	Very Poor	Idealized vector
Out standing	1	2	3	4	5	6	7	8	9	1.0000
Excellent	1/2	1	2	3	4	5	6	7	8	0.5616
Very Good	1/3	1/2	1	2	3	4	5	6	7	0.5095
Above Average	1/4	1/3	1/2	1	2	3	4	5	6	0.3519
Average	1/5	1/4	1/3	1/2	1	2	3	4	5	0.2411
Below Average	1/6	1/5	1/4	1/3	1/2	1	2	3	4	0.1652
Fair	1/7	1/6	1/5	1/4	1/3	1/2	1	2	3	0.1141
Poor	1/8	1/7	1/6	1/5	1/4	1/3	1/2	1	2	0.0802
Very Poor	1/9	1/8	1/7	1/6	1/5	1/4	1/3	1/2	1	0.0581

Table 14. Overall scores of 3PL providers

Criteria / Sub-Criteria	Local Weight (LW)	Global Weight (GW)	3PL (A)		3PL (B)		3PL (C)		3PL (D)	
			Rating scores	X GW	Rating scores	X GW	Rating scores	X GW	Rating scores	X GW
Service level	0.1117									
Reliability and timeliness	0.3202	0.0358	0.3519	0.0126	0.5616	0.0201	0.0581	0.0021	0.5616	0.0201
Quality of service	0.2042	0.0228	0.5616	0.0128	0.1141	0.0026	1.0000	0.0228	0.2411	0.0055
Flexibility and responsiveness	0.3249	0.0363	0.5095	0.0185	1.0000	0.0363	0.0802	0.0029	0.0581	0.0021
Employee satisfaction level	0.0000	0.0000	1.0000	0.0000	0.0581	0.0000	0.5616	0.0000	1.0000	0.0000
Agile service	0.1507	0.0168	0.5616	0.0095	0.5095	0.0086	0.2411	0.0041	0.3519	0.0059
Financial Security	0.0501									
Financial stability	0.2095	0.0105	0.3519	0.0037	0.5616	0.0059	0.5095	0.0053	0.0581	0.0006
Logistics costs	0.1961	0.0098	0.2411	0.0024	0.0802	0.0008	1.0000	0.0098	0.1652	0.0016
Market share	0.0018	0.0001	0.5616	0.0001	0.5095	0.0000	0.3519	0.0000	0.5616	0.0001
Total asset turnover	0.1741	0.0087	0.1652	0.0014	1.0000	0.0087	0.5616	0.0049	0.0581	0.0005
Profit growth rate	0.4185	0.0210	0.5095	0.0107	0.1652	0.0035	0.1652	0.0035	1.0000	0.0210
Capabilities	0.0233									
IT capability	0.1641	0.0038	0.0802	0.0003	0.5095	0.0019	0.1141	0.0004	0.5095	0.0019
Warehouse management capability	0.2909	0.0068	0.5616	0.0038	0.0802	0.0005	0.5095	0.0035	0.3519	0.0024
Optimization capability	0.1577	0.0037	0.3519	0.0013	0.0581	0.0002	0.1141	0.0004	1.0000	0.0037
Inventory replenishment	0.1702	0.0040	1.0000	0.0040	0.5095	0.0020	0.5616	0.0022	0.2411	0.0010
Logistics manpower	0.0852	0.0020	0.5616	0.0011	0.2411	0.0005	0.2411	0.0005	0.5616	0.0011
Enterprise ability	0.1319	0.0031	0.2411	0.0007	1.0000	0.0031	0.0802	0.0002	0.0802	0.0002
Location	0.1075									
Infrastructure	0.2677	0.0288	0.1141	0.0033	0.5616	0.0162	1.0000	0.0288	0.1652	0.0048
Control	0.1219	0.0131	0.5616	0.0074	0.0581	0.0008	0.0581	0.0008	1.0000	0.0131
Information level	0.1458	0.0157	0.3519	0.0055	0.2411	0.0038	0.5095	0.0080	0.5616	0.0088
Execution	0.1836	0.0197	0.5095	0.0101	1.0000	0.0197	0.3519	0.0069	0.1141	0.0023
Reporting	0.2809	0.0302	0.5616	0.0170	0.0802	0.0024	1.0000	0.0302	0.2411	0.0073
Global Presence	0.1003									
Geographic spread and access to retailers	0.1068	0.0107	0.0581	0.0006	0.5095	0.0055	0.5616	0.0060	0.3519	0.0038
Flexibility in operations and delivery	0.1196	0.0120	0.5616	0.0067	0.0581	0.0007	0.0581	0.0007	0.5095	0.0061
Response time against complaint	0.2600	0.0261	0.0802	0.0021	0.5616	0.0146	0.5095	0.0133	0.5616	0.0146
Environmental Awareness	0.2501	0.0251	0.1652	0.0041	0.2411	0.0060	0.1141	0.0029	1.0000	0.0251
Quality of management	0.2635	0.0264	0.5616	0.0148	1.0000	0.0264	0.3519	0.0093	0.1141	0.0030
Relationship Management	0.0816									
Trust and fairness	0.0578	0.0047	1.0000	0.0047	0.1652	0.0008	0.0581	0.0003	0.0581	0.0003
Benefit and risk sharing	0.5275	0.0430	0.5616	0.0242	0.5095	0.0219	0.5095	0.0219	1.0000	0.0430
Quality of management	0.0187	0.0015	0.0802	0.0001	1.0000	0.0015	0.5616	0.0009	0.1652	0.0003
Compatibility	0.2153	0.0176	0.5095	0.0090	0.5095	0.0090	0.1141	0.0020	0.3519	0.0062
Market knowledge	0.0465	0.0038	0.2411	0.0009	0.0581	0.0002	0.3519	0.0013	0.5616	0.0021

continued on following page

Table 14. Continued

Criteria / Sub-Criteria	Local Weight (LW)	Global Weight (GW)	3PL (A)		3PL (B)		3PL (C)		3PL (D)	
			Rating scores	X GW	Rating scores	X GW	Rating scores	X GW	Rating scores	X GW
Cost of relationship	0.0000	0.0000	0.5616	0.0000	0.2411	0.0000	1.0000	0.0000	0.5095	0.0000
Resiliency against Disaster	0.1342	0.0110	0.0581	0.0006	0.5095	0.0056	0.1652	0.0018	0.1652	0.0018
Client fulfilment	0.5256									
Effective communication	0.1397	0.0734	1.0000	0.0734	0.5616	0.0412	0.0802	0.0059	0.3519	0.0258
Service improvement	0.1432	0.0753	0.5616	0.0423	0.0581	0.0044	0.5616	0.0423	1.0000	0.0753
Cost saving	0.1432	0.0753	0.1141	0.0086	0.1652	0.0124	0.1141	0.0086	0.1141	0.0086
Overall working relations	0.0617	0.0324	0.0581	0.0019	0.5095	0.0165	0.5095	0.0165	0.5095	0.0165
Payment conditions	0.1390	0.0731	0.5616	0.0410	0.3519	0.0257	1.0000	0.0731	0.3519	0.0257
On-time delivery rate	0.1215	0.0639	0.3519	0.0225	0.0802	0.0051	0.2411	0.0154	0.5616	0.0359
Contract fulfilment on scheduled time	0.1374	0.0722	0.5095	0.0368	1.0000	0.0722	0.1652	0.0119	0.2411	0.0174
Information sharing	0.1143	0.0601	1.0000	0.0601	0.1141	0.0069	0.3519	0.0211	0.5095	0.0306
Overall scores			18.3045		17.1035		16.4487		17.8705	
Renormalized scores			**0.2625**		**0.2453**		**0.2359**		**0.2563**	

Chapter 6
Intuitionistic Fuzzy Set Theory with Fair Share CPU Scheduler:
A Dynamic Approach

Supriya Raheja
NorthCap University, India

ABSTRACT

CPU Scheduler is one of the crucial modules of any operating system. Operating system uses CPU schedulers based on the requirements. Fair Share CPU scheduler is preferable scheduler by many operating systems. It distributes fixed processor time amongst different parts of a system like groups, users etc. Fixed distribution of processor time can distress the performance of system. Moreover, traditional CPU schedulers ignore the impreciseness and uncertainty of the tasks. This work presents the Intuitionistic Fuzzy based Fair Share scheduler (IFFS). IFFS scheduler has extended the research in the field of the impreciseness of tasks. Intuitionistic Fuzzy Inference Systems, IFIS-DP and IFIS-DCS have been designed to share the CPU time dynamically and for assigning dynamic priority to each task. Experimental results prove the better performance of IFFS scheduler over FS scheduler.

CPU SCHEDULER: MAIN MODULE OF AN OPERATING SYSTEM

Operating System is a significant software component of any computer system. These software components are mainly divided into two categories,

- System software and
- Application software.

Operating system belongs to the system software category. The main task of system software is to run the application software. System software supports the running of programs which uses computer system to run. In modern era, computer system contains a number of different hardware components like:

DOI: 10.4018/978-1-5225-1848-8.ch006

- Processor (CPU),
- Hard disk drives,
- Monitors,
- Different memory,
- Keyboards and
- Other input and output devices.

The other main task of operating system is to manage these resources. The operating system has various modules that works together to carry out the resource management functions and there in between interactions. For example Task Manger, this creates and destroys the tasks, Memory Manager, which manages the usage of main memory (allocation and release of memory blocks), Device Drivers, which deals with hardware devices, File Manager that manages the file system etc (Galvin & Gagne, 2009). Another main module of an operating system is the CPU Scheduler which handles the scheduling of tasks.

In a multitasking environment, multiple tasks are simultaneously stored in main memory. Operating system requires CPU schedulers to provide the concurrent execution of tasks. When multiple tasks are stored in main memory, the CPU schedulers makes an illusion that each task has its own CPU. It is also known as Short Term scheduler. However, the CPU will be executing only one task at a time, but it switch from one task to another task to support concurrent execution (Dhamdhere, 2006; Stallings, 2014). The CPU scheduler at particular time selects one ready task from the main memory (Ready Queue) and allocates CPU to that task. The CPU scheduler decides whom to execute next and for how long. The performance of the system varies in accordance to the chosen criteria i.e. on scheduling technique since it changes the ready state of task to running state. When the task is in ready queue then it is in ready state and when CPU is assigned to it, it is in running or active state. An effective scheduling can enhance the performance; on the other hand an ineffective scheduling can lead to reduce the performance of the entire system. Like, with non-preemptive scheduling algorithm, a system can suffer with the convoy effect. When many tasks are in waiting state while one big task is executing it is called convoy effect. It degrades the performance as the one big task halts the execution of all other small tasks.

Different operating systems use different schedulers based on their requirements like the operating systems which handle the tasks based on their criticality prefer to use priority based schedulers. The CPU scheduler usually executes the highest-priority tasks first. For example, whenever a currently running task is interrupted by an event then the CPU scheduler must make some optimal adjustments and schedule the next highest-priority task to processor. Windows & Solaris based operating systems prefer to uses multilevel feedback queue scheduler whereas UNIX based operating systems usually use fair share scheduler. This chapter discusses Fair Share (FS) CPU scheduler.

FAIR SHARE SCHEDULER AND ITS STATIC APPROACH

FS CPU scheduler equally/statically distributes the processor time between groups, users and tasks, as opposed to the fixed distribution among tasks (Round Robin Schedulers). Fairness should be the major consideration for any scheduler. Traditional schedulers considered that every task belongs to a different user, therefore schedulers like Round Robin logically distributes equal CPU services/processor time to all tasks (Sabin et al.,1996; Nie et al., 2011). When several tasks need a processor, operating system often allots processor time slices on a round robin fashion i.e. scheduler distributes equal time to all tasks having same priority. Let us suppose if there are 4 tasks want to access the CPU for execution then each task will get only 1/4 of processor time on each cycle. Most of the operating systems define their own length for time slice, such as 20ms, 50 ms or any arbitrary number.

These schedulers perform well as per their functionalities because previously, the focus of designers was only tasks. However, system usually supports various groups of related tasks. But, a common criticism regarding scheduling algorithms discussed in literature so far is the relationship between user and task (Kay & Lauder, 1988). For example, UNIX and other multiuser systems make groups of tasks that belong to a particular user. As the time goes, it becomes clear that the scheduler should be fair between users as well as tasks. Fair Share scheduler has addressed this problem.

Fair share scheduler permits processor time to be shared fairly among system groups or users in a system (Stallings, 2014; Bui et al., 2010). It is the fraction of processor time that should be allocated to group of tasks that belongs to the same user. Let us consider a case where fair share scheduling approach can be applied. Suppose a society group whose members are using one multiuser system. These members are further divided into two groups. One is the society head and other is the secretaries of this group. The society head uses the system to perform important and intensive work. Whereas many secretaries use the system for less intensive work such as collecting information etc. The secretaries consumes more processor time as compare to the society head as they are many. But the society head performs the important task over other tasks. However, if the system allows only 30% of the processor time to secretaries' work and the remaining 70% for society head, the society head would not suffer. In this manner, fair share scheduling ensures the fair distribution of processor time.

Let us suppose three distinct users (U1, U2, and U3) in a system, where each is concurrently running one task. The scheduler logically divides the CPU time in such a way that each user can get 33.3% as its fair share. Suppose U3 starts another task, then the tasks running by U1 and U2 still get the same CPU share (33.3%) but the tasks of U3 now receive 16.7% of CPU share.

Further, fair share scheduling allows another layer of abstraction i.e. partition of users into groups and then assign the CPU share to groups as well. Scheduler first assigns the share of CPU time among groups then to users within group and later on among tasks of those users. When more groups become active then share of processor time among each group will be reduced. Consider two groups (G1 and G2) with 2 users (U11 and U12) and 3 users (U21, U22 and U23) respectively. The total CPU time is divided between groups and users as follows:

- **Total CPU Share:** 100%.
- **Groups (G1 and G2):** 100/2=50% each.
- **User (U11 and U12):** 50/2=25% each.
- **User (U21, U22 and U23):** 50/3= 16.67% each.

The common method of applying the fair share scheduling technique is to apply the round robin scheduling strategy at each level (Raheja, 2012). A scheduler must attain the following objectives to be considered fair

- Each user should get their pre assigned share of the CPU.
- A user consuming higher share as compared to the pre assigned share should receive a lower chance to occupy the CPU, while a user consuming less share should receive a higher chance to engage the CPU.

- If one user is not using its share of CPU time with in an assigned duration, all the other users can use that additional share of CPU time. It means a user can get more CPU time if another user is not using it.

In a non-preemptive multi-tasking system like UNIX, CPU scheduling technique adjusts the priority of each task in a way that size of time quantum changes accordingly (Park et al., 2008). However, an additional new arrived task will affect all the other ready tasks in a system, as the CPU resource will be redistributed between all tasks (Halim, 2003; Caprita et al., 2005). Kay and Lauder (1988) introduced a hierarchical Share, "A Fair Share Scheduler" that not only supports sharing between users but among the groups of users. But, this scheduler was basically designed for a student environment and was only applicable for a UNIX like multitasking systems, where all users with different tasks can share the CPU time. Moreover, this technique is not applicable for the pre-emptive scheduling.

There has been limited research on the fairness of non-preemptive scheduling techniques in terms of metrics like turnaround time, utilization, throughput etc. Sabin et al. (2004) has studied various scheduling policies. He proposed a technique to provide fairness in non-preemptive scheduling.

However, the static approach does not fulfil all the aforementioned objectives to be fair. The distribution of small processor time to a busy group may slow down the performance of a system. The performance of a scheduler can be improved with the dynamic approach of distribution rather than fixed approach. Moreover, operating system is not aware of exact attributes of tasks like burst time, priority of tasks, deadline etc (Alam et al., 2011; Raheja et al., 2014) before execution. There can be the probability of impreciseness present in the decisions of CPU scheduler. By considering the impreciseness associated with these attributes, performance of system can be further improved. Developments in Fair Share scheduler do not exactly dealing with this impreciseness and inexactness.

Fuzzy Logic introduced by Professor Zadeh (1965) was used by Xia and Feng (2005) to design feedback scheduler for embedded control system. Bashir Alam et al. (2009) has used the same theory to handle the uncertainty of tasks. They have proposed a fuzzy inference system to provide the solutions for the shortcomings of fair share scheduling algorithm like minimizing the overheads in terms of context switches. They have made use of round robin scheduling with in the group to schedule the tasks. But they have not illustrated the improvements in performance of system.

Bui et al. (2010) has used fair share scheduling for multiprocessor systems. Fair share scheduling has been used to manage the performance of system. They presented the detailed analysis of fair share scheduling techniques with respect to various parameters. Fair share scheduler is analysed under the workload of two different environments. There is no significant development in the literature which shows the handling of uncertainty. As per literature, one or two approaches provide the dynamic environment to CPU scheduler that makes decisions for CPU share (Alam, 2009; Raheja, 2015).

The main objective of this chapter is to discuss the intuitionistic fuzzy based fair share scheduler (IFFS), one application of intuitionistic fuzzy set theory. Here, intuitionistic fuzzy set theory deals with the impreciseness and inexactness of tasks. IFFS contains three modules:

- First module is intuitionistic fuzzy inference system IFIS-DCS,
- Second is intuitionistic fuzzy inference system IFIS-DP and
- The last one module is scheduling algorithm.

IFFS deals with the impreciseness and inexactness of tasks; Dynamically shares processor time amongst groups, users as well as tasks; and It also provides dynamic priority to each task of active users. Additionally, it also improves the performance of system over FS scheduler in terms of average response time, average normalized turnaround time and number of context switches.

DYNAMIC APPROACH: INTUITIONISTIC FUZZY BASED FAIR SHARE CPU SCHEDULER

This research work is grounded on one of the generalized form of fuzzy set theory named as intuitionistic fuzzy set theory (Atanassov, 1986). Although IFFS scheduler looks complicated but the deviation is not as much as it seems. The distinct function of this work is to share CPU time dynamically in between users and then among tasks by considering all the necessary parameters like:

- Burst time,
- Number of users,
- Total burst time per user, and
- Number of tasks per user.

Moreover, IFFS generates dynamic priorities for the tasks by considering the current situation of the user. The following section first discusses preliminaries of intuitionistic fuzzy set theory which is the core part of this chapter.

Intuitionistic Fuzzy Set Theory

The problem of incomplete and imprecise knowledge has been tackled by number of mathematicians and researchers. Professor Zadeh (1965) had introduced fuzzy logic; the most successful novel approach to the family of logics. In fuzzy logic each object has a single degree of membership in interval [0, 1].

Definition 1 (Fuzzy Set)

Let us suppose Z as the universe of discourse such that $Z = \{z_1, z_2, ..., z_n\}$. A fuzzy set F can be defined as $\{z, \mu_F(z)\}$ where $\mu_F(z) \rightarrow [0,1]$. The membership value $\mu_F(z)$ denotes the single degree of membership of element z in F. If the value of is more towards 1, the element z belongs more to the set (Zadeh, 1984; Zimmerman, 2001; Zadeh, 2008).

Number of extended forms of fuzzy set theory has been introduced in literature. Atanassov (1986) has familiarized with intuitionistic fuzzy set theory that divides the point based single fuzzy membership into two functions, One for degree of membership and Another for degree of non-membership but in the same interim. Besides, it cannot use these values at the same interval. Pawlak (1982) has defined a different approach to impreciseness/vagueness named rough set. Just like intuitionistic approach, in this approach author expressed impreciseness as a pair of sets which specifies the lower and upper approximation of the set. Lower approximation provides the information for the elements which undoubtedly belongs to the set whereas upper approximation tells about those elements which possibly belongs to the set. Professor Gau et al. (1993) has presented another extended approach of fuzzy set theory. They named it vague set theory. They have divided the concept of fuzzy single membership function/value $\mu_F(z)$ into two values. True value: value that is in favour and false value: value that is in against.

Definition 2 (Intuitionistic Fuzzy Set)

An intuitionistic fuzzy set *I* in the universe of discourse $Z = \{z_1, z_2, ..., z_n\}$ is defined with two membership functions (Li, 2008; Li et al., 2009); degree of membership D_I and non-degree of membership function N_I. Here, *I* in *Z* is an object representing as $I = \{< z, D_I(z), N_I(z) >: z \in Z\}$

Both functions must satisfy the following conditions $D_I(z) \in [0,1], N_I(z) \in [0,1]$. These two member functions can work independently but under some constraints. One of the constraints is that the total value of both functions should not go beyond 1 as mentioned in equation (1).

$$0 \leq D_I(z) + N_I(z) \leq 1 \tag{1}$$

Definition 3 (Intuitionistic Fuzzy Set Index)

The degree of indeterminacy of $z \in Z$ to the intuitionistic fuzzy set *I*. It is also called as hesitation margin of *z* in *I*. It can be calculated as

$$I_I(z) = 1 - D_I(z) - N_I(z) \text{ where } 0 \leq I_I(z) \leq 1 \tag{2}$$

It expresses the lack of knowledge about the belongingness of *z* to intuitionistic fuzzy set *I*.

Definition 4 (Intuitionistic Fuzzy Value)

The degree of membership of *z* in the intuitionistic fuzzy set *I* is bounded by a subinterval $< D_I(z), N_I(z) >$ of [0,1], where the interval $< D_I(z), N_I(z) >$ represents the intuitionistic fuzzy value (An & Wilfred, 2005; Bustince & Burillo, 1996) of *z* in *I*.

Let us suppose *Z* as universe of tasks where $Z \in$ 'burst time of the tasks'. Assume the intuitionistic fuzzy value for burst time of a task T_1 is given as $<T_1, 0.7, 0.2>$. This value specifies that the degree of membership of T_1 $(D_I(T1))$ is 0.7; non-degree of membership of T_1 $(N_I(T1))$ is 0.2 whereas left .1 denote the intuitionistic fuzzy index of T_1 $(I_I(T1))$ (i.e. the value which shows the degree of indeterminacy of T_1 in *I*.

Definition 5 (Evidence Membership Function)

Evidence membership function represents the overall evidence contained in an intuitionistic fuzzy value. It is defined as in equation (3).

$$M_E = \frac{(D_I + 1 + N_I)}{2}, 0 \leq M_E \leq 1 \tag{3}$$

When the intuitionistic fuzzy value is <1, 0>, M_E has highest value that represents the total belongingness of element, whereas at intuitionistic fuzzy value <0, 1>, M_E has the lowest value which represents the empty intuitionistic fuzzy value.

IFFS CPU Scheduler

IFFS Scheduler is having three modules as shown in Figure 1.

- IFIS-DCS
- IFIS-DP
- IFFS Scheduling Algorithm

IFIS-DCS module provides the dynamic share among the users within group whereas IFIS-DP supplies the dynamic priority to each task of user. IFFS scheduling algorithm, schedules the tasks in such a manner that each task will get its fair share of CPU time. The next sections will discuss all these modules in detail.

1. Intuitionistic Fuzzy Inference System for Dynamic CPU Share (IFIS-DCS)

Intuitionistic fuzzy inference system considers the current state of ready queue to look after the uncertainty and impreciseness of tasks. Authors have discussed in literature that when a new task come to the system, operating system is not capable to know the exact attributes of task (Raheja, 2013, 2014). So, these Intuitionistic fuzzy inference systems will handle the inexactness accordingly.

Figure 1. IFFS CPU Scheduler

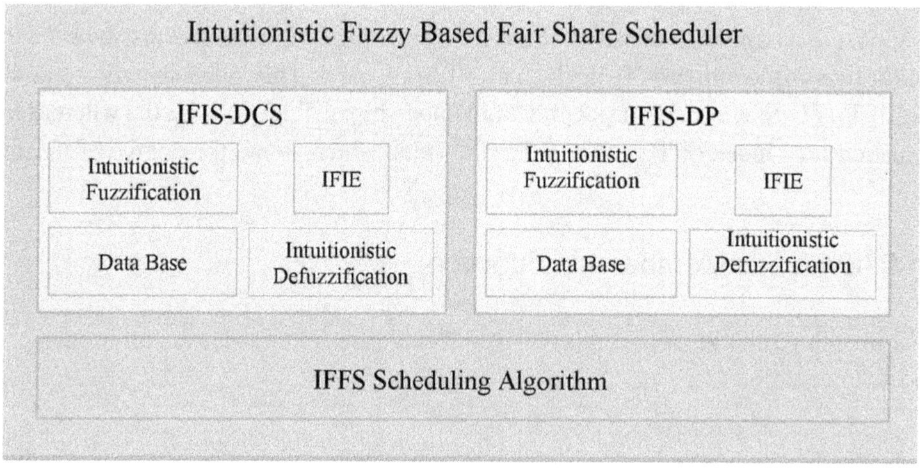

Figure 2. IFIS-DCS

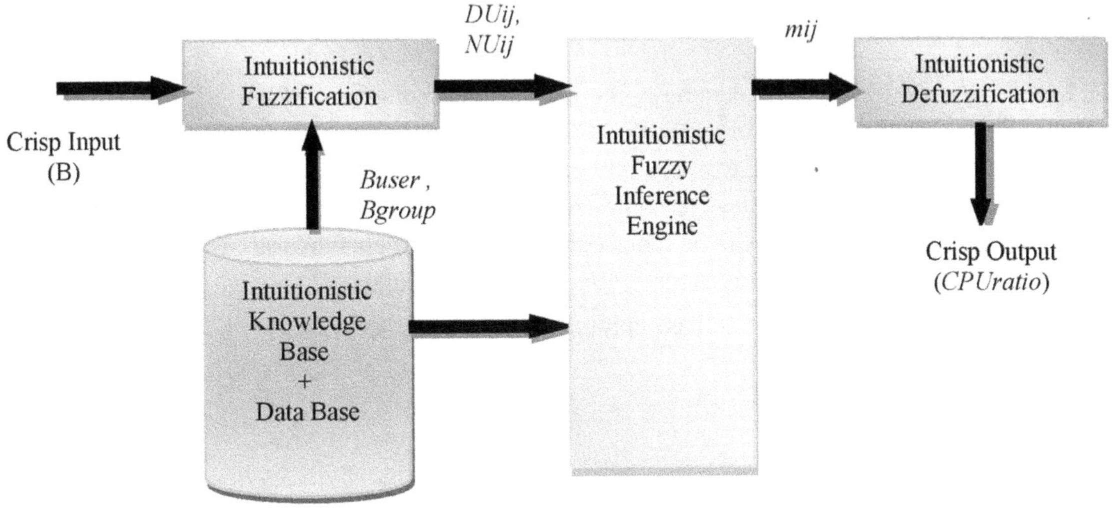

Proposed IFIS-DCS as shown in Figure 2 follows the following steps:

a. Intuitionistic Fuzzification
b. Intuitionistic Fuzzy Inference Engine
c. Intuitionistic Defuzzification

a. Intuitionistic Fuzzification

Intuitionistic Fuzzification process maps the inputs to the membership functions; degree of membership D_I and non-degree of membership function N_I. Finally, it converts each input into respective intuitionistic fuzzy value $< D_I, N_I >$. It receives the external input as well as data from the knowledgebase and data base. Based on these parameters it generates both the membership values.

Under IFIS-DCS, intuitionistic Fuzzification mainly handles the inexactness of burst time. It also consider the other parameters which affect the performance of task present in ready queue like the total number of users, total number of active tasks etc. In this way, impreciseness can be handled in more appropriately.

Let us assume a Uniprocessor System with *m* groups, *n* users in a group and *N* tasks performed by a user.

$$G = \{G_i \mid i \in 1...m\}, U = \{U_j \in 1...n\}, T = \{T_k \in 1...N\}$$

Intuitionistic Fuzzification process maps each input into intuitionistic fuzzy value using equation (4) and equation (5).

$$D_{Uij} = \frac{B_{user} * N_{user}}{B_{group} * N_{task\_user}} \tag{4}$$

$$N_{Uij} = 1 - sqrt(\frac{B_{user} * N_{user}}{B_{group} * N_{task\_user}}) \quad (5)$$

It extracts the values from the data base like B_{user} as total burst time of a user, B_{group} as total burst time of a group, N_{user} as number of users of active group and N_{task_user} as number of tasks of active user. Hence using all the knowledge of user and group, our Intuitionistic Fuzzification process assigns the degree of membership value D_{Uij} and non-degree of membership value N_{Uij} to each user *j* of group *i*. For example, user 2 of group 1 can be represented as U_{12} and their functions as D_{U12} & N_{U12} respectively. Finally the intuitionistic fuzzy value $< D_{Uij}, N_{Uij} >$ is calculated.

These membership functions should complying the following Axioms

Axiom 1: $0 \le D_{Uij} \le 1$
Axiom 2: $0 \le N_{Uij} \le 1$
Axiom 3: $0 \le D_{Uij} + N_{Uij} \le 1$

b. Intuitionistic Fuzzy Inference Engine (IFIE)

Intuitionistic fuzzy Inference Engine transforms the two membership values of intuitionistic fuzzy values into a single value as given in equation (6). Here, author uses the concept of evidence membership function discussed in section 3.1.

$$M_{Eij} = (D_{Uij} - N_{Uij} + 1) / 2 \quad (6)$$

Here M_{Eij} represents the overall evidence value of each user j of group i. Suppose the value of D_{U12} & N_{U12} is 0.5 and 0.3 respectively then M_{E12} is 0.6. Equation (6) must follow the axiom 4.

Axiom 4: $0 \le M_{Eij} \le 1$

c. Intuitionistic Defuzzification

Intuitionistic defuzzification process converts the multiple inputs into a single output value. Intuitionistic defuzzification process of IFIS-DCS takes the evidence value of each user *j* from 1 to *m*. Let us consider group 1 is having 3 users. Then it takes the evidence value for all three users such as M_{E11}, M_{E12} & M_{E13} from the intuitionistic fuzzy inference engine. Then it maps multiple intuitionistic fuzzy values into to a single output value CPU_{share}. It calculates the median of all evidence value w.r.t a group as given in equation (7). Later on this CPU_{share} is used to calculate the dynamic share of CPU time.

$$CPU_{share} = median(M_{Eij}), i \in 1...m \ \& \ j \in 1...n \quad (7)$$

2. Intuitionistic Fuzzy Inference System for Dynamic Priority (IFIS-DP)

Just like IFIS-DCS, IFIS-DP handles the inexactness and impreciseness of input: static priority of the task. As the priority of tasks depend on other factors like current state of the task and the state of the other tasks of a user; keeping in view IFIS-DP generates the dynamic priority by considering the static priority, number of users, number of tasks amongst user etc.

IFIS-DP works for each task T_k of each user U_j of each group G_i, where $k \in 1...N, j \in 1...n \; \& \; i \in 1...m$. Intuitionistic Fuzzification process takes P_{active}, static priority of active task T_k as input. It also extracts $P_{\max}$ (maximum priority) and $P_{\min}$ (minimum priority) values of an active user from the data base. By considering all these values, it converts the input priority into intuitionistic fuzzy value $< D_P, N_P >$ using equation (8). Just like mentioned in IFIS-DCS intuitionistic Fuzzification process, equation (8) must follow the entire axiom rules from axiom 1 to axiom 3.

$$< D_P, N_P >=< \frac{P_{\max} - P_{active}}{P_{\max} + P_{\min}}, \frac{P_{active} - P_{\min}}{P_{\max} + P_{\min}} > \quad (8)$$

Intuitionistic fuzzy inference engine computes the overall evidence value M_{Epk} for $\forall k$ as given in equation (9) where $k \in 1...N$.

$$M_{Epk} = (Dpk + 1 - Npk) \,/\, 2 \quad (9)$$

If user U_{12} is having three active tasks then the value of k varies from 1 to 3 and IFIE returns the overall evidence value as $M_{EP1}, M_{EP2}, \& M_{EP3}$ using equation (9).

$$M_{EP1} = (D_{P1} - N_{P1} + 1) \,/\, 2, M_{EP2} = (D_{P2} - N_{P2} + 1) \,/\, 2 \;\&\; M_{EP3} = (D_{P3} - N_{P3} + 1) \,/\, 2$$

Intuitionistic Fuzzification process of IFIS-DP further maps the overall evidence value of each task T_k into a single value named $P_{dynamic}$. System chooses the minimum value of M_{Epk} for $\forall k$ and assigns to $P_{dynamic}$ as given in equation (10).

$$P_{dynamic} = \min(M_{EPi}) \quad (10)$$

3. Intuitionistic Fuzzy Based Fair Share (IFFS) Scheduling Algorithm

An intuitionistic fuzzy based fair share scheduling policy allows the CPU time to be shared fairly among all tasks of users. A scheduling policy must accomplish the objectives aforementioned in order to be fair.

With these objectives, author proposes IFFS scheduling algorithm that consider both users and tasks. Moreover, it will be fair with all users and their associated tasks. It shares the CPU time at three levels:

- CPU_share_level1,
- CPU_share_level2 and
- CPU_share_level3 as shown in Figure 3.

Figure 3. Notations of CPU share

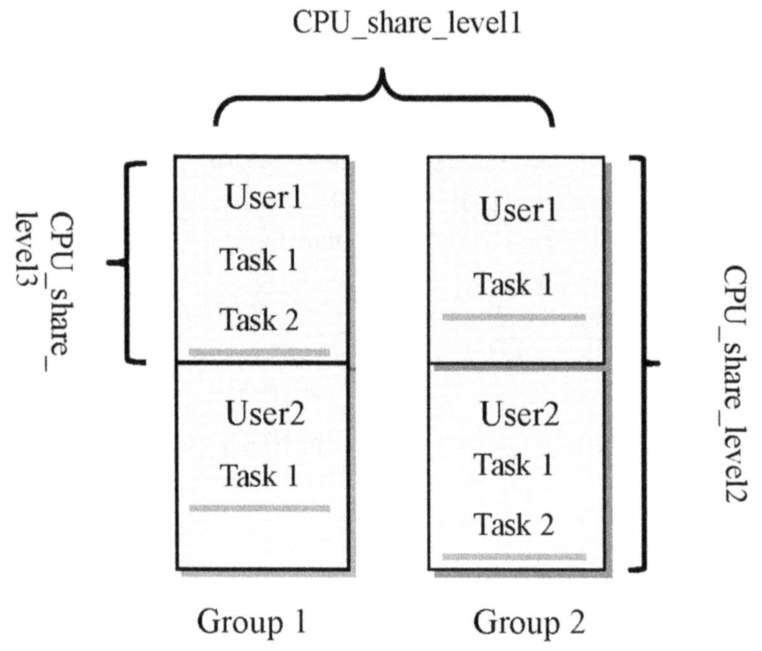

CPU_share_level1 distributes processor time among different groups in a system. Here author has assumed number of groups as 2. CPU_share_level2 distributes processor time between different users of a group whereas CPU_share_level3 distributes time among tasks of a user.

Let us assume uniprocessor system with 100% CPU share/processor time. Group 1 is always having high priority over group 2. In group 1, user U_i with i = 1 is having more priority over other users with i = 2….n. Priority is decreasing with increase in value of i.

a. Pseudo Code for CPU Share at level1

```
Begin
    Initialize variable N_group as number of groups. Here N_group = 2
    Determine the share of CPU time for each group using equation (11)
```

$$CPU_{share\_level1} = \frac{100}{N_{group}} \tag{11}$$

```
End
```

Now the system keeps track of the utilization of CPU time by each user and by each task of user.

b. Pseudo Code for CPU share at level2

```
Begin
    Do Loop
```

```
        Initialize variable $N_{user}$ number of active user
    Until $N_{group} = 0$
    Do Loop
        Initialize variable $N_{task\_user}$ as number of active tasks
    Until $N_{user} = 0$
    Do Loop
        Initialize variable $B_{task}$ as burst time of task
        Initialize variable $P_{task}$ as priority of task
        Until $N_{task\_user} = 0$
        Do Loop
        Do Loop
            Run the IFIS-DCS       //pseudo code given
            Calculate $CPU_{share}$ using equation (7)
                $CPU_{share} = median(M_E)$
            Determined the share of CPU time for each user using equation
(12)
                $CPU_{share\_level2} = CPU_{share\_level1} * CPU_{share}$       (12)
            Assign value of $CPU_{share\_level2}$ to the user with $\max(B_{user})$ $B_{user}$ is
the total burst time needed for a user.
                    Until $N_{user} = 0$ Until $N_{group} = 0$
  End
```

c. Pseudo Code for IFIS-DCS

```
  Begin
      Degree_membership ( $B_{user}, N_{user}, B_{group}, N_{task\_user}$ )
          Calculate degree of membership value $D_U$  $D_U = \frac{B_{user} * N_{user}}{B_{group} * N_{task\_user}}$
          return ( $D_U$ )
      endfunction
      Nondegree_membership ( $B_{user}, N_{user}, B_{group}, N_{task\_user}$ )
          Calculate non-degree of membership value $N_U$
$N_U = 1 - sqrt(\frac{B_{user} * N_{user}}{B_{group} * N_{task\_user}})$
          return ( $N_U$ )
        endfunction
      Calculate overall evidence value
          $M_E = (D_U - N_U + 1) / 2$
      return ( $M_E$ )
  End
```

d. Pseudo Code for CPU share at level3

At level 3 each task of user is scheduled in Round Robin fashion that's why CPU share is constituted as time quantum.

```
Begin
    Do Loop
        Initialize variable  t_q as the static time quantum initially as-
signed by system.
        Determine the dynamic time quantum dtq using equation (13)
```

$$dtq = \frac{B_{user}}{N_{task\_group} * t_q} \quad (13)$$

```
        Until N_group = 0
End
```

e. Pseudo Code for Dynamic Priority

```
Begin
        Initialize the task T_k, k = 1...N with static priority
      Run the IFIS-DP
      Assign the dynamic priority to each task
End
```

f. Pseudo Code for IFIS-DP

```
Begin
    Degree_membership (P_max, P_min, P_active)
            Calculate degree of membership value D_P D_P = (P_max − P_active)/(P_max + P_min)
        return (D_P)
    endfunction
    Nondegree_membership (P_max, P_min, P_active)
            Calculate false-membership value N_P N_P = (P_active − P_min)/(P_max + P_min)
              return (N_P)
       endfunction
    Calculate  overall evidence value
        M_EP = (D_P − N_P + 1) / 2
    Calculate dynamic priority
        P_dynamic = min(M_EPi)
End
```

g. Steps Followed

Step 1: Consider each task T_k for each user U_j of each group G_i.
Step 2: Allocate the processor time/CPU share to each group G_i by running code for CPU_share_level1
Step 3: Allocate the CPU share to each user U_j by running code for CPU_share_level2
Step 4: Assign the dynamic priorities to each task T_k by running code Dynamic Priority
Step 5: Treat task T_k with minimum dynamic priority value as the highest priority task.
Step 6: Compute the length of time quantum dtq for each group G_i using code for CPU_share_level3.
Step 7: Schedule the tasks as traditional fair share scheduling algorithm.

i. Schedule the highest priority task of user U_1 of G_1 with CPU for a calculated time period *dtq*.
ii. Repeat the process with other tasks of U_1 of G_1 until U_1 finishes its share of processor time.
iii. Repeat the same step (i) to step (ii) for the other users of U_j where *j*=2 to *n* until G_1 finishes its share of processor time.
iv. Repeat the process from step (i) to step (iii) with users of group G_2.

Step 8: Calculate average normalized turnaround time, average response time and total number of context switches.

IFFS scheduler also focuses on how to avoid the unnecessary context switches. Scheduler will check the task, if the remaining time of currently executing task is less than or equal to 1, it will not force the task to pre-empt but will allows the task to continue with its execution until it leaves the system. In this way IFFS scheduler reduces the number of context switches. Further, to reduce the computation time, author has considered that the IFFS scheduler will calculate the time quantum once for each group until the new task is started by the user.

h. Performance Metrics

In this work, author has analysed the three main performance metrics: average normalized turnaround time, average response time and the number of context switches. The performance of a scheduler can be improved by reducing the value of these parameters (Brinch, 1973). Author has discussed earlier that the response time is the important factor in the multitasking environment so, it plays an important role to analyse the performance of IFFS scheduler.

Definition 6 (Response Time)

Response Time is the time between the submission of request to the first response to that request. It represents the time when first time task has got the CPU. It is calculated as: $R_{time} = starttime - arrivaltime$

Average response time is calculated by as the average of response time of all the active tasks as given in equation (13). It is calculated as

$$ART = (\sum_{i=1}^{N} Rtime) \,/\, N \tag{13}$$

Definition 7 (Average Normalized Turnaround Time)

Normalized turnaround time represents the relative delay of a task. It is simply a ratio of turnaround time to burst time. It is calculated as

$$N_{time} = T_{time} / bursttime$$

Average normalized turnaround time is calculated as average of the normalized turnaround time of all the active tasks in ready queue as given in equation (14). It is calculated as

$$ANT = (\sum_{i=1}^{N} N_{time}) \, / \, N \qquad (14)$$

Definition 8 (Context Switch)

The total time between storing and loading the task is called as context switch time and this switching time is purely an overhead. Therefore, it can affect the performance of computer system as today's computers have lots of general purpose and status registers.

SIMULATION RESULTS

The overall number of tasks in the entire system can differ from 1 to 100. These tasks can be generated by system. The burst time of each task B_{task} is distributed from 10 to 100ms. Here, author assumed that the entire task sets are CPU bound i.e. all tasks execute only CPU related operations and will not execute I/O operations. Author has also assumed that the static value of time quantum t_q is always assigned by the system in *ms*.

In this chapter, author has explained step by step working of IFFS scheduler with the help of one small example with 2 groups, overall 5 users and 9 tasks. These 9 tasks with burst time *B*, priority *P* and the other associated information are divided among users and groups. Notations used throughout the work to denote groups, users and tasks are shown in Figure 4.

In group G_1, U_{11} represents the first user of G_1 whereas U_{12} represent second user of G_1. Author has already mentioned this terminology in section 3.2.1. Similar way T_{111} represents task1 of U_{11} and T_{211} represents task1 of U_{21}. For this particular sample task set given in Figure 5, author has assumed the length of static time quantum as 4ms. Through this example author has tried to discuss the working of scheduler in detail and compared the performance of IFFS scheduler with FS scheduler.

Working of FS Scheduler

The FS scheduler equally partitions the CPU time (100%) among aforementioned groups G_1 and G_2 (50% each at level 1). Then amongst group G_1 (at level 2), the scheduler equally distributes 25% of

Figure 4. Notations to illustrate groups, users and tasks

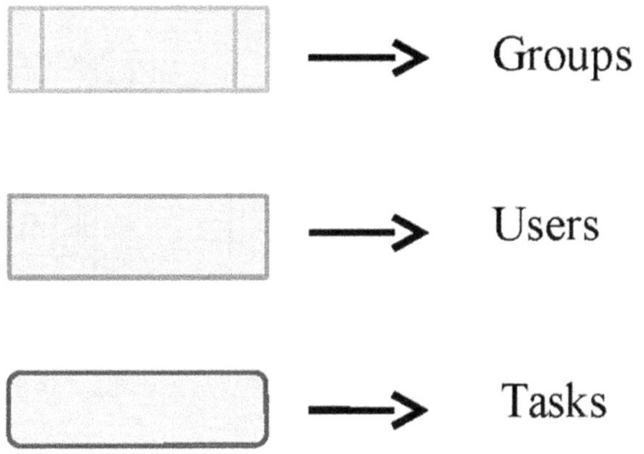

Figure 5. Example task set

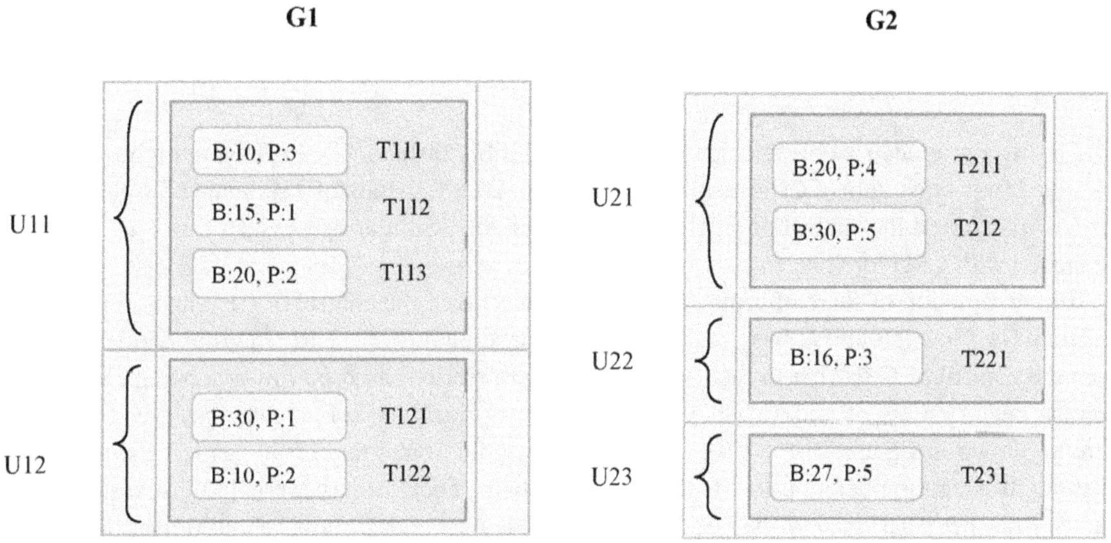

processor time to each user U_{11} and U_{12} . Whereas, in case of G_2, scheduler distributes 50% of processor time among three users U_{21}, U_{22} and U_{23} i.e. 16.67% share to each user as shown in Figure 6.

According to the FS scheduling approach, tasks of user U_{11} are scheduled using round robin approach. Since, at level 1, G_1 is having higher priority over G_2 and it is assumed at level 2, U_{11} is having more priority over U_{12} (user with lower number represents the higher priority). Amongst the tasks of U_{11}, T_{112} is having highest priority as shown in Figure 5 and therefore scheduled first with CPU. The detailed scheduling orders of tasks are illustrated with the help of Gantt chart.

1. First Cycle of CPU

Figure 7 (a) explains the execution of all tasks of user U_{11} and user U_{12} during first cycle of CPU. Both users have used their 25% of CPU share. Among 25% share, scheduler has assigned CPU to each

Figure 6. CPU share using FS scheduler at each level for Example Task Set

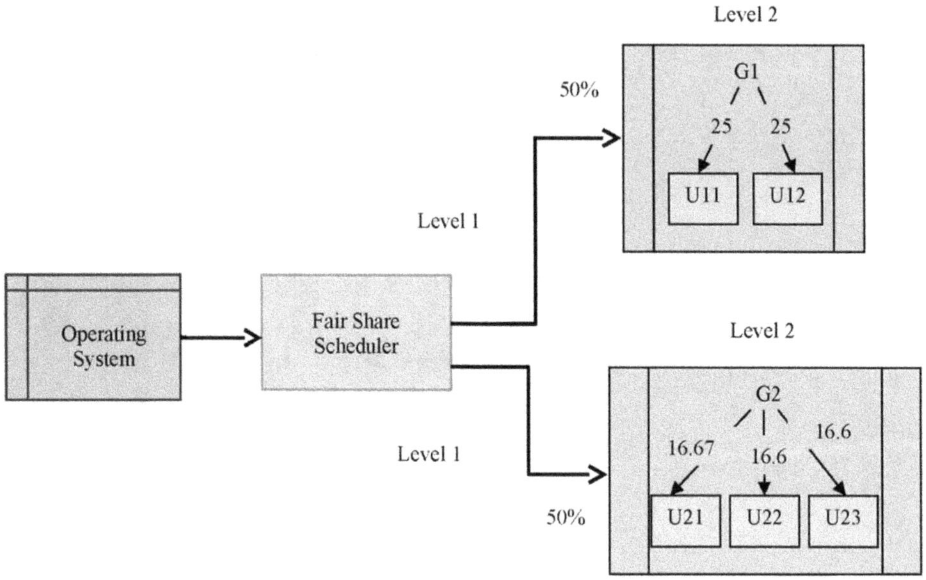

task for 4ms (assumed size of time quantum) in round robin fashion. T_{112} is having highest priority so it received first 4 quantum of CPU then T_{113} and so on. After finishing of 25% share of user U_{11}, scheduler has transferred the control of CPU to another user U_{12}. Similar way tasks of user U_{12} have been dispatched with CPU for 4ms.

After completion of share of group G_1, scheduler has taken the control of CPU and transferred it to the group G_2. Now, group G_2 has got the chance to consume its 50% share. In group G_2, scheduler has given the control of CPU first to the user U_{21} (highest priority based on lower subscript value). Task T_{211} has been first dispatched with CPU being a highest priority task. Each user has used their 16.67% share as shown in Figure 7(b). Since, user U_{22} needs only 16ms; task T_{221} has left the system after finishing its execution before utilizing the complete share. Therefore, the remaining share of 0.67% of user U_{22} has been assigned to the next user U_{23}. In this way both groups has used the 100% share of CPU time/processor time during first cycle.

2. Second Cycle of CPU

After successful completion of first cycle, CPU has started its second cycle with group G_1 as shown in Figure 7(c). Again group G_1 has been allotted for 50% share of processor time. But the group needs only 25ms more (U_{11} – 20ms, U_{12} – 5ms). So, both users have completed their execution before using their entire share (25% each).

Like group G_1, execution of group G_2 during second cycle of CPU is illustrated with the help of Gantt chart in Figure 7(d). Scheduler has assigned the CPU to user U_{21} up to its 25% processor time. After 25% of CPU share, almost tasks has left the system but still one task T_{212} has been remained with their burst time. Scheduler has transferred the CPU to user U_{23}.

3. Third Cycle of CPU

Figure 7. FS Gantt chart: (a) G_1 (first cycle) (b) G_2 (first cycle) (c) G_1 (second cycle) (d) G_2 (second cycle) (e) G2 (third cycle)

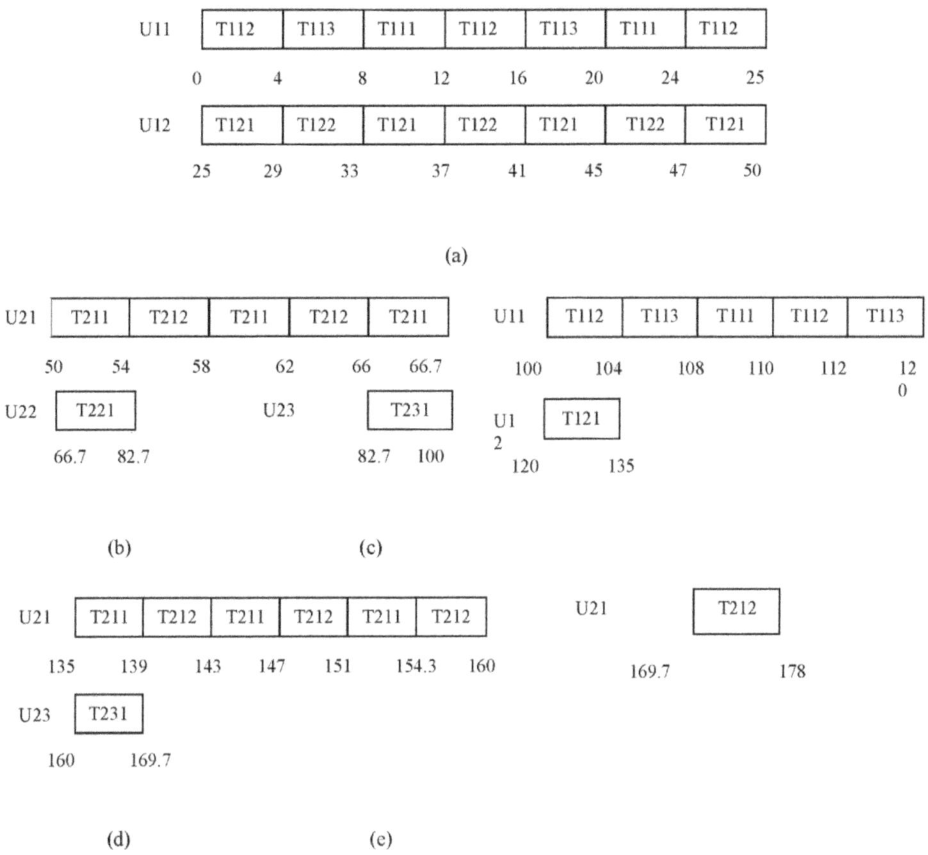

However, at the end of second cycle, group G_1 has finished and group G_2 is left with only one task T_{212} in ready queue. So, the whole CPU share of third cycle has been assigned to the group G_2 as shown in Figure 7(e).

On the basis of Gantt charts, author has calculated the normalized turnaround time and response time for each task.

Working of IFFS Scheduler

To compare the performance of FS scheduler with IFFS scheduler, author has applied IFFS scheduler on the same example set. As discussed in section 3, the first module 'IFIS-DCS' of IFFS scheduler distributes CPU time/processor time dynamically over the fixed/static distribution. It means the CPU share can vary based on the current state of ready queue. The dynamic distribution of example task set is shown in Figure 8.

IFIS-DCS module has partitioned the 100% share of CPU time into 50% each for groups G_1 and G_2 respectively. At level 1, it has distributed the same share as done by FS scheduler. But, further at level 2, it has distributed the 50% share of group G_1 among user U_{11} and U_{12} as 26.5% and 23.5% for respectively which was 25% to each user in case of FS scheduler. Likewise, it has distributed 50% of group G_2 into 30.5% to user U_{21}, 7.5% to user U_{22} and 12% to user U_{23} respectively which was again 16.67% to each user by FS scheduler. Figure 8 illustrates that the distribution of CPU time in case of IFFS scheduler is variable rather than fixed.

Afterwards IFFS scheduler has executed its second module *IFIS-DP* to generate the dynamic priority for each task amongst different users and is denoted by M_{Eijk} where i represent the group no, j represents the user no and k represents the task no. The computed dynamic priorities for both groups are mentioned in Table 1 and Table 2.

Figure 8. Intuitionistic Fuzzy Based CPU share at each level for Example Task set

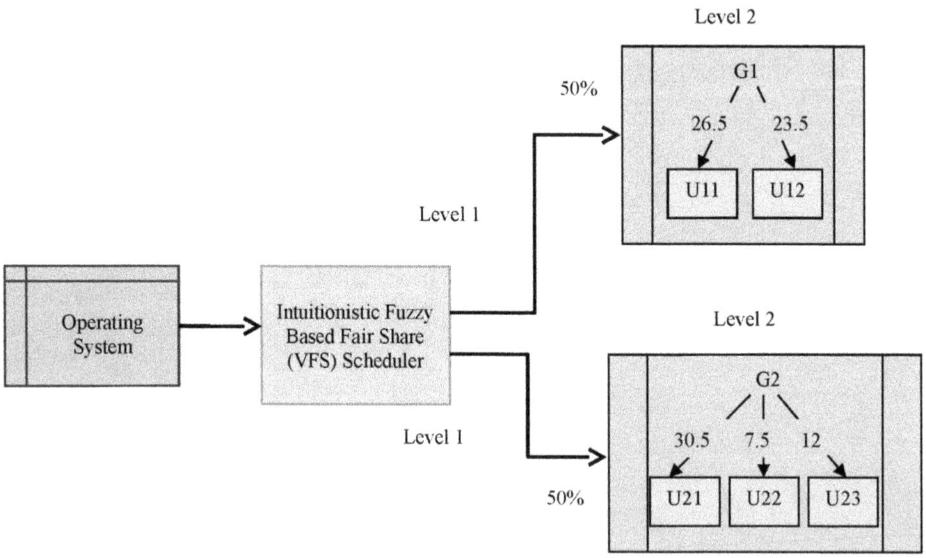

Table 1. Computed Dynamic Priority for G_1

Group G1(Dynamic Priority)				
User U_{11}			*User U_{12}*	
Task T_{111}	Task T_{112}	Task T_{113}	Task T_{121}	Task T_{122}
M_{E111}	M_{E112}	M_{E113}	M_{E121}	M_{E122}
0.25	0.75	0.5	0.67	0.33

Table 2. Computed Dynamic Priority for G_2

Group G2(Dynamic Priority)			
User U_{21}		*User U_{22}*	*User U_{23}*
Task T_{211}	Task T_{212}	Task T_{221}	Task T_{231}
M_{E211}	M_{E212}	No need to calculate priority as only one task	No need to calculate priority as only one task
0.55	0.44		

Module IFIS-DP has not calculated dynamic priority for those users who are executing only one task as mentioned in Table 2. IFFS scheduler has been considered the highest priority task among user which is with minimum value. Consider for user U_{11}, IFFS scheduler has chosen task T_{111} as highest priority task whereas FS scheduler has chosen task T_{112} as highest priority task.

At level 3, processor time has been shared in the form of time quantum to the tasks. FS scheduler has been allotted a fixed/static time quantum whereas IFFS scheduler has been generated a dynamic/variable time quantum for each group. It has returned the value of *dtq* for G_1 as 4.25ms and for G_2 as 5.8ms. For reducing computation time, scheduler will not recalculate time quantum at each cycle. It will be calculated only on the arrival of a new task for that particular user who has started the new task. Finally, IFFS scheduler has used its third and last module i.e. *fair share scheduling algorithm* over the example task set. Updated example task set with values of dynamic priority is shown in Figure 9.

1. First Cycle of CPU

Just like fair share scheduling algorithm, intuitionistic fuzzy based fair share scheduling algorithm has given the control of CPU to group G_1. Among group G_1, control has given to user U_{11}. IFFS scheduler has first dispatched the task T_{111} with CPU for 4.25 ms (calculated time quantum) since it is having minimum dynamic priority value.

As shown in Figure 10(a), each task of user U_{11} has been executed repeatedly until user finished its share of 26.5ms. Scheduler has transferred the control of CPU to another user U_{12}. The task T_{122} has been dispatched with CPU and similarly, all other tasks in a round robin fashion for a time slot of 4.25ms. At 45ms of CPU cycle, when task T_{121} got second time, control of CPU before finishing share of user U_{12} (23.5%), the task has continued its execution till the completion as its remaining burst time is less than 1.

Figure 9. Updated Example Task Set with Dynamic Priorities

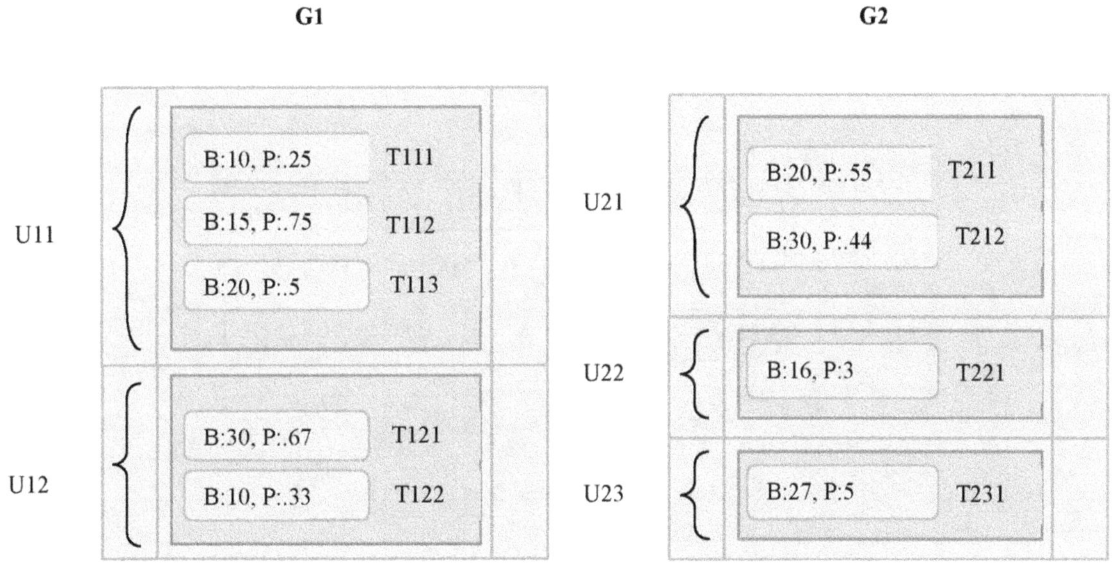

Figure 10. IFFS Gantt chart: (a) G_1 *(first cycle) (b)* G_2 *(first cycle) (c)* G_1 *(second cycle) (d)* G_2 *(second cycle)*

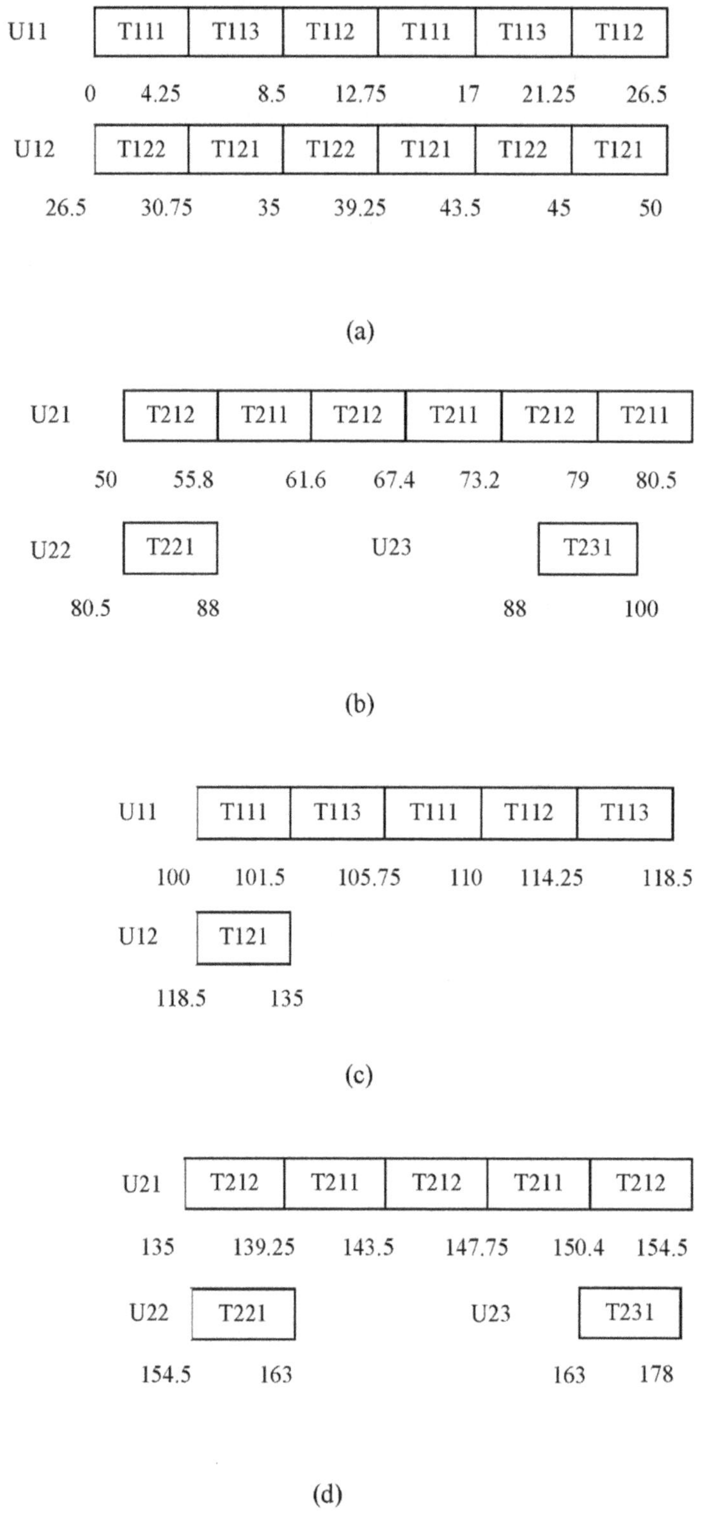

Likewise after completion of 50% share of group G_1, Scheduler has transferred the control of CPU to group G_2. The task T_{212} has been dispatched first with CPU as the lower dynamic priority task among all tasks of user U_{21}. Both task of user U_{11} has been executed repeatedly for a time slot of 5.8ms (calculated time quantum for G_2) until user finished its share of 30.5ms as shown in Figure 10(b). Similarly, scheduler transferred the control of CPU first to the user U_{22} and then to user U_{23}.

2. Second Cycle of CPU

IFFS scheduler has distributed the processor time/CPU time to both groups during the second cycle of CPU. The CPU share for group G_1 is given in Table 3.

The tasks of user U_{11} have been scheduled again for 5.8ms and have been completed their execution before using its complete share of 36%. So, remaining share of user U_{11} has been allotted to the next user U_{12} by IFFS scheduler as shown in Figure 10 (c). Since, the second user has also finished before utilizing full share, the remaining CPU share (15%) of user U_{12} has been allocated to group G_2.

Now the group G_2 has a total of 65% (50% its own and 15% from G_1) of CPU share. IFFS scheduler distributes the 65% share amongst user U_{21}, user U_{22} and user U_{23} as given in Table 4.

Similar ways, the tasks of user U_{21} have been scheduled with CPU. Afterwards, the tasks of user U_{22} and the tasks of user U_{23} have been dispatched with CPU as shown in Figure 10 (d).

As compare to FS scheduler, IFFS scheduler has finished the complete execution of example task set with in two cycles whereas FS scheduler has consumed three cycles of CPU. To analyse the performance of both schedulers, author have computed the normalized turnaround time and response time of each task using equation (14) and equation (13). Then author has calculated the average response time ART, average normalized turnaround time ANTT and the total number of context switches NCS (mentioned in section 4). Author has illustrated the analysis results of example task set with the help of graph shown in Figure 11 where on x-axis '1' represents the FS scheduler and '2' represents the IFFS scheduler.

From the graph, author has concluded that there is improvement in terms of average response time (represented by black colour) and the number of context switches (represented by light grey colour). But there is only a slight deviation in the normalized turnaround time (represented by dark grey colour).

Author has applied the same simulation process over multiple task sets with different number of users and their associated tasks, to better analyse the performance of both schedulers. In this chapter author has discussed the analysis results (all performance metrics) of 30 different task sets. Results of all three performance metrics are illustrated with the help of graphs.

Author has compared the different task sets in terms of average response time, average normalized turnaround time and number of context switches. Results are shown in Figure 12, Figure 13, and Figure 14 respectively. In the graphs of average response time and number of context switches, results of IFFS

Table 3. CPU share for G_1

G1 (50%)	
U11	U12
36%	14%

Table 4. CPU share for G_2

G2 (65%)		
U_{21}	U_{22}	U_{23}
50%	11%	4%

Figure 11. Performance IFFS vs. FS (Example Task Set)

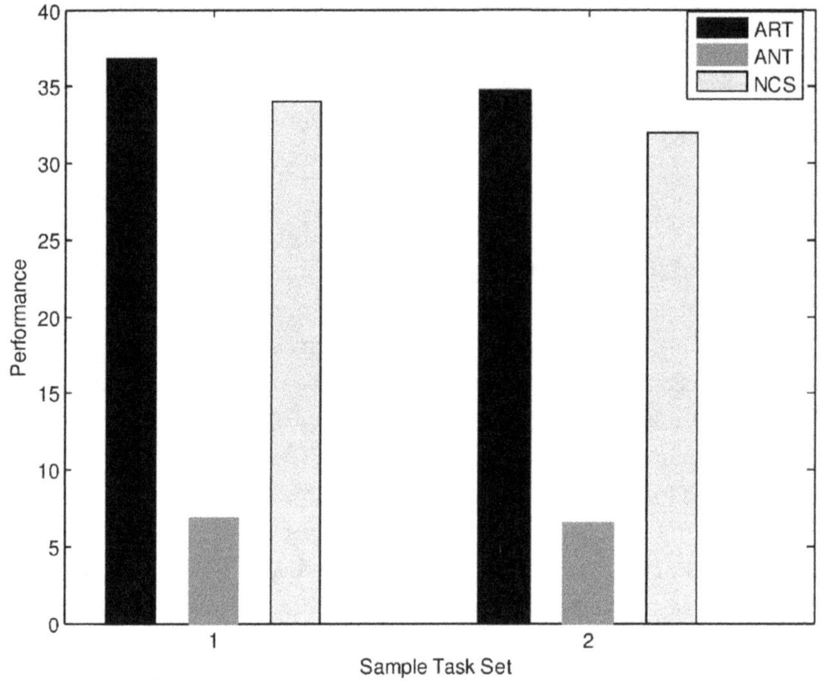

Figure 12. Performance (IFFS vs. FS) -average response time

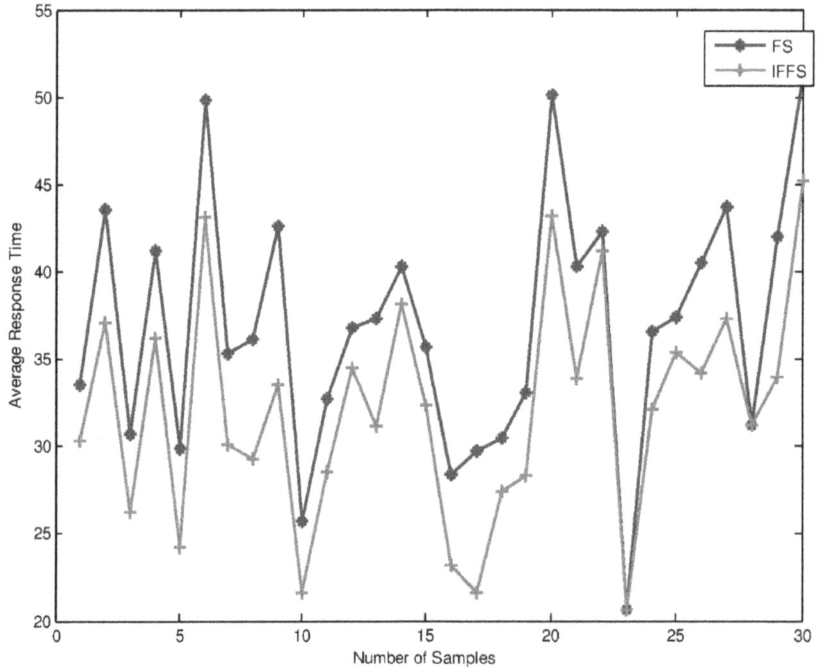

Figure 13. Performance (IFFS vs. FS) -average normalized turnaround time

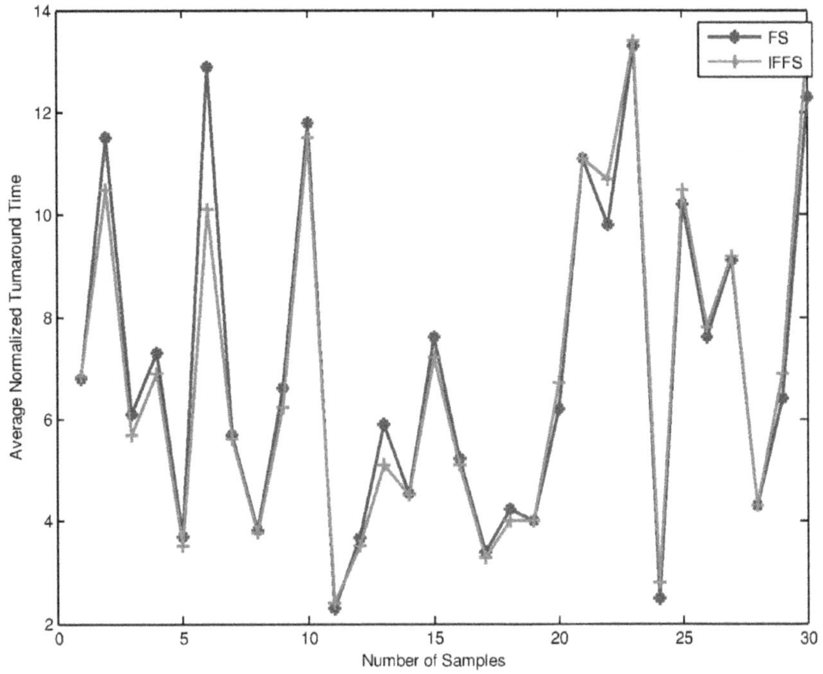

Figure 14. Performance (IFFS vs. FS) -number of context switch

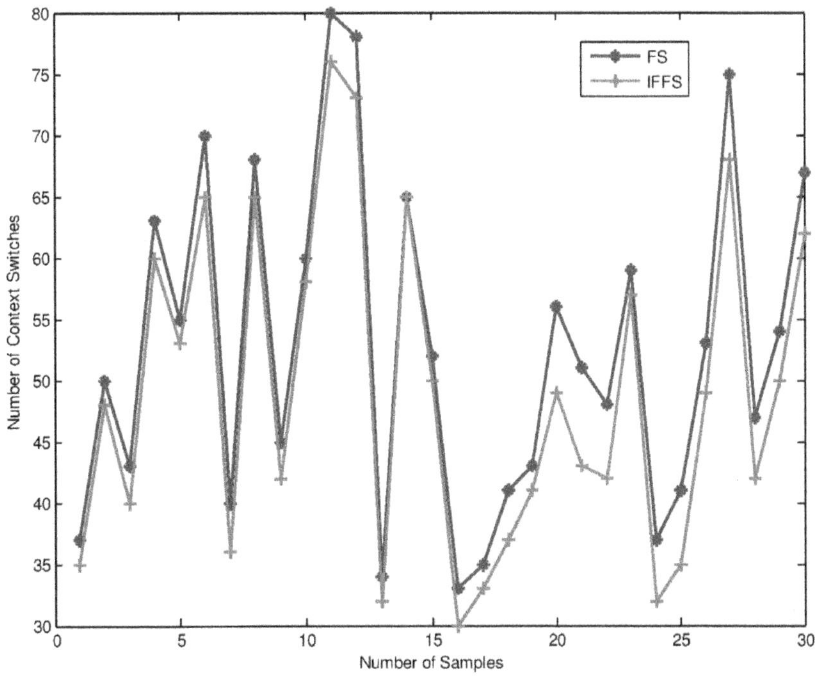

Figure 15. Improvement in performance of IFFS scheduler with FS Scheduler

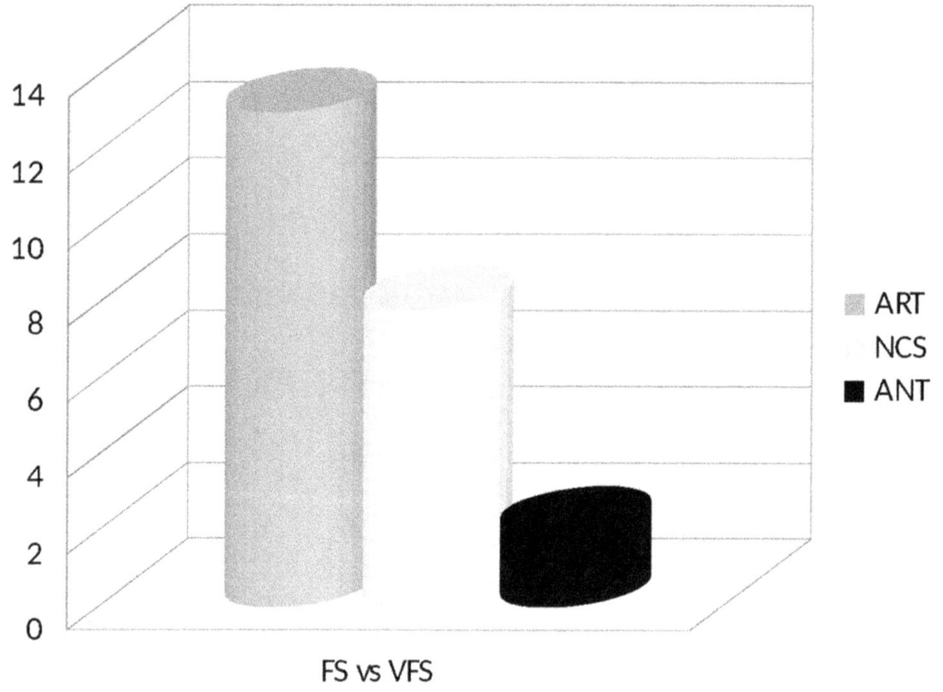

scheduler are at lower side as compare to FS scheduler. Reduction in both values illustrates the improvement in these two performance parameters. Whereas in the graph of average normalized turnaround time, at some places it illustrates little bit reduction in values and somewhere it shows no improvements. But as discussed earlier for multitasking environment the main factor is response time so other factors may be partially ignored.

IFFS scheduler performs 13% better in average response time, 2.5% in average normalized turnaround time and 7.8% with respect to number of context as compare to FS scheduler as shown in Figure 15.

Author has analysed from these graphs that IFFS scheduler provides better results over FS scheduler. So, finally author has concluded that IFFS scheduler has distributed the CPU time fairly amongst all users and tasks. Moreover, it has considered all the factors associated with tasks while distribution over number of tasks to handle the impreciseness and uncertainty of ready queue.

CONCLUSION

In this chapter, author has discussed an intuitionistic fuzzy approach to fair-share scheduler named as IFFS. IFFS scheduler has fairly distributed the CPU time/processor at three levels of system. It has been distributed time at group level, then at user level and finally, at task level. Moreover, at user level, this scheduler has also considered the priorities of each task. It has provided a variable/dynamic priority to each task. IFFS scheduler has used three modules to perform the aforementioned tasks. A module, IFIS-DCS has been used for allocation and distribution of CPU share at three levels whereas module IFIS-DP has been used to assign dynamic priories to each task. Last module, IFFS scheduling algorithm has been used for scheduling and dispatching of tasks. In the chapter, author has presented one example to show the detailed working of both FS and IFFS schedulers. Author has analysed the performance of both schedulers in terms of average response time, average normalized turnaround time and number of context switches and proved that IFFS scheduler gives better results.

REFERENCES

Abdurazzag, A. A., & Vladimir, M. (2008). Fuzzy Logic Based Algorithm for Uniprocessor Scheduling.*Proceedings of the International Conference on Computer and Communication Engineering* (pp. 499-504). IEEE.

Alam, B., Doja, M. N., & Biswas, R. (2009). Improving the performance of fair share scheduling algorithm using fuzzy logic.*International Conference on Advances in Computing, Communication and Control* (pp. 567-570). doi:10.1145/1523103.1523216

Alam, B., Doja, M. N., Biswas, R., & Alam, M. (2011). Fuzzy Priority CPU Scheduling Algorithm. *International Journal of Computer Science and Issues*, *8*(6), 386–390.

An, L., & Wilfred, N. (2005). Vague Sets or Intuitionistic Fuzzy Sets for Handling Vague Data: Which One Is Better? Lecture Notes in Computer Science, 376, 401–416.

Atanassov, K. (1986). Intuitionistic fuzzy sets. *Fuzzy Sets and Systems*, *20*(1), 87–96. doi:10.1016/S0165-0114(86)80034-3

Atanassov, K. (2000). Intuitionistic Fuzzy Sets: Theory and Applications. In Studies in Fuzziness and Soft Computing. Physica-Verlag.

Brinch Hansen, R. (1973). *Operating System Principles*. ACM Digital Library, Prentice Hall.

Bui, H., Emeneker, W., Apon, A., Hoffman, D., & Dowdy, L. (2010). Fair share Scheduling – a case study. In *Proceedings of 11th LCI Conference* (1-10).

Bustince, H., & Burillo, P. (1996). Vague sets are intuitionistic fuzzy sets. *Fuzzy Sets and Systems*, *79*(3), 403–405. doi:10.1016/0165-0114(95)00154-9

Caprita, B., Nieh, J., & Chan, W. C. (2005). Group Round-Robin: Improving the Fairness and Complexity of Packet Scheduling.*Proceedings of the 2005 ACM symposium on Architecture for networking and communications systems* (pp. 29-40). doi:10.1145/1095890.1095896

Dhamdhere, D. M. (2006). *Operating Systems A concept Based Approach* (2nd ed.). Tata: McGraw-Hill.

Galvin, S., & Gagne. (2009). *Operating systems concepts* (8th ed.). Wiley.

Gau, W. L., & Buehrer, D. J. (1993). Vague Sets. *IEEE Transactions on Systems, Man, and Cybernetics*, *23*(2), 610–614. doi:10.1109/21.229476

Halim Zaim, A. (2003). Design of A Scheduler: Comparison of Different Scheduling Algorithms. *Journal of Electronic and Electrical Engineering*, *3*(2), 859–877.

Kadhim, S. J., & Al-Aubidy, K. M. (2010). Design and Evaluation of a Fuzzy-Based CPU Scheduling Algorithm. *Information Processing and Management. Communications in Computer and Information Science*, *70*, 45–52. doi:10.1007/978-3-642-12214-9_9

Kay, J., & Lauder, P. (1988). A Fair Share Scheduler. *Communications of the ACM*, *31*(1), 44–55. doi:10.1145/35043.35047

Li, D.-F. (2008). Extension of the LINMAP for multiattribute decision making under Atanassovs intuitionistic fuzzy environment. *Fuzzy Optimization and Decision Making*, *7*(1), 17–34. doi:10.1007/s10700-007-9022-x

Li, D.-F., Wang, Y., Liu, S., & Shan, F. (2009). Fractional programming methodology for multi-attribute group decision-making using IFS. *Applied Soft Computing*, *9*(1), 219–225. doi:10.1016/j.asoc.2008.04.006

Lim, S., & Cho, S. (2007). Intelligent OS Process Scheduling Using Fuzzy Inference with User Models. *Lecture Notes in Computer Science*, *45*, 725–734. doi:10.1007/978-3-540-73325-6_72

Nie, B., Du, J., Guoliang, X., Liu, H., Yu, R., & Wen, Q. (2011). A New Operating System Scheduling Algorithm. *Advanced Research on Electronic Commerce, Web Application, and Communication, Springer*, *143*, 92–96. doi:10.1007/978-3-642-20367-1_15

Park, M., Jin, Y. H., Chae, J., & Kim, C. (2008). Quantum-Based Fixed Priority Scheduling. *International Journal of Computer and Information Sciences*, *11*(5), 341–356.

Raheja, S., & Dhadich, R. (2013). Many Valued Logics for Modeling Vagueness. *International Journal of Computers and Applications*, *61*(7), 35–39. doi:10.5120/9942-4579

Raheja, S., Dhadich, R., & Rajpal, S. (2012). An Optimum Time Quantum using Linguistic Synthesis for Round Robin Scheduling Algorithm. *International Journal of Soft Computing*, *3*(1), 57–66. doi:10.5121/ijsc.2012.3105

Raheja, S., Dhadich, R., & Rajpal, S. (2013). Vague Oriented Highest Response Ratio Next Scheduling Algorithm. *IEEE-International Conference on Fuzzy Systems* (pp. 1-7).

Raheja, S., Dhadich, R., & Rajpal, S. (2014). Designing of 2-Stage CPU Scheduler using Vague Logic. *Advances in Fuzzy Systems*, *2014*, 1–10. doi:10.1155/2014/841976

Raheja, S., Dhadich, R., & Rajpal, S. (2015). Designing of Vague Logic Based Fair-Share CPU Scheduler. *International Journal of Fuzzy System Applications, IGI Global*, *4*, 25–49.

Sabin, G., Kochhar, G., & Sadayappan, P. (2004). Job fairness in non-preemptive job scheduling.*4th International Conference on Parallel processing* (pp. 186-194).

Stallings, W. (2014). Operating Systems Internal and Design Principles (8th ed.). Academic Press.

Tanenbaum, S. A., & Woodfhull, S. A. (2008). Operating Systems Design and Implementation (2nd ed.). Academic Press.

Xia & Feng. (2005). Fuzzy logic based feedback scheduler for embedded control systems. In *Advances in Intelligent Computing*. Springer Berlin Heidelberg.

Zadeh, L. A. (1965). Fuzzy sets. *Information and Control*, *8*(3), 338–356. doi:10.1016/S0019-9958(65)90241-X

Zadeh, L. A. (1984). Making computers think like people. *IEEE Spectrum*, *8*(8), 26–32. doi:10.1109/MSPEC.1984.6370431

Zadeh, L. A. (2008). Is there a need for Fuzzy Logic. *Information Science*, *178*(13), 2751–2779. doi:10.1016/j.ins.2008.02.012

Zimmerman, J. (2001). *Fuzzy set theory And Its Application*. Norwell, MA: Kluwer Academic Publishers. doi:10.1007/978-94-010-0646-0

Chapter 7
MAGDM Problems with Correlation Coefficient of Triangular Fuzzy IFS

John Robinson P.
Bishop Heber College, India

Henry Amirtharaj E. C.
Bishop Heber College, India

ABSTRACT

Correlation coefficient of Intuitionistic Fuzzy Set (IFS), Interval valued IFS, Triangular IFS and Trapezoidal IFS are already present in the literature. This paper proposes the correlation coefficient for Triangular Fuzzy Intuitionistic Fuzzy set (TrFIFS). The method on uncertain Multiple Attribute Group Decision Making (MAGDM) problems based on aggregating intuitionistic fuzzy information is investigated for TrFIFSs. The Triangular Fuzzy Intuitionistic Fuzzy Ordered Weighted Averaging (TrFIFOWA) operator is proposed for TrFIFSs and the Triangular Fuzzy Intuitionistic Fuzzy Ordered Weighted Geometric (TrFIFOWG) operator is utilized for decision making models where expert weights are completely unknown. Based on these operators and the correlation coefficient defined for the TrFIFSs, new decision making models are proposed with numerical illustrations. Some comparisons are also made with existing ranking methods for validity.

INTRODUCTION

Decision making is the study of identifying and choosing alternatives based on the values and preferences of the decision maker. Making a decision implies that there are alternative choices to be considered, and in such cases many of these alternatives as possible should be identified and the one that best fits with our goals, objectives, desires, values, should be chosen (Kahraman, 2008). Decision making should start with the identification of the decision maker(s) and stakeholder(s) in the decision, reducing possible disagreement about problem definition, requirements, goals and criteria. It is very important to make a distinction between the cases where we have a single or multiple criteria. When a decision

DOI: 10.4018/978-1-5225-1848-8.ch007

problem has a single criterion or a single aggregate measure, then the decision can be made implicitly by determining the alternative with the best value of the single criterion or aggregate measure. When a decision problem has a finite number of criteria or multiple criteria, and the number of the feasible alternatives (the ones meeting requirements) is infinite, and then the decision problem belongs to the field of multiple criteria optimization. Also, techniques of multiple criteria optimization can be used when there are a finite number of feasible alternatives, but are given only in implicit form. This research work focuses on decision making problems when the number of the criteria (attribute) and alternatives is finite, and the alternatives are explicitly given. Problems of this type are called Multi Attribute Decision Making (MADM) problems.

Consider a MADM problem with m criteria and n alternatives. Let $\boldsymbol{C}_1,\ldots,\boldsymbol{C}_m$ and $\boldsymbol{A}_1,\ldots,\boldsymbol{A}_n$ denote the criteria and alternatives, respectively. A standard feature of MADM methodology is the decision table as shown in the following (See Figure 1). In this table each row belongs to a criterion and each column describes the performance of an alternative. The score a_{ij} describes the performance of the alternative $\boldsymbol{A}_j$ against the criterion $\boldsymbol{C}_i$. For the sake of simplicity it is assumed that a higher score value means a better performance, as any goal of minimization can easily be transformed into a goal of maximization. Here the weights w_i reflects the relative importance of criteria $\boldsymbol{C}_i$ to the decision, and is assumed to be positive. These weights are usually determined on subjective basis and they represent the opinion of a single decision maker or synthesize the opinions of a group of experts using a group decision technique. The values $x_1, \ldots, x_n$ are associated with the alternatives in the decision table, and are the final ranking values of the alternatives $\boldsymbol{A}_j$. Usually, a higher ranking value means a better performance of the alternative; hence the alternative with the highest ranking value is the best of the alternatives. Group decision is usually understood as aggregating different individual preferences on a given set of alternatives to a single collective preference. It is assumed that the individuals participating in making a group decision face the same common problem and are all interested in finding a solution. A group decision situation

Figure 1. A simple Decision Table

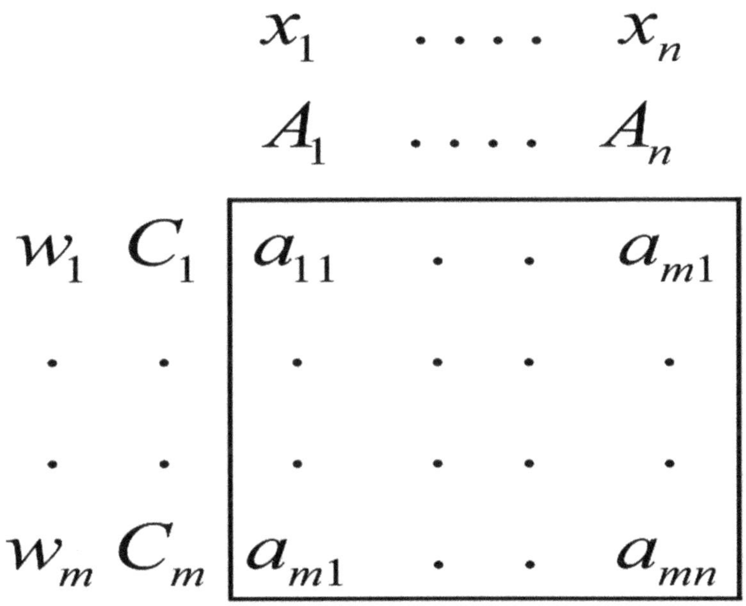

involves multiple actors (decision makers), each with different skills, experience and knowledge related to different aspects (criteria) of the problem. In a correct method for synthesizing group decisions, the competence of different actors to different professional fields has also to be considered. It is assumed that each actor considers the same sets of alternatives and criteria. It is also assumed that there is a special actor with authority to establish consensus rules and determine voting powers to group members on different criteria. Many researchers call this entity the Supra Decision Maker (SDM). The final decision is derived by aggregating (synthesizing) the opinions of group members according to rules and priorities defined by the SDM. There are several approaches to extend the basic MADM techniques for group decision. Consider a decision problem with l group members (decision makers) $\boldsymbol{D}_1,..,\boldsymbol{D}_l$, n alternatives $\boldsymbol{A}_1,...,\boldsymbol{A}_n$ and m criteria $\boldsymbol{C}_1,..,\boldsymbol{C}_m$. In case of a factual criterion, the evaluation scores must be identical for any alternative and any decision maker, while subjective (judgmental) criteria can be evaluated differently by each decision maker. In this work, Multiple Attribute Group Decision Making (MAGDM) problems with applications of different classes of aggregation operators, and correlation coefficient in triangular fuzzy IFS environment as the ranking method have been concentrated upon.

MAGDM problems are of importance in most kinds of fields such as engineering, economics and management. It is obvious that much knowledge in the real world is fuzzy rather than precise. Imprecision comes from a variety of sources such as unquantifiable information (Li & Nan, 2011). In many situations decision makers have imprecise/vague information about alternatives with respect to attributes. It is well known that the conventional decision making analysis using different techniques and tools has been found to be inadequate to handle uncertainty of fuzzy data. To overcome this problem, the concept of fuzzy approach (Azar, 2010) has been used in the evaluation of decision making systems. For a long period of time, efforts have been made in designing various decision making systems suitable for the arising day-to-day problems. MAGDM problems are wide spread in real life decision making situations and the problem is to find a desirable solution from a finite number of feasible alternatives assessed on multiple attributes, both quantitative and qualitative (Power, 2013). In order to choose a desirable solution, the decision maker often provides his/her preference information which takes the form of numerical values, such as:

- Exact values,
- Interval number values, and
- Fuzzy numbers.

However, under many conditions, numerical values are inadequate or insufficient to model real-life decision problems. Indeed, human judgments including preference information may be stated in intuitionistic fuzzy information, especially in triangular fuzzy intuitionistic fuzzy information. Hence, MAGDM problems under intuitionistic fuzzy or triangular fuzzy intuitionistic fuzzy environment are an interesting area of study for researchers in the recent days.

In the process of MAGDM problems with triangular fuzzy intuitionistic fuzzy information, sometimes, the attribute values take the form of Triangular Fuzzy Intuitionistic Fuzzy Number (TrFIFN). The information about attribute weights may sometimes be known or partially known or sometimes completely unknown. This may be because of the lack of knowledge of the data or the expert's limited expertise about the problem domain. The MAGDM model requires that the selection be made among the decision alternatives described by their attributes. Solving a MAGDM problem having a predetermined, limited number of decision alternatives involves sorting and ranking, and can be viewed as alternative methods for combining the information in a problem's decision matrix together with additional information from the decision maker to determine a final ranking or selection from among the alternatives. Besides the information contained in the decision matrix, all but the simplest MAGDM technique requires additional information from the decision matrix to arrive at a final ranking or selection.

Intuitionistic Fuzzy Sets (IFSs) proposed by Atanassov, (1986, 1994, 1999) is a generalization of the concept of fuzzy sets. Atanassov & Gargov, (1989) expanded the IFSs, using interval value to express membership and non-membership function of IFSs. Liu & Yuan, (2007) introduced the concept of fuzzy number IFSs as a further generalization of IFSs. Among the works done in IFSs, Atanassov (1986; 1994), Atanassov and Gargov (1989), Szmidt and Kacrzyk (2000, 2002, 2003), and Gerstenkorn and Manko (1991) can be mentioned. With best of our knowledge, Burillo et al., (1994) proposed the definition of intuitionistic fuzzy number (IFN) and studied the perturbations of IFN and the first properties of the correlation between these numbers. Many researchers have applied the IFS theory to the field of decision making. Recently some researches (Boren et al.., 2009; Chen et al., 2010; Liu & Yuan, 2007; Wei, 2010; Wang, 2008; Zhang & Liu, 2010) showed great interest in the fuzzy number IFSs and applied it to the field of decision making. Based on the arithmetic aggregation operators, Xu (2006), Xu & Chen (2007), and Wang, (2008) developed some new geometric aggregation operators and Intuitionistic Fuzzy Ordered Weighted Averaging (*IFOWA*) operator. Chen & Tan (1994) presented some products for dealing with MADM problems based on vague sets. Szmidt & Kacrzyk (2002) proposed some solution concepts in group decision making with intuitionistic (individual and social) fuzzy preference relations. Szmidt & Kacrzyk (2003) investigated the consensus-reaching process in group decision making based on individual intuitionistic fuzzy preference relations. Herrera et al. (1999) developed an aggregation process for combining numerical, interval valued and linguistic information, and then proposed different extensions of this process to deal with contexts in which can appear information such as IFSs or multi-granular linguistic information. Xu & Yager (2006) developed some geometric aggregation operators for MADM problems. Li (2005) investigated MADM problems with intuitionistic fuzzy information and constructed several linear programming models to generate optimal weights for attributes.

It is well known that the conventional correlation analysis using probabilities and statistics has been found to be inadequate to handle uncertainty of failure data and modeling. The method to measure the correlation between two variables involving fuzziness is a challenge to classical statistical theory. Bustince and Burillo (1995), Gernstenkorn and Manko (1991), Hong and Hwang (1995), and Yu (1993) defined a different correlation formula to measure the interrelation of intuitionistic fuzzy sets. In their definition, the correlation coefficient lies between 0 and 1, differing from the conventional range of [-1,1]. The work of Wang & Li (1999) suffers the same problem of the correlation lying between 0 and 1, in studying the correlation coefficient of interval valued fuzzy sets. Chiang and Lin (1999) took random samples from fuzzy sets, treating the membership grades to be crisp observations, to calculate the correlation lying in the interval [-1,1], where the sense of fuzziness is lost. Park et al. (2009) proposed the correlation coefficient of interval valued intuitionistic fuzzy sets. Robinson & Amirtharaj, (2011a, 2011b, 2012a, 2012b, 2014a, 2014b) proposed correlation coefficient for vague sets, interval vague sets, triangular and trapezoidal IFSs and a revised correlation coefficient for Triangular and Trapezoidal IFSs using graded mean integration representation.

In this paper, a novel method of correlation coefficient of Triangular Fuzzy Intuitionistic Fuzzy Sets (TrFIFS) is proposed and developed by taking into account the membership, non-membership and the hesitation degrees of TrFIFSs. TrFIFSs are advantageous than the usual intuitionistic fuzzy sets because in data modeling point of view, the interval valued data can add more precision to the real world problems.

Table 1. Comparison of MAGDM Models

Proposed Method of Correlation coefficient for *TrFIFNs*
Proposed Method of Normalized Distance Measure
Chen et al. (2010) Method of Ranking (Score function)
Chen et al. (2010) Method in combination with Proposed Correlation coefficient for *TrFIFNs*

The *OWA* operators for TrFIFSs is proposed in this paper for MAGDM problems. Aiming at the fuzzy number intuitionistic fuzzy information aggregating problems, this paper utilizes the Triangular Fuzzy Intuitionistic Fuzzy Weighted Geometric (*TrFIFWG*) operator and the Triangular Fuzzy Intuitionistic Fuzzy Ordered Weighted Geometric (*TrFIFOWG*) operator proposed by Wang (2008). Based on these operators, an approach is suggested to solve uncertain multiple attribute group decision making problems, where the attribute values are triangular fuzzy intuitionistic fuzzy numbers and weights of decision-makers are completely unknown. A new MAGDM algorithm is developed to solve the MAGDM problems in which the correlation coefficient of TrFIFSs is used for ranking alternatives. A distance function extended from the Hamming distance function is also proposed for TrFIFSs in ranking alternatives for MAGDM problems. Table 1 shows the list of numerical illustrations provided to demonstrate the developed approach and also a comparison made with existing methods (Chen et al., 2010).

ARITHMETIC OPERATIONS FOR TrFIFSs

In this section, some basic definitions and arithmetic operations of TrFIFSs are presented.

Definition 1

Triangular Fuzzy Number (TrFN)

$A=(a,b,c)$ is called a triangular fuzzy number, if the membership function $\mu_A : R \to [0,1]$ is expressed as:

$$\mu_A(x)=\begin{cases}\dfrac{x-a}{b-a} & a\le x\le b\\ \dfrac{x-c}{b-c} & b\le x\le c\\ 0 & otherwise\end{cases} \tag{1}$$

where $x\in R,\quad 0\le a\le b\le c\le 1.$

Definition 2

Triangular Fuzzy Intuitionistic Fuzzy Number (TrFIFN)

Let *X* be a non-empty set. Then $A = \left\{ \left\langle x, \mu_A(x), \gamma_A(x) \right\rangle \ / x \in X \right\}$ is called a Triangular Fuzzy Intuitionistic Fuzzy Number (TrFIFN) if $\mu_A(x) = \left(\mu_{A_1}(x), \mu_{A_2}(x), \mu_{A_3}(x) \right)$ and $\gamma_A(x) = \left(\gamma_{A_1}(x), \gamma_{A_2}(x), \gamma_{A_3}(x) \right)$ are triangular fuzzy numbers, which can express the membership degree and the non-membership degree of *x* in *X*, and fulfill $0 \le \mu_{A_3}(x) + \gamma_{A_3}(x) \le 1, \quad \forall x \in X$. An intuitionistic fuzzy number expressed on the basis of triangular fuzzy number is called as triangular fuzzy intuitionistic fuzzy number.

The hesitancy degree of the TrFIFN is given as follows:

$$\begin{aligned} \pi_{A_1}(x) &= 1 - \mu_{A_1}(x) - \gamma_{A_1}(x), \\ \pi_{A_2}(x) &= 1 - \mu_{A_2}(x) - \gamma_{A_2}(x), \\ \pi_{A_3}(x) &= 1 - \mu_{A_3}(x) - \gamma_{A_3}(x), \ \forall x \in X. \end{aligned}$$

If $A = \left\{ \left\langle x, \mu_A(x), \gamma_A(x) \right\rangle \ / x \in X \right\}, \ B = \left\{ \left\langle x, \mu_B(x), \gamma_B(x) \right\rangle \ / x \in X \right\}$ are intuitionistic fuzzy numbers, then the operation rules of the intuitionistic fuzzy numbers are as follows:

$$A + B = \left\{ \left\langle x, \mu_A(x) + \mu_B(x) - \mu_A(x).\mu_B(x), \gamma_A(x).\gamma_B(x) \right\rangle \ / x \in X \right\}, \tag{2}$$

$$A.B = \left\{ \left\langle x, \mu_A(x)\mu_B(x), \gamma_A(x) + \gamma_B(x) - \gamma_A(x).\gamma_B(x) \right\rangle \ / x \in X \right\}, \tag{3}$$

$$nA = \left\{ \left\langle x, 1 - [1 - \mu_A(x)]^n, [\gamma_A(x)]^n \right\rangle \ / x \in X \right\}, \tag{4}$$

$$A^n = \left\{ \left\langle x, [\mu_A(x)]^n, 1 - [1 - \gamma_A(x)]^n \right\rangle \ / x \in X, n > 0 \right\}. \tag{5}$$

Suppose,

$$\begin{aligned} A &= \left\{ \left\langle x, [\mu_{A_1}(x), \mu_{A_2}(x), \mu_{A_3}(x)], [\gamma_{A_1}(x), \gamma_{A_2}(x), \gamma_{A_3}(x)] \right\rangle \ / x \in X \right\}, \\ B &= \left\{ \left\langle x, [\mu_{B_1}(x), \mu_{B_2}(x), \mu_{B_3}(x)], [\gamma_{B_1}(x), \gamma_{B_2}(x), \gamma_{B_3}(x)] \right\rangle \ / x \in X \right\} \end{aligned}$$

are two TrFIFNs, then according to the above operation rules of intuitionistic fuzzy numbers, and the operation rules of triangular fuzzy numbers, the operation rules of TrFIFNs are as follows:

$$\begin{aligned} A + B = \Big\{ &\left\langle [\mu_{A_1}(x), \mu_{A_2}(x), \mu_{A_3}(x)], [\gamma_{A_1}(x), \gamma_{A_2}(x), \gamma_{A_3}(x)] \right\rangle + \\ &\left\langle [\mu_{B_1}(x), \mu_{B_2}(x), \mu_{B_3}(x)], [\gamma_{B_1}(x), \gamma_{B_2}(x), \gamma_{B_3}(x)] \right\rangle \Big\} \end{aligned}$$

$$= \left\{ \left\langle [\mu_{A_1}(x) + \mu_{B_1}(x) - \mu_{A_1}(x).\mu_{B_1}(x), \mu_{A_2}(x) + \mu_{B_2}(x) - \mu_{A_2}(x).\mu_{B_2}(x), \mu_{A_3}(x) + \mu_{B_3}(x) - \mu_{A_3}(x).\mu_{B_3}(x)], \right. \right.$$
$$\left. \left. [\gamma_{A_1}(x).\gamma_{B_1}(x), \gamma_{A_2}(x).\gamma_{B_2}(x), \gamma_{A_3}(x).\gamma_{B_3}(x)] \right\rangle \right\} \tag{6}$$

$$A.B =$$
$$\left\{ \left\langle [\mu_{A_1}(x), \mu_{A_2}(x), \mu_{A_3}(x)], [\gamma_{A_1}(x), \gamma_{A_2}(x), \gamma_{A_3}(x)] \right\rangle . \left\langle [\mu_{B_1}(x), \mu_{B_2}(x), \mu_{B_3}(x)], [\gamma_{B_1}(x), \gamma_{B_2}(x), \gamma_{B_3}(x)] \right\rangle \right\}$$

$$= \left\{ \left\langle [\mu_{A_1}(x).\mu_{B_1}(x), \mu_{A_2}(x).\mu_{B_2}(x), \mu_{A_3}(x).\mu_{B_3}(x)], \right. \right.$$
$$\left. \left. [\gamma_{A_1}(x) + \gamma_{B_1}(x) - \gamma_{A_1}(x).\gamma_{B_1}(x), \gamma_{A_2}(x) + \gamma_{B_2}(x) - \gamma_{A_2}(x).\gamma_{B_2}(x), \gamma_{A_3}(x) + \gamma_{B_3}(x) - \gamma_{A_3}(x).\gamma_{B_3}(x)] \right\rangle \right\} \tag{7}$$

$$\lambda A = \lambda \left([\mu_{A_1}(x), \mu_{A_2}(x), \mu_{A_3}(x)], [\gamma_{A_1}(x), \gamma_{A_2}(x), \gamma_{A_3}(x)] \right)$$
$$= \left\{ \left\langle [1-(1-\mu_{A_1}(x))^{\lambda}, 1-(1-\mu_{A_2}(x))^{\lambda}, 1-(1-\mu_{A_3}(x))^{\lambda}], [(\gamma_{A_1}(x))^{n}, (\gamma_{A_2}(x))^{n}, (\gamma_{A_3}(x))^{n}] \right\rangle \right\}, \tag{8}$$

$$A^{\lambda} = \left([\mu_{A_1}(x), \mu_{A_2}(x), \mu_{A_3}(x)], [\gamma_{A_1}(x), \gamma_{A_2}(x), \gamma_{A_3}(x)] \right)^{\lambda}$$
$$= \left\{ \left\langle [(\mu_{A_1}(x))^{n}, (\mu_{A_2}(x))^{n}, (\mu_{A_3}(x))^{n}], [1-(1-\gamma_{A_1}(x))^{\lambda}, 1-(1-\gamma_{A_2}(x))^{\lambda}, 1-(1-\gamma_{A_3}(x))^{\lambda}] \right\rangle \right\}, \quad \lambda \geq 0. \tag{9}$$

For the above operation rules, the following are true:

i) $A + B = B + A$
ii) $A.B = B.A$
iii) $\lambda(A + B) = \lambda A + \lambda B, \quad \lambda \geq 0$
iv) $\lambda_1 A + \lambda_2 A = (\lambda_1 + \lambda_2)A, \quad \lambda_1, \lambda_2 \geq 0$
v) $A^{\lambda_1}.A^{\lambda_2} = (A)^{\lambda_1 + \lambda_2}, \quad \lambda_1, \lambda_2 \geq 0$

AGGREGATION OPERATORS IN DECISION MAKING

Different types of aggregation operators are found in literature for aggregating the information (Herrera & Herrera-Viedma, 2000; Li et al., 2009; Li, 2008, 2010a, 2010b, 2010c, 2010d; Li & Nan, 2011). A very common aggregation method is the Ordered Weighted Averaging (*OWA*) operator introduced by Yager, (1988), which provides a parameterized family of aggregation operators that include - as special cases - the maximum, the minimum and the average criteria. Since its appearance, the *OWA* operator has been used in varied applications in decision making problems. Karayiannis (2000) and Yager, (2004) suggested a generalization of the *OWA* operator by using generalized means, known as the generalized *OWA* (*GOWA*) operator. With this generalization, it is possible to include the special cases found in the *OWA* operator like the maximum and the minimum, and the special cases found in the generalized mean such as the geometric and the harmonic mean.

When using the *OWA* operator, it is assumed that the available information is exact numbers or singletons. However, this may not be the real situation found in the decision making problem. Sometimes, available information is vague or imprecise and thus it is not possible to analyze it with exact numbers. Then, it is necessary to use another approach that can assess the uncertainty such as the use of Fuzzy Numbers (FNs). With the use of FNs, the best and worst possible scenario can be analyzed with the possibility that the internal values of the fuzzy interval will occur. The main characteristic of this approach is that it uses uncertain information in the aggregation represented by FNs. The Fuzzy *OWA* (FOWA) operator has been studied by different authors like Merigo & Casanovas, (2008, 2009) and Yager, (2004). Recently, Merigo and Gil-Lafuente (2009a) suggested a new model that unifies the *OWA* operator and the weighted average (*WA*) in the same formulation. The main advantage of this new model against the previous ones (Xu & Da, 2003) is that it considers the degree of importance of each concept in aggregation. Hence, it is easy to deal with situations where the *WA* or the *OWA* is very relevant. This new approach is known as the Ordered Weighted Average – Weighted Average (*OWAWA*) operator. Note that more recently Merigo & Casanovas (2009) have suggested a new model that uses FNs in the decision analysis, calling it the Fuzzy *OWAWA* (*FOWAWA*) operator.

Xu, (2007a; 2007b; 2007c; 2007d) defined some new intuitionistic preference relations, such as:

- The consistent intuitionistic preference relation,
- Incomplete intuitionistic preference relation, and
- Acceptable intuitionistic preference relation, and studied their properties and also investigated the intuitionistic fuzzy MADM with the information about attribute weights completely unknown.

Different families of *OWA* operators can be used by choosing a different manifestation of the weighting vector (Merigo & Gil-Lafuente, 2009a; Merigo & Casanovas, 2008; 2009; Xu, 2005; Yager, 1988, 1993, 2007). The Generalized *OWA* (*GOWA*) operator was introduced by Karayiannis, (2000) and Yager, (2004). It generalizes a wide range of aggregation operators that includes the *OWA* operator with its particular cases, the Ordered Weighted Geometric (*OWG*) operator (Chiclana et al., 2004; Herrera et al., 2003; Xu & Da, 2003), the Ordered Weighted Harmonic Averaging (*OWHA*) operator and the Ordered Weighted Quadratic Averaging (*OWQA*) operator (Yager, 2004).

- Li, (2010d, 2014, 2016),
- Li & Nan, (2014),
- Li& Liu, (2015),
- Li & Yang, (2015),
- Li et al., (2010),
- Nan et al., (2010; 2014a; 2014b),
- Yang et al., (2016a; 2016b),
- Liu & Jin, (2012), Liu & Shi, (2015), and
- Liu et al., (2011, 2012, 2016) have proposed different methods using varieties of operators to the field of decision making.

DIFFERENT CLASSES OF AGGREGATION OPERATORS

Ordered Weighted Averaging (*OWA*) Operators

The *OWA* operator provides a parameterized family of aggregation operators which are used in many applications. The definition of the *OWA* operator was introduced by Yager (1988).

Definition 3

An *OWA* operator of dimension n is a mapping $OWA: R^n \rightarrow R$ that has an associated weighting vector $w = (w_1, w_2, ..., w_n)^T$ of dimension n having the properties, $w_j \in [0,1]$, $\sum_{j=1}^{n} w_j = 1$ and such that

$$OWA(a_1, a_2, ..., a_n) = \sum_{j=1}^{n} w_j b_j \tag{10}$$

where b_j is the jth largest of the a_i, for all i.

Example 1: Assume $w = (0.4, 0.3, 0.2, 0.1)^T$. Then,

$$OWA\ (0.7, 1.0, 0.2, 0.6) = \\ (0.4 \times 1.0) + (0.3 \times 0.7) + (0.2 \times 0.6) + (0.1 \times 0.2) = 0.75$$

A fundamental aspect of this operator is the reordering of the arguments, based on their values. That is, the weights rather than being associated with a specific argument, as in the case of the usual weighted average, are associated with a particular position in the ordering.

Geometric Mean (*GM*) Operator

Definition 4

A Geometric Mean (*GM*) operator of dimension m is a mapping $GM: R^m \rightarrow R$ and is defined as:

$$GM(a_1, a_2, ..., a_m) = \prod_{j=1}^{m} \left(a_j\right)^{1/m} \tag{11}$$

Ordered Weighted Geometric (*OWG*) Operators

The Ordered Weighted Geometric (*OWG*) operator is based on the *OWA* operator and the Geometric Mean (*GM*) operator and provides a parameterized family of aggregation operators used in many applications (Herrera et al., 2003). The definition of the *OWG* operator is as follows:

Definition 5

An *OWG* operator of dimension *m* is a mapping $OWG:\ R^m \rightarrow\ R$ that has an associated weighting vector $w=(w_1,w_2,\ldots,w_m)^T$ of dimension *m* having the properties, $w_j \in [0,1]$, $\sum_{j=1}^{m} w_j = 1$ and such that

$$OWG(a_1,a_2,\ldots,a_m)=\prod_{j=1}^{m} b_j^{\,w_j} \tag{12}$$

where b_j is the jth largest of the a_i.

Example 4: Assume $w=(0.4,0.3,0.2,0.1)^T$. Then

$$OWG\ (0.7,1.0,0.2,0.6)=(0.1^{0.4})\times(0.7^{0.3})\times(0.6^{0.2})\times(0.2^{0.1})=0.69$$

DECISION MAKING USING *TrFIFOWA* OPERATOR AND CORRELATION COEFFICIENT OF TRIANGULAR FUZZY INTUITIONISTIC FUZZY SETS (TrFIFSs)

Based on the methods of correlation coefficient of different higher order IFSs proposed by Robinson & Amirtharaj (2011a,2011b,2012a, 2012b), and Park et al. (2009) (given in Appendix), in this paper, a new method is proposed for correlation coefficient of TrFIFSs.

Correlation Coefficient of TrFIFS

Let $X = \{x_1, x_2, \ldots, x_n\}$ be the finite universal set and $A, B \in$ TrFIFS(X) be given by:

$$A=\left\{\left\langle x,[\mu_{A_1}(x),\mu_{A_2}(x),\mu_{A_3}(x)],[\gamma_{A_1}(x),\gamma_{A_2}(x),\gamma_{A_3}(x)]\right\rangle\ / x\in X\right\},$$
$$B=\left\{\left\langle x,[\mu_{B_1}(x),\mu_{B_2}(x),\mu_{B_3}(x)],[\gamma_{B_1}(x),\gamma_{B_2}(x),\gamma_{B_3}(x)]\right\rangle\ / x\in X\right\}$$

Then the Informational Energy of Triangular Fuzzy Intuitionistic Fuzzy Sets (TrFIFSs) *A* and *B* are defined as follows:

$$E_{TrFIFS}(A)=\frac{1}{3}\sum_{i=1}^{n}\left[\mu_A^2(x_i)+\gamma_A^2(x_i)+\pi_A^2(x_i)\right]$$
$$=\frac{1}{3}\sum_{i=1}^{n}\left[\left(\mu_{A_1}^2(x_i)+\mu_{A_2}^2(x_i)+\mu_{A_3}^2(x_i)\right)+\left(\gamma_{A_1}^2(x_i)+\gamma_{A_2}^2(x_i)+\gamma_{A_3}^2(x_i)\right)+\left(\pi_{A_1}^2(x_i)+\pi_{A_2}^2(x_i)+\pi_{A_3}^2(x_i)\right)\right] \tag{13}$$

$$E_{TrFIFS}(B) = \frac{1}{3}\sum_{i=1}^{n}\left[\mu_B^2(x_i) + \gamma_B^2(x_i) + \pi_B^2(x_i)\right]$$
$$= \frac{1}{3}\sum_{i=1}^{n}\left[\left(\mu_{B_1}^2(x_i) + \mu_{B_2}^2(x_i) + \mu_{B_3}^2(x_i)\right) + \left(\gamma_{B_1}^2(x_i) + \gamma_{B_2}^2(x_i) + \gamma_{B_3}^2(x_i)\right) + \left(\pi_{B_1}^2(x_i) + \pi_{B_2}^2(x_i) + \pi_{B_3}^2(x_i)\right)\right] \tag{14}$$

Then the correlation of Triangular Fuzzy Intuitionistic Fuzzy Sets (TrFIFSs) *A* and *B* is defined as follows:

$$\begin{aligned} C_{TrFIFS}(A,B) &= \frac{1}{3}\sum_{i=1}^{n}\left[\mu_A(x_i)\mu_B(x_i) + \gamma_A(x_i)\gamma_B(x_i) + \pi_A(x_i)\pi_B(x_i)\right] \\ &= \frac{1}{3}\sum_{i=1}^{n}\Big\{\left[\left(\mu_{A_1}(x_i)\mu_{B_1}(x_i) + \mu_{A_2}(x_i)\mu_{B_2}(x_i) + \mu_{A_3}(x_i)\mu_{B_3}(x_i)\right)\right] \\ &\quad + \left[\left(\gamma_{A_1}(x_i)\gamma_{B_1}(x_i) + \gamma_{A_2}(x_i)\gamma_{B_2}(x_i) + \gamma_{A_3}(x_i)\gamma_{B_3}(x_i)\right)\right] \\ &\quad + \left[\left(\pi_{A_1}(x_i)\pi_{B_1}(x_i) + \pi_{A_2}(x_i)\pi_{B_2}(x_i) + \pi_{A_3}(x_i)\pi_{B_3}(x_i)\right)\right]\Big\} \end{aligned} \tag{15}$$

The correlation coefficient between Triangular Fuzzy Intuitionistic Fuzzy Sets (TrFIFSs) *A* and *B* is given as:

$$K_{TrFIFS}(A,B) = \frac{C_{TrFIFS}(A,B)}{\sqrt{E_{TrFIFS}(A).E_{TrFIFS}(B)}}, \qquad 0 \le K_{TrFIFS}(A,B) \le 1. \tag{16}$$

Proposition 1: For *A*, *B* ∈ TrFIFS(*X*), the following are true:

1. $0 \le K_{TrFIFS}(A,B) \le 1$,
2. $C_{TrFIFS}(A,B) = C_{TrFIFS}(B,A)$,
3. $K_{TrFIFS}(A,B) = K_{TrFIFS}(B,A)$.

The following theorems are essential for the correlation coefficient of TrFIFSs.

Theorem 1: For *A*, *B* ∈ TrFIFS(*X*), then $0 \le K_{TrFIFS}(A,B) \le 1$.
Proof: Since $C_{TrFIFS}(A,B) \ge 0$, it can be proved that $K_{TrFIFS}(A,B) \le 1$.

For any arbitrary real number ξ, the following relation is true:

$$0 \le \sum_{i=1}^{n} \left\{ \left(\mu_{A_1}(x_i) - \xi\mu_{B_1}(x_i)\right)^2 + \left(\mu_{A_2}(x_i) - \xi\mu_{B_2}(x_i)\right)^2 + \left(\mu_{A_3}(x_i) - \xi\mu_{B_3}(x_i)\right)^2 \right.$$
$$+ \left(\gamma_{A_1}(x_i) - \xi\gamma_{B_1}(x_i)\right)^2 + \left(\gamma_{A_2}(x_i) - \xi\gamma_{B_2}(x_i)\right)^2 + \left(\gamma_{A_3}(x_i) - \xi\gamma_{B_3}(x_i)\right)^2$$
$$\left. + \left(\pi_{A_1}(x_i) - \xi\pi_{B_1}(x_i)\right)^2 + \left(\pi_{A_2}(x_i) - \xi\pi_{B_2}(x_i)\right)^2 + \left(\pi_{A_3}(x_i) - \xi\pi_{B_3}(x_i)\right)^2 \right\}$$

$$= \sum_{i=1}^{n} \left\{ \left(\mu_{A_1}^2(x_i) + \gamma_{A_1}^2(x_i) + \pi_{A_1}^2(x_i)\right) + \left(\mu_{A_2}^2(x_i) + \gamma_{A_2}^2(x_i) + \pi_{A_2}^2(x_i)\right) + \left(\mu_{A_3}^2(x_i) + \gamma_{A_3}^2(x_i) + \pi_{A_3}^2(x_i)\right) \right.$$

$$- 2\xi\left(\mu_{A_1}(x_i)\mu_{B_1}(x_i) + \gamma_{A_1}(x_i)\gamma_{B_1}(x_i) + \pi_{A_1}(x_i)\pi_{B_1}(x_i)\right)$$
$$- 2\xi\left(\mu_{A_2}(x_i)\mu_{B_2}(x_i) + \gamma_{A_2}(x_i)\gamma_{B_2}(x_i) + \pi_{A_2}(x_i)\pi_{B_2}(x_i)\right)$$
$$- 2\xi\left(\mu_{A_3}(x_i)\mu_{B_3}(x_i) + \gamma_{A_3}(x_i)\gamma_{B_3}(x_i) + \pi_{A_3}(x_i)\pi_{B_3}(x_i)\right)$$

$$\left. +\xi^2\left(\mu_{B_1}^2(x_i) + \gamma_{B_1}^2(x_i) + \pi_{B_1}^2(x_i)\right) + \xi^2\left(\mu_{B_2}^2(x_i) + \gamma_{B_2}^2(x_i) + \pi_{B_2}^2(x_i)\right) + \xi^2\left(\mu_{B_3}^2(x_i) + \gamma_{B_3}^2(x_i) + \pi_{B_3}^2(x_i)\right)\right\}$$

Hence,

$$\left\{\sum_{i=1}^{n} \left[\left(\mu_{A_1}(x_i)\mu_{B_1}(x_i) + \mu_{A_2}(x_i)\mu_{B_2}(x_i) + \mu_{A_3}(x_i)\mu_{B_3}(x_i) + \gamma_{A_1}(x_i)\gamma_{B_1}(x_i) + \gamma_{A_2}(x_i)\gamma_{B_2}(x_i) \right.\right.\right.$$
$$\left.\left.\left. + \gamma_{A_3}(x_i)\gamma_{B_3}(x_i) + \pi_{A_1}(x_i)\pi_{B_1}(x_i) + \pi_{A_2}(x_i)\pi_{B_2}(x_i) + \pi_{A_3}(x_i)\pi_{B_3}(x_i)\right)\right]\right\}^2$$

$$\le \left(\sum_{i=1}^{n} \left\{\left(\mu_{A_1}^2(x_i) + \gamma_{A_1}^2(x_i) + \pi_{A_1}^2(x_i)\right) + \left(\mu_{A_2}^2(x_i) + \gamma_{A_2}^2(x_i) + \pi_{A_2}^2(x_i)\right) + \left(\mu_{A_3}^2(x_i) + \gamma_{A_3}^2(x_i) + \pi_{A_3}^2(x_i)\right)\right\} \times \right.$$
$$\left. \sum_{i=1}^{n} \left\{\left(\mu_{B_1}^2(x_i) + \gamma_{B_1}^2(x_i) + \pi_{B_1}^2(x_i)\right) + \left(\mu_{B_2}^2(x_i) + \gamma_{B_2}^2(x_i) + \pi_{B_2}^2(x_i)\right) + \left(\mu_{B_3}^2(x_i) + \gamma_{B_3}^2(x_i) + \pi_{B_3}^2(x_i)\right)\right\} \right)$$

The above inequality can be written as:

$$\left(\frac{\left\{\sum_{i=1}^{n} \left[\mu_{A_1}(x_i)\mu_{B_1}(x_i) + \mu_{A_2}(x_i)\mu_{B_2}(x_i) + \mu_{A_3}(x_i)\mu_{B_3}(x_i) + \gamma_{A_1}(x_i)\gamma_{B_1}(x_i) + \gamma_{A_2}(x_i)\gamma_{B_2}(x_i) + \gamma_{A_3}(x_i)\gamma_{B_3}(x_i) + \pi_{A_1}(x_i)\pi_{B_1}(x_i) + \pi_{A_2}(x_i)\pi_{B_2}(x_i) + \pi_{A_3}(x_i)\pi_{B_3}(x_i)\right]^2\right\}}{\left(\sum_{i=1}^{n} \left\{\left(\mu_{A_1}^2(x_i) + \gamma_{A_1}^2(x_i) + \pi_{A_1}^2(x_i)\right) + \left(\mu_{A_2}^2(x_i) + \gamma_{A_2}^2(x_i) + \pi_{A_2}^2(x_i)\right) + \left(\mu_{A_3}^2(x_i) + \gamma_{A_3}^2(x_i) + \pi_{A_3}^2(x_i)\right) \times \sum_{i=1}^{n}\left\{\left(\mu_{B_1}^2(x_i) + \gamma_{B_1}^2(x_i) + \pi_{B_1}^2(x_i)\right) + \left(\mu_{B_2}^2(x_i) + \gamma_{B_2}^2(x_i) + \pi_{B_2}^2(x_i)\right) + \left(\mu_{B_3}^2(x_i) + \gamma_{B_3}^2(x_i) + \pi_{B_3}^2(x_i)\right)\right\}\right)} \right) \le 1$$

Therefore $\dfrac{\left[C_{TrFIFS}(A,B)\right]^2}{E_{TrFIFS}(A).E_{TrFIFS}(B)} \le 1$, hence $K_{TrFIFS}(A,B) = \dfrac{C_{TrFIFS}(A,B)}{\sqrt{E_{TrFIFS}(A).E_{TrFIFS}(B)}} \le 1$.

Theorem 2: $K_{TrFIFS}(A,B) = 1 \Leftrightarrow A = B$.

Proof: Considering the inequality in the proof of theorem 1, then the equality holds if and only if the following satisfy:

1. $\mu_{A_1}(x_i) = \xi\mu_{B_1}(x_i)$, $\mu_{A_2}(x_i) = \xi\mu_{B_2}(x_i)$, $\mu_{A_3}(x_i) = \xi\mu_{B_3}(x_i)$,
2. $\gamma_{A_1}(x_i) = \xi\gamma_{B_1}(x_i)$, $\gamma_{A_2}(x_i) = \xi\gamma_{B_2}(x_i)$, $\gamma_{A_3}(x_i) = \xi\gamma_{B_3}(x_i)$,
3. $\pi_{A_1}(x_i) = \xi\pi_{B_1}(x_i)$, $\pi_{A_2}(x_i) = \xi\pi_{B_2}(x_i)$, $\pi_{A_3}(x_i) = \xi\pi_{B_3}(x_i)$, for some positive real ξ.

As

$$\mu_{A_1}(x_i) + \gamma_{A_1}(x_i) + \pi_{A_1}(x_i) = \mu_{B_1}(x_i) + \gamma_{B_1}(x_i) + \pi_{B_1}(x_i) = 1,$$

$$\mu_{A_2}(x_i) + \gamma_{A_2}(x_i) + \pi_{A_2}(x_i) = \mu_{B_2}(x_i) + \gamma_{B_2}(x_i) + \pi_{B_2}(x_i) = 1,$$

$$\mu_{A_3}(x_i) + \gamma_{A_3}(x_i) + \pi_{A_3}(x_i) = \mu_{B_3}(x_i) + \gamma_{B_3}(x_i) + \pi_{B_3}(x_i) = 1,$$

then it means $\xi = 1$, and therefore $A = B$.

Theorem 3: $C_{TrFIFS}(A,B) = 0 \Leftrightarrow A$ and B are non-fuzzy sets and satisfy the condition $\mu_A(x_i) + \mu_B(x_i) = 1$ or $\gamma_A(x_i) + \gamma_B(x_i) = 1$ or $\pi_A(x_i) + \pi_B(x_i) = 1$, $\forall x_i \in \mathrm{X}$.

Proof: For all $x_i \in \mathrm{X}$, the following are true:

$$\left(\mu_{A_1}(x_i)\mu_{B_1}(x_i) + \gamma_{A_1}(x_i)\gamma_{B_1}(x_i) + \pi_{A_1}(x_i)\pi_{B_1}(x_i)\right) \ge 0,$$

$$\left(\mu_{A_2}(x_i)\mu_{B_2}(x_i) + \gamma_{A_2}(x_i)\gamma_{B_2}(x_i) + \pi_{A_2}(x_i)\pi_{B_2}(x_i)\right) \ge 0,$$

$$\left(\mu_{A_3}(x_i)\mu_{B_3}(x_i) + \gamma_{A_3}(x_i)\gamma_{B_3}(x_i) + \pi_{A_3}(x_i)\pi_{B_3}(x_i)\right) \ge 0.$$

Hence,

$$\left\{\left(\mu_{A_1}(x_i)\mu_{B_1}(x_i) + \gamma_{A_1}(x_i)\gamma_{B_1}(x_i) + \pi_{A_1}(x_i)\pi_{B_1}(x_i)\right) + \left(\mu_{A_2}(x_i)\mu_{B_2}(x_i) + \gamma_{A_2}(x_i)\gamma_{B_2}(x_i) + \pi_{A_2}(x_i)\pi_{B_2}(x_i)\right) + \left(\mu_{A_3}(x_i)\mu_{B_3}(x_i) + \gamma_{A_3}(x_i)\gamma_{B_3}(x_i) + \pi_{A_3}(x_i)\pi_{B_3}(x_i)\right)\right\} \ge 0.$$

$$\{\mu_{A_1}(x_i)\mu_{B_1}(x_i)+\mu_{A_2}(x_i)\mu_{B_2}(x_i)+\mu_{A_3}(x_i)\mu_{B_3}(x_i) +\gamma_{A_1}(x_i)\gamma_{B_1}(x_i)+\gamma_{A_2}(x_i)\gamma_{B_2}(x_i)+\gamma_{A_3}(x_i)\gamma_{B_3}(x_i) +\pi_{A_1}(x_i)\pi_{B_1}(x_i)+\pi_{A_2}(x_i)\pi_{B_2}(x_i)+\pi_{A_3}(x_i)\pi_{B_3}(x_i)\} \geq 0.$$

If $C_{TrFIFS}(A,B)=0$ for all $x_i \in X$, then the following should be true:

$\mu_{A_1}(x_i)\mu_{B_1}(x_i)+\mu_{A_2}(x_i)\mu_{B_2}(x_i)+\mu_{A_3}(x_i)\mu_{B_3}(x_i)=0,$
$\gamma_{A_1}(x_i)\gamma_{B_1}(x_i)+\gamma_{A_2}(x_i)\gamma_{B_2}(x_i)+\gamma_{A_3}(x_i)\gamma_{B_3}(x_i)=0,$ and
$\pi_{A_1}(x_i)\pi_{B_1}(x_i)+\pi_{A_2}(x_i)\pi_{B_2}(x_i)+\pi_{A_3}(x_i)\pi_{B_3}(x_i)=0.$

If $\mu_{A_1}(x_i)=1$ then $\mu_{B_1}(x_i)=0$ and $\gamma_{A_1}(x_i)=\pi_{A_1}(x_i)=0,$
If $\mu_{A_2}(x_i)=1$ then $\mu_{B_2}(x_i)=0$ and $\gamma_{A_2}(x_i)=\pi_{A_2}(x_i)=0,$
If $\mu_{A_3}(x_i)=1$ then $\mu_{B_3}(x_i)=0$ and $\gamma_{A_3}(x_i)=\pi_{A_3}(x_i)=0.$
Also, If $\mu_{B_1}(x_i)=1$ then $\mu_{A_1}(x_i)=0$ and $\gamma_{B_1}(x_i)=\pi_{B_1}(x_i)=0,$
If $\mu_{B_2}(x_i)=1$ then $\mu_{A_2}(x_i)=0$ and $\gamma_{B_2}(x_i)=\pi_{B_2}(x_i)=0,$
If $\mu_{B_3}(x_i)=1$ then $\mu_{A_3}(x_i)=0$ and $\gamma_{B_3}(x_i)=\pi_{B_3}(x_i)=0.$
Hence $\mu_{A_1}(x_i)+\mu_{B_1}(x_i)=1,\ \mu_{A_2}(x_i)+\mu_{B_2}(x_i)=1,\ \mu_{A_3}(x_i)+\mu_{B_3}(x_i)=1,$
Conversely, When *A* and *B* are non-fuzzy sets and $\mu_{A_1}(x_i)+\mu_{B_1}(x_i)=1,\ \mu_{A_2}(x_i)+\mu_{B_2}(x_i)=1,$ $\mu_{A_3}(x_i)+\mu_{B_3}(x_i)=1,$
If $\mu_{A_1}(x_i)=1$ then $\mu_{B_1}(x_i)=0$ and $\gamma_{A_1}(x_i)=\pi_{A_1}(x_i)=0,$
If $\mu_{A_2}(x_i)=1$ then $\mu_{B_2}(x_i)=0$ and $\gamma_{A_2}(x_i)=\pi_{A_2}(x_i)=0,$
If $\mu_{A_3}(x_i)=1$ then $\mu_{B_3}(x_i)=0$ and $\gamma_{A_3}(x_i)=\pi_{A_3}(x_i)=0.$
Also, If $\mu_{B_1}(x_i)=1$ then $\mu_{A_1}(x_i)=0$ and $\gamma_{B_1}(x_i)=\pi_{B_1}(x_i)=0,$
If $\mu_{B_2}(x_i)=1$ then $\mu_{A_2}(x_i)=0$ and $\gamma_{B_2}(x_i)=\pi_{B_2}(x_i)=0,$
If $\mu_{B_3}(x_i)=1$ then $\mu_{A_3}(x_i)=0$ and $\gamma_{B_3}(x_i)=\pi_{B_3}(x_i)=0.$
Therefore $C_{TrFIFS}(A,B)=0$.
The cases $\gamma_A(x_i)+\gamma_B(x_i)=1$ and $\pi_A(x_i)+\pi_B(x_i)=1$ can be proved similarly.

Theorem 4: $C_{TrFIFS}(A,A)=1 \Leftrightarrow A$ is a non-fuzzy set.
Proof: If *A* is a non-fuzzy set, then $C_{TrFIFS}(A,A)=1$ is obvious.

Conversely, it can be proved by the method of contradiction. Assume *A* is not a non-fuzzy set. Then $0 \leq \mu_A(x_i) < 1,\ \ 0 \leq \gamma_A(x_i) < 1$ and $0 \leq \pi_A(x_i) < 1$, for some x_i.

Hence $\mu_A^{\,2}(x_i)+\gamma_A^{\,2}(x_i)+\pi_A^{\,2}(x_i)<1$.
That is,

$$\mu^2_{A_1}(x_i)+\gamma^2_{A_1}(x_i)+\pi^2_{A_1}(x_i)<1,\ \mu^2_{A_2}(x_i)+\gamma^2_{A_2}(x_i)+\pi^2_{A_2}(x_i)<1 \text{ and } \mu^2_{A_3}(x_i)+\gamma^2_{A_3}(x_i)+\pi^2_{A_3}(x_i)<1.$$

Also,

$$\mu^2_{A_1}(x_i)+\mu^2_{A_2}(x_i)+\mu^2_{A_3}(x_i)+\gamma^2_{A_1}(x_i)+\gamma^2_{A_2}(x_i)+\gamma^2_{A_3}(x_i)+\pi^2_{A_1}(x_i)+\pi^2_{A_2}(x_i)+\pi^2_{A_3}(x_i)<1.$$

Then

$$C_{TrFIFS}(A,A)=\frac{1}{3}\sum_{i=1}^{n}\left\{\mu^2_{A_1}(x_i)+\mu^2_{A_2}(x_i)+\mu^2_{A_3}(x_i)+\gamma^2_{A_1}(x_i)+\gamma^2_{A_2}(x_i)+\gamma^2_{A_3}(x_i)+\pi^2_{A_1}(x_i)+\pi^2_{A_2}(x_i)+\pi^2_{A_3}(x_i)\right\}<1.$$

This is contradictory, and so A is a non-fuzzy set.

OWA Operator for TrFIFNs

On the foundation of the operators discussed early in the paper, the weighted arithmetic operators for TrFIFNs are given as follows:

Definition 6

Let $\tilde{a}_j=\left\langle[\mu_{1j},\mu_{2j},\mu_{3j}],[\gamma_{1j},\gamma_{2j},\gamma_{3j}]\right\rangle,\ j=1,2,\ldots,n,$ be a collection of TrFIFN values. The Triangular Fuzzy Intuitionistic Fuzzy Weighted Averaging (*TrFIFWA*) operator, $TrFIFWA: Q^n\rightarrow Q$ is defined as:

$$TrFIFWA_\omega(\tilde{a}_1,\tilde{a}_2,\ldots,\tilde{a}_n)=\sum_{j=1}^{n}\omega_j\tilde{a}_j=\left(1-\prod_{j=1}^{n}(1-\mu_j)^{\omega_j},\prod_{j=1}^{n}(\gamma_j)^{\omega_j}\right)$$
$$=\left(\left[1-\prod_{j=1}^{n}(1-\mu_{1j})^{\omega_j},1-\prod_{j=1}^{n}(1-\mu_{2j})^{\omega_j},1-\prod_{j=1}^{n}(1-\mu_{3j})^{\omega_j}\right],\left[\prod_{j=1}^{n}(\gamma_{1j})^{\omega_j},\prod_{j=1}^{n}(\gamma_{2j})^{\omega_j},\prod_{j=1}^{n}(\gamma_{3j})^{\omega_j}\right]\right) \quad (17)$$

where $\omega=(\omega_1,\omega_2,\ldots,\omega_t)^T$ is the weight vector of $\tilde{a}_j=\left\langle[\mu_{1j},\mu_{2j},\mu_{3j}],[\gamma_{1j},\gamma_{2j},\gamma_{3j}]\right\rangle,\ j=1,2,\ldots,n,$ such that $\omega_j>0$ and $\sum_{j=1}^{n}\omega_j=1$.

Definition 7

Let $\tilde{a}_j = \left\langle [\mu_{1j}, \mu_{2j}, \mu_{3j}], [\gamma_{1j}, \gamma_{2j}, \gamma_{3j}] \right\rangle$, $j = 1,2,...,n$, be a collection of TrFIFN values. The Triangular Fuzzy Intuitionistic Fuzzy Ordered Weighted Averaging (*TrFIFOWA*) operator of dimension n is a mapping $TrFIFOWA : Q^n \to Q$ is defined as:

$$TrFIFOWA_w(\tilde{a}_1, \tilde{a}_2, ..., \tilde{a}_n) = \sum_{j=1}^{n} w_j \tilde{a}_{\sigma(j)} = \left(1 - \prod_{j=1}^{n}(1-\mu_{\sigma(j)})^{w_j}, \prod_{j=1}^{n}(\gamma_{\sigma(j)})^{w_j}\right)$$
$$- \left(\left[1 - \prod_{j=1}^{n}(1-\mu_{1,\sigma(j)})^{w_j}, 1 - \prod_{j=1}^{n}(1-\mu_{2,\sigma(j)})^{w_j}, 1 - \prod_{j=1}^{n}(1-\mu_{3,\sigma(j)})^{w_j}\right], \left[\prod_{j=1}^{n}(\gamma_{1,\sigma(j)})^{w_j}, \prod_{j=1}^{n}(\gamma_{2,\sigma(j)})^{w_j}, \prod_{j=1}^{n}(\gamma_{3,\sigma(j)})^{w_j}\right]\right) \quad (18)$$

where $w = (w_1, w_2, ..., w_m)^T$ is the associated weighting vector such that $w_j > 0$, $\sum_{j=1}^{n} w_j = 1$. Furthermore, $(\sigma(1), \sigma(2), ..., \sigma(n))$ is the permutation of (1,2,…,n) such that $\tilde{a}_{\sigma(j-1)} \geq \tilde{a}_{\sigma(j)}$, for all

j = 1,2,…,n.

Algorithm for MAGDM Using *TrFIFOWA* Operator and Correlation Coefficient of TrFIFS

Let $A = \{A_1, A_2, ..., A_m\}$ be a set of alternatives, $G = \{G_1, G_2, ..., G_n\}$ be the set of attributes, $\omega = (\omega_1, \omega_2, ..., \omega_n)$ is the weighting vector of the attribute G_j, j=1,2,…,n, where $\omega_j \in [0,1]$, such that $\sum_{j=1}^{n} \omega_j = 1$. Let $D = \{D_1, D_2, ..., D_t\}$ be the set of decision makers, $w = (w_1, w_2, ..., w_t)^T$ be the weighting vector of the decision makers, with $w_j > 0$, $\sum_{j=1}^{t} w_j = 1$.

Suppose that,

$$\tilde{R}_k = \left(\tilde{r}_{hij}^{(k)}\right)_{m \times n} = \left(\mu_{hij}^{(k)}, \gamma_{hij}^{(k)}\right)_{m \times n} = \left(\left[\mu_{1ij}^{(k)}, \mu_{2ij}^{(k)}, \mu_{3ij}^{(k)}\right], \left[\gamma_{1ij}^{(k)}, \gamma_{2ij}^{(k)}, \gamma_{3ij}^{(k)}\right]\right)_{m \times n}$$

be the triangular fuzzy intuitionistic fuzzy number decision matrix, where, $\left[\mu_{1ij}^{(k)}, \mu_{2ij}^{(k)}, \mu_{3ij}^{(k)}\right]$ is the degree of the membership value that the alternative A_i satisfies the attribute G_j given by the decision maker D_k and $\left[\gamma_{1ij}^{(k)}, \gamma_{2ij}^{(k)}, \gamma_{3ij}^{(k)}\right]$ is the degree of non-membership value for the alternative A_i, where $\mu_{hij}^{(k)}, \gamma_{hij}^{(k)} \subset [0,1]$ and $\mu_{3ij}^{(k)} + \gamma_{3ij}^{(k)} \leq 1$, i =1,2,…,m, j =1,2,…,n, k =1,2,…,t.

The developed model of MAGDM is given as follows:

Step 1: Utilize the decision information given in the matrix $\tilde{R}_k$, and the *TrFIFOWA* operator which has the associated weighting vector $w = (w_1, w_2, \ldots, w_t)^T$, and aggregate all the decision matrices into a collective decision matrix $\tilde{R} = (\tilde{r}_{h,ij})_{m\times n}$.

$$\tilde{r}_i^{(k)} = TrFIFOWA_w(\tilde{a}_1, \tilde{a}_2, \ldots, \tilde{a}_n) = \sum_{j=1}^{n} w_j \tilde{a}_{\sigma(j)} = \left(1 - \prod_{j=1}^{n}(1-\mu_{\sigma(j)})^{w_j}, \prod_{j=1}^{n}(\gamma_{\sigma(j)})^{w_j}\right)$$

$$= \left(\left[1 - \prod_{j=1}^{n}(1-\mu_{1,\sigma(j)})^{w_j}, 1 - \prod_{j=1}^{n}(1-\mu_{2,\sigma(j)})^{w_j}, 1 - \prod_{j=1}^{n}(1-\mu_{3,\sigma(j)})^{w_j}\right], \left[\prod_{j=1}^{n}(\gamma_{1,\sigma(j)})^{w_j}, \prod_{j=1}^{n}(\gamma_{2,\sigma(j)})^{w_j}, \prod_{j=1}^{n}(\gamma_{3,\sigma(j)})^{w_j}\right]\right)$$

Step 2: Utilize the *TrFIFWA* operator to aggregate the collective overall preferences given by the decision makers into a group integrated attribute value of the alternative A_i.

$$\tilde{r}_i = TrFIFWA_\omega(\tilde{a}_1, \tilde{a}_2, \ldots, \tilde{a}_n) = \sum_{j=1}^{n} \omega_j \tilde{a}_j = \left(1 - \prod_{j=1}^{n}(1-\mu_j)^{\omega_j}, \prod_{j=1}^{n}(\gamma_j)^{\omega_j}\right)$$
$$= \left(\left[1 - \prod_{j=1}^{n}(1-\mu_{1j})^{\omega_j}, 1 - \prod_{j=1}^{n}(1-\mu_{2j})^{\omega_j}, 1 - \prod_{j=1}^{n}(1-\mu_{3j})^{\omega_j}\right], \left[\prod_{j=1}^{n}(\gamma_{1j})^{\omega_j}, \prod_{j=1}^{n}(\gamma_{2j})^{\omega_j}, \prod_{j=1}^{n}(\gamma_{3j})^{\omega_j}\right]\right)$$

Step 3: Utilize the correlation coefficient (equations (13) to (16)) of TrFIFNs to derive the closeness between the overall group integrated attribute values $\tilde{r}_i$ and the TrFIFN positive ideal value $\tilde{r}^+$, where $\tilde{r}^+ = \langle[1,1,1],[0,0,0]\rangle$.

Step 4: Rank alternatives A_i, i = 1,2,…,m, and select the best in accordance with the highest closeness obtained from step-3.

NUMERICAL ILLUSTRATION

Let us suppose a student seeks admission in five different colleges (A_i, $i = 1,2,\ldots,5$), in a particular city, and wants to select the most desirable college for higher studies. In the process of choosing one among the five different colleges, four different attributes are considered, namely:

- G_1 – College popularity,
- G_2 – Teaching faculty strength,
- G_3 – Good placement records,
- G_4 – Good transport facility.

In the process of evaluating the colleges, the student seeks the help of three decision makers whose weighting vector is given as $w = (w_1, w_2, \ldots, w_t)^T = (0.35, 0.40, 0.25)^T$, and under the four attributes whose weighting vector is $\omega = (0.2, 0.1, 0.3, 0.4)^T$. The decision matrices (5x4) are given as follows:

$$\widetilde{R}_1 = \begin{pmatrix} \langle[0.2,0.3,0.4],[0.1,0.2,0.3]\rangle & \langle[0.3,0.38,0.5],[0.09,0.15,0.2]\rangle & \langle[0.1,0.15,0.2],[0.3,0.4,0.5]\rangle & \langle[0.07,0.09,0.1],[0.3,0.4,0.6]\rangle \\ \langle[0.4,0.5,0.6],[0.1,0.18,0.2]\rangle & \langle[0.3,0.4,0.6],[0.07,0.09,0.1]\rangle & \langle[0.3,0.5,0.6],[0.09,0.098,0.1]\rangle & \langle[0.1,0.2,0.3],[0.3,0.35,0.4]\rangle \\ \langle[0.3,0.4,0.5],[0.2,0.25,0.3]\rangle & \langle[0.2,0.29,0.4],[0.1,0.2,0.3]\rangle & \langle[0.1,0.3,0.4],[0.1,0.18,0.2]\rangle & \langle[0.1,0.3,0.5],[0.1,0.19,0.2]\rangle \\ \langle[0.5,0.6,0.7],[0.08,0.09,0.1]\rangle & \langle[0.1,0.4,0.5],[0.1,0.17,0.2]\rangle & \langle[0.1,0.14,0.2],[0.1,0.2,0.3]\rangle & \langle[0.05,0.08,0.1],[0.2,0.3,0.5]\rangle \\ \langle[0.2,0.4,0.5],[0.07,0.08,0.1]\rangle & \langle[0.1,0.2,0.3],[0.1,0.16,0.2]\rangle & \langle[0.2,0.4,0.6],[0.1,0.13,0.2]\rangle & \langle[0.1,0.2,0.4],[0.1,0.16,0.2]\rangle \end{pmatrix}$$

$$\widetilde{R}_2 = \begin{pmatrix} \langle[0.2,0.4,0.5],[0.1,0.3,0.4]\rangle & \langle[0.2,0.4,0.6],[0.1,0.2,0.3]\rangle & \langle[0.1,0.2,0.3],[0.4,0.5,0.6]\rangle & \langle[0.1,0.15,0.2],[0.5,0.6,0.7]\rangle \\ \langle[0.3,0.5,0.7],[0.1,0.2,0.3]\rangle & \langle[0.4,0.6,0.7],[0.1,0.18,0.2]\rangle & \langle[0.4,0.6,0.7],[0.1,0.14,0.2]\rangle & \langle[0.2,0.3,0.4],[0.2,0.4,0.5]\rangle \\ \langle[0.1,0.4,0.6],[0.2,0.3,0.4]\rangle & \langle[0.3,0.4,0.5],[0.1,0.2,0.4]\rangle & \langle[0.1,0.3,0.5],[0.1,0.2,0.3]\rangle & \langle[0.2,0.5,0.6],[0.2,0.25,0.3]\rangle \\ \langle[0.4,0.7,0.8],[0.08,0.09,0.1]\rangle & \langle[0.2,0.5,0.6],[0.1,0.2,0.3]\rangle & \langle[0.1,0.2,0.3],[0.1,0.3,0.4]\rangle & \langle[0.1,0.17,0.2],[0.4,0.5,0.6]\rangle \\ \langle[0.2,0.3,0.6],[0.1,0.19,0.2]\rangle & \langle[0.2,0.3,0.4],[0.1,0.19,0.3]\rangle & \langle[0.4,0.5,0.6],[0.1,0.16,0.2]\rangle & \langle[0.1,0.4,0.5],[0.1,0.2,0.3]\rangle \end{pmatrix}$$

$$\widetilde{R}_3 = \begin{pmatrix} \langle[0.1,0.2,0.4],[0.2,0.4,0.5]\rangle & \langle[0.2,0.3,0.5],[0.1,0.2,0.4]\rangle & \langle[0.1,0.13,0.2],[0.4,0.5,0.7]\rangle & \langle[0.05,0.08,0.1],[0.4,0.7,0.8]\rangle \\ \langle[0.2,0.4,0.6],[0.1,0.3,0.4]\rangle & \langle[0.4,0.5,0.6],[0.1,0.2,0.3]\rangle & \langle[0.4,0.5,0.6],[0.1,0.2,0.3]\rangle & \langle[0.1,0.2,0.3],[0.3,0.5,0.6]\rangle \\ \langle[0.1,0.3,0.4],[0.2,0.4,0.5]\rangle & \langle[0.2,0.3,0.4],[0.1,0.3,0.5]\rangle & \langle[0.1,0.2,0.3],[0.3,0.35,0.4]\rangle & \langle[0.1,0.4,0.5],[0.2,0.3,0.4]\rangle \\ \langle[0.2,0.5,0.7],[0.1,0.16,0.2]\rangle & \langle[0.2,0.4,0.5],[0.1,0.2,0.4]\rangle & \langle[0.1,0.17,0.2],[0.1,0.4,0.5]\rangle & \langle[0.05,0.06,0.1],[0.2,0.5,0.7]\rangle \\ \langle[0.3,0.4,0.5],[0.1,0.2,0.3]\rangle & \langle[0.1,0.2,0.3],[0.2,0.3,0.4]\rangle & \langle[0.2,0.3,0.6],[0.1,0.19,0.2]\rangle & \langle[0.2,0.3,0.4],[0.1,0.2,0.4]\rangle \end{pmatrix}$$

Step 1: Utilize the decision information given in the matrix $\widetilde{R}_k$, and the *TrFIFOWA* operator, and the weighting vector of the decision makers is given as $w = (w_1, w_2, ..., w_t)^T = (0.35, 0.40, 0.25)^T$, and aggregate all the decision matrices into a collective decision matrix $\widetilde{R} = (\widetilde{r}_{h,ij})_{m\times n}$.

$$\widetilde{r}_i^{(k)} = TrFIFOWA_w(\widetilde{a}_1, \widetilde{a}_2, ..., \widetilde{a}_n) = \sum_{j=1}^{n} w_j \widetilde{a}_{\sigma(j)}$$

$$= \left(\left[1-\prod_{j=1}^{n}(1-\mu_{1,\sigma(j)})^{w_j}, 1-\prod_{j=1}^{n}(1-\mu_{2,\sigma(j)})^{w_j}, 1-\prod_{j=1}^{n}(1-\mu_{3,\sigma(j)})^{w_j}\right], \left[\prod_{j=1}^{n}(\gamma_{1,\sigma(j)})^{w_j}, \prod_{j=1}^{n}(\gamma_{2,\sigma(j)})^{w_j}, \prod_{j=1}^{n}(\gamma_{3,\sigma(j)})^{w_j}\right]\right)$$

Here,

$$\widetilde{r}_{h,ij} = \left(1-\prod_{j=1}^{n}(1-\mu_{h,\sigma(j)})^{w_j}, \prod_{j=1}^{n}(\gamma_{h,\sigma(j)})^{w_j}\right)$$

$$= \left(\left[1-\prod_{j=1}^{n}(1-\mu_{1,\sigma(j)})^{w_j}, 1-\prod_{j=1}^{n}(1-\mu_{2,\sigma(j)})^{w_j}, 1-\prod_{j=1}^{n}(1-\mu_{3,\sigma(j)})^{w_j}\right], \left[\prod_{j=1}^{n}(\gamma_{1,\sigma(j)})^{w_j}, \prod_{j=1}^{n}(\gamma_{2,\sigma(j)})^{w_j}, \prod_{j=1}^{n}(\gamma_{3,\sigma(j)})^{w_j}\right]\right)$$

$$\widetilde{r}_{h,11} = ([0.1761, 0.3142, 0.4371], [0.1189, 0.2741, 0.3769])$$

Similarly, all the entries of the collective decision matrix can be calculated. The collective decision matrix $\widetilde{R} = (\widetilde{r}_{h,ij})_{m\times n}$ is as follows:

$$\widetilde{R} = \begin{pmatrix} \langle[0.1761,0.3142,0.4371],[0.1189,0.2741,0.3769]\rangle & \langle[0.2416,0.3682,0.5375],[0.0958,0.1782,0.2741]\rangle & \langle[0.0999,0.1629,0.2365],[0.3565,0.4573,0.5797]\rangle & \langle[0.0757,0.1090,0.1363],[0.3855,0.5302,0.6805]\rangle \\ \langle[0.3195,0.4767,0.6383],[0.1000,0.2122,0.2741]\rangle & \langle[0.3618,0.5026,0.6383],[0.0867,0.1400,0.1677]\rangle & \langle[0.3618,0.5375,0.6383],[0.0958,0.1327,0.1677]\rangle & \langle[0.1363,0.2365,0.3367],[0.2603,0.4009,0.4786]\rangle \\ \langle[0.1860,0.3764,0.5159],[0.2000,0.2997,0.3769]\rangle & \langle[0.2365,0.3329,0.4371],[0.1000,0.2213,0.3769]\rangle & \langle[0.0999,0.2762,0.4149],[0.1316,0.2205,0.2741]\rangle & \langle[0.1363,0.4012,0.5375],[0.1515,0.2344,0.2741]\rangle \\ \langle[0.4006,0.6175,0.7397],[0.0846,0.1039,0.1189]\rangle & \langle[0.1614,0.4371,0.5375],[0.1000,0.1874,0.2741]\rangle & \langle[0.0999,0.1689,0.2365],[0.1000,0.2741,0.3769]\rangle & \langle[0.0678,0.1077,0.1363],[0.2549,0.4076,0.5797]\rangle \\ \langle[0.2262,0.3667,0.5375],[0.0867,0.1362,0.1677]\rangle & \langle[0.1363,0.2365,0.3367],[0.1189,0.1988,0.2741]\rangle & \langle[0.2766,0.4149,0.6000],[0.1000,0.1537,0.2000]\rangle & \langle[0.1261,0.3004,0.4371],[0.1000,0.1829,0.2741]\rangle \end{pmatrix}$$

Step 2: Utilize the *TrFIFWA* operator to aggregate the collective overall preferences given by the decision makers into a group integrated attribute value of the alternative A_i.

$$\tilde{r}_i = TrFIFWA_{\omega}(\tilde{a}_1, \tilde{a}_2, ..., \tilde{a}_n) = \sum_{j=1}^{n} \omega_j \tilde{a}_j = \left(1 - \prod_{j=1}^{n} (1-\mu_j)^{\omega_j}, \prod_{j=1}^{n} (\gamma_j)^{\omega_j}\right)$$
$$= \left(\left[1 - \prod_{j=1}^{n}(1-\mu_{1j})^{\omega_j}, 1 - \prod_{j=1}^{n}(1-\mu_{2j})^{\omega_j}, 1 - \prod_{j=1}^{n}(1-\mu_{3j})^{\omega_j}\right], \left[\prod_{j=1}^{n}(\gamma_{1j})^{\omega_j}, \prod_{j=1}^{n}(\gamma_{2j})^{\omega_j}, \prod_{j=1}^{n}(\gamma_{3j})^{\omega_j}\right]\right)$$

Where $\omega = (\omega_1, \omega_2, ..., \omega_t)^T = (0.2, 0.1, 0.3, 0.4)^T$ is the weight vector of the attributes of

$$\tilde{a}_j = \langle [\mu_{1j}, \mu_{2j}, \mu_{3j}], [\gamma_{1j}, \gamma_{2j}, \gamma_{3j}] \rangle, \quad j = 1, 2, ..., n, \text{and } \omega_j > 0, \quad \sum_{j=1}^{n} \omega_j = 1.$$

$$\tilde{r}_1 = \left(\left[1 - \prod_{j=1}^{n}(1-\mu_{1j})^{\omega_j}, 1 - \prod_{j=1}^{n}(1-\mu_{2j})^{\omega_j}, 1 - \prod_{j=1}^{n}(1-\mu_{3j})^{\omega_j}\right], \left[\prod_{j=1}^{n}(\gamma_{1j})^{\omega_j}, \prod_{j=1}^{n}(\gamma_{2j})^{\omega_j}, \prod_{j=1}^{n}(\gamma_{3j})^{\omega_j}\right]\right)$$

$$= \Big(\Big[1-(1-\mu_{11})^{\omega_1}.(1-\mu_{21})^{\omega_2}.(1-\mu_{31})^{\omega_3}.(1-\mu_{41})^{\omega_4}, 1-(1-\mu_{12})^{\omega_1}.(1-\mu_{22})^{\omega_2}.(1-\mu_{32})^{\omega_3}.(1-\mu_{42})^{\omega_4},$$
$$1-(1-\mu_{13})^{\omega_1}.(1-\mu_{23})^{\omega_2}.(1-\mu_{33})^{\omega_3}.(1-\mu_{43})^{\omega_4}\Big],$$
$$\Big[(\gamma_{11})^{\omega_1}(\gamma_{21})^{\omega_2}(\gamma_{31})^{\omega_3}(\gamma_{41})^{\omega_4}, (\gamma_{12})^{\omega_1}(\gamma_{22})^{\omega_2}(\gamma_{32})^{\omega_3}(\gamma_{42})^{\omega_4}, (\gamma_{13})^{\omega_1}(\gamma_{23})^{\omega_2}(\gamma_{33})^{\omega_3}(\gamma_{43})^{\omega_4}\Big]\Big)$$

$$\tilde{r}_1 = \Big\langle \Big[1-(1-0.1761)^{0.2}.(1-0.2416)^{0.1}.(1-0.0999)^{0.3}.(1-0.0757)^{0.4},$$
$$1-(1-0.3142)^{0.2}.(1-0.3682)^{0.1}.(1-0.1629)^{0.3}.(1-0.1090)^{0.4},$$
$$1-(1-0.4371)^{0.2}.(1-0.5375)^{0.1}.(1-0.2365)^{0.3}.(1-0.1363)^{0.4}\Big],$$
$$\Big[(0.1189)^{0.2}(0.0958)^{0.1}(0.3565)^{0.3}(0.3855)^{0.4}, (0.2471)^{0.2}(0.1782)^{0.1}(0.4573)^{0.3}(0.5302)^{0.4},$$
$$(0.3769)^{0.2}(0.2741)^{0.1}(0.5797)^{0.3}(0.6805)^{0.4}\Big]\Big\rangle$$

$$\tilde{r}_1 = \langle [0.1214, 0.1982, 0.2822], [0.2589, 0.3904, 0.5262] \rangle$$

Similarly, the remaining collective values are calculated as follows:

$$\tilde{r}_2 = \langle [0.2704, 0.4164, 0.5390], [0.1427, 0.1740, 0.2814] \rangle$$
$$\tilde{r}_3 = \langle [0.1464, 0.2521, 0.4892], [0.1473, 0.2403, 0.3016] \rangle$$
$$\tilde{r}_4 = \langle [0.1644, 0.2958, 0.3848], [0.1406, 0.2547, 0.3443] \rangle$$
$$\tilde{r}_5 = \langle [0.1950, 0.3443, 0.5034], [0.0989, 0.1650, 0.2260] \rangle$$

Step 3: Utilize the correlation coefficient of TrFIFNs to derive the closeness between the overall group integrated attribute values $\tilde{r}_i$ and the TrFIFN positive ideal value $\tilde{r}^+$, where $\tilde{r}^+ = \langle [1, 1, 1], [0, 0, 0] \rangle$. Using equations (17) to (20), the correlation coefficients can be calculated by the formula:

$$K_{TrFIFS}(A,B)=\frac{C_{TrFIFS}(A,B)}{\sqrt{E_{TrFIFS}(A).E_{TrFIFS}(B)}},\qquad 0\le K_{TrFIFS}(A,B)\le 1.$$

Hence the calculated values are given as follows:

$$C_{TrFIFS}(\tilde{r}_1,\tilde{r}^+)=0.2006,\ C_{TrFIFS}(\tilde{r}_2,\tilde{r}^+)=0.4086,\ C_{TrFIFS}(\tilde{r}_3,\tilde{r}^+)=0.2959,$$
$$C_{TrFIFS}(\tilde{r}_4,\tilde{r}^+)=0.2816,\ C_{TrFIFS}(\tilde{r}_5,\tilde{r}^+)=0.3475,$$
$$K_{TrFIFS}(\tilde{r}_1,\tilde{r}^+)=0.3151, K_{TrFIFS}(\tilde{r}_2,\tilde{r}^+)=0.6430,\ K_{TrFIFS}(\tilde{r}_3,\tilde{r}^+)=0.4504,$$
$$K_{TrFIFS}(\tilde{r}_4,\tilde{r}^+)=0.4408,\ K_{TrFIFS}(\tilde{r}_5,\tilde{r}^+)=0.5258$$

Step 4: Ranking alternatives A_i, $i = 1,2,\ldots,m$, from the highest closeness value (correlation coefficient) obtained from step-3, the result is as follows:

$$A_2 > A_5 > A_3 > A_4 > A_1.$$

Hence the best alternative is A_2 .

DECISION MAKING USING *TrFIFOWG* OPERATOR AND CORRELATION COEFFICIENT OF TrFIFS

The Triangular Fuzzy Intuitionistic Fuzzy Ordered Weighted Geometric (*TrFIFOWG*) operator is based on the *TrFIFOWA* operator and the Geometric Mean (*GM*) operator. An MAGDM model is developed here based on the *TrFIFOWG*, and the correlation coefficient of TrFIFS is utilized for ranking alternatives.

OWG Operator for TrFIFNs

Definition 8

Let $\tilde{a}_j=\left\langle[\mu_{1j},\mu_{2j},\mu_{3j}],[\gamma_{1j},\gamma_{2j},\gamma_{3j}]\right\rangle,\ \ j=1,2,\ldots,n,$ be a collection of TrFIFN values. The Triangular Fuzzy Intuitionistic Fuzzy Weighted Geometric (*TrFIFWG*) operator, $TrFIFWG: Q^n\to Q$ is defined as:

$$TrFIFWG_\omega(\tilde{a}_1,\tilde{a}_2,\ldots,\tilde{a}_n)=\sum_{j=1}^{n}\tilde{a}_j^{\,\omega_j}=\left(\prod_{j=1}^{n}(\mu_j)^{\omega_j},1-\prod_{j=1}^{n}(1-\gamma_j)^{\omega_j}\right)$$
$$=\left(\left[\prod_{j=1}^{n}(\mu_{1j})^{\omega_j},\prod_{j=1}^{n}(\mu_{2j})^{\omega_j},\prod_{j=1}^{n}(\mu_{3j})^{\omega_j}\right],\ \left[1-\prod_{j=1}^{n}(1-\gamma_{1j})^{\omega_j},1-\prod_{j=1}^{n}(1-\gamma_{2j})^{\omega_j},1-\prod_{j=1}^{n}(1-\gamma_{3j})^{\omega_j}\right]\right) \tag{19}$$

where $\omega = (\omega_1, \omega_2, ..., \omega_t)^T$ is the weight vector of $\tilde{a}_j = \langle [\mu_{1j}, \mu_{2j}, \mu_{3j}], [\gamma_{1j}, \gamma_{2j}, \gamma_{3j}] \rangle$, $j = 1, 2, ..., n$, such that $\omega_j > 0$ and $\sum_{j=1}^{n} \omega_j = 1$.

Definition 9

Let $\tilde{a}_j = \langle [\mu_{1j}, \mu_{2j}, \mu_{3j}], [\gamma_{1j}, \gamma_{2j}, \gamma_{3j}] \rangle$, $j = 1, 2, ..., n$, be a collection of TrFIFN values. The Triangular Fuzzy Intuitionistic Fuzzy Ordered Weighted Geometric (*TrFIFOWG*) operator of dimension *n* is a mapping $TrFIFOWG : Q^n \rightarrow Q$ is defined as:

$$TrFIFOWG_w(\tilde{a}_1, \tilde{a}_2, ..., \tilde{a}_n) = \sum_{j=1}^{n} \tilde{a}_{\sigma(j)}^{w_j} = \left(\prod_{j=1}^{n} (\mu_{\sigma(j)})^{w_j}, 1 - \prod_{j=1}^{n} (1 - \gamma_{\sigma(j)})^{w_j} \right)$$
$$= \left(\left[\prod_{j=1}^{n} (\mu_{1,\sigma(j)})^{w_j}, \prod_{j=1}^{n} (\mu_{2,\sigma(j)})^{w_j}, \prod_{j=1}^{n} (\mu_{3,\sigma(j)})^{w_j} \right], \left[1 - \prod_{j=1}^{n} (1 - \gamma_{1,\sigma(j)})^{w_j}, 1 - \prod_{j=1}^{n} (1 - \gamma_{2,\sigma(j)})^{w_j}, 1 - \prod_{j=1}^{n} (1 - \gamma_{3,\sigma(j)})^{w_j} \right] \right) \quad (20)$$

where $w = (w_1, w_2, ..., w_m)^T$ is the associated weighting vector such that $w_j > 0$, $\sum_{j=1}^{n} w_j = 1$. Furthermore, $(\sigma(1), \sigma(2), ..., \sigma(n))$ is the permutation of (1,2,…,n) such that $\tilde{a}_{\sigma(j-1)} \geq \tilde{a}_{\sigma(j)}$, for all

$j = 1,2,…,n$.

Central to this operator is the reordering of the arguments, based on their values, in particular an argument $\tilde{a}_j$ is not associated with a particular weight w_j but rather a weight w_j is associated with a particular ordered position *i* of the arguments. One important issue of the *TrFIFOWG* operator is to determine its associated weights. Xu, (2005) reviewed existing main methods, and developed a normal distribution – based method to determine the *OWA* weights, which is determined by:

$$w_j = \frac{e^{-\left[(j-\mu_n)^2 / 2\sigma_n^2\right]}}{\sum_{i=1}^{n} e^{-\left[(j-\mu_n)^2 / 2\sigma_n^2\right]}} \quad (21)$$

where μ_n is the mean of the collection of 1,2,…,n and σ_n is the standard deviation of the collection of 1,2,…,n. μ_n and σ_n are obtained by the following formulae respectively:

$$\mu_n = \frac{1+n}{2}, \quad \sigma_n = \sqrt{\frac{1}{n} \sum_{i=1}^{n} \left(i - \mu_n \right)^2} \quad (22)$$

This method can relieve the influence of unfair arguments on decision results by weighting arguments with small values.

Algorithm for MAGDM Using *TrFIFOWG* Operator and Correlation Coefficient of TrFIFS

The developed model of MAGDM is given below:

Step 1: Utilize the *TrFIFOWG* operator to aggregate all the integrated attribute values given by all the decision makers into the group integrated attribute value of alternative A_i:

$$TrFIFOWG_w(\tilde{a}_1,\tilde{a}_2,\ldots,\tilde{a}_n)=\sum_{j=1}^{n}\tilde{a}_{\sigma(j)}^{w_j}=\left(\prod_{j=1}^{n}(\mu_{\sigma(j)})^{w_j},1-\prod_{j=1}^{n}(1-\gamma_{\sigma(j)})^{w_j}\right)$$
$$=\left(\left[\prod_{j=1}^{n}(\mu_{1,\sigma(j)})^{w_j},\prod_{j=1}^{n}(\mu_{2,\sigma(j)})^{w_j},\prod_{j=1}^{n}(\mu_{3,\sigma(j)})^{w_j}\right],\left[1-\prod_{j=1}^{n}(1-\gamma_{1,\sigma(j)})^{w_j},1-\prod_{j=1}^{n}(1-\gamma_{2,\sigma(j)})^{w_j},1-\prod_{j=1}^{n}(1-\gamma_{3,\sigma(j)})^{w_j}\right]\right)$$

Step 2: Utilize the *TrFIFWG* operator to aggregate the collective overall preferences given by the decision makers into a group integrated attribute value of the alternative A_i.

$$\tilde{r}_i=TrFIFWG_\omega(\tilde{a}_1,\tilde{a}_2,\ldots,\tilde{a}_n)=\sum_{j=1}^{n}\tilde{a}_j^{\omega_j}=\left(\prod_{j=1}^{n}(\mu_j)^{\omega_j},1-\prod_{j=1}^{n}(1-\gamma_j)^{\omega_j}\right)$$
$$=\left(\left[\prod_{j=1}^{n}(\mu_{1j})^{\omega_j},\prod_{j=1}^{n}(\mu_{2j})^{\omega_j},\prod_{j=1}^{n}(\mu_{3j})^{\omega_j}\right],\left[1-\prod_{j=1}^{n}(1-\gamma_{1j})^{\omega_j},1-\prod_{j=1}^{n}(1-\gamma_{2j})^{\omega_j},1-\prod_{j=1}^{n}(1-\gamma_{3j})^{\omega_j}\right]\right)$$

Step 3: Calculate the correlation coefficient (equations (13) – (16)) between collective overall values $\tilde{r}_i=\langle[\mu_1,\mu_2,\mu_3],[\gamma_1,\gamma_2,\gamma_3]\rangle$ and triangular fuzzy intuitionistic positive ideal solution $\tilde{r}^{+}=\langle[\mu_1^{+},\mu_2^{+},\mu_3^{+}],[\gamma_1^{+},\gamma_2^{+},\gamma_3^{+}]\rangle=\langle[1,1,1],[0,0,0]\rangle$:

Step 4: Rank alternative A_i. (i=1,2,3) and select the best one(s) in accordance with the correlation coefficient obtained. The higher correlation coefficient, the better the alternatives A_i.

NUMERICAL ILLUSTRATION

Let us suppose there is an university composed of 4 colleges A_k (k = 1,2,3,4) which wants to select the most excellent one. In the process of choosing one of the colleges, three factors are considered (Wang, 2008):

I_1: Teaching, I_2: Scientific research, I_3: Service.

In order to evaluate the colleges, a committee composed of three decision makers is formed. Suppose that the weight vector of I_i (i=1,2,3) is ω = (0.3608,0.3091,0.3301) but the weights of three decision makers are completely unknown. The preference information of the colleges A_k (k = 1,2,3,4) are represented by the TrFIFNs listed below. Then, the proposed procedure is utilized to get the most desirable college(s).

$$\tilde{R}_1 = \begin{bmatrix} \langle[0.5,0.6,0.7],[0.1,0.2,0.3]\rangle, & \langle[0.6,0.7,0.8],[0.1,0.1,0.2]\rangle, & \langle[0.3,0.4,0.5],[0.2,0.3,0.4]\rangle \\ \langle[0.4,0.5,0.6],[0.1,0.2,0.3]\rangle, & \langle[0.5,0.6,0.6],[0.1,0.2,0.3]\rangle, & \langle[0.5,0.6,0.7],[0.1,0.2,0.3]\rangle \\ \langle[0.7,0.7,0.8],[0.1,0.1,0.2]\rangle, & \langle[0.5,0.6,0.7],[0.1,0.1,0.2]\rangle, & \langle[0.5,0.5,0.6],[0.2,0.3,0.3]\rangle \\ \langle[0.5,0.6,0.7],[0.1,0.2,0.2]\rangle, & \langle[0.3,0.3,0.4],[0.1,0.2,0.2]\rangle, & \langle[0.6,0.7,0.7],[0.1,0.1,0.2]\rangle \end{bmatrix}$$

$$\tilde{R}_2 = \begin{bmatrix} \langle[0.6,0.6,0.7],[0.1,0.2,0.3]\rangle, & \langle[0.7,0.8,0.9],[0.1,0.1,0.1]\rangle, & \langle[0.5,0.5,0.5],[0.2,0.3,0.4]\rangle \\ \langle[0.5,0.6,0.6],[0.2,0.3,0.4]\rangle, & \langle[0.7,0.7,0.7],[0.1,0.2,0.3]\rangle, & \langle[0.6,0.6,0.6],[0.1,0.2,0.3]\rangle \\ \langle[0.8,0.8,0.8],[0.1,0.1,0.2]\rangle, & \langle[0.6,0.7,0.8],[0.1,0.2,0.2]\rangle, & \langle[0.5,0.6,0.6],[0.2,0.3,0.3]\rangle \\ \langle[0.5,0.6,0.7],[0.2,0.2,0.2]\rangle, & \langle[0.4,0.5,0.6],[0.1,0.1,0.1]\rangle, & \langle[0.8,0.8,0.8],[0.2,0.2,0.2]\rangle \end{bmatrix}$$

$$\tilde{R}_3 = \begin{bmatrix} \langle[0.6,0.6,0.7],[0.1,0.2,0.3]\rangle, & \langle[0.7,0.7,0.7],[0.1,0.1,0.1]\rangle, & \langle[0.4,0.4,0.4],[0.1,0.1,0.1]\rangle \\ \langle[0.7,0.8,0.8],[0.1,0.2,0.2]\rangle, & \langle[0.6,0.6,0.6],[0.1,0.2,0.3]\rangle, & \langle[0.5,0.6,0.7],[0.1,0.1,0.1]\rangle \\ \langle[0.7,0.7,0.8],[0.1,0.1,0.1]\rangle, & \langle[0.7,0.8,0.9],[0.1,0.1,0.1]\rangle, & \langle[0.6,0.6,0.6],[0.2,0.2,0.2]\rangle \\ \langle[0.6,0.6,0.6],[0.1,0.1,0.1]\rangle, & \langle[0.5,0.5,0.5],[0.1,0.1,0.1]\rangle, & \langle[0.6,0.7,0.8],[0.1,0.1,0.1]\rangle \end{bmatrix}$$

Step 1: Utilize the *TrFIFOWG* operator to aggregate the attribute value of the kth row into the integrated attribute value $\tilde{R}$ which is given by the decision maker D_k.

By (22), μ_n is the mean of collection of 1,2,…,n, $\mu_n = \frac{1+n}{2}$, σ_n is the standard deviation of the collection of 1,2,3,…,n, $\sigma_n = \sqrt{\frac{1}{n}\sum_{i=1}^{n}\left(i-\mu_n\right)^2}$ and the *OWG* weight is determined by (21) as follows: $w_1 = 0.2429$, $w_2 = 0.5142$, $w_3 = 0.2429$. Then $w = (0.2429, 0.5142, 0.2429)^T$ is the weighting vector. Hence the matrix is obtained as follows:

$$\tilde{R} = \left(\begin{array}{l} \qquad\qquad I_1 \\ \langle[0.5463,0.6000,0.7000],[0.0999,0.2000,0.3000]\rangle, \\ \langle[0.4837,0.5858,0.6434],[0.1254,0.2255,0.3035]\rangle, \\ \langle[0.7231,0.7231,0.8000],[0.0999,0.0999,0.1768]\rangle, \\ \langle[0.5226,0.6000,0.6743],[0.1254,0.1768,0.1768]\rangle, \end{array}\right.$$

$$I_2$$

$$\begin{array}{l}\langle[0.6466,0.7231,0.7969],[0.0999,0.0999,0.1529]\rangle,\\ \langle[0.5671,0.6229,0.6229],[0.0999,0.2030,0.3000]\rangle,\\ \langle[0.5671,0.6679,0.7686],[0.0999,0.1254,0.1768]\rangle,\\ \langle[0.3642,0.3845,0.4659],[0.0999,0.1529,0.1529]\rangle,\end{array}$$

$$I_3$$

$$\left.\begin{array}{l}\langle[0.3642,0.4222,0.4736],[0.1768,0.2559,0.3379]\rangle\\ \langle[0.5226,0.6000,0.6743],[0.0999,0.1768,0.2559]\rangle\\ \langle[0.5226,0.5463,0.6000],[0.2000,0.2769,0.2769]\rangle\\ \langle[0.6434,0.7231,0.7469],[0.1254,0.1254,0.1768]\rangle\end{array}\right)$$

Step 2: Utilize the *TrFIFWG* operator to aggregate all integrated attribute values given by all decision makers into the group integrated attribute value of alternative A_i:

$$\tilde{r}_i = TrFIFWG_{\omega}(\tilde{a}_1, \tilde{a}_2, \ldots, \tilde{a}_n) = \sum_{j=1}^{n} \tilde{a}_j^{\,\omega_j} = \left(\prod_{j=1}^{3}(\mu_j)^{\omega_j}, 1-\prod_{j=1}^{3}(1-\gamma_j)^{\omega_j}\right)$$

$$= \left(\left[\prod_{j=1}^{3}(\mu_{1j})^{\omega_j}, \prod_{j=1}^{3}(\mu_{2j})^{\omega_j}, \prod_{j=1}^{3}(\mu_{3j})^{\omega_j}\right], \left[1-\prod_{j=1}^{3}(1-\gamma_{1j})^{\omega_j}, 1-\prod_{j=1}^{3}(1-\gamma_{2j})^{\omega_j}, 1-\prod_{j=1}^{3}(1-\gamma_{3j})^{\omega_j}\right]\right)$$

Hence,

$$\begin{array}{l}\tilde{r}_1 = \langle[0.5034, 0.5660, 0.6404], [0.1260, 0.1899, 0.2710]\rangle\\ \tilde{r}_2 = \langle[0.5212, 0.6017, 0.6469], [0.1092, 0.2027, 0.2870]\rangle\\ \tilde{r}_3 = \langle[0.6026, 0.6432, 0.7186], [0.1342, 0.1700, 0.2113]\rangle\\ \tilde{r}_4 = \langle[0.5006, 0.5561, 0.6221], [0.1176, 0.1527, 0.1695]\rangle\end{array}$$

Step 3: The correlation coefficient is given as: $K_{TrFIFS}(A,B) = \dfrac{C_{TrFIFS}(A,B)}{\sqrt{E_{TrFIFS}(A).E_{TrFIFS}(B)}}$.

Hence the calculated values are given as:

$$\begin{array}{ll}C(\tilde{r}_1, \tilde{r}^{+}) = 0.5699, & K_{TrFNIFS}(\tilde{r}_1, \tilde{r}^{+}) = 0.8611\\ C(\tilde{r}_2, \tilde{r}^{+}) = 0.5899, & K_{TrFNIFS}(\tilde{r}_2, \tilde{r}^{+}) = 0.8739\\ C(\tilde{r}_3, \tilde{r}^{+}) = 0.6548, & K_{TrFNIFS}(\tilde{r}_3, \tilde{r}^{+}) = 0.9279\\ C(\tilde{r}_4, \tilde{r}^{+}) = 0.5596, & K_{TrFNIFS}(\tilde{r}_4, \tilde{r}^{+}) = 0.8545\end{array}$$

Step 4: Ranking alternatives A_i, i = 1,2,…,m, from the highest closeness (correlation coefficient) obtained from step-3, the result is as follows:

$$A_3 > A_2 > A_1 > A_4.$$

Hence the best alternative is A_3.

COMPARISON OF PROPOSED MAGDM WITH EXISTING METHODS

Comparison with the Ranking of Normalized Distance Measure

Definition 10

Let

$$A = \left\langle [\mu_{A_1}(x), \mu_{A_2}(x), \mu_{A_3}(x)], [\gamma_{A_1}(x), \gamma_{A_2}(x), \gamma_{A_3}(x)] \right\rangle$$

and

$$B = \left\langle [\mu_{B_1}(x), \mu_{B_2}(x), \mu_{B_3}(x)], [\gamma_{B_1}(x), \gamma_{B_2}(x), \gamma_{B_3}(x)] \right\rangle$$

be two triangular fuzzy intuitionistic fuzzy numbers. Then the new normalized distance between A and B is defined as follows:

$$d(A,B) = \frac{1}{8}\left[\left|\left(1+\mu_{A_1}-\gamma_{A_1}\right)-\left(1+\mu_{B_1}-\gamma_{B_1}\right)\right| + \left|\left(1+\mu_{A_2}-\gamma_{A_2}\right)-\left(1+\mu_{B_2}-\gamma_{B_2}\right)\right| + \left|\left(1+\mu_{A_3}-\gamma_{A_3}\right)-\left(1+\mu_{B_3}-\gamma_{B_3}\right)\right| \right] \tag{23}$$

Replace step-3 and step-4 of the above decision making model with the new distance measure.

$$d\left(\tilde{r}_i, \tilde{r}^+\right) = \frac{1}{8}\left[\left|\left(1+\mu_1-\gamma_1\right)-\left(1+\mu_1^+-\gamma_1^+\right)\right| + \left|\left(1+\mu_2-\gamma_2\right)-\left(1+\mu_2^+-\gamma_2^+\right)\right| + \left|\left(1+\mu_3-\gamma_3\right)-\left(1+\mu_3^+-\gamma_3^+\right)\right| \right]$$

$$d\left(\tilde{r}_1, \tilde{r}^+\right) = 0.2293, \quad d\left(\tilde{r}_2, \tilde{r}^+\right) = 0.2271, \quad d\left(\tilde{r}_3, \tilde{r}^+\right) = 0.1761, \quad d\left(\tilde{r}_4, \tilde{r}^+\right) = 0.2123.$$

Rank alternative A_i (i=1,2, 3) and select the best one(s) in accordance with $d\left(\tilde{r}_i, \tilde{r}^+\right)$. The smaller $d\left(\tilde{r}_i, \tilde{r}^+\right)$, the better the alternatives A_i.

$$A_3 < A_4 < A_2 < A_1.$$

Thus the most desirable college is A_3 .

Comparison with the Ranking of Chen et al., (2010) Method

Definition 11

Let X be a non-empty set. Then $A = \{\langle x, \mu_A(x), \gamma_A(x)\rangle \ / x \in X\}$ is called a Triangular Fuzzy Intuitionistic Fuzzy Number if $\mu_A(x) = \left(\mu_{A_1}(x), \mu_{A_2}(x), \mu_{A_3}(x)\right)$ and $\gamma_A(x) = \left(\gamma_{A_1}(x), \gamma_{A_2}(x), \gamma_{A_3}(x)\right)$ are triangular fuzzy numbers, with $0 \le \mu_{A_3}(x) + \gamma_{A_3}(x) \le 1, \ \forall x \in X$. Then the Score function of A is defined as follows:

$$S(A) = E(\mu_A)\left\{\frac{1 + E(\mu_A) - E(\gamma_A)}{2}\right\}\left(1 - E(\mu_A) - E(\gamma_A)\right), \quad (24)$$

where $S(A) \in [0,1]$, $E(\mu_A)$ and $E(\gamma_A)$ are the fuzzy means for the TrFIFNs μ_A and γ_A respectively.

For any TrFN $A = \left(a, b, c\right) \in F(D)$, where $\mu_A(x)$ is defined as follows:

$$\mu_A(x) = \begin{cases} (x-a)/(b-a), & a \le x \le b \\ (c-x)/(c-b), & b \le x \le c \\ 0, & otherwise \end{cases}$$

Then, $E(A)$ is defined as $E(A) = \dfrac{a + 2b + c}{4}$.

Following the ranking method with the score function of Chen et al., (2010) the decision alternatives are ranked as follows:

$$A_3 < A_4 < A_1 < A_2.$$

Thus the most desirable college is A_3.

Comparison with Chen et al., (2010) Method combined with Proposed Correlation Coefficient of TrFIFNs

Following the decision making method of Chen et al. (2010) and replacing the score and variation function by the proposed correlation coefficient of TrFIFNs, the decision making method is as follows:

Step 1: Utilize the *TrFIFWG* operator to aggregate the attribute values, and the weighting vector is ω = (0.3608, 0.3091,0.3301).

Step 2: Utilize the *TrFIFOWG* operator to aggregate all the integrated attribute values given by all decision makers, where $w = (0.2429, 0.5142, 0.2429)^T$ into the group integrated attribute value of alternative A_i . Hence we get:

$$\tilde{r}_1 = \left\langle \left[0.5338, 0.5797, 0.6415\right], \left[0.1345, 0.1867, 0.2671\right] \right\rangle$$
$$\tilde{r}_2 = \left\langle \left[0.5358, 0.6161, 0.6539\right], \left[0.1166, 0.2093, 0.2869\right] \right\rangle$$
$$\tilde{r}_3 = \left\langle \left[0.6416, 0.6758, 0.7443\right], \left[0.1253, 0.1516, 0.1937\right] \right\rangle$$
$$\tilde{r}_4 = \left\langle \left[0.5179, 0.5755, 0.6452\right], \left[0.1245, 0.1522, 0.1608\right] \right\rangle$$

Step 3: Replacing the score and variation function of Chen et al., (2010) method with the correlation coefficient of TrFIFNs, the calculated values are as follows:

$$C(\tilde{r}_1, \tilde{r}^+) = 0.5850, \quad K_{TrFNIFS}(\tilde{r}_1, \tilde{r}^+) = 0.8787$$
$$C(\tilde{r}_2, \tilde{r}^+) = 0.6019, \quad K_{TrFNIFS}(\tilde{r}_2, \tilde{r}^+) = 0.8904$$
$$C(\tilde{r}_3, \tilde{r}^+) = 0.6872, \quad K_{TrFNIFS}(\tilde{r}_3, \tilde{r}^+) = 0.9450$$
$$C(\tilde{r}_4, \tilde{r}^+) = 0.5795, \quad K_{TrFNIFS}(\tilde{r}_4, \tilde{r}^+) = 0.8737$$

Step 4: Ranking the alternatives based on the correlation coefficient, it can be observed that:

$$A_3 > A_2 > A_1 > A_4.$$

Thus the most desirable college is A_3.

Table 2. Comparison Table

MAGDM Models	Ranking of Alternatives
Proposed Method of Correlation coefficient for *TrFIFNs*	$A_3 > A_2 > A_1 > A_4.$
Proposed Method of Normalized Distance Measure	$A_3 < A_4 < A_2 < A_1.$
Chen et al. (2010) Method of Ranking (Score function)	$A_3 < A_4 < A_1 < A_2.$
Chen et al. (2010) Method in combination with Proposed Correlation coefficient for *TrFIFNs*	$A_3 > A_2 > A_1 > A_4.$

CONCLUSION

In this paper two different MAGDM models were proposed and the first model was based on ordered weighted averaging operators. In the first model, the Triangular Fuzzy Intuitionistic Fuzzy Weighted Averaging (*TrFIFWA*) operator and the Triangular Fuzzy Intuitionistic Fuzzy Ordered Weighted Averaging (*TrFIFOWA*) operator were proposed and utilized for aggregating the information. Correlation coefficient of TrFIFS was proposed and used in the MAGDM problem for ranking alternatives. In the second model, some geometric aggregation operators (Wang, 2008) based on TrFIFS were utilized in the situation where the decision-makers' weights were completely unknown and the attribute values took the form of TrFIFSs. A numerical illustration was given to show the effectiveness of the proposed approach. A new distance function was defined, and the proposed method of decision making together with the correlation coefficient of TrFIFNs was compared with the method employing the distance function and the Chen et al., (2010) method. A variation was observed (Table 2) in the final ranking of decision variables in the second model, with having no change in the ranking of the most desired variable. In both models, the newly proposed correlation coefficient of TrFIFNs proved a better ranking procedure since it preserves the linear relationship between the variables.

REFERENCES

Atanassov, K. (1986). Intuitionistic fuzzy sets. *Fuzzy Sets and Systems*, *20*(1), 87–96. doi:10.1016/S0165-0114(86)80034-3

Atanassov, K. (1994). Operators over interval-valued intuitionistic fuzzy sets. *Fuzzy Sets and Systems*, *64*(2), 159–174. doi:10.1016/0165-0114(94)90331-X

Atanassov, K. (1999). Intuitionistic Fuzzy Sets: Theory and Applications. Physica-Verlag.

Atanassov, K., & Gargov, G. (1989). Interval-valued intuitionistic fuzzy sets. *Fuzzy Sets and Systems*, *31*(3), 343–349. doi:10.1016/0165-0114(89)90205-4

Azar, A. T. (2010). *Fuzzy Systems*. Vienna, Austria: IN-TECH.

Azar, A. T. (2010). Adaptive Neuro-Fuzzy Systems. In A. T. Azar (Ed.), *Fuzzy Systems*. Vienna, Austria: IN-TECH. doi:10.5772/7220

Azar, A. T. (2012). Overview of Type-2 Fuzzy logic systems. *International Journal of Fuzzy System Applications*, *2*(4), 1–28. doi:10.4018/ijfsa.2012100101

Boren, F. E., Gene, S., Kurt, M., & Akay, D. (2009). A multi-criteria intuitionistic fuzzy group decision making for supplier selsection with TOPSIS method. *Expert Systems with Applications*, *36*(8), 11363–11368. doi:10.1016/j.eswa.2009.03.039

Burillo, P., Bustince, H., & Mohedano, V. (1994). Some definitions of intuitionistic fuzzy number. Fuzzy based expert systems, fuzzy Bulgarian enthusiasts, Sofia, Bulgaria.

Bustince, H., & Burillo, P. (1995). Correlation of interval-valued intuitionistic fuzzy sets. *Fuzzy Sets and Systems*, *74*(2), 237–244. doi:10.1016/0165-0114(94)00343-6

Chen, D. F., Zhang, L., & Jiao, J. (2010). Triagular Fuzzy Number Intuitionistic Fuzzy Aggregation Operators and Their Group Decision Making. *LNAI*, *6320*, 350–357.

Chen, S. M., & Tan, J. M. (1994). Handling multi-criteria fuzzy decision making problems based on vague sets. *Fuzzy Sets and Systems*, *67*(2), 163–172. doi:10.1016/0165-0114(94)90084-1

Chiang, D. A., & Lin, N. P. (1999). Correlation of fuzzy sets. *Fuzzy Sets and Systems*, *102*(2), 221–226. doi:10.1016/S0165-0114(97)00127-9

Chiclana, F., Herrera, F., Herrera-Viedma, E., & Alonso, S. (2004). Induced ordered weighted geometric operators and their use in the aggregation of multiplicative preference relations. *International Journal of Intelligent Systems*, *19*(3), 233–255. doi:10.1002/int.10172

Gerstenkorn, T., & Manko, J. (1991). Correlation of intuitionistic fuzzy sets. *Fuzzy Sets and Systems*, *44*(1), 39–43. doi:10.1016/0165-0114(91)90031-K

Herrera, F., & Herrera-Viedma, E. (2000). Linguistic decision analysis: Steps for solving decision problems under linguistic information. *Fuzzy Sets and Systems*, *115*(10), 67–82. doi:10.1016/S0165-0114(99)00024-X

Herrera, F., Herrera-Viedma, E., & Chiclana, F. (2003). A study of the origin and uses of the ordered weighted geometric operator in multicriteria decision making. *International Journal of Intelligent Systems*, *18*(6), 689–707. doi:10.1002/int.10106

Herrera, F., Martinez, L., & Sanchez, P. J. (1999). Managing non-homogenous information in group decision making. *European Journal of Operational Research*, *116*, 115–132.

Hong, D. H., & Hwang, S. W. (1995). Correlation of intuitionistic fuzzy sets in probability spaces. *Fuzzy Sets and Systems*, *75*(1), 77–81. doi:10.1016/0165-0114(94)00330-A

Kahraman, C. (2008). Multi-Criteria Decision Making-Theory and Applications with Recent Develoments. Springer Optimization and its Applications, 16.

Karayiannis, N. (2000). Soft learning vector quantization and clustering algorithms based on ordered weighted aggregation operators. *IEEE Transactions on Neural Networks*, *11*(5), 1093–1105. doi:10.1109/72.870042 PMID:18249837

Li, D.-F. (2005). Multi attribute decision making models and methods using intuitionistic fuzzy sets. *Journal of Computer and System Sciences*, *70*(1), 73–85. doi:10.1016/j.jcss.2004.06.002

Li, D.-F. (2008). Extension of the LINMAP for multi attribute decision making under Atanassovs intuitionistic fuzzy environment. *Fuzzy Optimization and Decision Making*, *7*(1), 17–34. doi:10.1007/s10700-007-9022-x

Li, D.-F. (2010a). Linear programming method for MADM with interval-valued intuitionistic fuzzy sets. *Expert Systems with Applications*, *37*(8), 5939–5945. doi:10.1016/j.eswa.2010.02.011

Li, D.-F. (2010b). TOPSIS-based nonlinear-programming methodology for multiattribute decision making with interval-valued intuitionistic fuzzy sets. *IEEE Transactions on Fuzzy Systems*, *18*(2), 299–311.

Li, D.-F. (2010c). Multiattribute decision making method based on generalized OWA operators with intuitionistic fuzzy sets. *Expert Systems with Applications*, *37*(12), 8673–8678. doi:10.1016/j.eswa.2010.06.062

Li, D.-F. (2010d). A ratio ranking method of triangular intuitionistic fuzzy numbers and its application to MADM problems. *Computers & Mathematics with Applications (Oxford, England)*, *60*(6), 1557–1570. doi:10.1016/j.camwa.2010.06.039

Li, D.-F. (2014). *Decision and Game Theory in Management with Intuitionistic Fuzzy Sets*. Heidelberg, Germany: Springer. doi:10.1007/978-3-642-40712-3

Li, D.-F. (2016). *Linear Programming Models and Methods of Matrix Games with Payoffs of Triangular Fuzzy Numbers*. Berlin: Springer. doi:10.1007/978-3-662-48476-0

Li, D.-F., & Liu, J.-C. (2015). A parameterized non-linear programming approach to solve matrix games with payoffs of I-fuzzy numbers. *IEEE Transactions on Fuzzy Systems*, *23*(4), 885–896. doi:10.1109/TFUZZ.2014.2333065

Li, D.-F., & Nan, J.-X. (2011). Extension of TOPSIS for Multi-attribute group decision making under Atanassov IFS environments. *International Journal of Fuzzy System Applications*, *1*(4), 47–61. doi:10.4018/ijfsa.2011100104

Li, D.-F., & Nan, J.-X. (2014). An interval-valued programming approach to matrix games with payoffs of triangular intuitionistic fuzzy numbers. *Iranian Journal of Fuzzy Systems*, *11*(2), 45–57.

Li, D.-F., Nan, J.-X., & Zhang, M.-J. (2010). A ranking method of triangular intuitionistic fuzzy numbers and application to decision making. *International Journal of Computational Intelligence Systems*, *3*(5), 522–530. doi:10.1080/18756891.2010.9727719

Li, D.-F., Wang, Y. C., Liu, S., & Shan, F. (2009). Fractional programming methodology for multi attribute decision making using IFS. *Applied Soft Computing*, *9*(1), 219–225. doi:10.1016/j.asoc.2008.04.006

Li, D.-F., & Yang, J. (2015). A difference-index based ranking method of trapezoidal intuitionistic fuzzy numbers and application to multiattribute decision making. *Mathematical and Computational Applications*, *20*(1), 25–38. doi:10.19029/mca-2015-003

Liu, F., & Yuan, X.H. (2007). Fuzzy number intuitionistic fuzzy set. *Fuzzy System and Mathematics*, *21*, 88-91.

Liu, P. D., He, L., & Yu, X. C. (2016). Generalized Hybrid Aggregation Operators Based on the 2-Dimension Uncertain Linguistic Information for Multiple Attribute Group Decision Making. *Group Decision and Negotiation*, *25*(1), 103–126. doi:10.1007/s10726-015-9434-x

Liu, P. D., & Jin, F. (2012). A multi-attribute group decision-making method based on weighted geometric aggregation operators of interval-valued trapezoidal fuzzy numbers. *Applied Mathematical Modelling*, *36*(6), 2498–2509. doi:10.1016/j.apm.2011.09.006

Liu, P. D., Jin, F., Zhang, X., Su, Y., & Wang, M. (2011). Minghe Wang, Research on the multi-attribute decision-making under risk with interval probability based on prospect theory and the uncertain linguistic variables. *Knowledge-Based Systems*, *24*(4), 554–561. doi:10.1016/j.knosys.2011.01.010

Liu, P. D., & Shi, L. L. (2015). The Generalized Hybrid Weighted Average Operator Based on Interval Neutrosophic Hesitant Set and Its Application to Multiple Attribute Decision Making. *Neural Computing & Applications*, *26*(2), 457–471. doi:10.1007/s00521-014-1736-4

Liu, P. D., Zhang, X., & Jin, F. (2012). A multi-attribute group decision-making method based on interval-valued trapezoidal fuzzy numbers hybrid harmonic averaging operators. *Journal of Intelligent & Fuzzy Systems*, *23*(5), 159–168.

Merigo, J. M., & Casanovas, M. (2008). Generalized aggregation operators and fuzzy numbers in a unified Model between the weighted average and the OWA operator. *International Journal of Intelligent Systems*, *24*, 934–954.

Merigo, J. M., & Casanovas, M. (2009). Induced aggregation operators in decision making with Dempster-Shafer belief structure. *International Journal of Intelligent Systems*, *24*(8), 934–954. doi:10.1002/int.20368

Merigo, J. M., & Gil-Lafuente, A. M. (2008a). On the use of the OWA operator in the Euclidean distance. *International Journal on Computer Science and Engineering*, *2*, 170–176.

Merigo, J. M., & Gil-Lafuente, A. M. (2009a). OWA operators in generalized distances. *International Journal of Applied Mathematics and Computer Science*, *5*, 11–18.

Nan, J.-X., Li, D.-F., & Zhang, M.-J. (2010). A lexicographic method for matrix games with payoffs of triangular intuitionistic fuzzy numbers. *International Journal of Computational Intelligence Systems*, *3*(3), 280–289. doi:10.1080/18756891.2010.9727699

Nan, J.-X., Zhang, M.-J., & Li, D.-F. (2014a). Intuitionistic fuzzy programming models for matrix games with payoffs of trapezoidal intuitionistic fuzzy numbers. *International Journal of Fuzzy Systems*, *16*(4), 444–456.

Nan, J.-X., Zhang, M.-J., & Li, D.-F. (2014b). A methodology for matrix games with payoffs of triangular intuitionistic fuzzy number. *Journal of Intelligent & Fuzzy Systems*, *26*(6), 2899–2912.

Park, D. G., Kwun, Y. C., Park, J. H., & Park, I. Y. (2009). Correlation coefficient of interval-valued intuitionistic fuzzy sets and its application to multiple attribute group decision making problems. *Mathematical and Computer Modelling*, *50*(9-10), 1279–1293. doi:10.1016/j.mcm.2009.06.010

Power, D. J. (2013). *Engineering effective decision support technologies: New models and applications*. Hershey, PA: IGI Global. doi:10.4018/978-1-4666-4002-3

Robinson, J. P., & Amirtharaj, E. C. H. (2011a). A short primer on the Correlation coefficient of Vague sets. *International Journal of Fuzzy System Applications*, *1*(2), 55–69. doi:10.4018/ijfsa.2011040105

Robinson, J. P., & Amirtharaj, E. C. H. (2011b). Extended TOPSIS with correlation coefficient of Triangular Intuitionistic fuzzy sets for Multiple Attribute Group Decision Making. *International Journal of Decision Support System Technology*, *3*(3), 15–40. doi:10.4018/jdsst.2011070102

Robinson, J. P., & Amirtharaj, E. C. H. (2012a). Vague Correlation coefficient of Interval Vague sets. *International Journal of Fuzzy System Applications*, *2*(1), 18–34. doi:10.4018/ijfsa.2012010102

Robinson, J.P., & Amirtharaj, E.C.H. (2012b). A Search for the Correlation coefficient of Triangular and Trapezoidal intuitionistic Fuzzy sets for Multiple Attribute Group Decision Making. *Communications in Computer and Information Science 283,* 333-342.

Robinson, J. P., & Amirtharaj, E. C. H. (2014a). Efficient Multiple Attribute Group Decision Making models with Correlation coefficient of Vague sets. *International Journal of Operations Research and Information Systems, 5*(3), 27–49. doi:10.4018/ijoris.2014070102

Robinson, J. P., & Amirtharaj, E. C. H. (2014b). MAGDM-MINER: A new algorithm for mining trapezoidal intuitionistic fuzzy correlation rules. *International Journal of Decision Support System Technology, 6*(1), 34–59. doi:10.4018/ijdsst.2014010103

Szmidt, E., & Kacprzyk, J. (2000). Distances between intuitionistic fuzzy sets. *Fuzzy Sets and Systems, 114*(3), 505–518. doi:10.1016/S0165-0114(98)00244-9

Szmidt, E., & Kacprzyk, J. (2002). Using intuitionistic fuzzy sets in group decision making. *Control and Cybernetics, 31*, 1037–1053.

Szmidt, E., & Kacprzyk, J. (2003). A consensus-reaching process under intuitionistic fuzzy preference relations. *International Journal of Intelligent Systems, 18*(7), 837–852. doi:10.1002/int.10119

Wang, G., & Li, X. (1999). Correlation of information energy of interval valued fuzzy numbers. *Fuzzy Sets and Systems, 103*(1), 169–175. doi:10.1016/S0165-0114(97)00303-5

Wang, X. F. (2008). Fuzzy number intuitionistic fuzzy geometric aggregation operators and their application to decision making. *Control and Decision, 23*, 68–73.

Wei, G. (2010). Some arithmetic aggregation operators with intuitionistic trapezoidal fuzzy numbers and their application to group decision making. *Journal of Computers, 5*(3), 345–351. doi:10.4304/jcp.5.3.345-351

Xu, Z. S. (2005). An Overview of Methods for Determining OWA Weights. *International Journal of Intelligent Systems, 20*(8), 843–865. doi:10.1002/int.20097

Xu, Z. S. (2006). Induced uncertain linguistic OWA operators applied to group decision making. *Information Fusion, 7*(2), 231–238. doi:10.1016/j.inffus.2004.06.005

Xu, Z. S. (2007a). Models for multiple attribute decision-making with intuitionistic fuzzy information. *International Journal of Uncertainty, Fuzziness and Knowledge-based Systems, 15*(3), 285–297. doi:10.1142/S0218488507004686

Xu, Z. S. (2007b). Intuitionistic fuzzy aggregation operators. *IEEE Transactions on Fuzzy Systems, 15*(6), 1179–1187. doi:10.1109/TFUZZ.2006.890678

Xu, Z. S. (2007c). Multi-person multi-attribute decision making models under intuitionistic fuzzy environment. *Fuzzy Optimization and Decision Making, 6*(3), 221–236. doi:10.1007/s10700-007-9009-7

Xu, Z. S. (2007d). Intuitionistic preference relations and their application in group decision making. *Information Science, 177*(11), 2363–2379. doi:10.1016/j.ins.2006.12.019

Xu, Z. S., & Chen, J. (2007). On geometric aggregation over interval-valued intuitionistic fuzzy information. In *Fourth International Conference on Fuzzy Systems and Knowledge Discovery (FSKD 2007)*, (vol. 2, pp. 466–471). doi:10.1109/FSKD.2007.427

Xu, Z. S., & Da, Q. L. (2003). An overview of operators for aggregating the information. *International Journal of Intelligent Systems*, *18*(9), 953–969. doi:10.1002/int.10127

Xu, Z. S., & Yager, R. R. (2006). Some geometric aggregation operators based on Intuitionistic Fuzzy sets. *International Journal of General Systems*, *35*(4), 417–433. doi:10.1080/03081070600574353

Yager, R. R. (1988). On ordered weighted averaging aggregation operators in multi-criteria decision making. *IEEE Transactions on Systems, Man, and Cybernetics*, *18*(1), 183–190. doi:10.1109/21.87068

Yager, R. R. (1993). Families of OWA operators. *Fuzzy Sets and Systems*, *59*(2), 125–148. doi:10.1016/0165-0114(93)90194-M

Yager, R. R. (2004). Generalized OWA aggregation operators. *Fuzzy Optimization and Decision Making*, *3*(1), 93–107. doi:10.1023/B:FODM.0000013074.68765.97

Yager, R. R. (2007). Centered OWA operators. *Soft Computing*, *11*(7), 631–639. doi:10.1007/s00500-006-0125-z

Yager, R. R., & Filev, D. P. (1999). Induced ordered weighted averaging operators. *IEEE Transactions on Systems, Man, and Cybernetics. Part B, Cybernetics*, *29*(2), 141–150. doi:10.1109/3477.752789 PMID:18252288

Yang, J., Fei, W., & Li, D.-F. (2016b). Nonlinear programming approach to solve bi-matrix games with payoffs represented by I-fuzzy numbers. *International Journal of Fuzzy Systems*, *18*(3), 492–503. doi:10.1007/s40815-015-0052-1

Yang, J., Li, D.-F., & Lai, L.-B. (2016a). Parameterized bilinear programming methodology for solving triangular intuitionistic fuzzy number bi-matrix games. *International Journal of Fuzzy Systems*, *31*(1), 115–125. doi:10.3233/IFS-162125

Yu, C. (1993). Correlation of fuzzy numbers. *Fuzzy Sets and Systems*, *55*(3), 303–307. doi:10.1016/0165-0114(93)90256-H

Zeng, W., & Li, H. (2007). Correlation Coefficient of Intuitionistic Fuzzy sets. *Journal of Industrial Engineering International*, *3*, 33–40.

Zhang, X., & Liu, P. D. (2010). Method for Aggregating Triangular Fuzzy Intuitionistic Fuzzy Information and its Application to Decision making. *Technological and Economic Development of Economy. Baltic Journal of Sustainability*, *16*(2), 280–290.

APPENDIX

Correlation Coefficient of Intuitionistic Fuzzy Sets (IFSs)

Let $X = \{x_1, x_2,..., x_n\}$ be the finite universal set and IFS(X) denote the set of all IFSs in X. For each A, $B \in$ IFS(X), Gerstenkorn and Manko, (1991) defined the correlation of two IFSs as follows:

$$C_{GM}\left(A,\ B\right) = \sum_{i=1}^{n} \left[\mu_A\left(x_i\right).\ \mu_B\left(x_i\right) + \gamma_A\left(x_i\right).\ \gamma_B\left(x_i\right)\right]$$

Following this is given the method proposed by Zeng & Li, (2007) for calculating correlation coefficient of IFSs is used, taking only the membership and non-membership grades into account. For

$$A = \left\{\left(\mu_A\left(x\right), \gamma_A\left(x\right), \pi_A\left(x\right)\right)\ / x \in X\right\},\ B = \left\{\left(\mu_B\left(x\right), \gamma_B\left(x\right), \pi_B\left(x\right)\right)\ / x \in X\right\},$$

the correlation of the IFSs is defined as:

$$C_{ZL}\left(A, B\right) = \frac{1}{n}\sum_{i=1}^{n} \left[\mu_A\left(x_i\right)\mu_B\left(x_i\right) + \gamma_A\left(x_i\right)\gamma_B\left(x_i\right) + \pi_A\left(x_i\right)\pi_B\left(x_i\right)\right].$$

Furthermore, the correlation coefficient of the IFSs, A and B is given by:

$$\rho_{ZL}(A,\ B) = \frac{C_{ZL}(A,\ B)}{\sqrt{C_{ZL}(A, A)\ .\ C_{ZL}(B, B)}},\quad 0 \le \rho_{ZL}(A, B) \le 1.$$

Correlation Coefficient of Interval-Valued Intuitionistic Fuzzy Sets (IVIFSs)

The Bustince and Bulliro's, (1995) method for calculating the correlation coefficients between IVIFSs is given as follows:

Let $X = \{x_1, x_2, ..., x_n\}$ be the finite universal set and A, $B \in$ IVIFS(X) be given by

$$A = \left\{\left\langle x, \left[\ \mu_{AL}\left(x\right), \mu_{AU}\left(x\right)\right], \left[\ \gamma_{AL}\left(x\right), \gamma_{AU}\left(x\right)\right]\right\rangle / x \in X\right\},$$

$$B = \left\{\left\langle x, \left[\ \mu_{BL}\left(x\right), \mu_{BU}\left(x\right)\right], \left[\ \gamma_{BL}\left(x\right), \gamma_{BU}\left(x\right)\right]\right\rangle / x \in X\right\},$$

Bustince and Burillo, (1995) defined the correlation of A, $B \in$ IVIFS(X) as follows:

$$C_{BB}(A,B) = \frac{1}{2}\sum_{i=1}^{n} \begin{bmatrix} \mu_{AL}(x_i)\mu_{BL}(x_i) + \mu_{AU}(x_i)\mu_{BU}(x_i) \\ + \gamma_{AL}(x_i)\gamma_{BL}(x_i) + \gamma_{AU}(x_i)\gamma_{BU}(x_i) \end{bmatrix}$$

and defined the correlation coefficient between A and B by the formula:

$$K_{BB}(A,\ B) = \frac{C_{BB}(A,\ B)}{\sqrt{C_{BB}(A,\ A)\ .\ C_{BB}(B,\ B)}}$$

For A, $B \in$ IVIFS(X), the correlation coefficient satisfies:

1. If $A = B$ then $K_{BB}(A, B) = 1$.
2. $K_{BB}(A, B) = K_{BB}(B, A)$.
3. $0 \leq K_{BB}(A, B) \leq 1$.

The Park et al., (2009) method of correlation coefficient of IVIFSs is an improved method compared to the method proposed by Bustince and Burillo, (1995). Park et al. (2009) defined the correlation of A, $B \in$ IVIFS(X) including the hesitation degree as follows:

For each $A \in$ IVIFS(X), the informational intuitionistic energy of A is defined as:

$$E_{PK}(A) = \frac{1}{2}\sum_{i=1}^{n}\left[\mu^2_{AL}(x_i) + \mu^2_{AU}(x_i) + \gamma^2_{AL}(x_i) + \gamma^2_{AU}(x_i) + \pi^2_{AL}(x_i) + \pi^2_{AU}(x_i)\right].$$

The correlation of A and B is given by the formula:

$$C_{PK}(A,B) = \frac{1}{2}\sum_{i=1}^{n} \begin{bmatrix} \mu_{AL}(x_i)\mu_{BL}(x_i) + \mu_{AU}(x_i)\mu_{BU}(x_i) + \gamma_{AL}(x_i)\gamma_{BL}(x_i) \\ + \gamma_{AU}(x_i)\gamma_{BU}(x_i) + \pi_{AL}(x_i)\pi_{BL}(x_i) + \pi_{AU}(x_i)\pi_{BU}(x_i) \end{bmatrix}.$$

For A, $B \in$ IVIFS(X), the correlation has the following properties:

1. $C_{PK}(A, A) = E(A)$.
2. $C_{PK}(A, B) = C_{PK}(B, A)$.

Furthermore, the correlation coefficient of A and B is defined by the formula

$$K_{PK}(A,\ B) = \frac{C_{PK}(A,\ B)}{\sqrt{E_{PK}(A)\ .\ E_{PK}(B)}}$$

Correlation Coefficient of Vague Sets

The Robinson and Amirtharaj, (2011a) method for calculating the correlation coefficients between vague sets is given as follows:

Let $X = \{x_1, x_2, \ldots, x_n\}$ be the finite universal set and $A, B \in VS(X)$ be given by

$$A = \left\{\left\langle x, \left[t_A(x), 1 - f_A(x)\right]\right\rangle \ / x \in X\right\},$$

$$B = \left\{\left\langle x, \left[t_B(x), 1 - f_B(x)\right]\right\rangle \ / x \in X\right\}.$$

And the length of the vague values are given by

$\pi_A(x) = 1 - t_A(x) - f_A(x)$, $\pi_B(x) = 1 - t_B(x) - f_B(x)$.

Now for each $A, B \in VS(X)$, the informational vague energy is defined as follows:

$$E_{VS}(A) = \frac{1}{n}\sum_{i=1}^{n}\left[t_A^2(x_i) + \left(1 - f_A(x_i)\right)^2 + \pi_A^2(x_i)\right],$$

$$E_{VS}(B) = \frac{1}{n}\sum_{i=1}^{n}\left[t_B^2(x_i) + \left(1 - f_B(x_i)\right)^2 + \pi_B^2(x_i)\right],$$

The correlation of A and B is given by the formula:

$$C_{VS}(A, B) = \frac{1}{n}\sum_{i=1}^{n}\left[t_A(x_i)t_B(x_i) + \left(1 - f_A(x_i)\right)\left(1 - f_B(x_i)\right) + \pi_A(x_i)\pi_B(x_i)\right],$$

Furthermore, the correlation coefficient of A and B is defined by the formula:

$$K_{VS}(A, B) = \frac{C_{VS}(A, B)}{\sqrt{E_{VS}(A) \cdot E_{VS}(B)}},$$

where $0 \leq K_{VS}(A,B) \leq 1$.

Correlation Coefficient of Triangular and Trapezoidal IFS

The Robinson ans Amirtharaj (2012b) method for calculating the correlation coefficients between Triangular IFSs is given as follows:

Let $A = \left([a_1, b_1, c_1]; \mu_A, \gamma_A\right)$, $B = \left([a_2, b_2, c_2]; \mu_B, \gamma_B\right)$ be two triangular IFSs.

Then for each $A, B \in$ TrIFS(X), the informational triangular intuitionistic energy of A and B is defined as follows:

$$E_{TrIFS}(A) = \frac{1}{n}\sum_{i=1}^{n}\left[\frac{a_1 + 4b_1 + c_1}{6}\right]^2 \left(\mu_A^{\ 2}(x_i) + \gamma_A^{\ 2}(x_i) + \pi_A^{\ 2}(x_i)\right)$$

and

$$E_{TrIFS}(B) = \frac{1}{n}\sum_{i=1}^{n}\left[\frac{a_2 + 4b_2 + c_2}{6}\right]^2 \left(\mu_B^{\ 2}(x_i) + \gamma_B^{\ 2}(x_i) + \pi_B^{\ 2}(x_i)\right).$$

The correlation of A and B is defined as:

$$C_{TrIFS}(A, B) = \frac{1}{n}\sum_{i=1}^{n}\left[\frac{a_1 + 4b_1 + c_1}{6}\right]\left[\frac{a_2 + 4b_2 + c_2}{6}\right]\left(\mu_A(x_i)\mu_B(x_i) + \gamma_A(x_i)\gamma_B(x_i) + \pi_A(x_i)\pi_B(x_i)\right)$$

Then the correlation coefficient between A and B is defined as:

$$K_{TrIFS}(A, B) = \frac{C_{TrIFS}(A, B)}{\sqrt{E_{TrIFS}(A).E_{TrIFS}(B)}},$$

where $0 \le K_{TrIFS}(A,B) \le 1$.

Similarly the correlation coefficient of trapezoidal intuitionistic fuzzy sets can be defined. Let $A = \left([a_1, b_1, c_1, d_1]; \mu_A, \gamma_A\right)$, $B = \left([a_2, b_2, c_2, d_2]; \mu_B, \gamma_B\right)$ be two trapezoidal IFSs (TzIFS). Then for each $A, B \in$ TzIFS(X), the informational trapezoidal intuitionistic energy of A and B is defined as follows:

$$E_{TzIFS}(A) = \frac{1}{n}\sum_{i=1}^{n}\left[\frac{a_1 + 2b_1 + 2c_1 + d_1}{6}\right]^2 \left(\mu_A^{\ 2}(x_i) + \gamma_A^{\ 2}(x_i) + \pi_A^{\ 2}(x_i)\right)$$

and

$$E_{TzIFS}(B) = \frac{1}{n}\sum_{i=1}^{n}\left[\frac{a_2 + 2b_2 + 2c_2 + d_2}{6}\right]^2 \left(\mu_B^{\ 2}(x_i) + \gamma_B^{\ 2}(x_i) + \pi_B^{\ 2}(x_i)\right).$$

Now the correlation of A and B is defined as:

$$C_{TzIFS}(A,B) = \frac{1}{n}\sum_{i=1}^{n}\left[\frac{a_1+2b_1+2c_1+d_1}{6}\right]\left[\frac{a_2+2b_2+2c_2+d_2}{6}\right]\left(\mu_A(x_i)\mu_B(x_i)+\gamma_A(x_i)\gamma_B(x_i)+\pi_A(x_i)\pi_B(x_i)\right)$$

Then the correlation coefficient between A and B is defined as:

$$K_{TzIFS}(A,B) = \frac{C_{TzIFS}(A,B)}{\sqrt{E_{TzIFS}(A).E_{TzIFS}(B)}},$$

where $0 \leq K_{TzIFS}(A,B) \leq 1$.

Chapter 8
Credibility Hypothesis Testing of Variance of Fuzzy Normal Distribution

S. Sampath
University of Madras, India

B. Ramya
University of Madras, India

ABSTRACT

This paper considers the problem of developing test procedures for testing credibility hypotheses about the variance of fuzzy normal distribution assuming the expected values of the distributions mentioned under null and alternative credibility hypotheses are known and equal. The cases where the underlying hypothesis is simple and composite (one sided) are considered. Tests have been derived with the help of the membership ratio criterion. Properties possessed by the developed tests, like best credibility rejection region and uniformly best rejection region have been studied. Examples are also given to illustrate the usage of the derived tests.

INTRODUCTION

A first study on Credibility Theory which is an analogue of Probability Theory to study subjective uncertainty has been carried out by Liu and Liu (2002). Credibility theory is considered as an important branch of mathematics. Further refinements on Credibility theory has been carried out by Liu (2008). Researchers have considered two different dimensions of Credibility theory, namely,

- Theoretical and
- Applied aspects.

DOI: 10.4018/978-1-5225-1848-8.ch008

Theoretical developments such as:

- Inequalities of fuzzy variables,
- Concept of convergence,
- Fundamental theorems,
- Independence of fuzzy variables,
- Fuzzy dominance and
- The concept of entropy, have been studied by various researchers.

On the other hand, the theory related to credibility theory has been applied in practical situations. For example, a study of:

- Modeling parallel machine schedule problem,
- Vehicle routing problem,
- Project scheduling problem,
- Quantity inventory model,
- Vendor selection problem in a supply chain were considered by a group of researchers.

Sampath and Ramya (2015) have a given a brief listing of works done in connection with applications of Credibility Theory. Sampath and Vaidyanathan (2009), Sampath and Kalaivani (2010) and Sampath and Senthilkumar (2014) applied credibility theory in mining of fuzzy datasets. Sampath (2009) and Sampath and Deepa (2013) considered the application of credibility theory in fuzzy statistical quality control.[1]Wan and Li (2013) and Wan et al. (2013) used Triangular intuitionistic fuzzy numbers to deal with ill-known quantities in decision data and decision making problems.

Huge volume of works has been done on theoretical aspects of credibility theory. Even though several studies have been made, one can feel the dearth of studies on inferential aspects of credibility theory. However, some studies have been made in Uncertainty theory which is a theory closely related to credibility theory. Estimation of the parameters involved in Uncertainty distributions have been considered by Wang and Peng (2012), and Wang et al. (2012). Wang et al. (2012a) proposed the concept of testing of hypothesis under uncertain situations. Further, Sampath and Ramya (2013) introduced a distance based approach for testing Uncertainty hypothesis. Sampath and Ramya (2015) developed membership ratio criterion for the problem of testing hypothesis about credibility distributions and studied important properties of test yielded by the membership ratio criterion with reference to the parameters of triangular credibility distribution. Sampath and Ramya (2014) considered tests for testing hypothesis about the expected value of fuzzy normal distributions and identified their optimal properties assuming the variances under null and alternative hypotheses are known and equal. In this chapter, the question of developing credibility tests for testing hypothesis about the variance of fuzzy normal distribution is considered when the expected values of the distributions under null and alternative credibility hypotheses are equal and known. The cases where the null and alternative credibility hypotheses are simple as well as composite (one sided) are considered.

BACKGROUND

Traditionally researchers who attempt to model a given situation treat tools meant for the purpose of reasoning, modelling and computing are crisp, deterministic and precise in character. Crispness means there is no state in between true and false regarding the status of an entity which is being considered thus leaving no scope for some state which lies between true and false. For example, while studying a manufacturing system, crispness assumes each one of the produced items should be clearly declared as either defective or non defective without leaving any scope for assigning a state in between defective and non defective. Preciseness assumes the parameters involved in the modelling process represent exactly the given system without any ambiguity. Deterministic property eliminates the scope for randomness leaving no doubt regarding values assumed by the appropriate entities. However, in real life situations these assumptions are not justified. In the opinion of Zadeh, "As the complexity of a system increases, our ability to make precise and yet significant statements about its behavior diminishes until a threshold is reached beyond which precision and significance (or relevance) become almost mutually exclusive characteristics". Probability theory has been widely used to model uncertain situations. However, many competing theories are available in the literature to model environments possessing uncertainties of different types. Fuzzy set theory is one such model based on classical set theory as well as dual logic. The first publication on fuzzy set theory was by Zadeh (1965). Fuzzy set theory provides a strict mathematical framework for vigorously studying vague conceptual phenomena. In the words of Zadeh (1965), "The notion of a fuzzy set provides a convenient point of departure for the construction of a conceptual framework which parallels in many respects the framework used in the case of ordinary sets, but is more general than the latter and, potentially, may prove to have a much wider scope of applicability, particularly in the fields of pattern classification and information processing. Essentially, such a framework provides a natural way of dealing with problems in which the source of imprecision is the absence of sharply defined criteria of class membership rather than the presence of random variables". In order to make rigorous studies on fuzzy events Zadeh (1978) introduced possibility measure using fuzzy set theory. This can be compared to the role played by measure theory in the development of probability theory. Possibility distributions and fuzzy variables are developed in the theory of possibility in a manner comparable with the development of probability distributions under random variables. Possibility theory was introduced to allow a reasoning to be carried out on imprecise or vague knowledge, making it possible to deal with uncertainties on this knowledge. Possibility is normally associated with some fuzziness, either in background knowledge on which possibility is based, or in the set for which possibility is asserted. Possibility measure lacks self-duality property which is essential for both theoretical and practical situations. Incorporating self-duality property, Liu and Liu (2002) proposed Credibility measure and developed a new theory called Credibility Theory through possibility theory. Further refinements on credibility theory were carried out by Liu (2008). In order to meet self duality requirement, Liu and Liu (2002) assigned for a non fuzzy set A, a measure called credibility measure based on the average of possibility and necessity measures.

The paper is organized as follows. The second section gives brief introduction to credibility theory. Certain definitions related to testing of credibility hypothesis are also presented in the same section. The third section describes the normal fuzzy distribution and reviews membership ratio criterion test developed by Sampath and Ramya (2014) for testing credibility hypothesis about the expected value. The fourth section is devoted for developing membership ratio criterion test for testing simple and composite credibility hypotheses about variance of normal credibility distribution assuming the expected value of the credibility distribution. Illustrative examples are also provided to illustrate the proposed test.

PRELIMINARIES ON CREDIBILITY THEORY AND TESTING OF CREDIBILITY HYPOTHESIS

In this section, a brief introduction to Credibility Theory (Liu and Liu, 2002; Liu, 2008) and also certain concepts related to the Testing of Credibility Hypotheses (Sampath & Ramya, 2015) are presented.

Let Θ be a nonempty set, P be the power set of Θ. Elements of P are called events. $Cr\left\{A\right\}$ indicates the credibility of occurrence of event A which is a number assigned to each event A that satisfies the axioms normality, monotonicity, self-duality and maximality.

$$Cr\left\{\Theta\right\}=1$$

$$Cr\left\{A\right\}\leq Cr\left\{B\right\} \text{ whenever } A\subset B$$

$$Cr\left\{A\right\}+Cr\left\{A^c\right\}=1 \text{ for any event } A$$

$$Cr\left\{\bigcup_i A_i\right\}=\sup_i\left(Cr\left\{A_i\right\}\right) \text{ for any events } \left\{A_i\right\} \text{ with } \sup_i\left(Cr\left\{A_i\right\}\right)<0.5$$

The triplet $\left(\Theta,\mathrm{P},Cr\right)$ is known as credibility measure space. A measurable function from the credibility measure space $\left(\Theta,\mathrm{P},Cr\right)$ to the set of real numbers is known as fuzzy variable.

Membership Function

The membership function μ of a fuzzy variable ξ defined on the credibility measure space $\left(\Theta,\mathrm{P},Cr\right)$ is defined as $\mu\left(x\right)=\left(2Cr\left\{\xi=x\right\}\right)\ \wedge\ 1,\ x\in R$

Credibility Distribution

The credibility distribution $\Phi:R\to\left[0,1\right]$ is defined as

$$\Phi\left(x\right)=Cr\left\{\theta\in\Theta\mid\xi\left(\theta\right)\leq x\right\}, \text{ for all } x\in R.$$

The support of the credibility distribution is the collection of all possible values assumed by the fuzzy variable with nonzero credibility.

In real life situations, assigning a suitable value to the credibility parameters (constants which appear in the credibility distribution of a fuzzy variable) in appropriate manner is a difficult task. Hence, Sampath and Ramya (2015) developed a mechanism for selecting a suitable value and proposed the credibility test for credibility parameters involved in triangular credibility distribution. Further, Sampath and Ramya (2014) identified the credibility rejection region for the expected value of the fuzzy normal distribution assuming variance is equal and known. Sampath and Ramya (2015a) investigated the properties of the test resulting from the application of membership ratio criterion when the competing hypotheses are statements about fuzzy exponential distribution.

To facilitate the readability of the paper, certain definitions, used by Sampath and Ramya (2014, 2015) related to credibility hypothesis testing are presented below.

Credibility Hypotheses and Test

A statement about the credibility distribution of a fuzzy variable is known as credibility hypotheses.

If a credibility hypothesis completely describes a credibility distribution then it is said to be a simple credibility hypothesis. On the other hand, it is known as a composite credibility hypothesis.

Credibility test is a rule which helps to accept or reject a credibility null hypothesis once the experimental value(s) of the underlying fuzzy variable(s) is known.

Credibility Rejection Region

The Credibility Rejection Region C is the set of points which leads to rejection of the credibility null hypothesis. Mathematically, the credibility rejection region is defined as

$$C = \left\{\xi \in \Theta \mid H_0 \text{ is rejected}\right\}$$

Type I and Type II Errors

Type I error is the action of rejecting the credibility null hypothesis when it is true and Type II error is the action of accepting the credibility null hypothesis when it is false.

Level of Significance

The maximum allowable credibility to commit Type I error by a credibility test is known as its level of significance and is denoted by α. The exact credibility of committing Type I error is known as the size of the test.

Power of a Credibility Test

The power of the credibility test with rejection region C^* is defined as $Cr\left(\xi \in C^*\right)$.

Best Credibility Rejection Region

A credibility rejection region C^* is said to be best credibility rejection region of level α if its power is maximum under credibility alternative hypothesis among all credibility rejection regions of level α. That is, C^* is the best credibility rejection region, if

$$Cr\left(\xi \in C^* \mid H_1\right) \geq Cr\left(\xi \in C \mid H_1\right) \tag{1}$$

where C is any region satisfying the condition

$$Cr\left(\xi \in C \mid H_0\right) \leq \alpha$$

Best Credibility Test

The rule that identifies the best credibility rejection region is known as best credibility test.

Membership Ratio Criterion

For testing the simple credibility null hypothesis, $H_0 : \xi$ has credibility distribution ξ_0 against the simple credibility alternative hypothesis $H_1 : \xi$ has credibility distribution ξ_1, the membership ratio criterion is

"Reject the credibility null hypothesis if the observed value of the fuzzy variable

$$\xi \in \left\{ x \mid \frac{\mu_2(x)}{\mu_1(x)} > k \right\} \tag{2}$$

where μ_1 and μ_2 are the membership functions corresponding to the credibility distributions ξ_0 and ξ_1 respectively; k being a constant selected so that

$$Cr\left(\xi \in C \mid H_0\right) = \alpha \tag{3}$$

The above definitions hold good as long as the null and alternative hypotheses are simple. When they become composite, we need to propose the optimal test. The Uniformly Best Credibility Rejection Region, defined below can be used to identify the optimal test when the hypotheses are composite.

Uniformly Best Credibility Rejection Region

A credibility rejection region C^* is said to be uniformly best credibility rejection region of level α if its power is maximum for all distributions considered under credibility alternative hypothesis among all credibility rejection regions of level α. That is, C^* is the uniformly best credibility rejection region, if

$$Cr\left(\xi \in C^{*} \mid H_{1}\right) \geq Cr\left(\xi \in C \mid H_{1}\right) \tag{4}$$

holds good for all distributions defined under the composite alternative hypothesis H_1, where C is any region satisfying the condition

$Cr\left(\xi \in C \mid H_{0}\right) \leq \alpha$ for all distributions defined under the composite null hypothesis H_0.

The following section describes the fuzzy normal distribution and gives a brief discussion on the results obtained by Sampath and Ramya (2014) in connection with testing of credibility hypothesis about the expectation of fuzzy normal distribution assuming the variance is known.

NORMAL FUZZY VARIABLE AND ITS CREDIBILITY HYPOTHESIS TEST ABOUT THE EXPECTED VALUE

Membership functions describe the characteristics of the fuzzy variables and they determine their credibility distributions. Most popular and widely studied credibility distributions are equipossible, triangular, trapezoidal, exponential and normal. Normal membership functions find its application where the membership function attains its maximum when the fuzzy variable assumes the value exactly equal to a real number e (expected value of normal credibility distribution) and decreasingly smaller values as the fuzzy variable tends to either $-\infty$ or $+\infty$.

A fuzzy variable ξ is said to have normal credibility distribution if its membership function is defined as

$$\mu(x) = \frac{2}{1+e^{\frac{\pi|x-e|}{\sqrt{6}\sigma}}}, x, e \in R, \sigma > 0 \tag{5}$$

Figure 1 describes the shape of the normal membership function. The normal membership curve is symmetric about e, the expected value. That is, the spread of fuzzy normal curve is equal on both sides of e and its variance is denoted as σ^2. The notation ξ ~ $\mathrm{N}(e,\sigma)$ indicates that the fuzzy variable ξ has fuzzy normal distribution with expected value e and variance σ^2. The unknown constants are termed as credibility parameters (Sampath & Ramya,2014) involved in normal membership functions are e and σ. Here, the credibility parameters e and σ decides the position of peakedness and the shape of the normal membership curve respectively. Figure 1 describes the shape of the normal credibility distribution.

Sampath and Ramya (2014) derived some important propositions related to fuzzy normal distribution that are relevant in our study. The propositions are presented below for easy reference. The proofs of the stated propositions are provided in the appendix.

Figure 1. Fuzzy normal curve

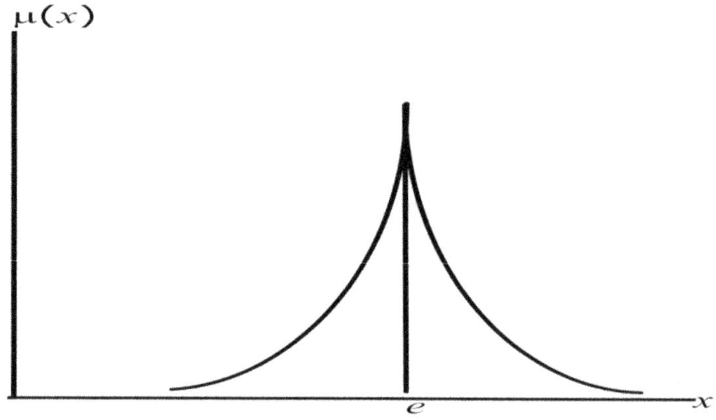

Proposition 1: If $\xi \sim N(e,\sigma)$ and $\alpha < \frac{1}{2}$ then all intervals of the form $\left(e + \frac{\sqrt{6}\sigma}{\pi}\ln\left(\frac{1-\alpha}{\alpha}\right), x\right)$ will have the credibility value α.

Proposition 2: If $\xi \sim N(e,\sigma)$ and $\alpha < \frac{1}{2}$ then all intervals of the form $\left(x, e - \frac{\sqrt{6}\sigma}{\pi}\ln\left(\frac{1-\alpha}{\alpha}\right)\right)$ will have credibility value α.

Proposition 3: If $\xi \sim N(e,\sigma)$ then all intervals of the form (e, x), $x > e$ will have credibility equal to $\frac{1}{2}$.

Proposition 4: If $\xi \sim N(e,\sigma)$ then all intervals of the form (x, e), $x < e$ will have credibility equal to $\frac{1}{2}$.

Proposition 5: If $\xi \sim N(e,\sigma)$ then all intervals of the form $\left(x, e + \frac{\sqrt{6}\sigma}{\pi}\ln\left(\frac{1-\alpha}{\alpha}\right)\right)$ where $x > e$ will have credibility value greater than α but less than $\frac{1}{2}$.

Proposition 6: If $\xi \sim N(e,\sigma)$ then all intervals of the form $\left(e - \frac{\sqrt{6}\sigma}{\pi}\ln\left(\frac{1-\alpha}{\alpha}\right), x\right)$ where $x < e$ will have credibility value greater than α but less than $\frac{1}{2}$.

Proposition 7: If $\xi \sim N(e,\sigma)$ then any interval of the form (x_0, x_1), $x_0 < e$ and $x_1 > e$ will have credibility greater than $\frac{1}{2}$.

Sampath and Ramya (2014) considered the problem of testing credibility hypothesis about the expected value e of the fuzzy normal distribution. They identified the credibility rejection region by using the membership ratio criterion for testing $H_0 : \xi \sim \mathrm{N}(e_1,\sigma)$ against $H_1 : \xi \sim \mathrm{N}(e_2,\sigma)$ when σ is known assuming $\alpha < \frac{1}{2}$. The credibility test identified by the membership ratio criterion for the cases $e_1 < e_2$ and $e_1 > e_2$ are

$$C^* = \left\{ x \mid x > e_1 + \frac{\sqrt{6}\sigma}{\pi} \ln\left(\frac{1-\alpha}{\alpha}\right) \right\}$$

and

$$C^* = \left\{ x \mid x < e_1 - \frac{\sqrt{6}\sigma}{\pi} \ln\left(\frac{1-\alpha}{\alpha}\right) \right\} \text{respectively.}$$

The optimal properties observed by the above credibility tests are

1. the credibility of the rejection region C^* under alternative hypothesis is always greater than α.
2. The credibility rejection region C^* is the best credibility rejection region of level α than any other credibility rejection region of level α.
3. The credibility rejection region C^* found to be one sided region for both the cases $e_1 < e_2$ and $e_1 > e_2$.

The following section discusses the properties of the tests obtained using the membership ratio criterion when testing hypothesis about the variance of fuzzy normal distribution.

TESTING OF VARIANCE OF NORMAL CREDIBILITY DISTRIBUTION

In this section, a test based on the membership ratio criterion for testing a hypothesis about the variance of fuzzy normal distribution is considered and its properties are examined. Consider the problem of testing the credibility null hypothesis that the fuzzy variable ξ has fuzzy normal with variance σ_1^2 against the credibility alternative hypothesis that the fuzzy variable ξ has fuzzy normal with variance σ_2^2. The following theorem gives the resulting credibility rejection region as well as its properties when membership ratio criterion is used under the case $\sigma_1 > \sigma_2$, assuming e is known.

Theorem 1: For testing the null credibility hypothesis $H_0 : \xi \sim \mathrm{N}\left(e,\sigma_1\right)$ against the alternative credibility hypothesis $H_1 : \xi \sim \mathrm{N}\left(e,\sigma_2\right)$, where $\sigma_1 > \sigma_2$ (when e is known)

Membership ratio criterion credibility rejection region of level $\alpha > \frac{1}{2}$ is given by

$$C^* = \left\{ x \mid e - \frac{\sqrt{6}\sigma_1}{\pi} \ln\left(\frac{\alpha}{1-\alpha}\right) < x < e + \frac{\sqrt{6}\sigma_1}{\pi} \ln\left(\frac{\alpha}{1-\alpha}\right) \right\}$$

b. $Cr\left(\xi \in C^* \mid H_1\right) > \alpha$ and

The credibility rejection region

$$C^* = \left\{ x \mid e - \frac{\sqrt{6}\sigma_1}{\pi} \ln\left(\frac{\alpha}{1-\alpha}\right) < x < e + \frac{\sqrt{6}\sigma_1}{\pi} \ln\left(\frac{\alpha}{1-\alpha}\right) \right\}$$

is the best credibility rejection region of level $\alpha > \frac{1}{2}$.

Proof:

For the case $\sigma_1 > \sigma_2$, the ratio $\frac{\mu_2(x)}{\mu_1(x)} = \frac{1 + e^{\frac{\pi|x-e|}{\sqrt{6}\sigma_1}}}{1 + e^{\frac{\pi|x-e|}{\sqrt{6}\sigma_2}}}$ is concave downwards in x and symmetric about e. Hence the set identified by membership ratio criterion, namely, $\left\{ x \mid \frac{\mu_2(x)}{\mu_1(x)} > k \right\}$ reduces to

$$C^* = \left\{ x \mid e - c < x < e + c \right\}$$

where c is selected such that

$$Cr\left(\xi \in C^* \mid H_0\right) = \alpha .$$

This implies,

$$\frac{1}{2}\left[\underset{x \in (e-c, e+c)}{Sup} \mu_1(x) + 1 - \underset{x \notin (e-c, e+c)}{Sup} \mu_1(x) \right] = \alpha$$

$$\Rightarrow 1 - \frac{1}{1 + e^{\frac{\pi c}{\sqrt{6}\sigma_1}}} = \alpha \tag{6}$$

Solving for c, we get

$$c = \frac{\sqrt{6}\sigma_1}{\pi} \ln\left(\frac{\alpha}{1-\alpha}\right)$$

Thus, (a) is proved.

b. The credibility of C^* under alternative hypothesis is

$$Cr\left(\xi \in C^* \mid H_1\right) = \frac{1}{2}\left[\underset{x\in(e-c,e+c)}{Sup}\mu_2\left(x\right) + 1 - \underset{x\notin(e-c,e+c)}{Sup}\mu_2\left(x\right)\right]$$

$$= \frac{1}{2}\left[1 + 1 - \frac{2}{1+e^{\frac{\pi c}{\sqrt{6}\sigma_2}}}\right]$$

$$= 1 - \frac{1}{1+e^{\frac{\pi c}{\sqrt{6}\sigma_2}}} \tag{7}$$

Using (5) and (6), we note that,

$$Cr\left(\xi \in C^* \mid H_1\right) - \alpha = 1 - \frac{1}{1+e^{\frac{\pi c}{\sqrt{6}\sigma_2}}} - 1 + \frac{1}{1+e^{\frac{\pi c}{\sqrt{6}\sigma_1}}}$$

$$= \frac{1}{1+e^{\frac{\pi(e-c_0)}{\sqrt{6}\sigma_1}}} - \frac{1}{1+e^{\frac{\pi(e-c_0)}{\sqrt{6}\sigma_2}}}$$

Thc assumption $\sigma_1 > \sigma_2$, implies $1 + e^{\frac{\pi c}{\sqrt{6}\sigma_1}} < 1 + e^{\frac{\pi c}{\sqrt{6}\sigma_2}}$

Therefore, $\frac{1}{1+e^{\frac{\pi c}{\sqrt{6}\sigma_1}}} > \frac{1}{1+e^{\frac{\pi c}{\sqrt{6}\sigma_2}}}$

Hence, $Cr\left(\xi \in C^* \mid H_1\right) > \alpha$.

In order to prove that

$$C^* = \left\{x \mid e - \frac{\sqrt{6}\sigma_1}{\pi}\ln\left(\frac{\alpha}{1-\alpha}\right) < x < e + \frac{\sqrt{6}\sigma_1}{\pi}\ln\left(\frac{\alpha}{1-\alpha}\right)\right\}$$

is the best credibility rejection region of level α, we must show that

$$Cr\left(\xi \in C^* \mid H_1\right) \geq Cr\left(\xi \in A^* \mid H_1\right) \tag{8}$$

for any A^* satisfying the condition

$$Cr\left(\xi \in A^* \mid H_0\right) \leq \alpha .$$

Now we shall identify possible regions of level α with the help of the propositions stated earlier.

1. Note that credibility of any subset of $\left\{x \mid e < x < \infty\right\}$ is always less than $\frac{1}{2}$. Hence any member of F_1 where F_1 is the collection of all intervals of the form $\left\{x \mid x \in \left(x_0, x_1\right), e < x_0 < x_1 < \infty\right\}$ will have size less than α.
2. Similarly, any member of F_2 which is the collection of all intervals of the form $\left\{x \mid x \in \left(x_0, x_1\right), -\infty < x_0 < x_1 < e\right\}$ will also have size less than α.
3. Let F_3 be the collection of all intervals of the form $\left\{x \mid x \in \left(x_0, x_1\right)\right\}$ where

$$e - \frac{\sqrt{6}\sigma_1}{\pi} \ln\left(\frac{\alpha}{1-\alpha}\right) < x_0 < x_1 < e + \frac{\sqrt{6}\sigma_1}{\pi} \ln\left(\frac{\alpha}{1-\alpha}\right).$$

By monotonocity of credibility measure, members of this class will have credibility value less than α.

Now we shall prove (8) by considering A^* of different forms as identified in F_1, F_2, F_3 and the unions of members of F_1, F_2 and/or F_3.

Case 1: $A^* \in F_1$

Consider an interval $\left(x_0, x_1\right) \in \left(e, \infty\right)$, as shown in the Figure 2. Then

$$Cr\left(\xi \in A^* \mid H_1\right) = Cr\left(\xi \in \left(x_0, x_1\right) \mid H_1\right)$$

$$= \frac{1}{2}\left[\underset{x \in \left(x_0, x_1\right)}{Sup}\, \mu_2\left(x\right) + 1 - \underset{x \notin \left(x_0, x_1\right)}{Sup}\, \mu_2\left(x\right)\right]$$

$$= \frac{1}{2}\left[\frac{2}{1 + e^{\frac{\pi\left(x_0 - e\right)}{\sqrt{6}\sigma_2}}} + 1 - 1\right]$$

Figure 2. Interval for the case A in F_1*

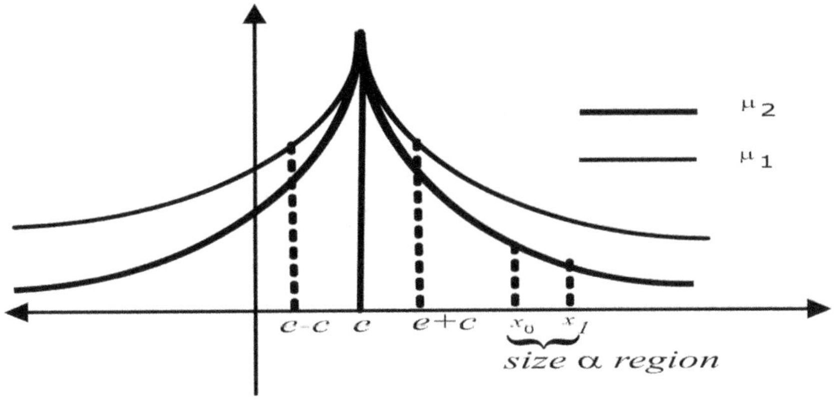

Therefore, $Cr\left(\xi \in A^* \mid H_1\right) = \dfrac{1}{1+e^{\frac{\pi(x_0-e)}{\sqrt{6}\sigma_2}}}$

Consider the difference, $Cr\left(\xi \in C^* \mid H_1\right) - Cr\left(\xi \in A^* \mid H_1\right)$

$$= 1 - \frac{1}{1+e^{\frac{\pi c}{\sqrt{6}\sigma_2}}} - \frac{1}{1+e^{\frac{\pi(x_0-e)}{\sqrt{6}\sigma_2}}}$$

Since $e < e + c$, we have $0 < \dfrac{\pi c}{\sqrt{6}\sigma_2}$

Hence, $2 < 1 + e^{\frac{\pi c}{\sqrt{6}\sigma_2}}$

$$\Rightarrow \frac{1}{1+e^{\frac{\pi c}{\sqrt{6}\sigma_2}}} < \frac{1}{2}$$

Further, the condition $e < x_0$, implies $\dfrac{1}{1+e^{\frac{\pi(e-x_0)}{\sqrt{6}\sigma_2}}} < \dfrac{1}{2}$.

Hence, we conclude that the second and third terms in the right hand side of the difference considered above are less than $\dfrac{1}{2}$.

Therefore,

$$Cr\left(\xi \in C^* \mid H_1\right) - Cr\left(\xi \in A^* \mid H_1\right) \geq 0$$

Figure 3. Interval for the case A^* in F_2

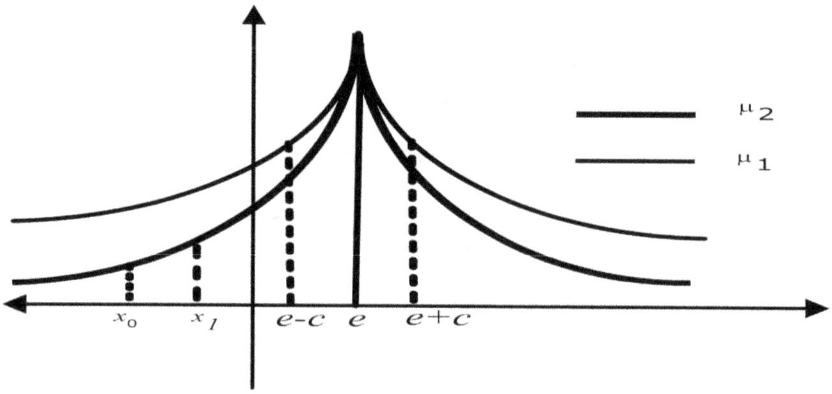

Case 2: $A^* \in F_2$

Consider an interval $\left(x_0, x_1\right) \in \left(-\infty, e\right)$, as shown in Figure 3.
In this case,

$$Cr\left(\xi \in A^* \mid H_1\right) = \frac{1}{2}\left[\underset{x \in \left(x_0, x_1\right)}{Sup}\ \mu_2\left(x\right) + 1 - \underset{x \notin \left(x_0, x_1\right)}{Sup}\ \mu_2\left(x\right)\right]$$

$$= \frac{1}{2}\left[\frac{2}{1 + e^{\frac{\pi\left(e - x_1\right)}{\sqrt{6}\sigma_2}}} + 1 - 1\right]$$

Therefore, $Cr\left(\xi \in A^* \mid H_1\right) = \dfrac{1}{1 + e^{\frac{\pi\left(e - x_1\right)}{\sqrt{6}\sigma_2}}}$

Consider the difference,

$$Cr\left(\xi \in C^* \mid H_1\right) - Cr\left(\xi \in A^* \mid H_1\right) = 1 - \frac{1}{1 + e^{\frac{\pi c}{\sqrt{6}\sigma_2}}} - \frac{1}{1 + e^{\frac{\pi\left(e - x_1\right)}{\sqrt{6}\sigma_2}}}$$

From Figure 3 we have, $x_1 < e - c < e$

$$\Rightarrow \frac{\pi\left(e - x_1\right)}{\sqrt{6}\sigma_2} > \frac{\pi c}{\sqrt{6}\sigma_2} > 0$$

Raising the power to e and adding 1 on both sides, we get

$$1+e^{\frac{\pi(e-x_1)}{\sqrt{6}\sigma_2}} > 1+e^{\frac{\pi c}{\sqrt{6}\sigma_2}} > 2$$

Taking the reciprocals we get

$$\frac{1}{1+e^{\frac{\pi(e-x_1)}{\sqrt{6}\sigma_2}}} < \frac{1}{1+e^{\frac{\pi c}{\sqrt{6}\sigma_2}}} < \frac{1}{2}$$

Therefore,

$$Cr\left(\xi \in C^* \mid H_1\right) - Cr\left(\xi \in A^* \mid H_1\right) \geq 0.$$

Case 3: $A^* \in F_3$

Consider an interval

$$\left(x_0, x_1\right) \in \left(e - \frac{\sqrt{6}\sigma_1}{\pi}\ln\left(\frac{\alpha}{1-\alpha}\right), e + \frac{\sqrt{6}\sigma_1}{\pi}\ln\left(\frac{\alpha}{1-\alpha}\right)\right),$$

as shown in Figure 4.

In this case,

$$Cr\left(\xi \in A^* \mid H_1\right) = Cr\left(\xi \in \left(x_0, x_1\right) \mid H_1\right)$$

Figure 4. Interval for the case A in F_3*

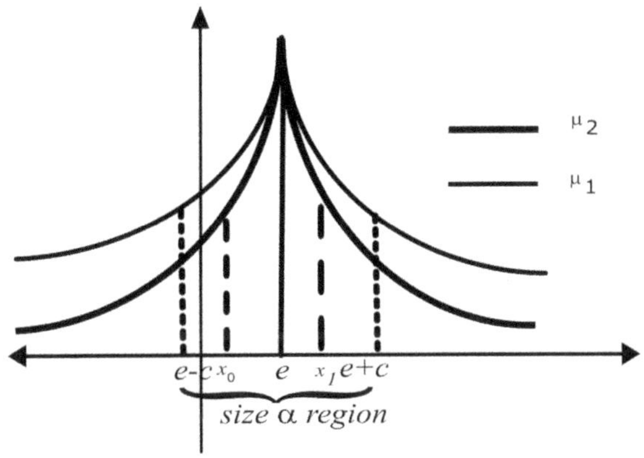

$$= \frac{1}{2}\left[\underset{x\in(x_0,x_1)}{Sup}\mu_2(x) + 1 - \underset{x\notin(x_0,x_1)}{Sup}\mu_2(x)\right]$$

$$= \frac{1}{2}\left[1 + 1 - \max\left\{\frac{2}{1+e^{\frac{\pi(e-x_0)}{\sqrt{6}\sigma_2}}}, \frac{2}{1+e^{\frac{\pi(x_1-e)}{\sqrt{6}\sigma_2}}}\right\}\right]$$

$$= 1 - \max\left\{\frac{1}{1+e^{\frac{\pi(e-x_0)}{\sqrt{6}\sigma_2}}}, \frac{1}{1+e^{\frac{\pi(x_1-e)}{\sqrt{6}\sigma_2}}}\right\}$$

Therefore,

$$Cr(\xi \in A^* \mid H_1) = \begin{cases} 1 - \dfrac{1}{1+e^{\frac{\pi(e-x_0)}{\sqrt{6}\sigma_2}}}, & \text{if } (e-x_0) < (x_1-e) \\ 1 - \dfrac{1}{1+e^{\frac{\pi(x_1-e)}{\sqrt{6}\sigma_2}}}, & \text{if } (x_1-e) < (e-x_0) \end{cases}$$

If $(e-x_0) < (x_1-e)$ then

$$Cr(\xi \in C^* \mid H_1) - Cr(\xi \in A^* \mid H_1) = \frac{1}{1+e^{\frac{\pi(e-x_0)}{\sqrt{6}\sigma_2}}} - \frac{1}{1+e^{\frac{\pi c}{\sqrt{6}\sigma_2}}}$$

Since $e - c < x_0$, we have

$$\frac{\pi c}{\sqrt{6}\sigma_2} > \frac{\pi(e-x_0)}{\sqrt{6}\sigma_2}$$

$$\Rightarrow \frac{1}{1+e^{\frac{\pi c}{\sqrt{6}\sigma_2}}} < \frac{1}{1+e^{\frac{\pi(e-x_0)}{\sqrt{6}\sigma_2}}}$$

Therefore,

$$Cr\left(\xi \in C^* \mid H_1\right) - Cr\left(\xi \in A^* \mid H_1\right) \geq 0$$

If $\left(e - x_0\right) > \left(x_1 - e\right)$ then

$Cr\left(\xi \in C^* \mid H_1\right) - Cr\left(\xi \in A^* \mid H_1\right)$ becomes

$$\frac{1}{1 + e^{\frac{\pi(x_1 - e)}{\sqrt{6}\sigma_2}}} - \frac{1}{1 + e^{\frac{\pi c}{\sqrt{6}\sigma_2}}}$$

Since $x_1 < e + c$, we have

$$\frac{\pi\left(x_1 - e\right)}{\sqrt{6}\sigma_2} < \frac{\pi c}{\sqrt{6}\sigma_2}$$

which gives

$$\frac{1}{1 + e^{\frac{\pi(x_1 - e)}{\sqrt{6}\sigma_2}}} > \frac{1}{1 + e^{\frac{\pi c}{\sqrt{6}\sigma_2}}}$$

Therefore,

$$Cr\left(\xi \in C^* \mid H_1\right) - Cr\left(\xi \in A^* \mid H_1\right) \geq 0$$

In the above discussion it is assumed that x_0 and x_1 are positioned respectively to the left and right of e. However, by monotonicity of credibility measure, the inequality given in (8) continues to hold good for any subinterval of

$$\left(e - \frac{\sqrt{6}\sigma_1}{\pi}\ln\left(\frac{\alpha}{1-\alpha}\right), e + \frac{\sqrt{6}\sigma_1}{\pi}\ln\left(\frac{\alpha}{1-\alpha}\right)\right).$$

Thus we have for A^* in F_1, F_2 and F_3 the condition $Cr\left(\xi \in C^* \mid H_1\right) \geq Cr\left(\xi \in A^* \mid H_1\right)$ is satisfied. By the maximality axiom of Credibility measures when the rejection region A^* of size α is constructed by taking the union of two or more regions belonging to F_1, F_2 and/or F_3, the inequality $Cr\left(\xi \in C^* \mid H_1\right) \geq Cr\left(\xi \in A^* \mid H_1\right)$ continues to hold good.

Now we shall consider the problem of developing the best credibility rejection region for testing credibility null hypothesis that the fuzzy variable ξ has normal credibility distribution with parameters e and σ_1 against the credibility alternative hypothesis that the fuzzy variable ξ has normal credibility distribution with parameter e and σ_2 assuming $\sigma_1 < \sigma_2$. The region identified by the membership ratio criterion is provided in Theorem 2 along with its optimal properties. Proof for the theorem follows the approach used for proving the earlier theorem.

Theorem 2: For testing the null credibility hypothesis $H_0 : \xi \sim \mathrm{N}\left(e, \sigma_1\right)$ against the alternative credibility hypothesis $H_1 : \xi \sim \mathrm{N}\left(e, \sigma_2\right)$, where $\sigma_1 < \sigma_2$ (when e is known)

Membership ratio criterion credibility rejection region of level $\alpha < \frac{1}{2}$ is given by

$$C^* = \left\{x \mid x < e - \frac{\sqrt{6}\sigma_1}{\pi} \ln\left(\frac{\alpha}{1-\alpha}\right), x > e + \frac{\sqrt{6}\sigma_1}{\pi} \ln\left(\frac{\alpha}{1-\alpha}\right)\right\}$$

b. $Cr\left(\xi \in C^* \mid H_1\right) > \alpha$ and

The credibility rejection region

$$C^* - \left\{x \mid x < e - \frac{\sqrt{6}\sigma_1}{\pi} \ln\left(\frac{\alpha}{1-\alpha}\right), x > e + \frac{\sqrt{6}\sigma_1}{\pi} \ln\left(\frac{\alpha}{1-\alpha}\right)\right\}$$

is the best credibility rejection region of level $\alpha < \frac{1}{2}$.

The following theorem provides the uniformly best credibility rejection region for the problem of testing the null credibility hypothesis $H_0 : \sigma^2 \geq \sigma_0^2$ against the alternative credibility hypothesis $H_1 : \sigma^2 < \sigma_0^2$ when e is known

Theorem 3: Let $\xi \sim \mathrm{N}\left(e, \sigma^2\right)$. For testing the null credibility hypothesis $H_0 : \sigma^2 \geq \sigma_0^2$ against the alternative credibility hypothesis $H_1 : \sigma^2 < \sigma_0^2$ when e is known, the credibility rejection region

$$C^* = \left\{x \mid e - \frac{\sqrt{6}\sigma}{\pi} \ln\left(\frac{\alpha}{1-\alpha}\right) < x < e + \frac{\sqrt{6}\sigma}{\pi} \ln\left(\frac{\alpha}{1-\alpha}\right)\right\}$$

is the uniformly best credibility rejection region of level whenever $\alpha > \frac{1}{2}$.

Proof: To start with, consider the problem of testing the simple credibility hypothesis $H_0^* : \sigma^2 = \sigma_0^2$ against the simple alternative $H_1^* : \sigma^2 = \sigma_*^2$ where $\sigma_0 < \sigma_*$, By part (a) of Theorem 1, the membership ratio criterion test of level $\alpha > \frac{1}{2}$ is given by the credibility rejection region

$$C^* = \left\{ x \mid e - \frac{\sqrt{6}\sigma_0}{\pi} \ln\left(\frac{\alpha}{1-\alpha}\right) < x < e + \frac{\sqrt{6}\sigma_0}{\pi} \ln\left(\frac{\alpha}{1-\alpha}\right) \right\}$$

By part (c) of Theorem 1, the credibility rejection region identified above has maximum credibility under the alternative $H_1^* : \sigma^2 = \sigma_*^2$. It is pertinent to note that the same region continues to be the best credibility rejection region of level $\alpha > \frac{1}{2}$ for testing $H_0^* : \sigma^2 = \sigma_0^2$ against all alternatives of the form $H_1^* : \sigma^2 = \sigma_*^2$ as long as σ_*^2 satisfies $\sigma_0^2 < \sigma_*^2$.

That is,

$$Cr\left(\xi \in C^* \mid \sigma^2\right) \geq Cr\left(\xi \in C \mid \sigma^2\right), \quad \forall\ \sigma^2 < \sigma_0^2$$

among all credibility rejection regions satisfying the condition

$$Cr\left(\xi \in C \mid \sigma^2 = \sigma_0^2\right) \leq \alpha. \tag{9}$$

In order to prove C^* is the uniformly best credibility rejection region of level α, we must show that

$$Cr\left(\xi \in C^* \mid \sigma^2\right) \geq Cr\left(\xi \in C \mid \sigma^2\right), \quad \forall\ \sigma^2 < \sigma_0^2$$

for any C satisfying the condition

$$Cr\left(\xi \in C \mid \sigma^2\right) \leq \alpha, \quad \forall\ \sigma^2 \geq \sigma_0^2 \tag{10}$$

Since the class of all credibility rejection regions satisfying (9) is a subclass of all credibility rejection regions satisfying (10) we conclude that

$$Cr\left(\xi \in C^* \mid \sigma^2\right) \geq Cr\left(\xi \in C \mid \sigma^2\right), \quad \forall\ \sigma^2 < \sigma_0^2$$

for any C satisfying the condition

$$Cr\left(\xi \in C \mid \sigma^2\right) \leq \alpha, \quad \forall\, \sigma^2 \geq \sigma_0^2.$$

The proof becomes complete once we prove that

$$Cr\left(\xi \in C^* \mid \sigma^2\right) \leq \alpha, \quad \forall\, \sigma^2 \geq \sigma_0^2$$

Note that for any $\sigma^2 \geq \sigma_0^2$,

$$Cr\left(\xi \in C^* \mid \sigma^2\right) = \frac{1}{2}\left[\underset{x\in(e-c,e+c)}{Sup} \frac{2}{1+e^{\frac{\pi|x-e|}{\sqrt{6}\sigma}}} + 1 - \underset{x\notin(e-c,e+c)}{Sup} \frac{2}{1+e^{\frac{\pi|x-e|}{\sqrt{6}\sigma}}}\right]$$

$$= 1 - \frac{1}{1+e^{\frac{\pi c}{\sqrt{6}\sigma}}} \tag{11}$$

Since the right hand side is a decreasing function of σ and it takes the value α when $\sigma = \sigma_0$. we conclude that $1 - \frac{1}{1+e^{\frac{\pi c}{\sqrt{6}\sigma}}} < \alpha$

Thus we have proved C^* satisfies the condition

$$Cr\left(\xi \in C^* \mid \sigma^2\right) \leq \alpha, \quad \forall\, \sigma^2 \geq \sigma_0^2$$

Hence the proof.

The following theorem gives the uniformly best credibility rejection region for testing the null credibility hypothesis $H_0 : \sigma^2 \leq \sigma_0^2$ against the alternative credibility hypothesis $H_1 : \sigma^2 > \sigma_0^2$ when e is known for fuzzy normal distribution.

Theorem 4: Let $\xi \sim \mathrm{N}\left(e, \sigma^2\right)$. For testing the null credibility hypothesis $H_0 : \sigma^2 \leq \sigma_0^2$ against the alternative credibility hypothesis $H_1 : \sigma^2 > \sigma_0^2$ when e is known, the credibility rejection region

$$C^* = \left\{x \mid x < e - \frac{\sqrt{6}\sigma_1}{\pi}\ln\left(\frac{1-\alpha}{\alpha}\right), x > e + \frac{\sqrt{6}\sigma_1}{\pi}\ln\left(\frac{1-\alpha}{\alpha}\right)\right\}$$

is the uniformly best credibility rejection region of level whenever $\alpha < \frac{1}{2}$.

Proof for the theorem follows the approach used in proving Theorem 4.

Illustration

A tyre manufacturer claims that the life length of tyres manufactured has fuzzy normal distribution with expected value 24,000 kms and variance 1000 kms when used in a city. Based on the past history and change in production technology, the company feels that the distribution has changed with expected value 24,000 kms and variance 900 kms.

The above assertions can be evaluated by considering this as a credibility hypothesis testing problem explained in this chapter. The null and alternative credibility hypotheses can be taken as $H : \xi \sim \mathrm{N}\left(24000, 1000\right)$ and $K : \xi \sim \mathrm{N}\left(24000, 900\right)$ with significance level 0.55.

From the data, we have $e = 24,000$, $\sigma_1 = \sqrt{1000}$, $\sigma_2 = \sqrt{900}$ and $\alpha = 0.55$. Therefore,

$$\frac{\sqrt{6}\sigma_1}{\pi}\ln\left(\frac{\alpha}{1-\alpha}\right) = \frac{\sqrt{6}\sqrt{1000}}{22/7}\ln\left(\frac{0.55}{0.45}\right) = 4.947772$$

Therefore, the best credibility rejection region is given by

$$\left\{x \mid e - \frac{\sqrt{6}\sigma_1}{\pi}\ln\left(\frac{\alpha}{1-\alpha}\right) < x < e + \frac{\sqrt{6}\sigma_1}{\pi}\ln\left(\frac{\alpha}{1-\alpha}\right)\right\}$$

$$= \left\{x \mid 23995.05 < x < 24004.95\right\}.$$

The power of the test is

$$Cr\left(\xi \in \left(23984.366, 24015.6339\right) \mid H_1\right) = \frac{1}{2}\left[1 + 1 - \frac{2}{1 + e^{\frac{\pi(4.947772)}{\sqrt{6}\sqrt{900}}}}\right]$$

$$= 0.5526851$$

Note that the power of the test is greater than the level 0.55.

It may be noted that, if the problem explained above is treated as testing the composite null credibility hypothesis $H_0 : \sigma^2 \geq 900$ against the composite alternative credibility hypothesis $H_1 : \sigma^2 < 900$ the interval obtained above continues to be the uniformly best credibility rejection region with significance level 0.55. The power of the test is

$$Cr\left(\xi \in \left(23995.31, 24004.69\right) \mid \sigma^2 < 900\right) = \frac{1}{2}\left[1 + 1 - \frac{2}{1 + e^{\frac{\pi(4.6939)}{\sqrt{6}\sigma}}}\right]$$

Figure 5. Shape of the power function

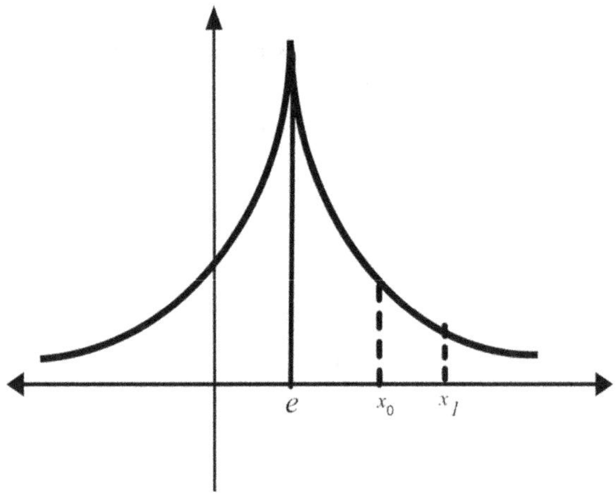

The power of the test is computed using the above power function with different values of σ^2 ranges between $\left(0, 900\right)$.

Figure 5 shows the shape of the power function for different value of σ^2.

The power curve in Figure 5 is decreasing when σ^2 tends to 900. The power of the function is 1 at $\sigma^2 = 0$ and is 0.55265 at $\sigma^2 = 900$. Note that the power of the test is always greater than the level 0.55.

CONCLUSION

In this chapter, the problem of testing the validity of hypotheses about the true value of the variance of a fuzzy normal variable is considered assuming the expected value of its credibility distribution is known.

Initially, using the membership ratio criterion credibility rejection regions have been identified for testing simple hypotheses about the variance of fuzzy normal distribution. It has been shown that the tests yielded by the membership ratio criterion always have credibility values under the alternative hypotheses exceeding the given level of significance. The credibility rejection regions derived using the membership ratio criterion are shown to be the best credibility rejection regions among all rejection regions of a given level α.

In the later part, the problem of developing uniformly best credibility rejection region for testing a one sided composite credibility hypothesis of the form $H_0 : \sigma^2 \geq \sigma_0^2$ against the alternative credibility hypothesis $H_1 : \sigma^2 < \sigma_0^2$ is also considered assuming e known and optimal tests have been identified. Uniformly best credibility rejection regions for the dual problem is also given.

It is interesting to note that the rejection regions corresponding to the best credibility test as well as uniformly best credibility critical regions are intervals, unlike the one sided best rejection regions associated with the those associated with testing hypotheses about the variance of normal distribution in probability theory when the mean is known. Hypothetical applications of the tests derived in this chapter are also provided for the purpose of illustration.

The authors are working in the direction of simultaneous testing of credibility hypotheses about the mean and variance of fuzzy normal distribution as well as other forms of testing problems.

REFERENCES

Liu, B. (2008). Uncertainty Theory (3rd ed.). Retrieved from http://orsc.edu.cn/liu/ut.pdf

Liu, B., & Liu, Y. K. (2002). Expected value of fuzzy variable and fuzzy expected models. *IEEE Transactions on Fuzzy Systems*, *4*(10), 445–450.

Sampath, S. (2009). Hybrid Single Sampling Plan. *World Applied Sciences Journal*, *12*(6), 1685–1690.

Sampath, S., & Deepa, S. P. (2013). Determination of Optimal Chance Double Sampling Plan using Genetic Algorithm. *Model Assisted Statistics and Applications*, *8*(4), 265–273.

Sampath, S., & Kalaivani, R. (2010). Clustering of fuzzy data using credibilistic critical values.*Proceedings of the IEEE Xplore, ICSIP*, (pp. 227 – 232). doi:10.1109/ICSIP.2010.5697474

Sampath, S., & Ramya, B. (2013). Distance-based test for uncertainty hypothesis testing. *Journal of Uncertainty Analysis and Applications*, *1*(1), 4. doi:10.1186/2195-5468-1-4

Sampath, S., & Ramya, B. (2014). Credibility Hypothesis Testing of Expectation of Fuzzy Normal Distribution. *Journal of Intelligent and Fuzzy Systems*, *27*(4), 1837–1847.

Sampath, S., & Ramya, B. (2015). Credibility Hypothesis Testing of Fuzzy Triangular Distributions. *Journal of Uncertain Systems, 9*(1), 62-74.

Sampath, S., & Ramya, B. (2015a). Credibility hypothesis testing of fuzzy Exponential distributions. *International Journal of Fuzzy Computation and Modelling*, *1*(3), 287–311. doi:10.1504/IJFCM.2015.069954

Sampath, S., & Senthilkumar, R. (2014). clustering of fuzzy data using credibilistic expected and critical values.*Proceedings of the IEEE Xplore, ICCCS*. doi:10.1109/ICCCS.2014.7068189

Sampath, S., & Vaidyanathan, V. S. (2009). *Credibilistic ISODATA algorithm. In Mathematical and Computational models recent trends*. New Delhi, India: Narosha Publishing House.

Wan, S. P., & Li, D.-F. (2013). Possibility mean and variance based method for multi-attribute decision making with triangular intuitionistic fuzzy numbers. *Journal of Intelligent and Fuzzy Systems*, *24*(4), 743–754.

Wan, S. P., Li, D.-F., & Rui, Z. F. (2013). Possibility mean, variance and covariance of triangular intuitionistic fuzzy numbers. *Journal of Intelligent and Fuzzy Systems*, *24*(4), 847–858.

Wang, X., Gao, Z., & Guo, H. (2012). Delphi method for estimating uncertainty distributions. *An International Interdisciplinary Journal*, *12*(2), 449–460.

Wang, X., Gao, Z., & Guo, H. (2012a). Uncertain hypothesis testing for two experts empirical data. *Mathematical and Computer Modelling*, *55*(3-4), 1478–1482. doi:10.1016/j.mcm.2011.10.039

Wang, X., & Peng, Z. (2012). *Method of moments for estimating uncertainty distributions*. Retrieved from http://www.orsc.edu.cn/online/ 100408.pdf

Zadeh, L. A. (1965). Fuzzy Sets. *Information and Control*, *8*(3), 338–353. doi:10.1016/S0019-9958(65)90241-X

Zadeh, L. A. (1978). Fuzzy sets as a basis for a theory of possibility. *Fuzzy Sets and Systems*, *1*(1), 3–28. doi:10.1016/0165-0114(78)90029-5

APPENDIX

Proposition 1: If $\xi \sim \mathrm{N}(e,\sigma)$ and $\alpha < \frac{1}{2}$ then all intervals of the form $\left(e + \frac{\sqrt{6}\sigma}{\pi}\ln\left(\frac{1-\alpha}{\alpha}\right), x\right)$ will have the credibility value α.

Proof: For any interval (x_0, x_1) as shown in Figure 6, where $e < x_0 < x_1$, we have

$$Cr\left[\xi \in (x_0, x_1)\right] = \frac{1}{2}\left[\underset{x\in(x_0,x_1)}{Sup}\ \mu(x) + 1 - \underset{x\notin(x_0,x_1)}{Sup}\ \mu(x)\right]$$

$$= \frac{1}{2}\left[\mu(x_0) + 1 - 1\right]$$

If $Cr\left[\xi \in (x_0, x_1)\right] = \alpha$ then

$$\frac{1}{2}\left[\frac{2}{1 + e^{\frac{\pi(x_0 - e)}{\sqrt{6}\sigma}}}\right] = \alpha \qquad \left[\because x_0 > e\right]$$

$$1 + e^{\frac{\pi(x_0 - e)}{\sqrt{6}\sigma}} = \frac{1}{\alpha}$$

Figure 6. Interval considered in Proposition 1

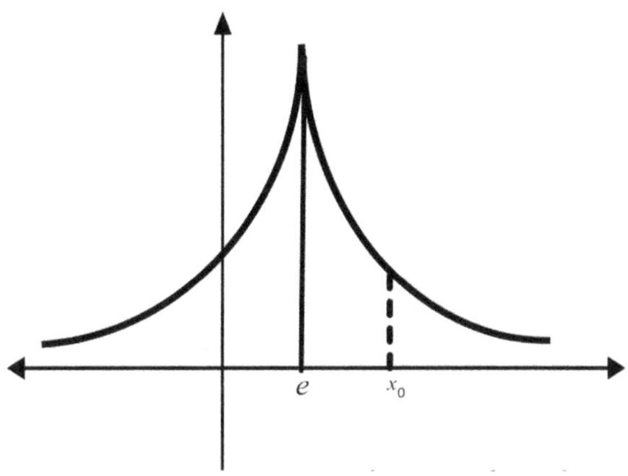

$$x_0 = e + \frac{\sqrt{6}\sigma}{\pi}\ln\left(\frac{1-\alpha}{\alpha}\right)$$

The above solution does not depend on x_1 and hence we conclude that all intervals of the form $\left(e + \frac{\sqrt{6}\sigma}{\pi}\ln\left(\frac{1-\alpha}{\alpha}\right), x\right)$ will have the credibility value α. Hence the property. ■

Remarks:

1. It may be noted that $\left(e + \frac{\sqrt{6}\sigma}{\pi}\ln\left(\frac{1-\alpha}{\alpha}\right), \infty\right)$ is the longest interval with credibility α.
2. The assumption $\alpha < \frac{1}{2}$ ensures the existence of an interval of the form to the right of e.

Proceeding as in the case of Proposition 1 the following can be proved.

Proposition 2: If $\xi \sim \mathrm{N}(e,\sigma)$ and $\alpha < \frac{1}{2}$ then all intervals of the form $\left(x, e - \frac{\sqrt{6}\sigma}{\pi}\ln\left(\frac{1-\alpha}{\alpha}\right)\right)$ will have the credibility value α.

Proposition 3: If $\xi \sim \mathrm{N}(e,\sigma)$ then all intervals of the form (e, x), $x > e$ will have the credibility equal to $\frac{1}{2}$.

Proof: Consider the interval (e, x_0), as given in Figure 7. The credibility of fuzzy variable in interval (e, x_0) is

$$Cr\left[\xi \in (e, x_0)\right] = \frac{1}{2}\left[\underset{x\in(e,x_0)}{Sup}\,\mu(x) + 1 - \underset{x\notin(e,x_0)}{Sup}\,\mu(x)\right] = \frac{1}{2}$$

Figure 7. Interval considered in Proposition 3

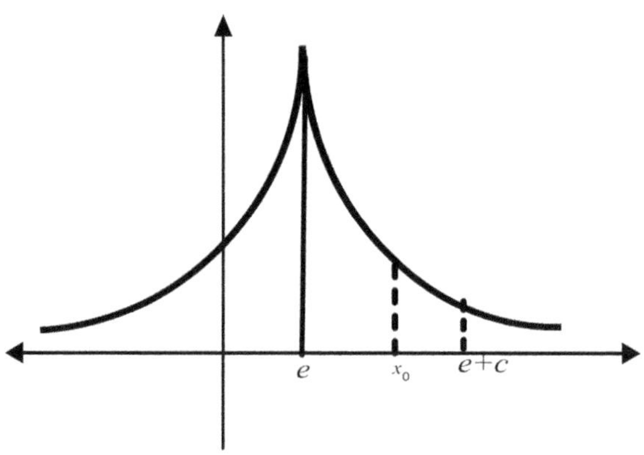

Hence the property. ■

The following Proposition can be proved using the above direction.

Proposition 4: If $\xi \sim \mathrm{N}(e,\sigma)$ then all intervals of the form (x,e), $x<e$ will have the credibility equal to $\frac{1}{2}$.

Proposition 5: If $\xi \sim \mathrm{N}(e,\sigma)$ then all intervals of the form $\left(x, e+\frac{\sqrt{6}\sigma}{\pi}\ln\left(\frac{1-\alpha}{\alpha}\right)\right)$ where $x>e$ will have the credibility value greater than α but less than $\frac{1}{2}$.

Proof: For any $S=\left\{x \mid x_0<x<e+\frac{\sqrt{6}\sigma}{\pi}\ln\left(\frac{1-\alpha}{\alpha}\right)\right\}$, where $x_0>e$ is as shown in Figure 8, we have

$$Cr\left[\xi \in S\right]=\frac{1}{2}\left[\underset{x\in S}{Sup}\,\mu(x)+1-\underset{x\notin S^c}{Sup}\,\mu(x)\right]$$

$$=\left[\frac{1}{1+e^{\frac{\pi(x_0-e)}{\sqrt{6}\sigma}}}\right]$$

Since $e<x_0<e+\frac{\sqrt{6}\sigma}{\pi}\ln\left(\frac{1-\alpha}{\alpha}\right)$, we have

$$0<\frac{\pi(x_0-e)}{\sqrt{6}\sigma}<\ln\left(\frac{1-\alpha}{\alpha}\right)$$

$$\frac{1}{2}>\left[\frac{1}{1+e^{\frac{\pi(x_0-e)}{\sqrt{6}\sigma}}}\right]>\alpha$$

Therefore, $\alpha<Cr\left[\xi\in S\right]<\frac{1}{2}$ ■

The following proposition can be easily proved using the above line of argument.

Proposition 6: If $\xi \sim \mathrm{N}(e,\sigma)$ then all intervals of the form $\left(e-\frac{\sqrt{6}\sigma}{\pi}\ln\left(\frac{1-\alpha}{\alpha}\right), x\right)$ where $x<e$ will have the credibility value greater than α but less than $\frac{1}{2}$.

Proposition 7: If $\xi \sim \mathrm{N}(e,\sigma)$ then any interval of the form (x_0, x_1), $x_0 < e$ and $x_1 > e$ will have credibility greater than $\frac{1}{2}$.

Proof: Consider the interval of the form (x_0, x_1). The credibility for the interval (x_0, x_1) is

$$Cr\left[\xi \in (x_0, x_1)\right] = \frac{1}{2}\left[\underset{x\in(x_0,x_1)}{Sup}\ \mu(x) + 1 - \underset{x\notin(x_0,x_1)}{Sup}\ \mu(x)\right]$$

$$= \frac{1}{2}\left[1 + 1 - \max\left(\mu(x_0), \mu(x_1)\right)\right]$$

$$= 1 - \max\left(\frac{1}{1+e^{\frac{\pi(e-x_0)}{\sqrt{6}\sigma}}}, \frac{1}{1+e^{\frac{\pi(x_1-e)}{\sqrt{6}\sigma}}}\right)$$

Since $\max\left(\frac{1}{1+e^{\frac{\pi(e-x_0)}{\sqrt{6}\sigma}}}, \frac{1}{1+e^{\frac{\pi(x_1-e)}{\sqrt{6}\sigma}}}\right) < \frac{1}{2}$,

we have $Cr\left[\xi \in (x_0, x_1)\right] > \frac{1}{2}$.

Chapter 9
Application of Fuzzy Soft Set in Patients' Prioritization

Samira Abbagholizadeh Rahimi
Université Laval, Canada

ABSTRACT

Based on studies, access to healthcare services and long waiting time is one of the main issues in many countries including Canada and United States. Healthcare organizations can't increase their limited resources nor treat all patients simultaneously. Then, patients' access to these services should be prioritized in a way that best uses the scarce resources and insures patients' safety. Prioritization is essential and inevitable not only because of resource shortage, which have not been improved during years, but also because it is a crucial issue that could contribute to the capability and stability of the healthcare systems, and most importantly to patients' safety. On the other hand, inappropriate prioritization of patients waiting for treatment, could affect directly on inefficiencies in healthcare delivery, quality of care, and most importantly on patients' safety and their quality of life and satisfaction. Inspired by these facts, in this chapter the importance of patients' prioritization and using fuzzy logic in this area will be discussed, and a novel hybrid framework using fuzzy soft sets for patients' prioritization will be proposed. The proposed framework may have a significant impact on patients' safety, and on both medical community and the public's faith in justice and equity.

INTRODUCTION

Medical knowledge and clinical practices are always associated with considerable amounts of uncertainty and about everything in medicine is inevitably vague (Sadegh-Zadeh, 2012). Many complicated problems like patients' prioritization problem involves such uncertainties. Despite this issue's importance, (to the best of our knowledge) no valid tool has been proposed in the literature to prioritize patients for medical treatment considering these uncertainties, and associated risks all together. To this end, this chapter will give useful impulses to face these major challenges in patients' prioritization, by developing a novel integrated framework which covers the current drawbacks and will provide theoretical solutions for them. This chapter focuses on uncertainties in clinicians' decisions and involving associated risks that could threaten patients while they are waiting for treatment, and proposes a novel integrated framework to deal with these crucial issues.

DOI: 10.4018/978-1-5225-1848-8.ch009

This problem, cannot be solved using classical mathematic methods. There are several well-known theories (such as theory of probability, theory of fuzzy sets, theory of vague sets, theory of interval mathematics, and etc.) to describe uncertainty, but all of these theories have their inherit difficulties as Molodtsov (1999) mentioned in his paper. The reason for these difficulties is, possibly, the inadequacy of the parameterization tool of the theories (Celik & Yamak, 2013). To overcome these difficulties, Molodtsov initiated the concept of soft sets as a new mathematical tool for dealing with uncertainties (Molodtsov, 1999). This so-called soft set theory seems to be free from the difficulties affecting the existing methods (Maji et al., 2003).

Maji et al. (2001) defined various operators for soft set theory. and in 2010, Ali & Shabir (2010) made some improvements of the introduced operations by Maji et al. Fundamental properties of soft sets has been defined by Aktas and Cagman (2007). They combined the soft set theory and group theory to defined soft groups. Some new operation were studied in Ali & Shabir (2010), Ali et al. (2009), and Feng and Li (2013). They gave the new concept of the soft product in soft set theory and discussed generalized decision making schemes, and many other researchers such as Li (2014), Li and& Ren(2015), and Yu and Li (2014) have studied soft set theory in different aspects.

Recently, research works on soft sets in different industries are very active and progressing rapidly. Applications of fuzzy soft set theory in many disciplines and real life situations have been studied by many researchers but, its application in healthcare industry is in its infancy stages. To the best of our knowledge this is the first time in literature that such novel integrated framework is introduced for patients' prioritization. This chapter focuses on developing an interdisciplinary, systematic and innovative prioritization framework which is inspired by Celik and Yamak (2013) work on medical diagnosis. The proposed framework considers uncertainty, multiple criteria, risks and their inherent interactions to prioritize patients' access to healthcare services. In this chapter, Analytical Network Process (ANP) is used to find relative importance weights of criteria and risks considering their possible interactions. Then, by using the notion of a fuzzy soft set together with arithmetic operations on fuzzy number, authors introduce how fuzzy soft set technology could be used for patients' prioritization.

The rest of the chapter is organized as follows. In the next two sections prioritization of patients' access to healthcare services, the related literature review and shortcomings are discussed in details. Basics of Fuzzy soft set theory are explained briefly after. Then the proposed framework is discussed. To illustrate the application and effectiveness of the proposed framework, a numerical example in surgery ward is illustrated. In order to demonstrate robustness of the proposed method, sensitivity analysis under various criteria/risks-weight-change scenarios is performed and the results are discussed. Finally, the proposed framework is compared with some of well-known Multi Criteria Decision Making (MCDM) methods in the comparison section to show its benefits and advantages. This chapter concludes with a summary of the chapter, a discussion on the major contribution of this framework and future directions.

PRIORITIZING PATIENTS ACCESS TO HEALTHCARE SERVICES

Patients in the Organization for Economic Co-operation and Development (OECD) countries (including United States and Canada) continue to wait too long to receive medically necessary treatment. Waiting times' situation not only have not been improved during years till 2016, but also they have gotten slightly worse (Barua, 2015). One of the main reasons for long waiting times is imbalance between demand and availability of scarce resources in healthcare organizations. Clinicians can't treat all patients simultaneously. Due to high costs, managers can't increase number of their limited resources either. Then, patients' access to these services should be prioritized in a way that best uses existence scarce resources, and insure patients' safety.

However, prioritization may not be the only option responding to limited resources in health care, but it may be one of the best options in identification of high risk patients and increasing safety of patients' waiting for treatment. There are other alternative like increasing efficiency or the overall amount spent for health care, and rationing by delay. But, rationing implicitly goes along with problems in principle (Nagel & Lauerer, 2016), and is not sufficiently narrowing the gap that occurs between demand and supply (Williams et al., 2012). Rationing regularly refers to actual withholding of health services, while prioritization describes a systematic and comprehensive approach to analyze what is more and what is less important. It leads to a ranking order (Meyer & Raspe, 2012) and a precise selection of patients' for treatment. Prioritization in healthcare domain is preferable approach over implicit approaches like rationing when tight budgets force clinicians to make allocation decisions (Nagel & Lauerer, 2016). Prioritization helps to allocate scare resources fairly and transparently (Nagel & Lauerer, 2016).

Currently, in healthcare systems after patients are referred to treatment, their situation is examined. If patients have non-life threatening condition, they will be enrolled on a first-come, first serve basis (Abbasgholizadeh Rahimi et al., 2016). But if their conditions are life-threatening they will be registered two or three emergency groups/priority levels. Higher priority-patients will be selected for service prior to those with a lower priority, regardless of when they are placed on the list (Barua, 2015). In the literature review, the comprehensive explanation of the different developed/used prioritization tools will be explained in details.

Inappropriate prioritization of patients waiting for treatment, affect directly on inefficiencies in health care delivery, quality of care, and most importantly on patients' medical conditions. Reports regarding the harms related to long waiting times and inappropriate prioritization of patients on waiting list are increasing. These harms include:

- Poorer medical results from care and decreased quality of life,
- Reduction in effectiveness of treatment (Day, 2013),
- Increased pain, and risk of adverse events (Barua & Esmail, 2013),
- Mortality (Prentice & Pizer, 2007), and so more.

An appropriate patients' prioritization can play an important role in diminishing these undesirable outcomes. Besides, it can have significant impact on both medical community and public's faith in justice and equity.

An excellent prioritization in health system requires not only good recognition of the system, expertise to analyze and understand the information (related to patients, clinicians and other resources) but also to consider and organize that information regularly in a way to make an adequate decision and to promote a collaboration among key stakeholders (e.g. managers, clinicians, patients).

There are different major challenging aspects in prioritization of patients' access to healthcare services as (Abbasgholizadeh Rahimi et al., 2016):

- How to involve confronting trade-offs between multiple, often conflicting, factors?
- How can groups of clinicians' opinions be simultaneously involved in the decision making procedure?

- How can patients and their families be involved in the decisions?
- What are the main criteria for prioritizing patients?
- What are the risks that could threaten patients on waiting list?
- How important are selected risks and criteria compared to each other?
- How can we handle uncertainties in clinicians' decisions in the procedure be handled?
- How dynamic nature of the health system could be considered in prioritization of patients?

Currently, access to healthcare services are not prioritized in a way that best uses limited resources and to ensure a transparent, equal and accurate rank of patients on waiting lists. There is no reliable and comprehensive method of assessing the relative priority of patients on waiting lists, and some researchers (Peacock et al., 2009;Domènech et al., 2013;Abbasgholizadeh Rahimi et al., 2014, 2016; Mullen, 2003; Russell et al., 2003; Abbasgholizadeh Rahimi & Jamshidi, 2014; Comas et al., 2008) in their studies stressed the need for an interdisciplinary and collaborative research to explore systematic and precise prioritization framework. Although some prioritization approaches have been developed for this aim but, they have several major shortcomings (Abbasgholizadeh Rahimi et al., 2016). following, we will discuss on how patients are prioritized currently and what are their main shortcomings.

LITERATURE REVIEW

The earliest publications on 'priorities in medicine' are from England (Butterfield, 1968; Godber, 1970) and Denmark (Pornak & Raspe, 2015), Norway was the first country worldwide that considered prioritization on a national level (Raspe, 2016).

Patients' prioritization is a complex decision making process and according to Hadorn (2000), "the normative basis for prioritization is that patients with more urgent conditions should receive services ahead of those with less urgent conditions, and patients with approximately the same degree of urgency should wait about the same length of time". Broadly, patients' priority can be defined by the position of patient in the waiting queue (leading to e.g., first-come, first-served (FCFS)), patient specific characteristics or the contribution to an objective function or to a combination of factors (Riet & Demeulemeester, 2015).

Some authors (Davis & Johnson, 1998; Fordyce & Phillips, 1970; Hadorn, 2000; Lack et al., 2000), (Domènech et al., 2013; Mariotti et al., 2014; Dowseya et al., 2014; Montoya et al., 2014) proposed various prioritization scoring systems to assist health professionals to make better decisions in determining which patient receive treatment sooner than others. A scoring system or points system consists of a method for deciding patients' relative priorities for treatment. In this type of prioritization system, a weighted set of criteria is proposed and each patient is assessed with respect to every criterion. The sum of all the values gives a "total score" for each patient, which is used to rank patients between them.

Pioneer scoring systems introduced in 1990s were criticized for being arbitrary and resulting in significant numbers of patients being mistakenly denied treatment (sometimes with fatal consequences). However, scoring systems have been proposed and are largely in use in:

- Italy,
- Sweden,
- New Zealand,
- United Kingdom,

- Norway,
- Germany,
- Canada and other OECD countries.

For instance, currently used scoring tools in Canada are in the following five main clinical areas:

- i.e. Cataract surgery,
- General surgery procedures,
- Hip and knee replacement,
- Magnetic resonance imaging scanning,
- Children's mental health.

Prioritization formulae were also developed (mostly by considering specific criteria like waiting time or emergency) with the aim of reducing waiting times and improving access of patients to healthcare services, Mullen (2003) gave a complete review of developed prioritization formulas. A summary of Mullen's review is shown in Table 1.

Naylor et al. (1993) suggested to assign to each patient, at referral's moment, a priority code based on an Urgency Rating Scale (URS). Prioritization based on other considerations has also been proposed for situations where the risk of death is low (Hadorn, 2000). In Dolan and Cookson's (2000) work the different principles of priority-setting decision makings have been focused qualitatively (e.g. need, equity and fairness principles). Nagel and Lauerer (2016) comprehensively discussed the different prioritization factors in their book and MacCormick et al. (2003) thoroughly discussed the different prioritization factors and their weighing, in a review study on patients' prioritization systems.

The prioritization has been rarely evaluated and considered within scheduling problem for surgical patients as well. However, in Testi et al. (2008) and Valente et al. (2009), the subset of patients to be operated on among the patients on waiting list and their order of admission (based on the prioritization system) was introduced which were based on both the waiting time of the patient since its referral and its urgency status. Valente et al. (2009) developed a model to prioritize access to elective surgery on the basis of clinical urgency and waiting time. While, Testi et al. (2008) in their research emphasized the importance of using both urgency related groups (URGs) and scoring system for scheduling patients' admissions in an explicit and transparent way, and a similar approach was also used in Min and Yih (2010) . Most recently, in 2015, Addis et al. considered the problem of selecting a set of patients among a given waiting list of elective patients and assigning them to a set of available operating room blocks. In their study, each patient was prioritized by a recommended maximum waiting time.

In emergency departments (EDs) upon their arrival to the ED, patients have a triage interview, then the triage nurse prioritizes them to three or five clinically distinct levels based on their vital signs, present illnesses and past medical history. The contemporary Emergency Severity Index (ESI) triage system sorts patients into five clinically distinct levels in EDs. ESI level 1 or level 2 are assigned to the most acutely-ill patients, and ESI levels 3, 4, and 5 are assigned to the lowest acutely-ill patients. There are several serious limitation of using the ESI triage sorting system in EDs; physicians and nurses in current triage system are not able to determine how to proceed when they faced with the scenario of multiple same level patients in the waiting room (Tanabe et al, 2005; Andersson et al., 2006). And, the main challenge is how to prioritize patients and how to identify who should proceed first for the treatment (Andersson

Table 1. Developed prioritization formulae

Description	Proposed Prioritization Formula	Description
(Luckman et al., 1969)	$P = S^a DT^{w^b}$	where P= priority score; S =social factor (1–3);D= disability factor (1–3); w=deterioration factor(1–3); T = time on waiting list(weeks); a and b are constants.
(Fordyce & Phillips, 1970),	$AI = (aS)(bD)(T)c^w$	where S =social factor; D =disability factor (1–5); w =urgency of condition (0–5); T=time on waiting list (weeks); a, b and c are constants.
(Phoenix, 1972)	$P = DT^{\sqrt{w}}$	Luckman J. (1969) proposed formula simplified to this one
(Eltringham & Clare, 1973)	$P = 800(1 - e^{-kT})$	where P =patient's priority score; k = f(urgency);T =time on waiting list.
(Clare, 1973)	$P = \frac{c^w T}{\sqrt{L}}$	where P =priority score; L= expected length of stay; c =constant; w =urgency or deterioration factor; T= time on waiting list.
(Culyer & Cullis, 1976)	$P = \frac{c^w (T + 28)}{\sqrt{L}}$	
(Langham & Thorogood, 1996) ; (Soljak, 1997); (Hadorn & Holmes, 1997),	$P = \sum_{i=1}^{5} S_i$	where P= priority score; S_i = score for ith clinical factor
(Dennett & Parry, 1998),	$P = \sum_{i=1}^{3} A_i . \sum_{j=1}^{2} B_j$	where P= total priority score; A_i = score on i th element of the clinical severity criterion (i =1 suffering (physical or mental) (values 0, 3, 5 or 7);i = 2 disability (1, 3, 5 or 7); i = 3 clinical cost of delay (1, 3 or 6)); B_j =score on jth element of the capacity to benefit criterion (j = 1 degree of improvement anticipated; j = 2 likelihood of improvement).
(Seddon et al., 1999),	$P = E\sum_{i=1}^{5} S_i$	where E =(100 – age)/30 if age > 70
(Lack et al., 2000)	$P = \sum_{i=1}^{5} S_i . w_i + \left(\frac{5t}{m} - 1\right)^2$	where t = time already waited and m=waiting time of the longest waiter.

et al., 2006). Moreover, patient medical state might change while waiting for treatment, which means that decisions should change dynamically as well (Claudio & Okudan, 2010).

The use of the utility theory to prioritize ED patients assigned to the same acuity level was recently increased. In 2010, Claudio and Okudan illustrated the use of the multi-attribute utility theory (MAUT) in patient prioritization using a hypothetical example. They explained the choice of MAUT due to the inherent uncertainty in ED settings, and that MAUT accounts for uncertainty (Ashour & Kremer, 2016). Ashour and Okudan (2010) also prioritized patients in ED using MAUT, and considering patient age, gender, pain level, and the assigned ESI. These factors were not considered in the ESI algorithm nor in Claudio and Kudan's study (Claudio & Okudan, 2010).

Argon and Ziya (2009) in their study assumed that each customer arrives with a signal that can be used as an indicator for his/her identity. They considered three different prioritization policies i.e., highest-signal-first (HSF) policy, two-class policy, first-come-first-served (FCFS) policy, and the generalized cμ (GE-cμ) policy. Then, they compared these policies based on waiting cost. Their study showed that when the waiting cost is linear with time, HSF policy outperformed any finite class priority policy otherwise the two-class policy and the FCFS policy performed better than the HSF policy (Ashour & Kremer, 2016). Even though Argon and Ziya (2009) have described techniques to choose the better signal, they did not develop a way to prioritize patients who visit the ED.

Ashour and Kremer (2016) most recently developed a dynamic prioritization tool for EDs. They applied group technology (GT) concept to the triage process to develop a dynamic grouping and prioritization (DGP) algorithm. They implemented discrete event simulation (DES) to investigate the impact of the DGP algorithm on the performance measures of the ED system. Their study demonstrated that DGP algorithm outperforms the ESI algorithm by shortening patients' average length of stay (LOS), average time to bed (TTB), time in emergency room, and lowering the percentage of tardy patients and their associated risk in the system (Ashour & Kremer, 2016). Patients' prioritization has been considered also in other types of medical activities such as:

- Organ Transplant (Cholongitas et al., 2010; Cholongitas & Burroughs, 2012; Lin & Harris, 2012; Lavee & Brock, 2012; Abbasgholizadeh Rahimi & Jamshidi, 2014; Ahlert & Kliemt, 2013; Al-Ebbinia et al., 2016),
- Trauma (Fieldsa et al., 2013), (Andersson et al., 2006), and in
- Cardiothoracic ICU (Yang et al., 2013).

Despite all these efforts, some important points are still overlooked and major shortcomings in current prioritization systems need to be improved (Peacock et al., 2009; Domènech et al., 2013; Abbasgholizadeh Rahimi et al., 2014; Mullen, 2003; Russell et al., 2003). These shortcomings are explicitly stated in the following:

1. Current prioritization tools cannot ensure that the ranking results are robust to face uncertainty. Uncertainties are inherent to most of real-life decision making processes and particularly in medicine, where the lack of information, as well as its imprecision and conflicting nature are common facts (Noseworthy et al., 2003; Abbasgholizadeh Rahimi et al., 2016).
2. Associated risks that could threaten patients' health during the waiting time have been overlooked. In all patient-related procedures, presents' risks are associated with severe effects and unavoidable consequences which increase along with the waiting time. By considering these risks in the prioritization procedure, decision makers could make sure that patients with the highest risks (e.g. the risk of injury, stroke, disability or even death) will be selected first (Abbasgholizadeh Rahimi et al., 2016).

3. Lack of group decision making is another shortcoming of current prioritization tools which cause biased and inaccurate prioritization of patients (Abbasgholizadeh Rahimi et al., 2016). For instance, in surgery wards medical staff may have different priorities or concerns that could conflict with surgeons' ones. But, at the end, surgeon is the only one who prioritizes patients based on his/her opinion. This may cause bias in the prioritization procedure, and dissatisfaction of other medical staff as well. To benefit from all surgery team members' knowledge and experience in prioritization procedure, other team member's assessments should be considered. Albeit, in order to have an accurate prioritization, the surgery team members should have different weights.
4. Possible interrelationships among criteria have been overlooked (Abbasgholizadeh Rahimi et al., 2016). Real life situations usually confirm possible interrelationships among decision criteria that can influence the final results. However, this issue rarely has been considered in patients' prioritization procedures up to date.
5. Dynamic behavior of system has rarely been considered. Practically, waiting lists are dynamic, new patients arrive and others quit continuously. Moreover, patients' condition evolve in time. Therefore, any prioritization procedure must take into account this dynamic behavior of the system in order to adequately support decision makers (Abbasgholizadeh Rahimi et al., 2016).
6. Last but not least, there is no systematic and comprehensive framework for patients' prioritization on waiting list.

Based upon the increasing number of demand for healthcare services and mentioned shortcomings of existing prioritization systems, there is need for a quantitative and qualitative approach that dynamically identifies and assesses the patients' priority and considers uncertainties and risks. The approach should improve the capability of prioritization systems to make differentiations among mix of patients, while lowering the cognitive stress and load on medical staff, which resulted from the dynamic nature of prioritization process and the complex environment of healthcare organizations, and aiding them to make better decisions. Methodology part of the chapter focuses on our developed methodologies to deal with some of the mentioned shortcomings in prioritization systems.

FUZZY SOFT SETS

In this section, authors give some known and useful definitions and notations regarding a soft set and a fuzzy soft set. The definitions and notions in this part may be found in references (Molodtsov, 1999; Celik & Yamak, 2013; Maji et al., 2003; De et al., 2001; Chetia & Das, 2010; Sanchez, 1979). Let U be an initial universal set and E be a set of parameters. The power set of U is denoted by P(U) and A is a subset of E.

Definition 1: A pair (F, A) is called a soft set over U, where F is a mapping given by F: A→P (U) (Molodtsov, 1999; Ali et al., 2009).

Definition 2: A fuzzy subset μ of U is defined as a map from U to [0, 1]. The family of all fuzzy subsets of U is denoted by F (U). Let μ, $\nu \in$ F (U) and x $\in$ U. Then the union and intersection of μ and ν are defined in the following way (Zadeh, 1965):

$(\mu \vee \nu)\,(x) = \mu(x) \vee \nu(x),$

$(\mu \wedge \nu)(x) = \mu(x) \wedge \nu(x)$,

$\mu \leq \nu$ if and only if $\mu(x) \leq \nu(x)$ for all $x \in U$.

Definition 3: Let U be a common universe, E be a set of parameters and $A \subseteq E$. Then a pair (F, A) is called a fuzzy soft set over U, where F is a mapping given by F: A→F (U) (Maji et al., 2003).

Definition 4: For two fuzzy soft sets (F, A) and (G, B) over a common universe U, we say that (F, A) is a fuzzy soft subset of (G, B) if: (i) $A \subseteq B$, (ii) $F(a) \leq G(a)$ for all $a \in A$. In this case, we write $(F, A) \subseteq (G, B)$ [26].

Proposition: Let (F, A) and (G, B) be two fuzzy soft sets over a common universe *U*, (Maji et al., 2001; Ali & Shabir, 2010) Then:

1. $(F, A) \cup (F, A) = (F, A)$,
2. $(F, A) \cap (F, A) = (F, A)$,
3. $\left((F, A) \cup (G, B)\right)^c = (F, A)^c \cap (G, B)^c$,
4. $\left((F, A) \cap (G, B)\right)^c = (F, A)^c \cup (G, B)^c$,
5. $\left((F, A) \vee (G, B)\right)^c = (F, A)^c \wedge (G, B)^c$,
6. $\left((F, A) \wedge (G, B)\right)^c = (F, A)^c \vee (G, B)^c$.

Definition 5:

1. A fuzzy subset μ on the universe of discourse $\mathbb{R}$ (the set of all real numbers) is convex if and only if for $a, b \in U$ $\mu(\alpha a + \beta b) \geq \mu(a) \wedge \mu(b)$, where $\alpha + \beta = 1$ (Maji et al., 2003; Kaufmann & Gupta, 1991).
2. A fuzzy subset μ on the universe of discourse U is called a normal fuzzy subset if there exist $a_i \in U$ such that $\mu(a_i) = 1$ (Maji et al., 2003; Kaufmann & Gupta, 1991).
3. A fuzzy number is a fuzzy subset defined on the universe of discourse $\mathbb{R}$ which is both convex and normal. For more information on fuzzy numbers readers could refer to (Li, 2014; Li & Ren, 2015; Li & Liu, 2015).

Fuzzy number can take various forms such as Triangular Fuzzy Number (TFN), Trapezoidal Fuzzy Number, and Gaussian Fuzzy Number. TFN is the most popular form due to its simple membership function which is represented by three parameters as (a, b, c). The parameters a, b, and c denote the smallest possible value, the most promising value, and the largest possible value of the fuzzy number, respectively. And the membership function is defined as below (Maji et al., 2003; Kaufmann & Gupta, 1991).

$$\mu(u) = \begin{cases} 0 & if\, u < a, \\ \dfrac{u-a}{b-a} & if\ a \leq u \leq b, \\ \dfrac{c-u}{c-b} & if\, b \leq u \leq c, \\ 0 & if \quad u \geq c\,. \end{cases}$$

If the membership function μ (u) is piecewise linear, then μ is said to be a trapezoidal fuzzy number. Now let μ and β be two triangular fuzzy numbers parameterized by the triplet $b_1 = (a_1, b_1, c_1)$ and $b_2 = (a_2, b_2, c_2)$ respectively.

Then addition and multiplication of μ and β are as below (Maji et al., 2003; Kaufmann & Gupta, 1991):

- $\mu \oplus \beta = \tilde{a}_2 \oplus \tilde{b}_2 = (a_1, a_2, a_3) \oplus (b_1, b_2, b_3) = (a_1 + b_1, a_2 + b_2, a_3 + b_3)$
- $\mu \otimes \beta = \tilde{a}_2 \otimes \tilde{b}_2 = (a_1, a_2, a_3) \otimes (b_1, b_2, b_3) = (a_1 \times b_1, a_2 \times b_2, a_3 \times b_3)$

It is required to reduce a given fuzzy number into a single crisp representative value. This is called defuzzification operation. Then in the next step authors give the defuzzification method of a triangular fuzzy number. The defuzzification value t of a triangular fuzzy number (l, m, u) is equal to (Maji et al., 2003):

$$t = \frac{l + m + m + u}{4} \tag{1}$$

METHODOLOGY: FUZZY SOFT SET FOR PATIENTS' PRIORITIZATION

In Figure 1 an algorithm of our proposed framework for patients' prioritization using ANP and fuzzy arithmetic operations is presented. This approach focuses on shortcomings related to uncertainty, risks and interrelationship among criteria and risks in prioritizing patients' access to healthcare services.

Assume that there is a set of m patients, $P = \{p_1, p_2, p_3, \ldots, p_m\}$, set of n criteria $C = \{c_1, c_2, c_3, \ldots, c_n\}$ (these criteria could be medical, social or other factors as explained in the previous chapter) related to a set of k risks $R = \{R_1, R_2, R_3, \ldots, R_k\}$. Authors apply fuzzy soft set theory to distinguish which patient's risks is high and is in priority to receive care and treatment considering uncertainty. For this, a fuzzy soft set (F, P) over C where F is a mapping $F: P \rightarrow F(C)$ is constructed.

This fuzzy soft set gives a relation matrix PC, called patient-criteria matrix, where the entries are fuzzy numbers $\tilde{p}$ parameterized by a triplet $(p - 1, p, p + 1)$. For instance, $\tilde{a}_{11}$ shows a fuzzy number for the first patient's situation considering first criterion. Then construct another fuzzy soft set (G, C) over R, where G is a mapping $G: C \rightarrow F(R)$. This fuzzy soft set gives a relation matrix CR, called criteria-risk matrix, where each element denotes the impact of the criteria for a certain risk. For instance, $\tilde{b}_{11}$ shows a fuzzy number for the first criterion's impact on first risk. These elements are also taken as triangular fuzzy numbers. Thus the general form of PC *matrix* is shown in Figure 1.

$$\begin{array}{c} c_1\left(W_{C_1}\right) c_2\left(W_{C_2}\right) \cdots \cdots c_n\left(W_{C_n}\right) \end{array}$$
$$PC = \begin{bmatrix} \tilde{a}_{11} & \tilde{a}_{12} & \cdots & \cdots & \tilde{a}_{1n} \\ \tilde{a}_{21} & \tilde{a}_{22} & \cdots & \cdots & \vdots \\ \vdots & \vdots & \cdots & \cdots & \vdots \\ \vdots & \vdots & \cdots & \cdots & \vdots \\ \tilde{a}_{m1} & \tilde{a}_{m1} & \cdots & \cdots & \tilde{a}_{mn} \end{bmatrix}$$

Figure 1. Systematic algorithm of proposed prioritization method

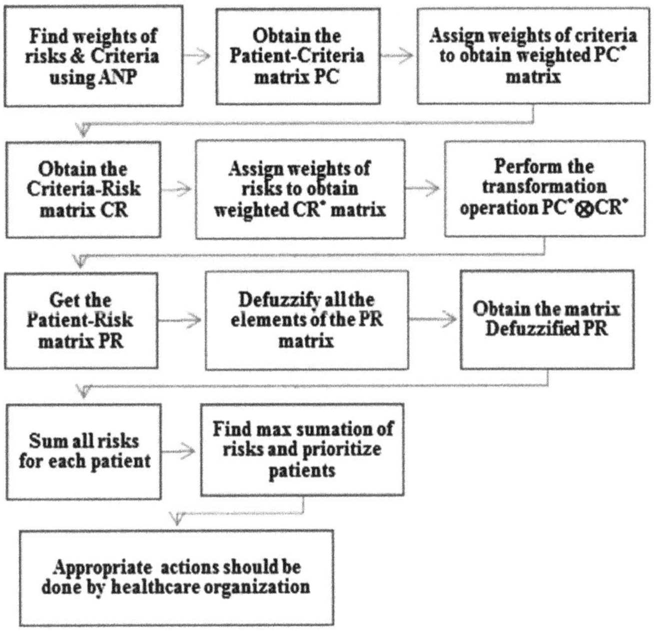

And the general form of *CR matrix* is:

$$
\begin{array}{c} \\ \\ CR = \end{array}
\begin{array}{c} \\ c_1 \\ c_2 \\ \vdots \\ \\ c_n \end{array}
\begin{array}{c} R_1\left(W_{R_1}\right) R_2\left(W_{R_2}\right) \cdots \cdots R_k\left(W_{R_n}\right) \\
\begin{bmatrix} \tilde{b}_{11} & \tilde{b}_{12} & \cdots & \cdots & \tilde{b}_{1k} \\ \tilde{b}_{21} & \tilde{b}_{22} & \cdots & \cdots & \tilde{b}_{2k} \\ \vdots & \vdots & \cdots & \cdots & \vdots \\ \vdots & \vdots & \cdots & \cdots & \vdots \\ \tilde{b}_{n1} & \tilde{b}_{n2} & \cdots & \cdots & \tilde{b}_{nk} \end{bmatrix} \end{array}
$$

Since some risks have a higher impact than others, it's better to treat all risks uniquely. The risks that can cause the biggest losses, should have the biggest importance weights. It's the same for criteria, each criterion has a relative importance weight comparing to other criteria.

For finding each criterion's importance weight (i.e. W_C) and each risk's importance weight (i.e. W_R), medical experts should compare each risk to all other risks and also each criterion to all other criteria using a questionnaire which has been developed based on ANP method. The reason of using ANP technique in this framework is that, ANP enables us to consider all possible interrelationships among these criteria and risks. For further information on ANP approach please refer to Saaty & Özdemir (2005). Now each of k risk and n criterion have weights, which indicates risk's/criterion's importance in comparison to others.

The numbers inside parenthesis in PC and CR matrices are weights obtained using ANP methodology. In PC and CR matrices each criterion and each risk's weight should be multiplied in its related column to get the weighted PC and CR matrices which are shown by PC* and CR*. For instance, W_{C_1} (Weight of first criteria, C_1) should be multiplied in first column of CP matrix.

Now, after obtaining weighted PC (i.e. PC*) and weighted CR (i.e. CR*) matrices, the transformation operation PC*⊗CR* is performed, the Patient-Risk matrix (PR) would be as follows:

$$PR = \begin{array}{c} \\ p_1 \\ p_2 \\ \vdots \\ \\ p_m \end{array} \overset{\begin{array}{cccccc} R_1 & R_2 & \cdots & \cdots & R_k \end{array}}{\begin{bmatrix} \tilde{d}_{11} & \tilde{d}_{12} & \cdots & \cdots & \tilde{d}_{1k} \\ \tilde{d}_{21} & \tilde{d}_{22} & \cdots & \cdots & \tilde{d}_{2k} \\ \vdots & \vdots & \cdots & \cdots & \vdots \\ \vdots & \vdots & \cdots & \cdots & \vdots \\ \tilde{d}_{m1} & \tilde{d}_{m2} & \cdots & \cdots & \tilde{d}_{mk} \end{bmatrix}}$$

where:

$$\tilde{d}_{il} = \left(\sum_{j=1}^{n} \left(a_{ij} - 1\right).\left(b_{jl} - 1\right),\ \sum_{j=1}^{n} a_{ij}.b_{jl},\ \sum_{j=1}^{n} \left(a_{ij} + 1\right).\left(b_{jl} + 1\right) \right) \tag{2}$$

Then, defuzzifying each element of the above matrix by equation (1), and the defuzzified Patient-Risk (PR) matrix would be as below:

$$Defuzzified\, PR =$$

$$\begin{array}{c} \\ p_1 \\ p_2 \\ \vdots \\ \\ p_m \end{array} \overset{\begin{array}{cccccc} R_1 & R_2 & \cdots & \cdots & R_k \end{array}}{\begin{bmatrix} v_{11} & v_{12} & \cdots & \cdots & v_{1k} \\ v_{21} & v_{22} & \cdots & \cdots & v_{2k} \\ \vdots & \cdots & \cdots & \cdots & \vdots \\ \vdots & \cdots & \cdots & \cdots & \vdots \\ v_{m1} & v_{m2} & \cdots & \cdots & v_{mk} \end{bmatrix}}$$

Now to get the final prioritization, the summation of each row should be calculated $Sv_m = \sum_{j=1}^{k} v_{mj}$ for example for the first patient $Sv_1 = \sum_{j=1}^{k} v_{1j}$. Then the final matrix *sum* (which is the result of summation) would be:

$$Sum = \begin{array}{c} p_1 \\ p_2 \\ \vdots \\ p_m \end{array} \begin{bmatrix} Sv_1 \\ Sv_2 \\ \vdots \\ Sv_m \end{bmatrix}$$

In this matrix Sv_1 is the priority number of the first patient and Sv_m is the priority number of mth patient.

Now patients can be ranked based on their priority score. The patient with the largest overall priority number should be the one selected for treatment first. This proposed framework can be applied in different clinical settings. Fields such as organ transplant, surgery, cancer treatment or other critical care.

In numerical example, application of proposed framework will be explained for prioritizing surgical patients.

NUMERICAL EXAMPLE

In this section, to illustrate the proposed framework, a hypothetical numerical example is presented. Suppose that in a hospital's surgery ward, there are four surgical patients $\{p_1, p_2, p_3, p_4\}$ on waiting lists. Surgeons reached to consensus to consider three criteria $\{C_1, C_2, C_3\}$ to prioritize these patients. They also selected three risks of R1 (Risk of death), R2 (Risk of comorbidity), and R3 (Risk of infection) $\{R_1, R_2, R_3\}$ which may threaten patients' health if patients' treatment is delayed. Authors used ANP technique and medical experts' opinion to obtain each criterion's and each risk's relative importance weights.

The ANP approach four phases (Saaty & Özdemir, 2005) (Chung et al., 2005), and detailed explanation of ANP can be found in Saaty (Saaty & Özdemir, 2005):

Phase 1: *Construction of model and problem structuring*: In this step the problem should be stated clearly and decomposed into a rational system like a network.

Phase 2: *Pairwise comparisons and priority vectors:* In ANP, like (AHP) pairs of decision elements at each cluster are compared with respect to their importance towards their criteria. In addition, interdependencies among criteria of a cluster must also be examined in pairs. The relative importance values are determined with Saaty's scale.

Phase 3: *Formation of Super-matrix*: A super-matrix is a partitioned matrix, where each matrix part represents a relationship between two clusters in a system. The super-matrix concept is similar to the Markov chain process. To obtain global priorities in a system with interdependent influences, the local priority vectors should be entered in the appropriate columns of a matrix.

Phase 4: *Synthesizing the criteria/risks*: The priority weights of the criteria/risks can be found in the normalized super-matrix.

In this study we used Super Decisions software version number 2.0.8. (which implements ANP) to obtain the relative weights of the risks and criteria. For more details on process of obtaining relative weights of criteria/risks for patients' prioritization using AHP/ANP in real practice, please refer to (Abbasgholizadeh Rahimi et al., 2016; Abbasgholizadeh Rahimi & Jamshidi, 2014). The calculated weights are as followings:

Criteria's Weights = {0.08, 0.68, 0.24},

Risks' Weights = {0.79, 0.15, 0.06},

Suppose that surgeons assign following numbers to multiple criteria for each of four patients:

$F(p_1) = \{C_1/3,\ C_2/5,\ C_3/6\},$

$F(p_2) = \{C_1/4,\ C_2/7,\ C_3/9\},$

$F(p_3) = \{C_1/5,\ C_2/8,\ C_3/2\},$

$F(p_4) = \{C_1/1,\ C_2/3,\ C_3/4\},$

Then the fuzzy soft set (F, P) is a parameterized family of all fuzzy sets over C and gives a collection of an approximate description of the Patient-Criteria in the hospital. This fuzzy soft set (F,P) represents the relation matrix Patient-Criteria (PC) and is:

$$C_1(0.79)\ \ C_2(0.15)\ \ C_3(0.06)$$

$$PC = \begin{matrix} p_1 \\ p_2 \\ p_3 \\ p_4 \end{matrix} \begin{bmatrix} \tilde{3} & \tilde{5} & \tilde{6} \\ \tilde{4} & \tilde{7} & \tilde{9} \\ \tilde{5} & \tilde{8} & \tilde{2} \\ \tilde{1} & \tilde{3} & \tilde{4} \end{bmatrix}$$

And the relation matrix of Criteria-Risk (CR) is as followings:

$$CR = \begin{matrix} \\ c_1 \\ c_2 \\ c_3 \end{matrix} \begin{matrix} R_1(0.79) & R_2(0.15) & R_3(0.06) \\ \tilde{4} & \tilde{2} & \tilde{3} \\ \tilde{6} & \overset{\frown}{5} & \tilde{7} \\ \tilde{9} & \tilde{8} & \tilde{1} \end{matrix}$$

Then, after assigning weights to relative columns in PC and CR matrices, and performing the transformation operation PC*⊗CR* is performed and authors get the Patient-Risk matrix PR as:

$$PR = \begin{matrix} \\ p_1 \\ p_2 \\ p_3 \\ p_4 \end{matrix} \begin{matrix} R_1 & R_2 & R_3 \\ \widetilde{27.1} & \widetilde{4.3} & \widetilde{1.5} \\ \widetilde{38.9} & \widetilde{6.2} & \widetilde{2.1} \\ \widetilde{30.4} & \widetilde{4.7} & \widetilde{2.3} \\ \widetilde{16.7} & \widetilde{2.7} & \widetilde{0.9} \end{matrix}.$$

where:

$$\widetilde{27.1} = (18.7,\ 27.1,\ 37), \widetilde{4.3} = (2.9,\ 4.3,\ 6), \widetilde{1.5} = (0.99,\ 1.5,\ 2.2),$$

$$\widetilde{38.9} = (28.8,\ 38.9,\ 50.6), \widetilde{6.2} = (4.5,\ 6.2,\ 8.3), \widetilde{2.1} = (1.49,\ 2.1,\ 2.9),$$

$$\widetilde{30.4} = (21.07,\ 30.4,\ 41.4), \widetilde{4.7} = (3.1,\ 4.7,\ 6.6), \widetilde{2.3} = (1.75,\ 2.3,\ 3.13),$$

$$\widetilde{16.7} = (9.9,\ 16.7,\ 25.1), \widetilde{2.7} = (1.57,\ 2.7,\ 4.1), \widetilde{0.9} = (0.48,\ 0.92,\ 1.48),$$

Above numbers are gained using equation (2). Now, after defuzzifying the above numbers using equation (1), the defuzzified PR will be as following:

$$Defuzzified\,PR =$$

$$\begin{array}{c} \\ p_1 \\ p_2 \\ p_3 \\ p_4 \end{array} \begin{array}{c} \begin{array}{ccc} R_1 & R_2 & R_3 \end{array} \\ \begin{bmatrix} 19.5 & 40.25 & 15.45 \\ 28.1 & 58.25 & 21.45 \\ 17.5 & 33.75 & 22.35 \\ 11.9 & 25.25 & 8.85 \end{bmatrix} \end{array}$$

After summation of each row what is obtained will be summation of risks for each patient. The final matrix *Sum* is:

$$\rightarrow Sum = \begin{array}{c} p_1 \\ p_2 \\ p_3 \\ p_4 \end{array} \begin{bmatrix} 33.5 \\ 47.8 \\ 38.1 \\ 20.8 \end{bmatrix}$$

Considering the obtained priority scores in the *Sum* matrix, the second patient's (p_2) situation is more risky than others and this patient should be the first priority for surgery. Other patients' prioritization would be as following: $p_3 > p_1 > p_4$.

ROBUSTNESS ANALYSIS

The proposed framework is deemed robust if it obtains the same result under different (uncertain) operating conditions. Will the same patient be recommended for treatment if the criteria and risks' weights vary? This analysis is important to validate model's reliability. This is especially important for critical problems, where the margin of error could mean fewer life years saved for waiting patients. Figure 2 illustrates the sensitivity analysis results in multiple scenarios when the weight of the most important criterion (i.e. C_2) and the most important risk (i.e. R_1) varied by ±10% and ±20% of its original value. Scenarios 1-4 are related to criteria and scenarios 5-8 are related to risks.

From the experiments, authors can conclude that the proposed framework is robust under all eight scenarios, and the decision makers should be confident that patient 2 is the optimal choice, even though the derived criteria/risk's importance weights may be imperfect, and the probability of choosing the wrong patient for treatment is therefore almost zero.

COMPARISON

For illustrating the advantages and contributions of the proposed framework, Table 2 is provided. This table compares the proposed integrated method with four other well-known methods (namely: Analytic

Figure 2. Sensitivity analysis under eight different scenarios
For a more accurate representation of this figure, please see the electronic version.

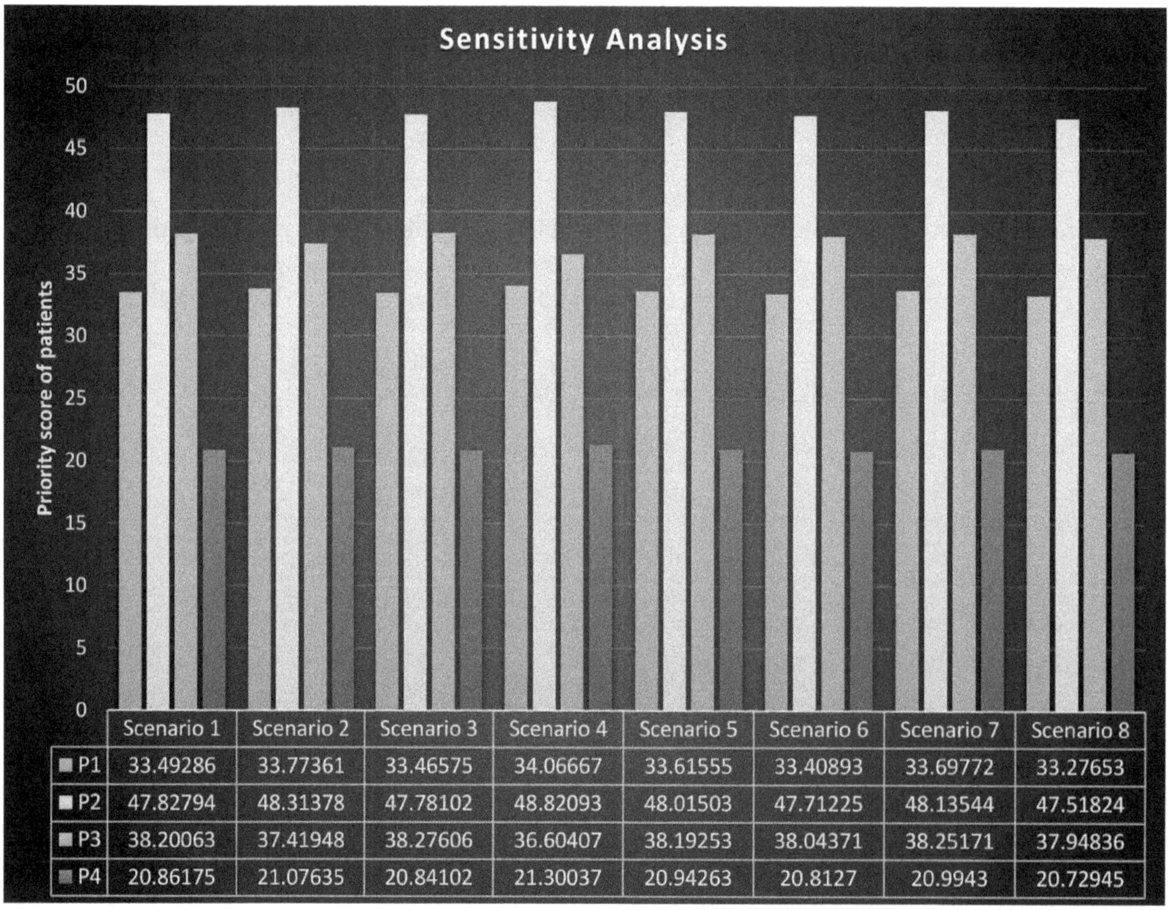

Table 2. Comparison of the proposed framework with other we-known methods

	AHP	ANP	TOPSIS	DEA	Proposed Method
Consider uncertainty					✓
Involve associated risks					✓
Involve multiple criteria	✓	✓	✓	✓	✓
Involve group of Decision Makers (DMs)	✓	✓		✓	✓
Consider importance weight of each expert (based on their knowledge)	✓	✓			✓
Assign importance weights for criteria	✓	✓	✓	✓	✓
Consider dependency among criteria		✓			✓
Easiness of the procedure (simplicity)	✓		✓		✓
Time-efficiency in large scale					✓

Hierarchy Process (AHP), Analytic Network Process (ANP), Technique for Order of Preference by Similarity to Ideal Solution (TOPSIS), and Data Envelopment Analysis (DEA)) from different perspectives.

In brief, none of the mentioned methods consider the associated uncertainties in patients' prioritization procedure except the proposed framework. Besides, in the proposed framework the authors involved the associated risks in the procedure as an important factor while this important issue has been overlooked in others. All of the methods including the proposed framework consider multiple criteria and we can assign importance weights for these criteria, but none of them except ANP and proposed framework considers the relationships among criteria. This framework can consider group of decision makers' opinion like other methods. From the time point of view, the required time for prioritizing patients in large scale in the proposed framework is less than others. This comparison has been done based on experts' opinion however, for more precise comparison further studies in practice is required. This could be done by conducting a clinical studies, involving clinicians in the procedure, and implementing each methodology in practice and comparing the results and each of the mentioned aspects in different methodologies.

CONCLUSION

Access to healthcare services and long waiting time is one of the main issues in most of the countries including Canada and United States, and based on recent reports, this situation not only have not been improved during years till 2016, but also they have gotten slightly worse. Healthcare organizations can't increase their limited resources nor treat all patients simultaneously. Then, patients' access to these services should be prioritized in a way that best use the scarce resources and most importantly insure safety of patients who are waiting for treatment.

Prioritization is essential and inevitable not only because of resource shortage, which have not been improved during years, but also because it may be one of the best options in identification of high risk patients and increasing safety of patients' waiting for treatment. Prioritization is a crucial issue that will contribute to the capability and stability of the systems as well. This decision making under complex, uncertain and dynamic environments is not simple as ABC so in order to reduce the workload of clinicians and impact on waiting times for healthcare services, development of prioritization tools are necessary.

In this chapter, the authors proposed a novel integrated and easy to use framework which prioritizes a diverse mix of patients in waiting lists into an order of importance considering the following strengths and main features:

1. The importance of each criterion and each risk is computed, considering group of medical experts' opinions,
2. Associated uncertainties are handled,
3. Importance of each decision maker's opinion has been considered by assigning weights to their knowledge and experience to obtain more precise results,
4. Possible interrelationships among criteria and also among risks are considered, and
5. Various risks that threaten patients' safety have been considered as an important and distinguished factor.

This is an original and innovative framework and the above features, distinguish it from currently used methods. To exhibit the simplicity, and applicability of the proposed framework, in this chapter authors provided an example of surgical patients' prioritization. The proposed framework not only considers uncertainties and risks in determining patients' priorities, but also remains noticeably robust. Finally, the proposed method is compared with some of the well-known methods in the comparison section to show its main benefits. The results of this study may increase safety of patients' waiting for treatment, and could have a significant impact on both medical community and the public's faith in justice and equity.

Although the authors believe this study provides real value, there are areas for future enhancements. The main weakness of the proposed model is that patients were not involved in the prioritization procedure. Besides, dynamic nature of the healthcare system has not been considered in this study, involving these aspects could increase the validity and efficacy of the results. Another interesting future research in this regard would be to study whether the proposed framework yield a fruitful result in real practice in different clinical settings.

REFERENCES

Abbasgholizadeh Rahimi, S., & Jamshidi, A. (2014a). *Prioritization of Organ Transplant Patients using Analytic Network Process.* Montreal, Canada: Academic Press.

Abbasgholizadeh Rahimi, S., Jamshidi, A., Ruiz, A. & Ait-Kadi, D. (2014b). *Applied Methods in Prioritization of Patients in Surgery Waiting Lists.* Montreal, Canada: Academic Press.

Abbasgholizadeh Rahimi, S., Jamshidi, A., Ruiz, A., & Ait Kadi, D. (2016). A new dynamic integrated framework for surgical patients prioritization considering risks and uncertainties. *Decision Support Systems*, *88*, 112–120. doi:10.1016/j.dss.2016.06.003

Addis, B., Carello, G., Grosso, A., & Tanfani, E. (2015). *Operating room scheduling and rescheduling: a rolling horizon approach.* Flex Serv Manuf J.

Ahlert, M., & Kliemt, H. (2013). Problems of priority change in kidney allocation and beyond. *The European Journal of Health Economics*, *14*(3), 383–390. doi:10.1007/s10198-012-0382-y PMID:22358456

Aktas, H., & Cagman, N. (2007). Soft sets and soft groups. *Inf Sci*, *177*(13), 2726–2735. doi:10.1016/j.ins.2006.12.008

Al-Ebbinia, L., Oztekin, A., & Chen, Y. (2016). FLAS: Fuzzy lung allocation system for US-based transplantations. *European Journal of Operational Research*, *248*(3), 1051–1065. doi:10.1016/j.ejor.2015.08.001

Ali, M., Feng, F., Liu, X., Min, W. K., & Shabir, M. (2009). On some new operations in soft set theory. *Computers & Mathematics with Applications (Oxford, England)*, *57*(9), 1547–1553. doi:10.1016/j.camwa.2008.11.009

Ali, M., & Shabir, M. (2010). Comments on De Morgan's law in fuzzy soft sets. *J. Fuzzy Math.*, *18*(3), 679–686.

Andersson, A., Omberg, M., & Svedlund, M. (2006). Triage in the emergency department—a qualitative study of the factors which nurses consider when making decisions. *Nursing in Critical Care*, *11*(3), 136–145. doi:10.1111/j.1362-1017.2006.00162.x PMID:16719019

Argon, N., & Ziya, S. (2009). Priority Assignment Under Imperfect Information on Customer Type Identities. *Manuf Serv Oper Manag*, *11*, 674–693.

Ashour, O., & Kremer, G. (2016). Dynamic patient grouping and prioritization: A new approach to emergency department flow improvement. *Health Care Management Science*, *19*(2), 192–205. doi:10.1007/s10729-014-9311-1 PMID:25487711

Ashour, O. & Okudan, G. (2010). *Patient sorting through emergency severity index and descriptive variables' utility.* Academic Press.

Barua, B. (2015). *Waiting Your Turn: Wait Times for Health Care in Canada, 2015 Report.* Fraser institute.

Barua, B., & Esmail, N. (2013). *Why are we waiting so long? Health care wait times nearly double over the past two decades to 18.2 weeks.* Fraser Institute.

Butterfield, W. (1968). *Priorities in medicine.* London: The Nuffield Provincial Hospitals Trust.

Celik, Y., & Yamak, S. (2013). Fuzzy soft set theory applied to medical diagnosis using fuzzy arithmetic operations. *Journal of Inequalities and Applications*, *82*, 1–9.

Chetia, B., & Das, P. (2010). An application of interval valued fuzzy soft set in medical diagnosis. *Int. J. Contemp. Math. Sci.*, *5*(38), 1887–1894.

Cholongitas, E., & Burroughs, A. (2012). The evolution in the prioritization for liver transplantation. *Annals of Gastroenterology*, *25*(1). PMID:24713804

Cholongitas, E., Germani, G., & Burroughs, A. K. (2010). Prioritization for liver transplantation. *Gastroenterology & Hepatology Nature Reviews*, *7*(12), 659–668. doi:10.1038/nrgastro.2010.169 PMID:21045793

Chung, S., Lee, A., & Pearn, W. (2005). Analytic network process (ANP) approach for product mix Planning in semiconductor fabricator. *International Journal of Production Economics*, *96*(1), 15–36. doi:10.1016/j.ijpe.2004.02.006

Clare, P. (1973). *Waiting Lists: Priority Formulae.* Birmingham, UK: Notes taken at a meeting of the Midlands Health Services Discussion Group.

Claudio, D., & Okudan, G. (2010). Utility function based patient prioritization in the emergency department. *European Journal of Industrial Engineering*, *4*(1), 59–77. doi:10.1504/EJIE.2010.029570

Comas, M., Castells, X., Hoffmeister, L., Román, R., Cots, F., Mar, J., & Espallargues, M. et al. (2008). Discrete-Event Simulation Applied to Analysis of Waiting Lists.Evaluation of a Prioritization System for Cataract Surgery. *Value in Health*, *11*(7), 1203–1213. doi:10.1111/j.1524-4733.2008.00322.x PMID:18494754

Culyer, A., & Cullis, J. (1976). Some economics of hospital waiting lists in the NHS. *Journal of Social Policy*, *5*(3), 239–264. doi:10.1017/S0047279400004748

Davis, B., & Johnson, S. (1998). Real-time priority scoring system must be used for prioritisation on waiting lists. *British Medical Journal*. PMID:10373189

Day, B. (2013). *Reducing Wait Times for Health Care*. Fraser Institute.

De, S., Biswas, R., & Roy, A. (2001). An application of intuitionistic fuzzy sets in medical diagnosis. *Fuzzy Sets and Systems*, *117*(2), 209–213. doi:10.1016/S0165-0114(98)00235-8

Dennett, E., & Parry, B. (1998). Generic surgical priority criteria scoring system: The clinical reality. *The New Zealand Medical Journal*, *111*, 163–166. PMID:9612483

Dolan, P., & Cookson, R. (2000). A qualitative study of the extent to which health gain matters when choosing between groups of patients. *Health Policy (Amsterdam)*, *51*(1), 19–30. doi:10.1016/S0168-8510(99)00079-2 PMID:11010223

Domènech, M., Adam, P., Tebé, C., & Espallargues, M. (2013). Developing a universal tool for the prioritization of patients waiting for elective surgery. *Journal of Health Policy*, *113*(1-2), 118–126. doi:10.1016/j.healthpol.2013.07.006 PMID:23932414

Dowseya, M., Gunn, J., & Choong, P. (2014). Selecting those to refer for joint replacement: Who will likely benefit and who will not? *Best Practice & Research. Clinical Rheumatology*, *28*(1), 157–171. doi:10.1016/j.berh.2014.01.005 PMID:24792950

Eltringham, D., & Clare, P. (1973). *Waiting List Management by Computer*. OR Unit Regional Hospital Board.

Feng, F., & Li, Y. (2013). Soft subsets and soft product operations. *Inf Sci*, *232*, 44–57. doi:10.1016/j.ins.2013.01.001

Fieldsa, E., Okudana, G., & Ashour, O. (2013). Rank aggregation methods comparison: A case for triage prioritization. *Expert Systems with Applications*, *40*(4), 1305–1311. doi:10.1016/j.eswa.2012.08.060

Fordyce, A., & Phillips, R. (1970). Waiting list management by computer. *The Hospital*, *66*(9), 303–305.

Godber, G. (1970). Priorities in medicine. *The Journal of the Royal College of General Practitioners*, *20*, 313–322. PMID:5500451

Hadorn, D. (2000). Setting priorities for waiting lists: Defining our terms. *Canadian Medical Association Journal*, *163*(7), 857–860. PMID:11033717

Hadorn, D., & Holmes, A. (1997). The New Zealand priority criteria project Part 2: Coronary artery bypass graft surgery. *British Medical Journal*, *314*(7074), 135–138. doi:10.1136/bmj.314.7074.135 PMID:9006478

Kaufmann, A., & Gupta, M. (1991). Introduction to Fuzzy Arithmetic Theory & Applications. Van Nostrand.

Lack, A., Edwards, R., & Boland, A. (2000). Weights for waits: Lessons from Salisbury. *Journal of Health Services Research & Policy*, *5*(2), 83–88. PMID:10947552

Langham, S., & Thorogood, M. (1996). Waiting lists, Bypassing the time. *The Health Service Journal*, *106*(33). PMID:10156829

Lavee, J., & Brock, D. (2012). Prioritizing registered donors in organ allocation: An ethical appraisal of the Israeli organ transplant law. *Current Opinion in Critical Care*, *18*(6), 707–711. doi:10.1097/MCC.0b013e328357a2e2 PMID:22914426

Li, D.-F. (2014). *Decision and Game Theory in Management With Intuitionistic Fuzzy Sets*. Springer. doi:10.1007/978-3-642-40712-3

Li, D.-F., & Liu, J.-C. (2015). A parameterized non-linear programming approach to solve matrix games with payoffs of I-fuzzy numbers. *IEEE Transactions on Fuzzy Systems*, *23*(4), 885–896. doi:10.1109/TFUZZ.2014.2333065

Li, D.-F., & Ren, H.-P. (2015). Multi-attribute decision making method considering the amount and reliability of intuitionistic fuzzy information. *Journal of Intelligent & Fuzzy Systems*, *28*, 1877–1883.

Lin, C., & Harris, S. (2012). A Unified Framework for the Prioritization of Organ Transplant Patients: Analytic Hierarchy Process, Sensitivity and Multifactor Robustness Study. *Journal of Multi-Criteria Decision Analysis*, *20*(3-4), 157–172. doi:10.1002/mcda.1480

Luckman, J., Mackenzie, M., & Stringer, J. (1969). *Management Policies for Large Ward Units*. Coventry, UK: Institute for Operational Research.

MacCormick, A., Collecutt, W., & Parry, B. (2003). Prioritizing patients for elective surgery: A systematic review. *ANZ Journal of Surgery*, *73*(8), 633–642. doi:10.1046/j.1445-2197.2003.02605.x PMID:12887536

Maji, P., Biswas, R., & Roy, A. (2001). Fuzzy soft set. *J. Fuzzy Math.*, *9*(3), 677–692.

Maji, P., Biswas, R., & Roy, A. (2003). *Soft set theory*. Comp.&Math.Appl.

Mariotti, G., Siciliani, L., Rebba, V., Fellini, R., Gentilini, M., Benea, G., & Liva, C. et al. (2014). Waiting time prioritisation for specialist services in Italy: The homogeneous waiting time groups approach. *Health Policy (Amsterdam)*, *117*(1), 54–63. doi:10.1016/j.healthpol.2014.01.018 PMID:24576498

Meyer, T., & Raspe, H. (2012). Priority setting: Priorisierung: Was ist das und wie geht das? *Rehabilitation, 51*(2), 73–80. doi:10.1055/s-0032-1306288 PMID:22570153

Min, D., & Yih, Y. (2010). An elective surgery scheduling problem considering patient priority. *Computers & Operations Research, 37*(6), 1091–1099. doi:10.1016/j.cor.2009.09.016

Molodtsov, D. (1999). Soft set theory first results. *Computers & Mathematics with Applications (Oxford, England), 37*(4-5), 19–31. doi:10.1016/S0898-1221(99)00056-5

Montoya, S., González, M., & López, S. et al.. (2014). Study to Develop a Waiting List Prioritization Score for Varicose Vein Surgery. *Annals of Vascular Surgery, 28*(2), 306–312. doi:10.1016/j.avsg.2012.11.017 PMID:24084264

Mullen, P. (2003). Prioritizing waiting lists: How and why? *European Journal of Operational Research, 150*(1), 32–45. doi:10.1016/S0377-2217(02)00779-8

Nagel, E., & Lauerer, M. (2016). *Prioritization in Medicine An International Dialogue.* Springer International. doi:10.1007/978-3-319-21112-1

Nagel, E., & Lauerer, M. (2016). *Prioritization in Medicine: An International Dialogue.* Springer International Publishing. doi:10.1007/978-3-319-21112-1

Naylor, C. (1993). Waiting for coronary revascularization in Toronto: 2 years' experience with a regional referral office. *Canadian Medical Association Journal, 149*(7), 955–962. PMID:8402424

Noseworthy, T., McGurran, J., & Hadorn, D. C. (2003). Waiting for scheduled services in Canada: Development of priority-setting scoring systems. *Journal of Evaluation in Clinical Practice, 9*(1), 23–31. doi:10.1046/j.1365-2753.2003.00377.x PMID:12558699

Peacock, S., et al. (2009). Overcoming barriers to priority setting using interdisciplinary methods. *Health Policy, 92*(2-3), 124-32. doi: 10.1016/j.healthpol.2009.02.006

Phoenix, C. J. (1972). Waiting list management and admission scheduling. Butterworth.

Pornak, S., & Raspe, H. (2015). Die dänische Debatte über Priorisierung und Posteriorisierung in der Medizin – Was gibt es Neues? *Gesundheitswesen (Bundesverband der Arzte des Offentlichen Gesundheitsdienstes (Germany)), 77*, 1–7.

Prentice, J., & Pizer, S. (2007). Delayed Access to Health Care and Mortality. *Health Services Research, 42*(2), 644–662. doi:10.1111/j.1475-6773.2006.00626.x PMID:17362211

Raspe, H. (2016). Prioritisation: (At Least) Two Normative. In *Prioritization in Medicine: An International Dialogue* (pp. 85–99). Springer International Publishing. doi:10.1007/978-3-319-21112-1_7

Riet, C., & Demeulemeester, E. (2015). Trade-offs in operating room planning for electives and emergencies: A review. *Operations Research for Health Care, 7*, 52–69. doi:10.1016/j.orhc.2015.05.005

Russell, C., Roberts, M., Williamson, T. G., McKercher, J., Jolly, S. E., & McNeil, J. (2003). Clinical categorization for elective surgery in Victoria. *ANZ Journal of Surgery*, *73*(10), 839–842. doi:10.1046/j.1445-2197.2003.02797.x PMID:14525579

Saaty, T., & Özdemir, M. (2005). *The Encyclion: A Dictionary of Decisions with Dependence and Feedback based on the ANP*. RWS Publications.

Sadegh-Zadeh, K. (2012). *Handbook of analytic philosophy of medicine*. New York: Springer Dordrecht Heidelberg. doi:10.1007/978-94-007-2260-6

Sanchez, E. (1979). Inverse of fuzzy relations, application to possibility distributions and medical diagnosis. *Fuzzy Sets and Systems*, *2*(1), 75–86. doi:10.1016/0165-0114(79)90017-4

Seddon, M., French, J. K., Amos, D. J., Ramanathan, K., McLaughlin, S. C., & White, H. D. (1999). Waiting times and prioritisation for coronary artery bypass survey in New Zealand. *Heart (British Cardiac Society)*, *81*(6), 586–592. doi:10.1136/hrt.81.6.586 PMID:10336915

Soljak, M. (1997). *The Use of Scoring Systems in the NHS: Cardiac Surgery in London. From Waiting Lists to Booking Systems HSMC*. University of Birmingham.

Tanabe, P., Travers, D., Gilboy, N., Rosenau, A., Sierzega, G., Rupp, V., & Adams, J. G. et al. (2005). Refining emergency severity index (ESI) triage criteria. *Academic Emergency Medicine*, *12*(6), 497–501. doi:10.1111/j.1553-2712.2005.tb00888.x PMID:15930399

Testi, A., Tanfani, E., Valente, R., Ansaldo, G. L., & Torre, G. C. (2008). Prioritizing surgical waiting lists. *Journal of Evaluation in Clinical Practice*, *14*(1), 59–64. doi:10.1111/j.1365-2753.2007.00794.x PMID:18211645

Valente, R., Testi, A., Tanfani, E., Fato, M., Porro, I., Santo, M., & Ansaldo, G. et al. (2009). A model to prioritize access to elective surgery on the basis of clinical urgency and waiting time. *BMC Health Services Research*, *9*(1), 1. doi:10.1186/1472-6963-9-1 PMID:19118494

Williams, I., Robinson, S., & Dickinson, H. (2012). *Rationing in health care: the theory and practice of priority setting*. Bristol: The Policy Press.

Yang, M., Fry, M., & Raikhelkar, J. (2013). A Model to Create an Efficient and Equitable Admission Policy for Patients Arriving to the Cardiothoracic ICU. *The Clinical Investigator*, *41*(2), 414–422. PMID:23263573

Yu, D.-J., & Li, D.-F. (2014). Dual hesitant fuzzy multi-criteria decision making and its application to teaching quality assessment. *Journal of Intelligent and Fuzzy Systems*, *27*(4), 1679–1688.

Zadeh, L. (1965). Fuzzy sets. *Information and Control*, *8*(3), 338–353. doi:10.1016/S0019-9958(65)90241-X

KEY TERMS AND DEFINITIONS

First-Come First-Served: A basic prioritization rule mentioning the first arrived client should be the first to receive the service.

Group Decision Making: Group decision making give an opportunity to group of experts and professionals to express their decisions and opinions and considers the values that each point of views as important.

Multi Criteria Decision Analysis: A valuable tool that could solve complex decision making problems, usually involve the problems that are characterized as a choice among alternatives.

Prioritization: Deciding which clients are perceived or measured important or significant and ranking them based on their importance or significance called prioritization.

Risks: A thing that may happen or be the cause of loss, injury, disease, or death.

Soft Set: In order to deal with uncertainty in a parametric manner, (in 1999) Molodtsov proposed a generalization of fuzzy set theory called Soft set theory.

The Organization for Economic Co-Operation and Development (OECD): OECD is a forum consisting of the governments of 34 democracies with market economies work with each other to promote economic growth, prosperity, and sustainable development.

Waiting List: A waiting list is a queue of clients who are assumed to need a service that is in short supply.

Chapter 10
Performance Estimation of Firms by G-L-A Supply Chain under Imperfect Data

Anoop Kumar Sahu
J.K.I.E., India

Nitin Kumar Sahu
Guru Ghasidas Vishwavidyalaya, India

Atul Kumar Sahu
Guru Ghasidas Vishwavidyalaya, India

ABSTRACT

Amongst various proposed SC philosophies, Lean, Green and Agile strategies of SC worked seminal to solve many problems of firms. Performance measurement acts as a tool to quantify overall efficiency via G-L-A SC activities of firm, where Green SC agenda is to control and diminish pollution and Lean SC helps to reduce waste of firm, while Agile SC target to manage SC more cum quick responding to clients. This chapter proposed a MCDM performance appraisement module (module constituted by mixing the segregated chain of green-lean-agile logistic activities and corresponding their interrelated metrics) conjunctive with Fuzzy Performance Index model in purpose to estimation the overall performance of individual firm. Furthermore, a centroid method coupled with fuzzy number set is proposed for classifying ill and strong G-L-A metrics of firm, so that managers could escalate their firms performance in case of non desirable performance. A hypothetical case research of two firms i.e. gear and shaft manufacturing firms are shown to measure their performance under G-LA supply chain.

HISTORICAL BACKGROUND AND SCM

The philosophy of Supply Chain (SC) was articulated by Tom Peter in front of Ford firm. He elaborated that SC network, is counted as an incomplete network as long as links aren't established amongst:

DOI: 10.4018/978-1-5225-1848-8.ch010

- The vendor,
- Manufacturing firm,
- Warehouse,
- Wholesaler,
- Retailer and end users,

Each element of said network is known as agent of SC. In SC, entire agents are allied with each other through an inter relationship, communication and collaboration for its own mutual profit. As a practical example, he also considered the convention centre of ford firm as a particle application, and explained that the convention centre of ford firm is built and not into use. Ford firm must use this convention centre for sheep to supply wool for its own automobile upholstery. Moreover, he also discussed that the degradation in efficiency and effectiveness of any agent of SC not solely leverages the performances of SC agents survive at upstream and downstream level of SC, but also disgrace the reputation of each agent.

During the half spanning of 19th century, the philosophy of supply chain is impounded by several industries, as it worked for a great success, growth and excellence in business of firms. In the mid of 19th century, proposed concept is accepted as a heart of each firm. Many definitions are proposed in regards to supply chain, expressed here.

Definitions of Supply Chain

SC is a network of organizations that are involved, through upstream and downstream linkages, in the different processes and activities, which produce value in the forms of products and services in the hands of the ultimate customer (Christopher, 1998). SC is considered as a network that performs the procurement of raw material, transformation of materials into intermediate and end-products, distribution and selling of the end-products to end customers (Petrovic, Roy, & Petrovic, 1999). SC is an integrated process, which involves the processing of raw materials into finished goods and making them available to the end user (Beamon, 1999). SC is the upstream fraction of the value chain activities aiming to ensure that the right materials, services and technology are purchased from the right sources, at right time, at right quantity. SC is a continuous process, from raw materials to finished goods, via each traditional distinct function such as forecasting, purchasing, manufacturing, distribution, and sales and marketing (Chan, Qi, Chan, & Lau, 2003).

SC comprises a worldwide network of suppliers, factories, warehouses, distribution centre's and retailers through which raw materials are acquired, transformed and delivered to the end user (Lin, Chiu, & Tseng, 2006). SC is considered as a value-adding relationship between partially discrete, yet interdependent entities that cooperatively procure and transform raw materials into finished products through sequential network structures (Hult, Ketchen, & Arrelti, 2007). SC is a combined system which comprises planning, sourcing, making and development of processes with its constituent parts to include:

- Material suppliers,
- Production facilities,
- Distribution centers and customers linked together through the feed forward flow of material,
- As well as feedback flow of information (Olugu & Wong, 2009).

Later, the concept of SC management was introduced on account of several below said grounds / reasons.

First, companies have become more specialized in searching their suppliers who may provide them materials i.e. stuffs, semi finished goods etc. on feasible price with best quality under lead time. It became imperative for companies to select the best supplier under overall supply chains.

Second, partial stems from increased national and international competition. Companies had multiple sources to choose their supplier for satisfying customer demands, locating products throughout the distribution channel for maximum customer accessibility at minimum cost. Previously, companies looked at solving the distribution problems through maintaining inventory at various locations throughout the supply chain. However, the dynamic nature of the marketplace makes holding inventory a risky and potentially unprofitable business. Customers' buying habits are constantly changing, and competitors are continuously adding and deleting products. The cost of holding any inventory also means most companies cannot provide a low cost product when funds are tied up in inventory.

Third, companies indeed necessity to be negotiated with supplier firms for purchasing materials with respect to low price under lead time. Therefore, the necessity to develop the decision making models as well approaches for electing supplier became essential. The definition of Supply chain management is discussed here:

SCM integrate suppliers, manufacturers, warehouses and stores so as to produce and distribute products and services to the customers at right quantities, at right locations at right time, minimizing the system wide cost while satisfying the service level requirements. SCM is concerned to integrate its logistic, procurement, operations and marketing functions with other supply chain members, in order that material, information, component/parts and finished product may flow seamless from point of origin to final customer at the low unit cost and high level services. (Peter & Thomas, 2000)

Figure 1 describes the parameters, used to manage the supply chain of firms.

Figure 1. Supply chain management

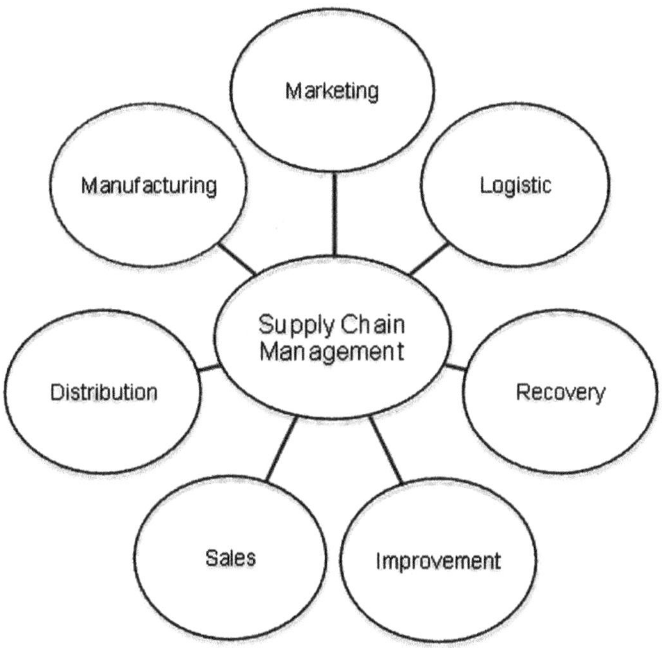

The verdict making is found so difficult task, which drags the brain of peoples into implications. Right verdict help the firms for achieving their unified goals such as enhancing productivity, customer satisfaction, complete the demand and maximize profit (minimize losses). In last decade, the decision making has found out its applications in the vast range of in managing effective supply chain. In today's era, MCDM process compensated the need of firm, as it aids the boss of the firm for carrying out the decision under contradicting supply chain activities. MCDM help the boss of firm to manage supply chain performance of own firm.

In last year, many decision support systems are developed and helped the managers of supplier selection and too balance the supply chain of their firms. But, it is indeed necessarily to develop the strong and upgraded decision support models and systems.

GREEN SUPPLY CHAIN

Across the boundary of 1990, in industries, managers individually paid attention at arm's length of quantity and quality cum SC revolution and weren't viewed the SC actions / activities of their firms coupled with environmental (green) issues. Merely a separate department was responsible to handle and control ongoing environmental pollution. That department significantly focused to control the green issues caused by individually manufacturing activities such as material handling, production and process etc. Firms weren't concentrated to control the green issues caused by SC core agent such as suppliers (Srivastava, 2007; Toke, & Dandekar, 2010). But when, so many industries began to establish and integrate their business over world on account of the customized goods demands as in market. Nonetheless, firms were steady with same department (Simpson & Samson, 2008). So, loose concentrations of firms towards managing the green activities in extent of traditional SC created the causes of rapid deterioration of environment, which resulted the undesirable vast pollutions i.e.:

- Ill-biological particles,
- Fossil fuel,
- Hazard particle,
- Ill-particles,
- Toxic gasses etc. In environment or overcast.

To control or obtain the pollution at zero level in industrial realm, the industrialists, researchers and ecologists; scholars implemented the consistence rules & legislation, several reactive environmental policies, productive enterprise's green programs and environmentally goods policies in arena of traditional SC network known as Green Supply Chain (GSC) (Garvin, 1993; Min & Galle, 1997). Therefore, several definitions have been proposed by several researchers and ecologist scholars in regard to green supply chain in previews published manuscripts; defined in below section.

Green supply chain refers to the implementation of an environmental moral and thinking at the each instance of traditional supply chain networking. Green supply chain is considered as a way to drive out the unwanted and undesirable pollutions,

- e.g. Ill-biological particles,
- Fossil fuel,

- Hazard particle,
- Ill-particles,
- Toxic gasses etc., from the each subsystem/agents of traditional SC circuit.

Green supply chain deals with suppliers, factories, warehouses, distribution centers and retailers through, which raw materials are acquired, transformed and delivered to the end users.

Green supply chain consists of the purchasing function's involvement in activities that include reduction, recycling, reuse and the substitution of materials (Narasimhan & Carter, 1998). Figure 2 defined the green supply chain management resources to control the pollution in environment.

During the journey of last years, Multi Criterion Decision Making (MCDM) found out its application undertaking the problem in relation to green supply chain management. Last years, several DSSs, consist of multi criterion hierarchical modules coupled with MCDM methods, tool, software etc, in order to bring the decision to end in GSCM discipline. Currently, it is indeed necessity to develop momentous novel DSS in the arena of green supply chain management discipline in order to undertaking complicated decision making problems.

Figure 2. Green supply chain management

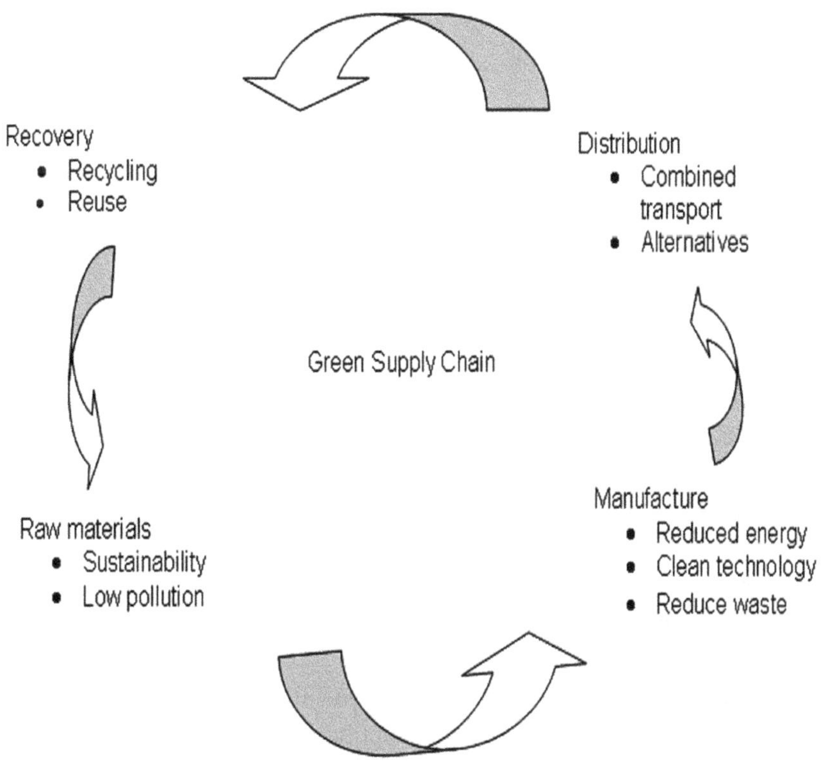

AGILE AND LEAN SUPPLY CHAIN

Agile supply chain deals with alters in dissimilar facets i.e. alter in customer demands, scientific advancements and unbalanced business environment (Sharifi, Colquhoun, Barclay, & Dann, 2001).

Agile supply chain is defined as an agility index, while is a mixture for measuring the strength levels of agility criterions and sub-criterions by exploration of multi criterion decision making methods. For measuring agility in agility supply chain, it is indeed to place those criterions into index, which have the recommendation of expert's panel. Agility supply chain help the manager to measure the responses towards the clients of their firms.

Agile Supply chain is an evaluation model based on fuzzy logic and the multi-criteria Decision-Making (MCDM), which provides a path for measuring supply chain agility and also recognizing the major barriers to improving the agility levels of defined firms. Agile supply chain is considered to be structure under the goals of satisfying clients and workers where each firm can design its own business policy, progression and information systems. The structure is built by four main beliefs:

- Mastering change and uncertainty,
- Pioneering management structures and practical organization,
- Supportive relationships, and
- Elastic and clever expertise and machineries.

Agile supply chain needs to be extremely plastic and to be able to be reconfigured rapidly in answer to alter in the business setting. Thus, an ASC can be thought of as an active network of constituent companies, whose constituents and construction are likely to need to alter frequently. The winning process of an ASC mainly depends upon the firm's ability to select the most suitable potential buddies in any given condition.

The main necessary ideas to develop a structure to analyze the mutual performance of corporations have come into view from information exchanged between buddies and exploitation of this distribution, mixing of activities, maturity degrees of partnership, performance indicators associated to high-quality joint practices. In last decade, the philosophy of agility has found out its applications in the vast range of in managing effective supply chain. MCDM in the context of agile supply chain help the boss of firm to manage supply chain performance of own firm and too draw.

In last year, many decision support systems are developed and helped the managers of supplier selection and too balance the supply chain of their firms. But, it is indeed necessarily to develop the strong and upgraded decision support models and systems. Figure 3 defined parameters for managing the leanness in supply chain.

The Lean is a team-based approach to nonstop upgrading technologies / approaches/ paths towards reducing the unworthily added activities or waste from the perspective of the clients. Lean idea originated since many years and approximately ran over most of the manufacturing firm. Few years ago, lean concept has been applied merged with supply chain & logistics management realm Called Lean supply chain Management.

Lean supply chain offers explanations of both basic and advanced Lean tools as well as specific Lean implementation occasions. Lean SC deals with the supplier election and balancing the supply chain of firms. Lean supply chain believes to create the waste elimination environment in industries. Lean supply chain implementation methodology with critical success factors for reducing the waste of firms. In last

Figure 3. Lean supply chain management

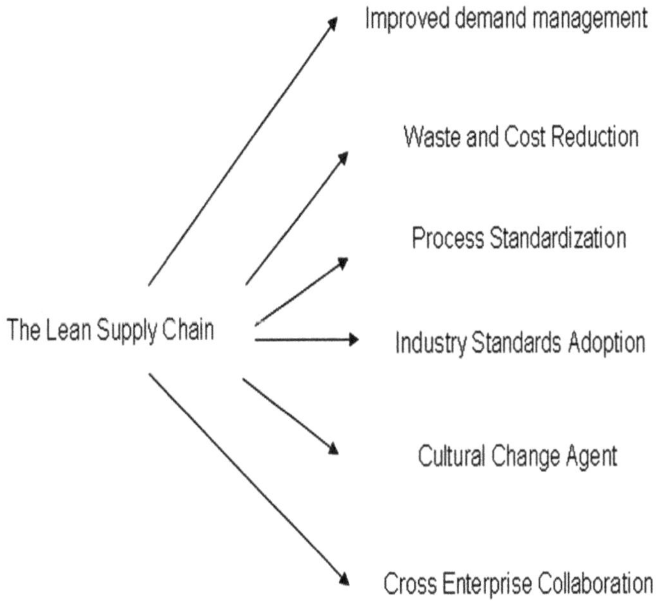

year, numerous lean supply chain based decision support systems are developed and helped the managers of supplier selection under lean strategy of SC. But, it is indeed realized by authors via literature survey, there is still need to develop the robust decision support module and systems for balancing the supply chain of firm. Figure 4 defined flow chart to manage the firm's performance for agile supply chain.

Figure 4. Agile supply chain management

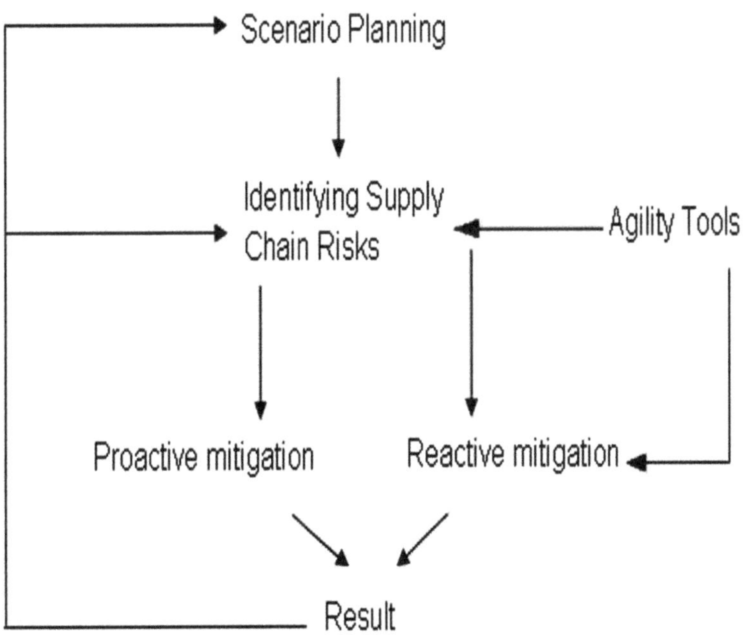

MOTIVATION

In last decades, each industry strikes to equilibrium own operational chains as per market demands, where the service, productivity and green concerns had low significance ratio than others i.e. rapid production, quality in manufacturing goods etc. But, from the last decade, customer satisfactions, productivity and green issues in industrial SC have become chief concerns to win competitive merit as well as customer satisfaction. Green and Lean SC have an objective to manage eco-friendly environment and reduce waste (manage standard productivity) throughout industrial supply chain, respectively, whereas agile has an objective to provide rapid service to clients. Therefore, it is investigated that there is no multi criteria hierarchical module, which can measure the performances of defined alternative firms under Green-Lean-Agile (G-L-A) supply chain. This issue has motivated the authors to introduce the notion of G-L-A supply chain, in order that managers could measure the performances of defined alternatives under green, lean and agile supply chains.

SC PERFORMANCES MEASUREMENT

Performance Measurement (PM) materializes the efficiency and effectiveness of industry against their practices, processes associated with SC network. Performance measurement describes the feedback or information on actions of supply chain towards accomplishing customer expectations and strategic objectives. It reflects the need for improvement in areas of unsatisfactory performance. By using of performance measurement tool, the efficiency and quality of firm can be escalated.

Performance measurement is counted as vital constituent of effective planning and control as well as verdict making. It provides essential criticisms or information to expose growth, augment in enthusiasm and identify problems. Performance measurement is consider as a part of methods, metrics, courses and systems, used in firms to:

- Explain strategies into tactics,
- Observe implementation, and
- Supply insight to get better financial and operations.

It reflects the tactical, incorporated development of commerce cleverness to hold up the organization courses. Performance measurement is concerted as the process of quantifying the efficiency and effectiveness of issues. Effectiveness is the extent to which customers' requirements are met while efficiency measures how economically a firm's resources are utilized (Neely, 2005; Kurien & Qureshi, 2011).

Last decade, performance measurement of individual or preferred firms under actions of SC become prominent very task, which create tension in brain of manager. MCDM (Multi Criterion Decision Making) realm effectively overcame managers' implications and aids them for measuring the performances of their firms under SC actions.

The performance measurement split up into two types (Li, 2010, 2016, 2014; 2009; Li & Sun, 2007a, 2007b; Cabral, Grilo & Machado, 2012; Chen & Chen, 2007; Kainuma & Tawara, 2006; Chen & Hsieh, 1999; Chen & Hwang, 1992; Knadel, 1986; Chen, 1985; Li & Yang, 2015; Sahu, Datta & Mahapatra, 2012, 2013, 2014, 2015; Samantra, Sahu, Datta & Mahapatra, 2013;; Li, Nan & Zhang, 2010):

1. Quantitative measurement.
2. Qualitative measurement.
    - **Quantitative Measurement:** The quantitative measurement dealt with defined or crisp value, which is associated with non-partial information (objective information). Quantitative measurement delivers the accurate result. As, the appropriateness ratings and importance weights are assigned in the forms of crisp value against criterion (consist of criteria wear certain information).
    - **Qualitative Measurement:** The qualitative measurement dealt with vagueness, incompleteness, and impreciseness; associates the partial or linguistic information (subjective value).

Qualitative measurement delivers the inaccurate result, as appropriateness ratings and importance weights are assigned in the form of non crisp value against criterions. It determined that in MCDM problem, measures, indices and metrics plays a very dominant role during the performance measurement of individual or preferred organization under actions of SC indices. Where, Measures are considered as main criterion and metrics is counted as inter-related criterions of criterion. Last years, MCDM proposed several modules combined with MCDM methods and mathematical equation for measuring the performance of firms in arena of supply chain management strategies.

LITERATURE REVIEW

The relevant literature review is conducted to determine the research gaps, few of them elaborated below:

Huiyu and Weiwei (2010), articulated that inadequate numbers of papers are published in relation to measure the performance of firm under green supply chain management strategies. The more definitions are indeed to build for better expressing the green supply chain management. According to Green, Morton and New (1998), Green supply chain manages the firm for environmental issues and too influence the relationships between supply chain management and the ordinary environment. It is also referred in form of modernism in supply chain management. Srivastava (2007), articulated that GSC (Green Supply Chain) is an integration of green thinking into a supply chain management, including product drawing, stuff resourcing and evaluation, built-up processes, delivery of thc final products to the shoppers as well as end-of-life. Hsu and Hu (2008), green supply chain management is an approach to escalate the performance of the manufacturing processes and goods as per the requirements of the green policies. Noted by Toke, Gupta and Dandekar (2010), Green supply chain management is increasing attention of researchers of operations and supply chain management for preserving the green environment in world. The growing significance of GSCM is determined mainly by the improving environment e.g. reducing raw material resources, spilling over waste sites and rising level of non-pollution.

Sharifi and Zhang (1999), explained the idea in regards to developing the methodologies in order to achieve agility. Gunasekaran (1999) made endeavors to review the literature of agile manufacturing to identify key strategies and techniques of agile manufacturing. Moreover, the authors proposed a number of future research directions and too develop a structure for the development of agile manufacturing systems accompanied with four input dimensions,

- i.e. Strategies,
- Technologies,
- Systems and
- Peoples.

Yusuf, Sarhadi and Gunasekaran (1999), recognized the criterions of agility and talked about the collection of competitive advantages that amalgamated over time as a result of the altering requirements of manufacturing. The need to attain the competitive merits of manufacturing in synergy and without tradeoffs was proposed as a basic tool to the agile. The authors also appraised the meaning of agility from dissimilar perspectives and recommended a comprehensive definition, which could be adopted as a working definition by practitioners. Meredith and Francis (2000), discussed about a variety of issues on agility, which were found by Agile Manufacturing Research Group (AMRG). The authors too recognized the exact policies and practices of supported agility. Sharifi and Zhang (2001), the authors built a technique for attaining agility in manufacturing organizations. The technique was applied in two manufacturing companies and information was collected from the applications to authenticate the technique. Moreover, the authors provided a brief synopsis of the technique and details of implementation and validation in the two case examples.

Panizzolo (1998), dealt with defies proposed by lean production values for operations management. The authors proposed a model to precisely describe the operational chain of lean production. The model represented a conceptualization of lean production, constituted by:

- Process and equipment,
- Manufacturing planning and control,
- Human resources,
- Product design,
- Supplier relationships,
- Customer relationships.

Arbos (2002), proposed a method for implementation of lean management in a services production system. Lean management demonstrated as a case study of telecommunication services. Pavnaskar, Gershenson and Jambekar (2003), proposed a classification system to dish up as a link between manufacturing waste problems and lean manufacturing tools. Shah and Ward (2003), examined the upshots of three companies taking into account fourteen criterions i.e. plant size, plant age and unionization status, on the likelihood of implementing twenty two manufacturing practices that were key facets of lean production systems. Furthermore, the authors postulate four inter-related and internally consistent criterions; these were related to:

- Just-in-time (JIT),
- Total quality management (TQM),
- Total preventive maintenance (TPM), and
- Human resource management (HRM).

A hypothetical case research carried out to investigate criterion upshots on operational performance. Kainumaa and Tawara (2006) proposed the multiple attribute utility theory technique for assessing a supply chain. The authors considered said technique to manage the lean and green supply chain. Hult, Ketchen and Arrelti (2007) examined the strategic supply chain management and improved the performance of this through a culture of competitiveness and knowledge development. Sahu, Sahu and Sahu (2016a), developed interval-valued fuzzy number set and merged this with modified MCDM technique TOPSIS for benchmarking supplier firms under agile supply chain actions. Sahu, Sahu and Sahu (2016b), used a revised ranking approach accompanied with fuzzy performance important index to identify the barriers/ill-strong measures in Agile Supply Chain Management.

After conducting mixed literature review on applications of MCDM towards tackling the performance measurement problems of firms under SC strategies, the authors determined that MCDM field ought to be smarter towards undertaking double or tipple level hierarchical module of DSS. Moreover, the need of novel MCDM-algorithmic models is too essential.

Theory of Generalized Trapezoidal Fuzzy Numbers

A fuzzy set $\tilde{A}$ in a universe of discourse X is characterized by a membership function $\mu_{\tilde{A}}(x)$ which associates with each element x in X a real number in the interval [0, 1]. The function value $\mu_{\tilde{A}}(x)$ is termed the grade of membership of x in $\tilde{A}$. A trapezoidal fuzzy number can be defined as $\tilde{A} = (a_1, a_2, a_3, a_4; w_{\tilde{A}})$, as shown in Figure 5 and the membership function $\mu_{\tilde{A}}(x) : R \to [0,1]$, is defined as follows:

$$\mu_{\tilde{A}}(x) = \begin{cases} \dfrac{x - a_1}{a_2 - a_1} \times w_{\tilde{A}}, & x \in (a_1, a_2) \\ w_{\tilde{A}}, & x \in (a_2, a_3) \\ \dfrac{x - a_4}{a_3 - a_4} \times w_{\tilde{A}}, & x \in (a_3, a_4) \\ 0, & x \in (-\infty, a_1) \cup (a_4, \infty) \end{cases} \tag{1}$$

Here, $a_1 \le a_2 \le a_3 \le a_4$ and $w_{\tilde{A}} \in (0,1)$

Figure 5. Trapezoidal fuzzy number $\tilde{A}$

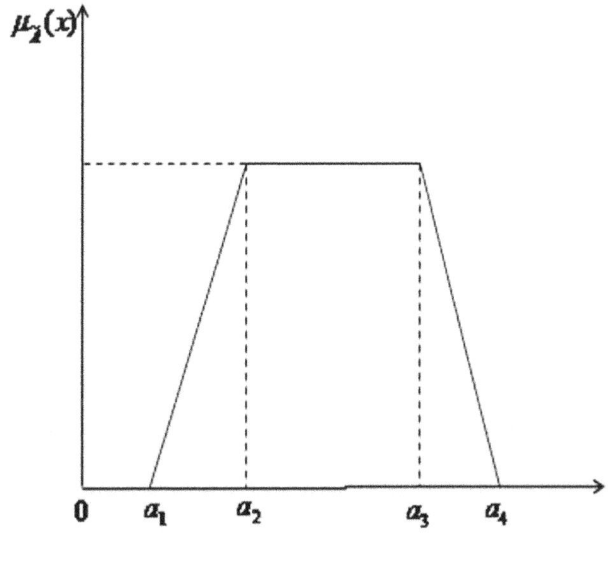

Suppose that $\tilde{a} = \left(a_1, a_2, a_3, a_4; w_{\tilde{A}}\right)$ and $\tilde{b} = \left(b_1, b_2, b_3, b_4; w_{\tilde{B}}\right)$ are two trapezoidal fuzzy numbers, then the operational rules of the trapezoidal fuzzy numbers $\tilde{a}$ and $\tilde{b}$ are shown as follows:

- Addition Operation

$$\tilde{a} \oplus \tilde{b} = \left(a_1, a_2, a_3, a_4; w_{\tilde{A}}\right) \oplus \left(b_1, b_2, b_3, b_4; w_{\tilde{B}}\right) =$$

$$\left(a_1 + b_1, a_2 + b_2, a_3 + b_3, a_4 + b_4; \min\left(w_{\tilde{A}}, w_{\tilde{B}}\right)\right) \tag{2}$$

- Substation Operation

$$\tilde{a} - \tilde{b} = \left(a_1, a_2, a_3, a_4; w_{\tilde{A}}\right) - \left(b_1, b_2, b_3, b_4; w_{\tilde{B}}\right) =$$

$$\left(a_1 - b_4, a_2 - b_3, a_3 - b_2, a_4 - b_1; \min\left(w_{\tilde{A}}, w_{\tilde{B}}\right)\right) \tag{3}$$

- Multiplication Operation

$$\tilde{a} \otimes \tilde{b} = \left(a_1, a_2, a_3, a_4; w_{\tilde{A}}\right) \otimes \left(b_1, b_2, b_3, b_4; w_{\tilde{B}}\right) =$$

$$\tilde{a} \otimes \tilde{b} = \left(a_1 \times b_1,\ a_2 \times b_2,\ a_3 \times b_3,\ a_4 \times b_4; \min\left(w_{\tilde{A}}, w_{\tilde{B}}\right)\right) \tag{4}$$

- Division Operation

$$\tilde{a} \,/\, \tilde{b} = \left(a_1, a_2, a_3, a_4; w_{\tilde{A}}\right) \Big/ \left(b_1, b_2, b_3, b_4; w_{\tilde{B}}\right)$$

$$= \left(a_1 / b_4, a_2 / b_3, a_3 / b_2, a_4 / b_1; \min\left(w_{\tilde{A}}, w_{\tilde{B}}\right)\right) \tag{5}$$

Ranking of Fuzzy Numbers

Fuzzy logic is a tool used to transform the oral information into mathematical data (Sahu et al., 2014, 2015a, 2015b, 2015c). The ranking of fuzzy numbers lead a pertinent job for approximate reasoning, optimization, forecasting, choice making, arrangement and analysis chain of criterions related to defined alternative. This chapter dealt with a ranking method proposed by Thorani, Rao and Shankar (2012). In this method, the centroid of a trapezoid is considered as the balancing tip of the trapezoid. Split out the trapezoid into three plane figures. These three plane figures are a triangle (APB), a rectangle

Figure 6. Trapezoidal fuzzy number

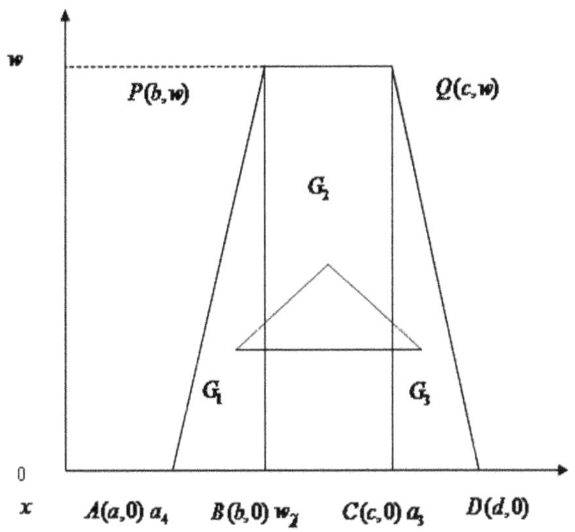

(BPQC), and a triangle (CQD), respectively. Let the centroid of the three plane figures be G_1, G_2 and G_3 respectively. The in center of these Centroid G_1, G_2 and G_3 is counted as the point of reference to describe the ranking of generalized trapezoidal fuzzy numbers set. The cause for selecting this point as a point of reference is that each centroid point are balancing points of each individual plane figure, and the in centre of these centroid points is a much more balancing point for a generalized trapezoidal fuzzy number. Therefore, this point would be a better reference point rather than the centroid point of the trapezoid, shown in Figure 6.

Consider a generalized trapezoidal fuzzy number $\tilde{A} = (a, b, c, d; w)$. The Centroid of the three plane figures are respectively.

$$G_1 = \left(\frac{a + 2b}{3}, \frac{w}{3}\right) \quad (6)$$

$$G_2 = \left(\frac{b + c}{3}, \frac{w}{3}\right) \quad (7)$$

$$G_3 = \left(\frac{2c + d}{3}, \frac{w}{3}\right) \quad (8)$$

Equation of the line $\overleftrightarrow{G_1G_3}$ is $y = \frac{w}{3}$ and G_2 does not lie on the line $\overleftrightarrow{G_1G_3}$. Therefore G_1, G_2 and G_3 are non-collinear and they form a triangle. We define the in centre $I_{\tilde{A}}(\overline{x}_0, \overline{y}_0)$ of the triangle with vertices G_1, G_2 and G_3 of the generalized trapezoidal fuzzy number as $\tilde{A} = (a, b, c, d; w)$ as

$$I_{\tilde{A}}(\overline{x}_0, \overline{y}_0) = \left(\frac{\alpha\left(\frac{a+2b}{3}\right) + \beta\left(\frac{b+c}{2}\right) + \gamma\left(\frac{2c+d}{3}\right)}{\alpha+\beta+\gamma}, \frac{\alpha\left(\frac{w}{3}\right) + \beta\left(\frac{w}{2}\right) + \gamma\left(\frac{w}{3}\right)}{\alpha+\beta+\gamma} \right) \tag{9}$$

where

$$\alpha = \frac{\sqrt{(c-3b+2d)^2 + w^2}}{6}, \beta = \frac{\sqrt{(2c+d-a-2b)^2 + w^2}}{3}, \gamma = \frac{\sqrt{(3c-2a-b)^2 + w^2}}{6}$$

As a special case, for triangular fuzzy number $\tilde{A} = (a, b, c, d; w)$ i.e. c = b the in centre of centroid is given by

$$I_{\tilde{A}}(\overline{x}_0, \overline{y}_0) = \left(\frac{x\left(\frac{a+2b}{3}\right) + yb + z\left(\frac{2b+d}{3}\right)}{x+y+z}, \frac{x\left(\frac{w}{3}\right) + y\left(\frac{w}{2}\right) + z\left(\frac{w}{3}\right)}{x+y+z} \right) \ldots\ldots\ldots\ldots\ldots \tag{10}$$

where

$$x = \frac{\sqrt{(2d-2b)^2 + w^2}}{6}, y = \frac{\sqrt{(d-a)^2 + w^2}}{3}, z = \frac{\sqrt{(2b-2a)^2 + w^2}}{6}$$

The ranking function of the generalized trapezoidal fuzzy number $\tilde{A} = (a, b, c, d; w)$ which maps the set of all fuzzy numbers to a set of real numbers

$$R(\tilde{A}) = (\overline{x}_0 \times \overline{y}_0) = \left(\frac{x\left(\frac{a+2b}{3}\right) + yb + z\left(\frac{2b+d}{3}\right)}{x+y+z}, \frac{x\left(\frac{w}{3}\right) + y\left(\frac{w}{2}\right) + z\left(\frac{w}{3}\right)}{x+y+z} \right) \ldots\ldots\ldots\ldots \tag{11}$$

This is the area between the in centers of centroid method, $I_{\tilde{A}}\left(\overline{x}_0, \overline{y}_0\right)$ as defined in Equation 10 and the original point.

HYPOTHETICAL CASE RESEARCH

The practical steps for measuring the performances of two manufacturing firms i.e. gear and shaft under G-L-A supply chain presented below.

Step 1: Construction of a group of expert's panel for evaluating the overall G-L-A supply chain performance.

Step 2: Selection of appropriate linguistic scale in terms of ratings and importance grades against evaluation criteria.

Step 3: Assignment of performance ratings as well as importance grade against criterions associated with module by a group of expert's panel.

Step 4: Transform the linguistic terms into Generalized Trapezoidal Fuzzy Numbers (GTFNs) and then aggregate the assigned linguistic terms (as rating and weights) convert into single responses.

Step 5: Applied fuzzy performance index model to compute the rating of 1st level criteria.

Step 6: Estimation of overall performance of firm under G-L-A supply chain.

Step 7: Identification of ill-performing areas which need future improvement.

Two hypothetical case researches are carried out, where G-L-A supply chain based appraisement module is constructed in purpose to measure the performances of two i.e. gear and shaft manufacturing firms. This chapter presented a decision support system (consist of multi criterion hierarchical module coupled with fuzzy performance index approach) in order to measure the supply chain performance of firm under green, agile, lean supply chain management strategies. In proposed module, green supply chain is undertook the several crucial MCDM metrics such as:

- Green service,($C_{1,1}$),
- Green advertisement,($C_{1,2}$),
- Green delivery,($C_{1,3}$),
- Emotional purchase,($C_{1,4}$),
- Routine purchase,($C_{1,5}$),
- Expansion of product market to a global level,($C_{1,6}$),
- Demand from customers,($C_{1,7}$),
- Collaboration with customers,($C_{1,8}$),
- Competitive advantage in adopting green strategies,($C_{1,9}$),
- Green strategy of substitute product producers,($C_{1,10}$),
- Export potential of the green product,($C_{1,11}$),
- Establishing company's green image,($C_{1,12}$),
- Cost of environmentally friendly goods,($C_{1,13}$),
- Cost of environmentally friendly packages,($C_{1,14}$).

Agile supply chain is extended into its interrelated metrics such as:

- Flexibility in volume of product,($C_{2,1}$),
- Sourcing responsiveness,($C_{2,2}$),
- Adaptability of delivery time by suppliers,($C_{2,3}$),
- Suppliers delivery time,($C_{2,4}$),
- Supplier relation management,($C_{2,5}$),
- Manufacturing responsiveness,($C_{2,6}$),
- Time of establishment and changing parts,($C_{2,7}$),
- Responsiveness level to the market, changes,($C_{2,8}$),
- Delivery responsiveness,($C_{2,9}$), Competency,($C_{2,10}$),

- Cooperation and interval external balance,($C_{2,11}$),
- Cooperation and interval internal balance,($C_{2,12}$).

Lean supply chain undertook the numerous crucial metrics such as:

- Organizational structure,(C_{31}),
- Smooth information flow,($C_{3,2}$),
- Team management for decision-making,($C_{3,3}$),
- Interchange-ability personnel,($C_{3,4}$),
- Nature of management,($C_{3,5}$),
- Clearly known management goals,($C_{3,6}$),
- Management involvement,($C_{3,7}$),
- Transparency in information sharing,($C_{3,8}$),
- Manufacturing management leanness,($C_{3,9}$),
- Customer response adaptation,($C_{3,10}$),
- Prevalence of continuous improvement culture,($C_{3,11}$),
- Employee involvement,($C_{3,12}$).

In aforementioned appraisement module is shown in Table 1 (See Appendix). An appropriate linguistic scale is elected, shown in Table 2 (See Appendix), which facilitated the experts to state their subjective judgment in terms of importance grade as well as ratings against evaluation criteria. For computing importance grade of criteria available at different hierarchical levels, a committee of four expert's panel, $DM_1, DM_2, DM_3 and\ DM_4$ is formed to express their ratings preferences in linguistic terms against 2nd level criteria, shown in Tables 3-4 (See Appendix) for Alternative-1 (gear manufacturing) and Alternative-2 (shaft manufacturing) firms.

Similarly, Expert's panel (E) expressed their importance grade in linguistic terms against 2nd level criteria for Alternative-1-2, shown in Table 5 (See Appendix). Moreover, expert's panel to express their importance grade in linguistic terms against 1st level criteria for Alternative-1-2, shown in Table 6 (See Appendix).

By using Equations 1-5, the fuzzy ratings and importance grade against individual 2nd level criteria for alternative-1-2 are aggregated, depicted in Tables 7-8 (See Appendix). Next same Equations 1-5 is used to compute importance grade against individual 1st level criteria as shown in Table 9 (See Appendix). Considering a G-L-A 2nd level appraisement hierarchical module, included criteria at 1st and 2nd level followed below said notations to compute the performances of Alternative-1 (gear manufacturing) and Alternative-2 (shaft manufacturing) firms.

C_i = i^{th} 1st level evaluation index; $i = 1, 2, ..., m.$

C_{ij} = j^{th} 2nd level evaluation index which is under i^{th} 1st level evaluation index C_i; $j = 1, 2, ..., n.$

The computed fuzzy rating of individual 1st level evaluation criteria can be calculated as (Equation 12).

$$U_i = \frac{\sum_{j=1}^{n} \left(w_{ij} \otimes U_{ij} \right)}{\sum_{j=1}^{n} w_{ij}} \tag{12}$$

Here U_{ij} represents aggregated fuzzy performance measure (rating) and w_{ij} represent aggregated fuzzy importance grade corresponding to attributes C_{ij} at 2nd level. Also, U_i represents the computed fuzzy performance measure (rating) corresponding to the index C_i at 1st level. Thus, overall fuzzy performance index $U\left(FPI\right)$ can be obtained as follows.

$$U\left(FPI\right) = \frac{\sum_{i=1}^{m} \left(w_i \otimes U_i \right)}{\sum_{i=1}^{m} w_i} \tag{13}$$

Here U_i = rating of ith 1st level evaluation index C_i; w_i = Importance grade of ith 1st level evaluation index C_i.

The FPI becomes (0.508179, 0.639028, 1.085557; 1.356745) for Alternative-1 gear manufacturing firm, While, The FPI becomes (0.498267, 0.631272, 1.102439; 1.39133) for Alternative-2 shaft manufacturing firm. After evaluating FPI, simultaneously it is felt indeed necessary to recognize ill and strong performing sub-criteria or metrics, so that managers could augment the performances of their firms under G-L-A supply chain SC. A Fuzzy Performance Importance Index (FPII) may be used to identify these ill-performing areas. FPII combines the performance rating and importance grade of various 2nd level sub-criteria. The higher the FPII of a factor, the higher is the contribution. The concept of FPII was introduced by (Lin, Chiu, & Tseng 2006).

$$FPII_{ij} = w'_{ij} \otimes U_{ij} \tag{14}$$

Here, $w'_{ij} = \left[\left[\left(1,1,1,1;1 \right) \right] - w_{il} \right]$

U_{ij} is the rating and w_{ij} is the importance weight of j_{th} index (at 2nd level).

U_{ij} is the rating and w_{ij} is the importance weight of j_{th} index (at 2nd level).

Fuzzy Performance Importance Index (FPII) has been computed against each of the 2nd level evaluation sub-criteria and FPII values are shown in Table 10 (See Appendix). After evaluation the FPII values, the crisps scores corresponding to FPII of individual 2nd level sub-criteria computed by exploring the concept of in centre of centroid method (Equation 9), ranking order has been shown in shown in Table 11 (See Appendix) for Alternative-1-2, where high crisp value doesn't require improvement ranking order.

CONCLUSION

Performance measurement is a tool, used for materializing the efficiency and effectiveness of firms. Effectiveness is extent to which client requirements are met, while efficiency measures how cheaply a firm's capitals are brought into use. During the journey of last five years, Multi Criterion Decision Making (MCDM) is found out activated towards undertaking industrial management problems in the pitch of supply chain management strategies such as green, agile, lean. Adequate numbers of literature surveys were conducted in the context of MCDM applications in the arena of supply chain management strategies. Then, it was determined that module of DSS is thus far suffering thought an unwell or poor improvement mode, it is still soliciting the novel dynamic effective as well as non-complicated mathematical approach. So, the need to develop new approach or DSS in extent of supply chain management strategies took a place.

This chapter presented a decision support system (consist of multi criterion hierarchical module coupled with fuzzy performance index approach) in order to measure the supply chain performance of firm under green, agile, lean supply chain management strategies. In proposed module, green supply chain is undertook the several crucial MCDM metrics such as:

- Green service,($C_{1,1}$),
- Green advertisement,($C_{1,2}$),
- Green delivery,($C_{1,3}$),
- Emotional purchase,($C_{1,4}$),
- Routine purchase,($C_{1,5}$),
- Expansion of product market to a global level,($C_{1,6}$),
- Demand from customers,($C_{1,7}$),
- Collaboration with customers,($C_{1,8}$),
- Competitive advantage in adopting green strategies,($C_{1,9}$),
- Green strategy of substitute product producers,($C_{1,10}$),
- Export potential of the green product,($C_{1,11}$),
- Establishing company's green image,($C_{1,12}$),
- Cost of environmentally friendly goods,($C_{1,13}$),
- Cost of environmentally friendly packages,($C_{1,14}$).

Agile supply chain is extended into its interrelated metrics such as:

- Flexibility in volume of product,($C_{2,1}$),
- Sourcing responsiveness,($C_{2,2}$),
- Adaptability of delivery time by suppliers,($C_{2,3}$),
- Suppliers delivery time,($C_{2,4}$),
- Supplier relation management,($C_{2,5}$),
- Manufacturing responsiveness,($C_{2,6}$),
- Time of establishment and changing parts,($C_{2,7}$),
- Responsiveness level to the market, changes,($C_{2,8}$),
- Delivery responsiveness,($C_{2,9}$), Competency,($C_{2,10}$),
- Cooperation and interval external balance,($C_{2,11}$),
- Cooperation and interval internal balance,($C_{2,12}$).

Lean supply chain undertook the numerous crucial metrics such as:

- Organizational structure,(C_{31}),
- Smooth information flow,($C_{3,2}$),
- Team management for decision-making,($C_{3,3}$),
- Interchange-ability personnel,($C_{3,4}$),
- Nature of management,($C_{3,5}$),
- Clearly known management goals,($C_{3,6}$),
- Management involvement,($C_{3,7}$),
- Transparency in information sharing,($C_{3,8}$),
- Manufacturing management leanness,($C_{3,9}$),
- Customer response adaptation,($C_{3,10}$),
- Prevalence of continuous improvement culture,($C_{3,11}$),
- Employee involvement,($C_{3,12}$).

The vice results of this chapter presented here:

- The built DSS permits the personnels to share their subjective information against defined G-L-A supply chain and its interrelated criteria associated with hierarchy of MCDM appraisement module and also enabled to undertake objective information.
- The presented fuzzy based DSS, which could be powered by Microsoft Windows XP or manual computation under fewer time.
- A corridor to identify the ill and strong sub-criteria or metrics is proposed. By exploration of proposed model, the managers could augment the performances of their firms under G-L-A supply chain.

REFERENCES

Arbos, L. C. (2002). Design of a rapid response and high efficiency service by lean production principles: Methodology and evaluation of variability of performance. *International Journal of Production Economics*, *80*(2), 169–183. doi:10.1016/S0925-5273(02)00316-X

Beamon, B. M. (1999). Designing the green supply chain. *Logist. Inform. Manage*, *12*(4), 332–342. doi:10.1108/09576059910284159

Cabral, I., Grilo, A., & Machado, V. C. (2012). A decision-making model for lean, agile, resilient and green supply chain management. *International Journal of Production Research*, *50*(17), 4830–4845. doi:10.1080/00207543.2012.657970

Chan, F. T. S., Qi, H. J., Chan, H. K., Lau, H. C. W., & Ip, R. W. L. (2003). A conceptual model of performance measurement for supply chains. *Management Decision*, *41*(7), 635–642. doi:10.1108/00251740310495568

Chen, S. H., & Hsieh, C. H. (1999). Ranking generalized fuzzy number with graded mean integration representation. In *Proceedings of the Eighth International Conference of Fuzzy Sets and Systems Association World Congress.*

Chen, S. J. (1985). Operations on fuzzy numbers with function principal. *Tamkang Journal of Management Sciences*, *6*(1), 13–25.

Chen, S. J., & Chen, S. M. (2007). Fuzzy risk analysis based on the ranking of generalized trapezoidal fuzzy numbers. *Applied Intelligence*, *26*(1), 1–11. doi:10.1007/s10489-006-0003-5

Chen, S. J., & Hwang, C. L. (1992). *Fuzzy multiple attribute decision making*. Berlin: Springer-Verlag. doi:10.1007/978-3-642-46768-4

Christopher, M. (1998). Logistics and supply chain management: Strategies for reducing cost and improving service. *International Journal of Logistics Research and Applications: A Leading Journal of Supply Chain Management, 2*(1), 103-104.

Garvin, A. O. (1993). The 12 commandments of environmental compliance,'. *Industrial Engineering (American Institute of Industrial Engineers)*, *25*(9), 18–22.

Green, K., Morton, B., & New, S. (1998). Green purchasing and supply policies: Do they improve companies environmental performance. *Supply Chain Management*, *3*(2), 89–95. doi:10.1108/13598549810215405

Gunasekaran, A. (1999). Agile manufacturing: A framework for research and development. *International Journal of Production Economics*, *62*(1-2), 87–105. doi:10.1016/S0925-5273(98)00222-9

Hsu, C. W., & Hu, A. H. (2008). Green supply chain management in the electronic industry'. *International Journal of Science and Technology*, *5*(2), 205–216.

Huiyu, C., & Weiwei, W. (2010). *Green supply chain management for a Chinese auto manufacturer.* Department of Technology and Built Environment, University of GAVLE.

Hult, G. T. M., Ketchen, D. J. K., & Arrelti, M. (2007). Strategic supply chain management: Improving performance through a culture of competitiveness and knowledge development. *Strategic Management Journal*, *28*(10), 1035–1052. doi:10.1002/smj.627

Kainuma, Y., & Tawara, N. (2006). A multiple attribute utility theory approach to lean and green supply chain management. *International Journal of Production Economics*, *101*(1), 99–108. doi:10.1016/j.ijpe.2005.05.010

Kainumaa, Y., & Tawara, N. (2006). A multiple attribute utility theory approach to lean and green supply chain management. *International Journal of Production Economics*, *101*(1), 99–108. doi:10.1016/j.ijpe.2005.05.010

Knadel, A. (1986). *Fuzzy mathematical techniques with applications*. Boston, MA: Addison-Wesley Longman Publishing Co.

Kurien, G. P., & Qureshi, M. N. (2011). Study of performance measurement practices in supply chain management. *International Journal of Business. Management and Social Sciences*, *2*(4), 19–34.

Li, D.-F. (2009). Multiattribute group decision making method using extended linguistic variables. *International Journal of Uncertainty, Fuzziness and Knowledge-based Systems*, *17*(6), 793–806. doi:10.1142/S0218488509006273

Li, D.-F. (2010). A new methodology for fuzzy multi-attribute group decision making with multi-granularity and non-homogeneous information. *Fuzzy Optimization and Decision Making*, *9*(1), 83–103. doi:10.1007/s10700-010-9071-4

Li, D.-F. (2010). A ratio ranking method of triangular intuitionistic fuzzy numbers and its application to MADM problems. *Computers & Mathematics with Applications (Oxford, England)*, *60*(6), 1557–1570. doi:10.1016/j.camwa.2010.06.039

Li, D.-F. (2014). *Decision and Game Theory in Management with Intuitionistic Fuzzy Sets*. Heidelberg, Germany: Springer. doi:10.1007/978-3-642-40712-3

Li, D.-F. (2016). *Linear Programming Models and Methods of Matrix Games with Payoffs of Triangular Fuzzy Numbers*. Berlin: Springer. doi:10.1007/978-3-662-48476-0

Li, D.-F.,, Nan, J. X., & Zhang, M. J. (2010). A ranking method of triangular intuitionistic fuzzy numbers and application to decision making. *International Journal of Computational Intelligence Systems*, *3*(5), 522–530. doi:10.1080/18756891.2010.9727719

Li, D.-F., & Sun, T. (2007a). Fuzzy linear programming approach to multiattribute decision making with linguistic variables and incomplete information. *Advances in Complex Systems*, *10*(4), 505–525. doi:10.1142/S0219525907001306

Li, D.-F., & Sun, T. (2007b). Fuzzy LINMAP method for multiattribute group decision making with linguistic variables and incomplete information. *International Journal of Uncertainty, Fuzziness and Knowledge-based Systems*, *15*(2), 153–173. doi:10.1142/S0218488507004509

Li, D.-F., & Yang, J. (2015). A difference-index based ranking method of trapezoidal intuitionistic fuzzy numbers and application to multiattribute decision making. *Mathematical and Computational Applications*, *20*(1), 25–38. doi:10.19029/mca-2015-003

Lin, C., Chiu, T. H., & Tseng, Y. H. (2006). Agility evaluation using fuzzy logic. *International Journal of Production Economics*, *101*(2), 353–368. doi:10.1016/j.ijpe.2005.01.011

Meredith, S., & Francis, D. (2000). Journey towards agility: The agile wheel explored. *The TQM Magazine*, *12*(2), 137–143. doi:10.1108/09544780010318398

Min, H., & Galle, W. P. (1997). Green purchasing strategies: Trends and implications. *International Journal of Purchasing and Materials Management*, *33*(2), 10–17. doi:10.1111/j.1745-493X.1997.tb00026.x

Nambirajan, T., & Ganeshkumar, C. (2010). Measurement of performance of supply chains and their impact on the competitiveness of manufacturing industries in union territory of Pondicherry (India). *International Research Conference on Business and Information*. University of Kelaniya.

Neely, A. (2005). The evolution of performance measurement research. *International Journal of Operations & Production Management*, *1*(1), 1264–1277. doi:10.1108/01443570510633648

Olugu, E. U., & Wong, K. Y. (2009). Supply Chain Performance Evaluation: Trends and Challenges. *American J. of Engineering and Applied Sciences*, *2*(1), 202–211. doi:10.3844/ajeas.2009.202.211

Panizzolo, R. (1998). Applying the lessons learned from 27 lean manufacturers: The relevance of relationships management. *International Journal of Production Economics*, *55*(3), 223–240. doi:10.1016/S0925-5273(98)00066-8

Pavnaskar, S. J., Gershenson, J. K., & Jambekar, A. B. (2003). Classification scheme for lean manufacturing tools. *International Journal of Production Research*, *41*(13), 3075–3090. doi:10.1080/0020754021000049817

Peter, C. B., & Thomas, W. S. (2000). Using the balanced scorecard to measure supply chain performance. *Journal of Business Logistics*, *21*(1), 75.

Petrovic, D., Roy, R., & Petrovic, R. (1999). Supply chain modeling using fuzzy sets. *International Journal of Production Economics*, *59*(1), 443–453. doi:10.1016/S0925-5273(98)00109-1

Sahu, A. K., Sahu, N. K., & Sahu, A. (2014). Appraisal of CNC machine tool by integrated MULTI MOORA-IGVN circumstances: An empirical study. *International Journal of Grey Systems: Theory and Application*, *4*(1), 104–123.

Sahu, A. K., Sahu, N. K., & Sahu, A. K. (2015b). Benchmarking CNC machine tool using hybrid fuzzy methodology a multi indices decision making approach. *International Journal of Fuzzy System Applications*, *4*(2), 28–46.

Sahu, A. K., Sahu, N. K., & Sahu, A. K. (2016a). Application of Integrated TOPSIS in ASC index: Partners Benchmarking perspective. *Benchmarking: An International Journal*, *23*(3), 540–563. doi:10.1108/BIJ-03-2014-0021

Sahu, A. K., Sahu, N. K., & Sahu, A. K. (2016b). Appraisal of Partner Enterprises under GTFNS Environment in Agile SC. *International Journal of Decision Support System Technology*, *8*(3), 1–19. doi:10.4018/IJDSST.2016070101

Sahu, A. K., Sahu, N. K., & Sahu, A. K. (2016c). Application of Modified MULTI-MOORA for CNC Machine Tool Evaluation in IVGTFNS Environment: An Empirical Study. *International Journal of Computer Aided Engineering and Technology*, *8*(3), 234–259. doi:10.1504/IJCAET.2016.077603

Sahu, N. K., Datta, S., & Mahapatra, S. S. (2012). Establishing green supplier appraisement platform using grey concepts. *International Journal of Grey Systems: Theory and Application*, *2*(3), 395–418.

Sahu, N. K., Datta, S., & Mahapatra, S. S. (2013). Decision making for selecting 3PL service provider using three parameter interval grey numbers. *International Journal of Logistic Systems and Management*, *14*(3), 261–297. doi:10.1504/IJLSM.2013.052061

Sahu, N. K., Datta, S., & Mahapatra, S. S. (2014). Green Supplier Appraisement in Fuzzy Environment. *Benchmarking: An International Journal*, *21*(3), 412–429. doi:10.1108/BIJ-06-2012-0042

Sahu, N. K., Datta, S. & Mahapatra, S. S. (2015). Fuzzy Based Appraisement Module for 3PL Evaluation and Selection. *Benchmarking, an International Journal, 22*(3), 354-392.

Sahu, N. K., Sahu A. K., & Sahu, A. K (2015a). Appraisement and Benchmarking of Third Party Logistic Service Provider by Exploration of Risk Based Approach. *Cogent Business and Management, 2,* 1-21.

Samantra, C., Sahu, N. K., Datta, S., & Mahapatra, S. S. (2013). Decision-Making in Selecting Reverse Logistics Alternative using Interval-Valued Fuzzy Sets Combined with VIKOR Approach. *International Journal of Services and Operations Management*, *14*(2), 175–196. doi:10.1504/IJSOM.2013.051828

Shah, R., & Ward, P. T. (2003). Lean manufacturing: Context, practice bundles, and performance. *Journal of Operations Management*, *21*(2), 129–149. doi:10.1016/S0272-6963(02)00108-0

Sharifi, H., Colquhoun, G., Barclay, I., & Dann, Z. (2001). Agile manufacturing: A management and operational framework. *Proceedings of the Institution of Mechanical Engineers. Part B, Journal of Engineering Manufacture*, *215*(6), 857–869. doi:10.1243/0954405011518647

Sharifi, H., & Zhang, Z. (2001). Agile manufacturing in practice: Application of a methodology. *International Journal of Operations & Production Management*, *21*(5/6), 772–794. doi:10.1108/01443570110390462

Sharifi, H., & Zhang, Z. A. (1999). Methodology for achieving agility in manufacturing organizations: An introduction. *International Journal of Production Economics*, *62*(1-2), 7–22. doi:10.1016/S0925-5273(98)00217-5

Sharp, J. M., Irani, Z., & Desai, S. (1999). Working towards agile manufacturing in the UK industry. *International Journal of Production Economics*, *62*(1), 155–169. doi:10.1016/S0925-5273(98)00228-X

Simpson, D., & Samson, D. (2008). *Developing strategies for green supply chain management. In Decision Line* (pp. 12–15). University of Melbourne.

Srivastava, S. K. (2007). Green supply-chain management: A state-of-the-art literature review. *International Journal of Management Reviews*, *9*(1), 53–80. doi:10.1111/j.1468-2370.2007.00202.x

Thorani, Y. L. P., Rao, P. P. B., & Shankar, N. R. (2012). Ordering generalized trapezoidal fuzzy numbers. *International Journal of Contemporary Mathematical Sciences*, *7*(12), 555–573.

Toke, L. K., Gupta, R. C., & Dandekar, M. (2010). Green supply chain management; critical research and practices. *Proceedings of the 2010 International Conference on Industrial Engineering and Operations Management*, (pp. 9-10).

Yusuf, Y. Y., Sarhadi, M., & Gunasekaran, A. (1999). Agile manufacturing: The drivers, concepts and attributes. *International Journal of Production Economics*, *62*(1-2), 33–43. doi:10.1016/S0925-5273(98)00219-9

APPENDIX

Table 1. G-L-A supply chain appraisement module

Goal (C)	Criteria, (C_i)	Interrelated Criteria, (C_{ij})
Fuzzy-Performance measurement of A_1 and A_2 firms under G-L-A supply chain, (C)	Green SC, (C_1)	Green service,($C_{1,1}$)
		Green advertisement,($C_{1,2}$)
		Green delivery,($C_{1,3}$)
		Emotional purchase,($C_{1,4}$)
		Routine purchase,($C_{1,5}$)
		Expansion of product market to a global level,($C_{1,6}$)
		Demand from customers,($C_{1,7}$)
		Collaboration with customers,($C_{1,8}$)
		Competitive advantage in adopting green strategies,($C_{1,9}$)
		Green strategy of substitute product producers,($C_{1,10}$)
		Export potential of the green product,($C_{1,11}$)
		Establishing company's green image,($C_{1,12}$)
		Cost of environmentally friendly goods,($C_{1,13}$)
		Cost of environmentally friendly packages,($C_{1,14}$)
	Agile SC, (C_2)	Flexibility in volume of product,($C_{2,1}$)
		Sourcing Responsiveness,($C_{2,2}$)
		Adaptability of delivery time by suppliers,($C_{2,3}$)
		Suppliers delivery time,($C_{2,4}$)
		Supplier relation management,($C_{2,5}$)
		Manufacturing Responsiveness,($C_{2,6}$)
		Time of establishment and changing parts,($C_{2,7}$)
		Responsiveness level to the market changes,($C_{2,8}$)
		Delivery Responsiveness,($C_{2,9}$)
		Competency,($C_{2,10}$)
		Cooperation and interval external balance,($C_{2,11}$)
		Cooperation and interval internal balance,($C_{2,12}$)
	Lean SC, (C_3)	Organizational structure,(C_{31})
		Smooth information flow,($C_{3,2}$)
		Team management for decision-making,($C_{3,3}$)
		Interchange-ability personnel,($C_{3,4}$)
		Nature of management,($C_{3,5}$)
		Clearly known management goals,($C_{3,6}$)
		Management involvement,($C_{3,7}$)
		Transparency in information sharing,($C_{3,8}$)
		Manufacturing management leanness,($C_{3,9}$)
		Customer response adaptation,($C_{3,10}$)
		Prevalence of continuous improvement culture,($C_{3,11}$)
		Employee involvement,($C_{3,12}$)

Table 2. Set of linguistic variables and their corresponding fuzzy representations

Linguistic Terms for Importance Grade	Linguistic Terms for Performance Rating	Fuzzy Representation
DL: Definitely low	DL: Definitely low	(0.0, 0.0, 0.0, 0.0; 1.0)
VL: Very low	VL: Very low	(0.0, 0.0, 0.02, 0.07; 1.0)
L: Low	L: Low	(0.04, 0.10, 0.18, 0.23; 1.0)
ML: More or less low	ML: More or less low	(0.17, 0.22, 0.36, 0.42; 1.0)
M: Middle	M: Middle	(0.32, 0.41, 0.58, 0.65; 1.0)
MH: More or less high	MH: More or less high	(0.58, 0.63, 0.80, 0.86; 1.0)
H: High	H: High	(0.72, 0.78, 0.92, 0.97; 1.0)
VH: Very high	VH: Very high	(0.93, 0.98, 1.0, 1.0; 1.0)
DH: Definitely high	DH: Definitely high	(1.0, 1.0, 1.0, 1.0; 1.0)

Table 3. Appropriateness ratings in terms of linguistic variable for 2nd level criteria individual against alternative A_1

Criteria, (C_{ij})	E_1	E_2	E_3	E_4
($C_{1,1}$)	DH	DH	DH	DH
($C_{1,2}$)	DH	DH	DH	DH
($C_{1,3}$)	DH	VL	VL	DH
($C_{1,4}$)	VL	VL	VL	DH
($C_{1,5}$)	VL	VL	L	DH
($C_{1,6}$)	VL	ML	L	DH
($C_{1,7}$)	ML	DH	DH	DH
($C_{1,8}$)	DH	DH	DH	DH
($C_{1,9}$)	DH	DH	DH	DH
($C_{1,10}$)	VL	VL	VL	DH
($C_{1,11}$)	VL	VL	VL	DH
($C_{1,12}$)	VL	L	VL	DH
($C_{1,13}$)	ML	L	ML	DH
($C_{1,14}$)	DH	DH	DH	DH
($C_{2,1}$)	DH	DH	DH	DH
($C_{2,2}$)	VL	VL	VL	DH
($C_{2,3}$)	MH	MH	MH	DH
($C_{2,4}$)	MH	MH	MH	DH
($C_{2,5}$)	H	H	MH	DH
($C_{2,6}$)	H	H	MH	DH
($C_{2,7}$)	VL	L	VL	DH
($C_{2,8}$)	ML	L	ML	VH

continued on following page

Table 3. Continued

Criteria, (C_{ij})	E_1	E_2	E_3	E_4
$C_{2,9}$)	DH	DH	DH	VH
($C_{2,10}$)	H	H	VH	VH
($C_{2,11}$)	DH	DH	VH	VH
($C_{2,12}$)	L	ML	VH	L
(C_{31})	DH	DH	VH	DH
($C_{3,2}$)	H	VH	VH	H
($C_{3,3}$)	DH	VH	VH	DH
($C_{3,4}$)	L	ML	VH	L
($C_{3,5}$)	DH	DH	VH	DH
($C_{3,6}$)	H	H	H	DH
($C_{3,7}$)	H	H	H	DH
($C_{3,8}$)	L	ML	VH	L
($C_{3,9}$)	DH	DH	VH	DH
($C_{3,10}$)	H	VH	VH	H
$C_{3,11}$)	DH	VH	VH	DH
($C_{3,12}$)	DH	DH	DH	DH

Table 4. Appropriateness ratings in terms of linguistic variable for 2nd level criteria individual against alternative A_2

Criteria, (C_{ij})	E_1	E_2	E_3	E_4
($C_{1,1}$)	MH	MH	MH	MH
($C_{1,2}$)	MH	MH	MH	MH
($C_{1,3}$)	MH	MH	MH	MH
($C_{1,4}$)	MH	MH	MH	MH
($C_{1,5}$)	MH	MH	MH	MH
($C_{1,6}$)	MH	MH	MH	MH
($C_{1,7}$)	MH	MH	MH	MH
($C_{1,8}$)	DH	DH	DH	DH
($C_{1,9}$)	DH	DH	DH	DH
($C_{1,10}$)	VL	VL	VL	DH
($C_{1,11}$)	VL	VL	VL	DH
($C_{1,12}$)	VL	L	VL	DH
($C_{1,13}$)	ML	L	ML	DH
($C_{1,14}$)	DH	DH	DH	DH
($C_{2,1}$)	DH	DH	DH	DH

continued on following page

Table 4. Continued

Criteria, (C_{ij})	E_1	E_2	E_3	E_4
($C_{2,2}$)	VL	VL	VL	DH
($C_{2,3}$)	MH	MH	MH	DH
($C_{2,4}$)	MH	MH	MH	DH
($C_{2,5}$)	H	H	MH	DH
($C_{2,6}$)	H	H	MH	DH
($C_{2,7}$)	VL	L	VL	DH
($C_{2,8}$)	ML	L	ML	VH
$C_{2,9}$)	DH	DH	DH	VH
($C_{2,10}$)	H	H	VH	VH
($C_{2,11}$)	DH	DH	VH	VH
($C_{2,12}$)	L	ML	VH	L
(C_{31})	DH	DH	VH	DH
($C_{3,2}$)	H	VH	VH	H
($C_{3,3}$)	DH	VH	VH	DH
($C_{3,4}$)	L	ML	VH	L
($C_{3,5}$)	DH	DH	VH	DH
($C_{3,6}$)	H	H	H	DH
($C_{3,7}$)	H	H	H	DH
($C_{3,8}$)	L	ML	VH	L
($C_{3,9}$)	DH	DH	VH	DH
($C_{3,10}$)	H	VH	VH	H
$C_{3,11}$)	DH	VH	VH	DH
($C_{3,12}$)	DH	DH	DH	DH

Table 5. Importance weights in terms of linguistic variable for 2nd level criteria individual against alternative A_1 and A_2

Criteria, (C_{ij})	E_1	E_2	E_3	E_4
($C_{1,1}$)	MH	MH	MH	MH
($C_{1,2}$)	MH	MH	MH	MH
($C_{1,3}$)	MH	MH	MH	MH
($C_{1,4}$)	VL	VL	VL	DH
($C_{1,5}$)	VL	VL	L	DH
($C_{1,6}$)	VL	ML	L	DH
($C_{1,7}$)	ML	DH	DH	DH
($C_{1,8}$)	DH	DH	DH	DH

continued on following page

Table 5. Continued

Criteria, (C_{ij})	E_1	E_2	E_3	E_4
($C_{1,9}$)	DH	DH	DH	DH
($C_{1,10}$)	VL	VL	VL	DH
($C_{1,11}$)	VL	VL	VL	DH
($C_{1,12}$)	VL	L	VL	DH
($C_{1,13}$)	ML	L	ML	DH
($C_{1,14}$)	DH	DH	DH	DH
($C_{2,1}$)	DH	DH	DH	DH
($C_{2,2}$)	VL	VL	VL	DH
($C_{2,3}$)	MH	MH	MH	DH
($C_{2,4}$)	MH	MH	MH	DH
($C_{2,5}$)	H	H	MH	DH
($C_{2,6}$)	H	H	MH	DH
($C_{2,7}$)	VL	L	VL	DH
($C_{2,8}$)	ML	L	ML	VH
$C_{2,9}$)	DH	DH	DH	VH
($C_{2,10}$)	H	H	VH	VH
($C_{2,11}$)	DH	DH	VH	VH
($C_{2,12}$)	L	ML	VH	L
(C_{31})	DH	DH	VH	DH
($C_{3,2}$)	H	VH	VH	H
($C_{3,3}$)	DH	VH	VH	DH
($C_{3,4}$)	L	ML	VH	L
($C_{3,5}$)	DH	DH	VH	DH
($C_{3,6}$)	H	H	H	DH
($C_{3,7}$)	H	H	H	DH
($C_{3,8}$)	L	ML	VH	L
($C_{3,9}$)	DH	DH	VH	DH
($C_{3,10}$)	H	VH	VH	H
$C_{3,11}$)	DH	VH	VH	DH
($C_{3,12}$)	DH	DH	DH	DH

Table 6. Importance weights in terms of linguistic variable for 1st level criteria individual against alternative A_1 and A_2

Criteria, (C_i)	Importance Weights			
($C_{1,1}$)	H	H	H	H
($C_{1,2}$)	H	H	H	H
($C_{1,3}$)	H	H	H	H

Table 7. Computed appropriateness ratings and importance weights for 2nd level criteria against alternatives A_1

Criteria, (C_{ij})	Computed Appropriateness Ratings	Computed Importance Weights
($C_{1,1}$)	(1.00,1.00,1.00,1.00;1.00)	(0.58,0.63,0.80,0.86;1.00)
($C_{1,2}$)	(1.00,1.00,1.00,1.00;1.00)	(0.58,0.63,0.80,0.86;1.00)
($C_{1,3}$)	(0.50,0.50,0.51,0.54;1.00)	(0.58,0.63,0.80,0.86;1.00)
($C_{1,4}$)	(0.25,0.25,0.27,0.30;1.00)	(0.25,0.25,0.27,0.30;1.00)
($C_{1,5}$)	(0.26,0.28,0.31,0.34;1.00)	(0.26,0.28,0.31,0.34;1.00)
($C_{1,6}$)	(0.30,0.33,0.39,0.43;1.00)	(0.30,0.33,0.39,0.43;1.00)
($C_{1,7}$)	(0.79,0.810.84,0.86;1.00)	(0.79,0.810.84,0.86;1.00)
($C_{1,8}$)	(1.00,1.00,1.00,1.00;1.00)	(1.00,1.00,1.00,1.00;1.00)
($C_{1,9}$)	(1.00,1.00,1.00,1.00;1.00)	(1.00,1.00,1.00,1.00;1.00)
($C_{1,10}$)	(0.25,0.25,0.27,0.30;1.00)	(0.25,0.25,0.27,0.30;1.00)
($C_{1,11}$)	(0.25,0.25,0.27,0.30;1.00)	(0.25,0.25,0.27,0.30;1.00)
($C_{1,12}$)	(0.26,0.28,0.31,0.34;1.00)	(0.26,0.28,0.31,0.34;1.00)
($C_{1,13}$)	(0.35,0.39,0.48,0.52;1.00	(0.35,0.39,0.48,0.52;1.00
($C_{1,14}$)	(1.00,1.00,1.00,1.00;1.00)	(1.00,1.00,1.00,1.00;1.00)
($C_{2,1}$)	(1.00,1.00,1.00,1.00;1.00)	(1.00,1.00,1.00,1.00;1.00)
($C_{2,2}$)	(0.25,0.25,0.27,0.30;1.00)	(0.25,0.25,0.27,0.30;1.00)
($C_{2,3}$)	(0.69,0.72,0.85,0.90;1.00)	(0.69,0.72,0.85,0.90;1.00)
($C_{2,4}$)	(0.69,0.72,0.85,0.90;1.00)	(0.69,0.72,0.85,0.90;1.00)
($C_{2,5}$)	(0.76,0.80,0.91,0.95;1.00)	(0.76,0.80,0.91,0.95;1.00)
($C_{2,6}$)	(0.76,0.80,0.91,0.95;1.00)	(0.76,0.80,0.91,0.95;1.00)
($C_{2,7}$)	(0.26,0.28,0.31,0.34;1.00)	(0.26,0.28,0.31,0.34;1.00)
($C_{2,8}$)	(0.33,0.38,0.48,0.52;1.00)	(0.33,0.38,0.48,0.52;1.00)
$C_{2,9}$)	(0.98,1.00,1.00,1.00;1.00)	(0.98,1.00,1.00,1.00;1.00)
($C_{2,10}$)	(0.83,0.88,0.96,0.99;1.00)	(0.83,0.88,0.96,0.99;1.00)
($C_{2,11}$)	(0.97,0.99,1.00,1.00;1.00)	(0.97,0.99,1.00,1.00;1.00)
($C_{2,12}$)	(0.30,0.35,0.43,0.47;1.00)	(0.30,0.35,0.43,0.47;1.00)
(C_{31})	(0.98,1.00,1.00,1.00;1.00)	(0.98,1.00,1.00,1.00;1.00)
($C_{3,2}$)	(0.83,0.88,0.96,0.99;1.00)	(0.83,0.88,0.96,0.99;1.00)
($C_{3,3}$)	(0.97,0.99,1.00,1.00;1.00)	(0.97,0.99,1.00,1.00;1.00)
($C_{3,4}$)	(0.30,0.35,0.43,0.47;1.00)	(0.30,0.35,0.43,0.47;1.00)
($C_{3,5}$)	(0.98,1.00,1.00,1.00,1.00)	(0.98,1.00,1.00,1.00,1.00)
($C_{3,6}$)	(0.79,0.84,0.94,0.98;1.00)	(0.79,0.84,0.94,0.98;1.00)
($C_{3,7}$)	(0.79,0.84,0.94,0.98;1.00)	(0.79,0.84,0.94,0.98;1.00)
($C_{3,8}$)	(0.30,0.35,0.43,0.47;1.00)	(0.30,0.35,0.43,0.47;1.00)
($C_{3,9}$)	(0.98,1.00,1.00,1.00;1.00)	(0.98,1.00,1.00,1.00;1.00)
($C_{3,10}$)	(0.83,0.88,0.96,0.99;1.00)	(0.83,0.88,0.96,0.99;1.00)
$C_{3,11}$)	(0.97,0.99,1.00,1.00;1.00)	(0.97,0.99,1.00,1.00;1.00)
($C_{3,12}$)	(1.00,1.00,1.00,1.00;1.00)	(1.00,1.00,1.00,1.00;1.00)

Table 8. Computed appropriateness ratings and importance weights for 2nd level criteria against alternatives A_2

Criteria, (C_{ij})	Computed Appropriateness Ratings	Computed Importance Weights
($C_{1,1}$)	(0.58,0.63,0.80,0.86;1.00)	(0.58,0.63,0.80,0.86;1.00)
($C_{1,2}$)	(0.58,0.63,0.80,0.86;1.00)	(0.58,0.63,0.80,0.86;1.00)
($C_{1,3}$)	(0.58,0.63,0.80,0.86;1.00)	(0.58,0.63,0.80,0.86;1.00)
($C_{1,4}$)	(0.58,0.63,0.80,0.86;1.00)	(0.25,0.25,0.27,0.30;1.00)
($C_{1,5}$)	(0.58,0.63,0.80,0.86;1.00)	(0.26,0.28,0.31,0.34;1.00)
($C_{1,6}$)	(0.58,0.63,0.80,0.86;1.00)	(0.30,0.33,0.39,0.43;1.00)
($C_{1,7}$)	(0.58,0.63,0.80,0.86;1.00)	(0.79,0.810.84,0.86;1.00)
($C_{1,8}$)	(1.00,1.00,1.00,1.00,1.00)	(1.00,1.00,1.00,1.00;1.00)
($C_{1,9}$)	(1.00,1.00,1.00,1.00,1.00)	(1.00,1.00,1.00,1.00;1.00)
($C_{1,10}$)	(0.25,0.25,0.27,0.30,1.00)	(0.25,0.25,0.27,0.30;1.00)
($C_{1,11}$)	(0.25,0.25,0.27,0.30,1.00)	(0.25,0.25,0.27,0.30;1.00)
($C_{1,12}$)	(0.26,0.28,0.31,0.34,1.00)	(0.26,0.28,0.31,0.34;1.00)
($C_{1,13}$)	(0.35,0.39,0.48,0.52,1.00)	(0.35,0.39,0.48,0.52;1.00
($C_{1,14}$)	(1.00,1.00,1.00,1.00,1.00)	(1.00,1.00,1.00,1.00;1.00)
($C_{2,1}$)	(1.00,1.00,1.00,1.00,1.00)	(1.00,1.00,1.00,1.00;1.00)
($C_{2,2}$)	(0.25,0.25,0.27,0.30,1.00)	(0.25,0.25,0.27,0.30;1.00)
($C_{2,3}$)	(0.69,0.72,0.85,0.90,1.00)	(0.69,0.72,0.85,0.90;1.00)
($C_{2,4}$)	(0.69,0.72,0.85,0.90,1.00)	(0.69,0.72,0.85,0.90;1.00)
($C_{2,5}$)	(0.76,0.80,0.91,0.95,1.00)	(0.76,0.80,0.91,0.95;1.00)
($C_{2,6}$)	(0.76,0.80,0.91,0.95,1.00)	(0.76,0.80,0.91,0.95;1.00)
($C_{2,7}$)	(0.26,0.28,0.31,0.34,1.00)	(0.26,0.28,0.31,0.34;1.00)
($C_{2,8}$)	(0.33,0.38,0.48,0.52,1.00)	(0.33,0.38,0.48,0.52;1.00)
$C_{2,9}$)	(0.98,1.00,1.00,1.00,1.00)	(0.98,1.00,1.00,1.00;1.00)
($C_{2,10}$)	(0.83,0.88,0.96,0.99,1.00)	(0.83,0.88,0.96,0.99;1.00)
($C_{2,11}$)	(0.97,0.99,1.00,1.00,1.00)	(0.97,0.99,1.00,1.00;1.00)
($C_{2,12}$)	(0.30,0.35,0.43,0.47,1.00)	(0.30,0.35,0.43,0.47;1.00)
(C_{31})	(0.98,1.00,1.00,1.00,1.00)	(0.98,1.00,1.00,1.00;1.00)
($C_{3,2}$)	(0.83,0.88,0.96,0.99,1.00)	(0.83,0.88,0.96,0.99;1.00)
($C_{3,3}$)	(0.97,0.99,1.00,1.00,1.00)	(0.97,0.99,1.00,1.00;1.00)
($C_{3,4}$)	(0.30,0.35,0.43,0.47,1.00)	(0.30,0.35,0.43,0.47;1.00)
($C_{3,5}$)	(0.98,1.00,1.00,1.00,1.00)	(0.98,1.00,1.00,1.00,1.00)
($C_{3,6}$)	(0.79,0.84,0.94,0.98,1.00)	(0.79,0.84,0.94,0.98;1.00)
($C_{3,7}$)	(0.79,0.84,0.94,0.98,1.00)	(0.79,0.84,0.94,0.98;1.00)
($C_{3,8}$)	(0.30,0.35,0.43,0.47,1.00)	(0.30,0.35,0.43,0.47;1.00)
($C_{3,9}$)	(0.98,1.00,1.00,1.00,1.00)	(0.98,1.00,1.00,1.00;1.00)
($C_{3,10}$)	(0.83,0.88,0.96,0.99,1.00)	(0.83,0.88,0.96,0.99;1.00)
$C_{3,11}$)	(0.97,0.99,1.00,1.00,1.00)	(0.97,0.99,1.00,1.00;1.00)
($C_{3,12}$)	(1.00,1.00,1.00,1.00,1.00)	(1.00,1.00,1.00,1.00;1.00)

Table 9. Computed importance weights for 1st level criteria against alternatives A_1 and A_2

Criteria, (C_i)	Computed Importance Weights
($C_{1,1}$)	(0.72, 0.78, 0.92, 0.97; 1.0)
($C_{1,2}$)	(0.72, 0.78, 0.92, 0.97; 1.0)
($C_{1,3}$)	(0.72, 0.78, 0.92, 0.97; 1.0)

Table 10. Computed fuzzy performance importance index against alternatives A_1 and A_2

Criteria, (C_{ij})	A_1	A_2
($C_{1,1}$)	(0.42,0.37,0.20,0.14;1.00)	(0.42,0.37,0.20,0.14;1.00)
($C_{1,2}$)	(0.42,0.37,0.20,0.14;1.00)	(0.42,0.37,0.20,0.14;1.00)
($C_{1,3}$)	(0.21,0.19,0.10,0.07;1.00)	(0.42,0.37,0.20,0.14;1.00)
($C_{1,4}$)	(0.19,0.19,0.19,0.21;1.00)	(0.75,0.75,0.73,0.69;1.00)
($C_{1,5}$)	(0.19,0.20,0.21,0.23;1.00)	(0.74,0.72,0.69,0.65;1.00)
($C_{1,6}$)	(0.21,0.22,0.24,0.25;1.00)	(0.69,0.67,0.61,0.57;1.00)
($C_{1,7}$)	(0.16,0.16,0.13,0.12;1.00)	(0.20,0.19,0.16,0.14;1.00)
($C_{1,8}$)	(0.00,0.00,0.00,0.00;1.00)	(0.00,0.00,0.00,0.00;1.00)
($C_{1,9}$)	(0.00,0.00,0.00,0.00;1.00)	(0.00,0.00,0.00,0.00;1.00)
($C_{1,10}$)	(0.19,0.19,0.19,0.21;1.00)	(0.19,0.19,0.19,0.21;1.00)
($C_{1,11}$)	(0.19,0.19,0.19,0.21;1.00)	(0.19,0.19,0.19,0.21;1.00)
($C_{1,12}$)	(0.19,0.20,0.21,0.23;1.00)	(0.19,0.20,0.21,0.23;1.00)
($C_{1,13}$)	(0.23,0.24,0.25,0.25;1.00)	(0.23,0.24,0.25,0.25;1.00)
($C_{1,14}$)	(0.00,0.00,0.00,0.00;1.00)	(0.00,0.00,0.00,0.00;1.00)
($C_{2,1}$)	(0.00,0.00,0.00,0.00;1.00)	(0.00,0.00,0.00,0.00;1.00)
($C_{2,2}$)	(0.19,0.19,0.19,0.21;1.00)	(0.19,0.19,0.19,0.21;1.00)
($C_{2,3}$)	(0.22,0.20,0.13,0.09;1.00)	(0.22,0.20,0.13,0.09;1.00)
($C_{2,4}$)	(0.22,0.20,0.13,0.09;1.00)	(0.22,0.20,0.13,0.09;1.00)
($C_{2,5}$)	(0.18,0.16,0.08,0.05;1.00)	(0.18,0.16,0.08,0.05;1.00)
($C_{2,6}$)	(0.18,0.16,0.08,0.05;1.00)	(0.18,0.16,0.08,0.05;1.00)
($C_{2,7}$)	(0.19,0.20,0.21,0.23;1.00)	(0.19,0.20,0.21,0.23;1.00)
($C_{2,8}$)	(0.22,0.240.25,0.25;1.00)	(0.22,0.24,0.25,0.25;1.00)
$C_{2,9}$)	(0.02,0.00,0.00,0.00;1.00)	(0.02,0.00,0.00,0.00;1.00)
($C_{2,10}$)	(0.14,0.11,0.04,0.01;1.00)	(0.14,0.11,0.04,0.01;1.00)
($C_{2,11}$)	(0.03,0.01,0.00,0.00;1.00)	(0.03,0.01,0.00,0.00;1.00)
($C_{2,12}$)	(0.21,0.23,0.25,0.25;1.00)	(0.21,0.23,0.25,0.25;1.00)
(C_{31})	(0.02,0.00,0.00,0.00;1.00)	(0.02,0.00,0.00,0.00;1.00)
($C_{3,2}$)	(0.14,0.11,0.04,0.01;1.00)	(0.14,0.11,0.04,0.01;1.00)
($C_{3,3}$)	(0.03,0.01,0.00,0.00;1.00)	(0.03,0.01,0.00,0.00;1.00)

continued on following page

Table 10. Continued

Criteria, (C_{ij})	A_1	A_2
($C_{3,4}$)	(0.21,0.23,0.25,0.25;1.00)	(0.21,0.23,0.25,0.25;1.00)
($C_{3,5}$)	(0.02,0.00,0.00,0.00;1.00)	(0.02,0.00,0.00,0.00;1.00)
($C_{3,6}$)	(0.17,0.14,0.06,0.02;1.00)	(0.17,0.14,0.06,0.02;1.00)
($C_{3,7}$)	(0.17,0.14,0.06,0.02;1.00)	(0.17,0.14,0.06,0.02;1.00)
($C_{3,8}$)	(0.21,0.23,0.25,0.25;1.00)	(0.21,0.23,0.25,0.25;1.00)
($C_{3,9}$)	(0.02,0.00,0.00,0.00;1.00)	(0.02,0.00,0.00,0.00;1.00)
($C_{3,10}$)	(0.14,0.11,0.04,0.01;1.00)	(0.14,0.11,0.04,0.01;1.00)
$C_{3,11}$)	(0.03,0.01,0.00,0.00;1.00)	(0.03,0.01,0.00,0.00;1.00)
($C_{3,12}$)	(0.00,0.00,0.00,0.00;1.00)	(0.00,0.00,0.00,0.00;1.00)

Table 11. Computed performance scores against alternatives A_1 and A_2

Criteria, (C_{ij})	$I_{\tilde{A}}(\bar{x}_0, \bar{y}_0)$	Performance Scores	$I_{\tilde{A}}(\bar{x}_0, \bar{y}_0)$	Performance Scores
($C_{1,1}$)	0.117	1.0	0.081	1.00
($C_{1,2}$)	0.117	2.0	0.081	1.00
($C_{1,3}$)	0.059	13.0	0.081	1.00
($C_{1,4}$)	0.080	10.0	0.215	2.00
($C_{1,5}$)	0.085	9.0	0.206	3.00
($C_{1,6}$)	0.095	8.0	0.186	4.00
($C_{1,7}$)	0.060	12.0	0.052	11.00
($C_{1,8}$)	0.000	19.0	0.000	17.00
($C_{1,9}$)	0.000	19.0	0.000	17.00
($C_{1,10}$)	0.080	10.0	0.080	9.00
($C_{1,11}$)	0.080	10.0	0.080	9.00
($C_{1,12}$)	0.085	9.0	0.085	8.00
($C_{1,13}$)	0.100	3.0	0.100	6.00
($C_{1,14}$)	0.000	19.0	0.000	17.00
($C_{2,1}$)	0.000	19.0	0.000	17.00
($C_{2,2}$)	0.080	10.0	0.080	9.00
($C_{2,3}$)	0.067	11.0	0.067	10.00

continued on following page

Table 11. Continued

Criteria, (C_{ij})	$I_{\tilde{A}}(\overline{x}_0, \overline{y}_0)$	Performance Scores	$I_{\tilde{A}}(\overline{x}_0, \overline{y}_0)$	Performance Scores
($C_{2,4}$)	0.067	11.0	0.067	10.00
($C_{2,5}$)	0.050	14.0	0.050	12.00
($C_{2,6}$)	0.050	14.0	0.050	12.00
($C_{2,7}$)	0.085	9.0	0.085	8.00
($C_{2,8}$)	0.100	4.0	0.100	5.00
$C_{2,9}$)	0.001	18.0	0.001	16.00
($C_{2,10}$)	0.030	16.0	0.030	14.00
($C_{2,11}$)	0.003	17.0	0.003	15.00
($C_{2,12}$)	0.097	5.0	0.097	7.00
(C_{31})	0.001	18.0	0.001	16.00
($C_{3,2}$)	0.030	16.0	0.030	14.00
($C_{3,3}$)	0.003	17.0	0.003	15.00
($C_{3,4}$)	0.097	6.0	0.097	7.00
($C_{3,5}$)	0.001	18.0	0.001	16.00
($C_{3,6}$)	0.040	15.0	0.040	13.00
($C_{3,7}$)	0.040	15.0	0.040	13.00
($C_{3,8}$)	0.097	7.0	0.097	7.00
($C_{3,9}$)	0.001	18.0	0.001	16.00
($C_{3,10}$)	0.030	16.0	0.030	14.00
$C_{3,11}$)	0.003	17.0	0.003	15.00
($C_{3,12}$)	0.000	19.0	0.000	17.00

Chapter 11
Appraise the Economic Values of Logistic Handling System under Mixed Information

Anoop Kumar Sahu
J.K.I.E., India

Nitin Kumar Sahu
Guru Ghasidas Central Vishwavidyalaya, India

Atul Kumar Sahu
Guru Ghasidas Central Vishwavidyalaya, India

ABSTRACT

Logistics activities are globally recognized as a core heart of each industry, as major content of productivity of industries exclusively depend on this. The system being used to manage logistic activities of firms is called as logistic system i.e. JCBs, robots, mono-trucks, trolleys conveyors etc. In recent years, assessments of logistic activities of firms are found as historic topic in the realm of industrial management. Each parameter leads the trait role in relation to materialize the economic values of material handling system. It is observed that in the context of decision making, MCDM problems under concerns of observed data rather than verbal data is resolved with short computation task by exploration of algorithmic techniques. It became indeed imperative to aid the managers from soft computing algorithmic technique for making decision under subjective or objective or subjective cum objective information. The present chapter flaunted an application of Multi Level Multi Criterion Decision Making (MLMCDM) algorithmic technique for appraising the economic values of MHS candidate choices under subjective (linguistic information provided by personnel) or objective or subjective cum objective information. A hypothetical research work is conveyed to end for aiding the firms from a DSS, in order that they could manage the logistic systems of their firms.

DOI: 10.4018/978-1-5225-1848-8.ch011

AT A GLANCE

This chapter regards the problems in relation to quantifying the economic values of Material Handling System (MHS). This chapter dealt with the fuzzy cum non-fuzzy information embedded with Multi Level Multi Criterion Decision Making (MLMCDM) algorithmic technique for appraising the economic values of MHS candidate alternatives. The proposed chapter objective is to crack the logistic system evaluation problems under mixed information. This chapter has vast scope amongst practitioners, academicians, research scholars, as they can relate shared technique for solving their professional problems. Furthermore, proposed technique can be implemented in order to resolve miscellaneous quandaries of industries (i.e. performance measurement of supply chain management, benchmarking of advanced manufacturing technologies, port selection, shipping logistic route evaluation, layout evaluation, banking sector evaluation, hotel election, etc.) This chapter displays a flexible decision support system for quantifying the economic values of MHS candidate alternatives.

INTRODUCTION

In today's emerging technology, MHS (Material Handling System) has been determined as boon for traditional and advanced manufacturing firms. MHS leads a significant function to complete the production tasks under scheduled time. MHS is found active to perform repetitious, thorny as well as precarious production tasks with great accuracy. The material handling system transfers the materials such as stuffs, parts, and finished goods from one place to another through pre-planned path or layout and enabling parts and material transportation between various stages of processing. It is responsible for:

- Ensuring the right product,
- Location,
- Condition,
- Quantity, and
- Timing of materials delivery.

Material handling systems are those systems, where parts are stacked onto pallets to send to a warehouse for storage, prior to be finally picked and assembled into customer orders for distribution and delivery. Askin and Standridge (1993), articulated that material handling systems are characterized by a high intensity of material flows and storage to dispatch to a prominent place.

Recently, several MHS,

- i.e. Powered trucks,
- Non-powered trucks,
- Dock equipment,
- Truck restraining devices,
- Fixed path conveying,
- Gravity conveyors,
- Horizontal powered conveyors,
- Vertical conveyors,

- Monorails,
- Horizontal,
- Automatic Guided Vehicles (AGV),
- Robot,
- Vertical carousels etc., are adopted by traditional and advanced manufacturing firms, depending upon the material and applications.

In present era, growth of automation technology enabled many material handling systems to performance hazardous tasks under efficient manners. Where, robot has been seen as a trait MHS for performing the several manufacturing operations with flexibility. Robots are self-powered MHS, work on automations and riding on planned path. Robot provides flexible transport of material without using valuable floor space. They are being used increasingly in automation of fabrication, finishing, and assembly operations; but usually these systems are used in batch operations.

Evaluation and selection of robot MHS for advanced manufacturing is found critical task, as appropriate robot MHS not only escalate the production rate, but also enhance overall productivity of firms. The few definition of robot MHS articulated here: Robot MHS follows standard programming language designed specifically for transferring the material with high-precision. Robot MHS is valuable for precision automation system uses programs to automatically execute a series of operations. Robot MHS increases productivity and flexibility.

Robot MHS selection problem for a particular industrial application is challenging tasks. This problem has become more complex in recent years due to rapid production under limited period of time. With wide variety of availability of robot types and models with distinct features, increase complexity and the diversity in their application areas offered by different robotic products. The selection of the most appropriate robot amongst several robots is become more difficult for companies. It became indeed necessity to select the best robot under consideration of robot's common subjective cum objective criterion in empire of Multi Criteria Decision Making (MCDM) problem. In the present chapter, a MLMCDM approach is applied for solving the selection problems of robot under robot's common subjective cum objective criterion in empire of MCDM. Entire measures dealt with:

- (+) positive attitude beside depreciation,
- Requirement of service,
- Purchasing cost and maintenance cost,
- Dealt with (-) negative attitude.

A hypothetical case research has been presented to justify the applications of proposed approach.

LITERATURE REVIEW

Webster and Reed (1971), developed a heuristics procedure, which elected MHE to perform a given set of moves for minimizing the system's costs associated with handling. Ekere and Hannam (1989), used several criteria describing programming features, model development characteristics, experimental and reporting features, and commercial and technical features were specified for the evaluation of manufacturing simulation software. Hlupic and Paul (1999), provides guidelines for selecting manufacturing simulation software according to the intended purpose of software use. The basic criteria to be examined in the software evaluation process are listed together with the level of their importance for particular types of users. Sharp, Wan, McGinnis, Goetschalckx, Bodner, Govindaraj, Ram and Everette (2001), explained that MHS selection usually does not address the possibility for electing different types of technology, such as forklift truck and AGVS. It addresses the possibility for electing technology as per material handling tasks and assigning them to technology applications. Chan and Lau (2001), integrates expert system with the analytical hierarchy process. Chu and Lin (2003), proposed a fuzzy-TOPSIS method for robot selection. Bhangale, Agrawal and Saha (2004), attempted to create exhaustive files by recognizing maximum possible number of attributes for robot manipulators and also illustrated the coding system and the mathematical, graphical and selection procedures. Sujono and Lashkari (2005), expressed that Material Handling Equipments (MHEs) are capable for performing the required material handling operations within the constraints operating on the manufacturing system.

Caputo and Pelagagge (2008), expressed few pertinent issues in the design and operations of material handling and storage systems (MH&SS) characterized by compound material flows and high-traffic intensity. Karsak (2008), introduced a decision model for robot selection based on quality function deployment (QFD) and fuzzy linear regression. Hassan (2010), Material handling (MH) equipments are important for looking after manufacturing and logistics actions, as they impacts upon operating cost, productivity, quality, and flexibility of operations of firms. Sahu, Datta and Mahapatra (2013), proposed an efficient decision-support system for electing industrial robot. The authors utilized grey set theory coupled with MULTIMOORA algorithmic approach in order to compute the performance score against industrial robots. Tansel, Yurdakul and Dengiz (2013), developed a two-phase robot selection decision support system, namely ROBSEL, to help the decision makers in their robot selection decisions. Hassan (2014), suggested that developers of expert's system ought to consider the efficiency of expert's system in exploring data and generation of equipments in respect to its functions and approaches. Sahu, Sahu and Sahu (2014), utilized MUTI-MOORA approach along with an application of grey number to evaluate the performances score against CNC machine tool alternatives under uncertain information. Sahu, Sahu and Sahu (2015a), integrated a hairy-based approach and applied said approach on novel developed multi hierarchical third-party logistics (3PL) service provider's appraisement module for evaluating the 3PL provider performance score. Sahu, Sahu and Sahu (2015b), employed TOPSIS approach to make the decision in regards to elect the best advanced manufacturing CNC machine tool under subjective information against criterion. Sahu, Sahu and Sahu (2016a), developed interval-valued fuzzy number set and merged this with modified MCDM technique TOPSIS for benchmarking supplier firms under agile supply chain actions. Sahu, Sahu and Sahu (2016b), used a revised ranking approach accompanied with fuzzy performance important index to identify the barriers/ill-strong measures in Agile Supply Chain Management. Sahu, Sahu and Sahu (2016c), explored Ratio system, Reference point, and full Mutification form coupled with aggregated interval valued trapezoidal fuzzy number set to evaluate the CNC machine tool from preferred options.

APPLICATION OF MLMCDM METHOD

The applications of ML-MCDM is not limited to solve the material handling appraisement and selection problems. The proposed method is also active to solve several below said real life problems (Li, 2007a, 2007b, 2009, 2011; Li & Nan, 2011; Chu & Lin, 2003; Sahu, Datta & Mahapatra, 2016; Ren, Chen, Fei

& Li, 2016; Li & Wan, 2014a, 2014b; Halgamuge, 1998; Van & Kerre, 1999; Runkler & Glesner, 1993; Kauffman & Gupta, 1991; Mizumoto, 1989; Laarhoven & Pedrycz, 1983; Tong, 1978; Li & Cheng, 2004). Material handling system selection for a given engineering application:

- Evaluation of alternative product designs.
- Evaluation and selection of modern machining methods.
- Evaluation and selection of flexible manufacturing systems.
- Machine group selection in a flexible manufacturing cell.
- Selection of automated inspection systems.
- Selection of a rapid prototyping process in rapid product development.
- Selection of software for design and manufacturing applications.
- Selection of the most appropriate welding machine for a given job.
- Evaluation of metal stamping layouts.
- Evaluation of environmentally conscious manufacturing programs.
- Environmental impact assessment of manufacturing processes.
- Evaluation of aggregate risk in green manufacturing.
- Selection of best product end-of-life scenario.
- Integrated project evaluation and selection.
- Facility location selection.
- Supplier evaluation and election.
- Election of buses, cars, trucks etc.
- Election of the best ship for given route.
- Election of route.
- Appraisement and Flexible manufacturing system.
- Supply chain strategic performance evaluation and election.
- Shipping port and maritime Engg problem.

MULTI-CRITERIA DECISION MAKING (MCDM)

MCDM constructing a global preference in relation to a set of choices evaluated under numerous criteria, which reflects the selection of the best actions from a set of choices, each criterion is called as conflicting criteria.

MCDM is allied with structuring and solving decision and planning problems involving multiple criteria. The aim of MCDM is to support personnel's to stand against industrial problems,

- i.e. Evaluation and selection of flexible manufacturing systems,
- Machine group,
- Automated inspection systems,
- Rapid prototyping process in rapid product development,
- Software for design and manufacturing applications,
- Welding machine for a given job,
- Metal stamping layouts,
- Environmentally conscious manufacturing programs,

- Bus,
- Truck,
- Shipping or etc.

Normally, MCDM doesn't deliver optimal solution for said problems due to the collection of information from personnel.

The solution, which is being obtained in domain of MCDM, can be interpreted in different paths. It could correspond to elect the optimum choice from preferred alternatives. MCDM problems consist of a finite number of alternatives, explicitly known in the beginning of the solution process. Each alternative is represented by its performance score in empire of chain of multiple criteria. The previous research scholars defined the MCDM as a decision making tool for finding the best alternative in accordance with the subjective perceptions of personnels. Decision Making (DM) splits out into several categories (Sahu, Datta & Mahapatra, 2012, 2013 2014, 2015; Samantra, Sahu, Datta & Mahapatra, 2013).

- Multiple-Criterion Decision Making: Commodity selection/ Personnel selection/ facility location/ Public facility selection.
- Multiple-Objective Decision Making: System design/ Production and manufacturing/ Capital budgeting/ Manpower planning/ Network Flows/ Socioeconomic problems.
- Group Decision Making: Political voting/ Commodity selection/ Project evaluation.

VALIDATION OF FUZZY SET THEORY

Manufacturing is the heart of any processing industry. It is a part for augmenting the world significance. It converts the raw material, inventory and semi-inventory into finish goods with taking the interaction between processing equipments; machine tools & human being. It encompasses around 30 to 35% value of all goods and services, which is produced. In today's advanced manufacturing organizations, many problems take place, which requires the verdict, brainpower, analytical view & strategic vision of the expert's panels. In order to answer the defies of competitive market, in domain of industries, the expert's panel performs a great role to select suitable manufacturing strategies, product designs, manufacturing processes, work piece and tool materials, machinery and MHS etc. from given set of alternatives. The verdict making is a so tricky task, as it requires the analytical ability, judgment power to make out the current and future actions and issues, may leverage the complete manufacturing processes. Decision makers in manufacturing industries regularly look the problems for assessing a broad range of alternatives / choices, and selecting one based on a set of contradictory criteria.

When the managers of organizations begin to purchase assets from their partner organizations, at that moment, expert's panel felt need to utilize the decision support systems or algorithmic models or pertinent approaches, which could aid them to elect the feasible or more economic asset from set of assets. When, managers decide to buy assets in accordance with technical data or information, in that case, traditional tool and approach would aid the managers to come out problems, but when managers decide to buy assets in accordance with non-technical data or information (oral data), in that case traditional tools and methods would not aid the managers for same problem. Then problem become more complicated and requires the utility of fuzzy logic theory.

Fuzzy logic starts with a set of partial data being supplied by human's language. The fuzzy system converts these human's language into their mathematical equivalents. This simplifies the job of the system designer and the computer, and results in much more accurate representations of the way systems behave in the real world. Additional benefits of fuzzy logic include its simplicity and its flexibility. Fuzzy logic can handle problems with imprecise and incomplete data. Fuzzy Logic is one of the best tools to model our imprecise and blurred world. The real world is too complicated for precise descriptions to be obtained; therefore approximations (or fuzziness) must be introduced in order to obtain a reasonable, yet traceable, model. Fuzzy logic is the tool for transforming human knowledge and its decision-making ability into a mathematical formula. In other words, it provides us with meaningful and powerful representation of measurement uncertainties and also with meaningful representation of vague concepts expressed in natural language.

In present chapter, a DSS (Decision Support System) has been proposed, which handled the uncertainty associated with Multi-Criteria Decision Making hierarchical module towards benchmarking the candidate robots. A multi-level multi criterion decision making method coupled with fuzzy set theory in associating with COA 'Center of Area approach has utilized to overcome ambiguity as well as vagueness associated with subjective (linguistic) information of expert panel, helped to undertake uncertainty of criteria. An example has been presented to justify the application of proposed approach.

FUZZY SET THEORY

Fuzzy Sets

A fuzzy set A can be denoted by $A = \left\{\left(x, f_A\left(x\right)\right) \middle| x \in U\right\}$, where U is the universe of discourse, x is an element in U, A is a fuzzy set in U, $f_A\left(x\right)$ is the membership function of A at x (Koulouriotis & Ketipi, 2011; Bairagi, Dey, Sarkar & Sanyal, 2012; Huang & Oliver, 1994; Li, 2013, 2016; Semwal, Chakraborty & Nandi, 2015; Li & Liu 2015). The larger $f_A\left(x\right)$, the stronger the grade of membership for x in A.

Fuzzy Numbers

A real fuzzy number A is described as any fuzzy subset of the real line R with membership function f_A which possesses the following properties:

f_A is a continuous mapping from R to $\left[0,1\right]$;

1. $f_A\left(x\right) = 0, \forall x \in \left(-\infty, a\right|$;

f_A is strictly increasing on $\left[a, b\right]$;

2. $f_A\left(x\right) = 1, x \in \left[b, c\right]$;

f_A is strictly deceasing on $\left[c, d\right]$;

3. $f_A = 0, \forall x \in \left|d, \infty\right)$;

Here a, b, c, d are real numbers. We may let $a = -\infty$, or $a = b$, or $b = c$, or $c = d$, or $d = +\infty$. Unless elsewhere specified, it is assumed that A is convex, normal and bounded, i.e. $-\infty < a, d < \infty$. For convenience, fuzzy number A can be denoted by $A = [a, b, c, d]$. The opposite of A can be given by $-A = [-d, -c, -b, -a; 1]$ x. Fuzzy number A is a triangular fuzzy number, denoted by (a, b, c), if, its membership function f_A is given by:

$$f_A(x) = \begin{cases} (x-a)/(b-a), & a \le x \le b, \\ (x-c)/(b-c), & b \le x \le c, \\ 0, & Otherwise. \end{cases} \tag{1}$$

Linguistic Values

A linguistic variable is a variable whose values are expressed in linguistic terms. Linguistic variable is a very helpful concept for dealing with situations which are too complex or not well-defined to be reasonably described by traditional quantitative expressions. It is assumed that decision makers have fully understood the meanings of these linguistic values and their corresponding fuzzy numbers before they assign these values to criteria.

Defuzzifying Triangular Fuzzy Numbers With COA

The following formulas are developed to defuzzify triangular fuzzy numbers based on dividing the area under the membership function in half. The defuzzification formulas for fuzzy number A in Equation (1) by using COA, i.e. $I_L(A) = I_R(A)$, are presented in the following three situations:

1. If $\overline{ab} > \overline{bc}$, then e is derived from "$I_L(A) = I_R(A)$" as

$$e = a + \frac{1}{2}\left[2a^2 - 2ab - 2ac + 2bc\right]^{\frac{1}{2}} \tag{2}$$

2. If $\overline{ab} < \overline{bc}$, then e is derived from "$I_L(A) = I_R(A)$" as

$$e = c - \frac{1}{2}\left[2c^2 + 2ab - 2ac - 2bc\right]^{\frac{1}{2}} \tag{3}$$

3. If $\overline{ab} = \overline{bc}$, then e is derived from "$I_L(A) = I_R(A)$" as

$$e = \frac{1}{2}(a + c) \tag{4}$$

MODEL DEVELOPMENT

Notations

Some important mathematical notations used in the proposed model are defined as follows:

D_v : Denotes decision maker $v, v = 1,..., q$;
A_i : Denotes fuzzy numbers used to evaluate the importance of the importance of the criteria, $i = 1,..., n$;
B_i : Denotes fuzzy numbers used to evaluate the suitability of alternatives versus qualitative criteria, $i = 1,..., n$;
$e(A_i)$: Denotes the defuzzified value of A_i through COA;
$e(B_i)$: Denotes the defuzzified value of B_i through COA;
$f_{x_1x_2...x_i...x_n}$: Denotes the n level (general) hierarchy structure to depict the relationship amongst criteria;
$m_{x_1x_2...x_{(i-1)}}$: Denotes number of sub-criteria for criterion $f_{x_1x_2...x_i}$;
$w_{x_1x_2...x_iv}$: Denotes the weight given by the vth decision-maker to the $x_1x_2...x_i{}^{th}$ criterion, $1 \leq v \leq q$;
W : Denotes vector;
M : Denotes matrix;
$r_{x_1x_2...x_itv}$: Denotes the suitability given by the vth decision-maker to the $x_1x_2...x_i{}^{th}$ criterion for alternative t;
$R_{m_{x_1x_2...x_{(i-1)}}\times p}$: Denotes $m_{x_1x_2...x_{(i-1)}} \times p$ matrix of the $m_{x_1x_2...x_{(i-1)}}$ suitability values of sub-criteria of the criterion $f_{x_1x_2...x_{(i-1)}}$ from p alternatives.

The MLMCDM Model

In this section, the proposed COA defuzzification method is applied to establish a MLMCDM (multiple levels multiple criteria decision making) model under fuzzy environment. Suppose the importance weights of different criteria and the ratings of various alternatives under qualitative criteria in the model are assessed in linguistic terms represented by triangular fuzzy numbers. Further suppose a set of linguistic terms represented by positive triangular fuzzy numbers $A_i, i = 1,.., n$, are applied by decision-makers D_v, $v = 1,..., q$, to evaluate the importance of the criteria. Also a set of linguistic terms represented by positive triangular fuzzy numbers $B_i, i = 1,.., n$, are applied by decision makers to evaluate the suitability of alternatives versus qualitative criteria. By applying Equations. 2 and 4, we obtain the values of COA of these fuzzy numbers as $e(A_i)$ and $e(B_i)$, respectively. The proposed model is developed by the following procedure.

Establish a Multiple Levels Hierarchy Structure for Criteria

A general hierarchical structure to depict criteria is presented as follows:

$$F_{x_i} = \left\{ f_{x_1 x_2 \ldots x_i \ldots x_n} \right\} \tag{5}$$

For example, f_{x_1} represents the first level criteria of evaluated alternatives, $f_{x_1 x_2}$ represents second level criteria of f_{x_1}, and the number of the second level criteria is m_{x_1}. Herein, the criteria in the hierarchical structure are assumed to be independent.

Decide the Weights

When decision makers assign weights to criteria, they must understand the meanings of the linguistic weights and their corresponding fuzzy numbers; in other words, we assume that decision makers' understanding of the concept of ''importance'' is in full compliance with the way that weights are used in the model.

The average weights associated with n level hierarchical structure are developed the following equation:

$$w_{x_1 x_2 \ldots x_i} = \frac{1}{q} \left\{ w_{x_1 x_2 \ldots x_i 1} + w_{x_1 x_2 \ldots x_i 2} + \ldots + w_{x_1 x_2 \ldots x_i v} + \ldots + w_{x_1 x_2 \ldots x_i q} \right\} \tag{6}$$

Here $w_{x_1 x_2 \ldots x_i v}$ is a defuzzified triangular fuzzy number from $e\left(A_i\right)$. Also $w_{x_1 x_2 \ldots x_i}$ represents the weight of criterion $f_{x_1 x_2 \ldots x_i}$.

Average Alternative Suitability vs. Qualitative Criteria

The average suitability of alternative t, $t = 1, \ldots, p$, versus each subjective criterion associated with n level hierarchy structure is presented as follows:

$$r_{x_1 x_2 \ldots x_i t} = \frac{1}{q} \left\{ r_{x_1 x_2 \ldots x_i t1} + r_{x_1 x_2 \ldots x_i t2} + \ldots + r_{x_1 x_2 \ldots x_i tv} + \ldots + r_{x_1 x_2 \ldots x_i tq} \right\} \tag{7}$$

Here $r_{x_1 x_2 \ldots x_i tv}$ is a defuzzified triangular fuzzy number from $e\left(B_i\right)$ and $r_{x_1 x_2 \ldots x_i t}$ represents the average suitability of alternative t versus criterion $f_{x_1 x_2 \ldots x_i}$.

Normalization of Alternative Suitability vs. Qualitative Criteria

Values (or suitability) of alternatives versus different quantitative criteria need to be normalized because they have different units. If only benefit (or cost) qualitative criteria are used, normalization can be omitted. In this model, suitability of alternatives versus quantitative criteria can be classified into benefit $\left(B\right)$ and cost $\left(C\right)$ ones. The normalization of the suitability can be accomplished by applying the following two formulas:

$$r_{x_1x_2\ldots x_it} = \frac{S_{x_1x_2\ldots x_it}}{\max_t\left\{S_{x_1x_2\ldots x_it}\right\}}, \tag{8}$$

$$r_{x_1x_2\ldots x_it} = \frac{\min_t\left\{S_{x_1x_2\ldots x_it}\right\}}{S_{x_1x_2\ldots x_it}}. \tag{9}$$

Here $r_{x_1x_2\ldots x_it}$ denotes the normalized value of $S_{x_1x_2\ldots x_it}$. Also $S_{x_1x_2\ldots x_it}$ denotes the suitability value of alternative t versus criterion $f_{x_1x_2\ldots x_i}$.

Synthetic Evaluation

The additive weighted evaluation matrices in the structure can be obtained by using multiplication and addition to aggregate the evaluation matrices and their corresponding weights matrices as follows:

$$M_{x_1x_2\ldots x_{(i-1)}} = W_{x_1x_2\ldots x_{(i-1)}} \times R_{m_{x_1x_2\ldots x_{(i-1)}}\times p}$$

$$= \begin{bmatrix} \sum_{x_i=1}^{m_{x_1x_2\ldots x_{(i-1)}}} w_{x_1x_2\ldots x_i}.r_{x_1x_2\ldots x_i1} \sum_{x_i=1}^{m_{x_1x_2\ldots x_{(i-1)}}} w_{x_1x_2\ldots x_i}.r_{x_1x_2\ldots x_i2}\cdots \sum_{x_i=1}^{m_{x_1x_2\ldots x_{(i-1)}}} w_{x_1x_2\ldots x_i}.r_{x_1x_2\ldots x_it}\cdots\cdots \\ \sum_{x_i=1}^{m_{x_1x_2\ldots x_{(i-1)}}} w_{x_1x_2\ldots x_i}.r_{x_1x_2\ldots x_ip} \end{bmatrix}$$

$$= \left[r_{x_1x_2\ldots x_{(i-1)}1} r_{x_1x_2\ldots x_{(i-1)}2}\cdots r_{x_1x_2\ldots x_{(i-1)}t}\cdots r_{x_1x_2\ldots x_{(i-1)}p}\right] \tag{10}$$

Here, $M_{x_1x_2\ldots x_{(i-1)}}$ is a $1\times p$ vector with the additive weighted evaluations of the p alternatives over the criteria set $f_{x_1x_2\ldots x_i}$, $W_{x_1x_2\ldots x_{(i-1)}}$ is the vector of the corresponding criteria weights and $R_{m_{x_1x_2\ldots x_{(i-1)}}\times p}$ is a matrix with the suitability of the alternatives on the criteria. $w_{x_1x_2\ldots x_{(i-1)}}$ is derived by Equation 6, t represents alternative t. $r_{x_1x_2\ldots x_it}$ is defined from Equation 7 when $f_{x_1x_2\ldots x_i}$ is a qualitative criterion with no sub-criteria, from Equation 8 and Equation (9) when $f_{x_1x_2\ldots x_i}$ is a quantitative criterion with no sub-criteria, or from $\sum_{x_{(i+1)}=1}^{m_{x_1x_2\ldots x_i}} w_{x_1x_2\ldots x_{(i+1)}}.r_{x_1x_2\ldots x_{(i+1)}t}$ when $f_{x_1x_2\ldots x_i}$ is not further analyzed into lower-level sub-criteria. $\sum_{x_i=1}^{m_{x_1x_2\ldots x_{(i-1)}}} w_{x_1x_2\ldots x_i}.r_{x_1x_2\ldots x_it}$ denotes the additive weighted evaluation value, $r_{x_1x_2\ldots x_{(i-1)}t}$, of sub-criterion $f_{x_1x_2\ldots x_{(i-1)}}$

of $f_{x_1x_2\ldots x_{(i-2)}}$ from alternative t, and is the corresponding element of the $x_{(i-1)}th$ row and the tth column in $R_{m_{x_1x_2\ldots x_{(i-2)}}\times p}$. The aggregation at every level of the hierarchy is done similarly to Equation 10. The final additive weighted evaluation matrix can then be derived by Equation 11 based on the rule of back propagation as follows:

$$M = W \times R_{m\times p} = \left[\sum_{x_1=1}^{m} w_{x_1}.r_{x_1 1}\sum_{x_1=1}^{m} w_{x_1}.r_{x_1 2}\ldots\sum_{x_1=1}^{m} w_{x_1}.r_{x_1 t}\ldots\sum_{x_1=1}^{m} w_{x_1}.r_{x_1 p}\right]$$

$$= \left[r_1 r_2 \ldots r_t \ldots r_p\right] \quad (11)$$

HYPOTHETICAL CASE RESEARCH

The robot MHS is a mechanical device, which is explored in industries in purpose to performance the hazardous tasks. A hypothetical case research is conducted in a well known advanced manufacturing organization, situated at western part of India, producing goods. In hypothetical case research, proposed DSS is examined for selecting the robot MHS. To select the most feasible alternative amongst preferred robot MHS alternatives, a committee of five expert's panel (decision makers), DM_1, DM_2, DM_3, DM_4 and DM_5 is formed from cross functional departments such as:

- Production,
- Marketing,
- Environmental,
- Assembly,
- Quality assurance of a firm.

The decision making committee assessed the performances of defined alternatives as per structured module (criteria hierarchy evaluated by DM), shown in Table 1 (see Appendix) and formulated problem has solved by MLMCDM (Chu & Varma, 2012).

The structured module dealt with thirty seven contradicting criteria, where:

- Effectiveness, (C_1),
- Efficiency, (C_2),
- Responsiveness towards scheduled production, (C_4),
- Overall Logistic Capability, (C_5),
- Quantity flexibility, (C_6),
- Volume flexibility, (C_7),
- Reliability, (C_8),
- Programming flexibility, (C_9),
- Programming features, (C_{10}),
- Productivity, (C_{11}),

- Coding aspects, (C_{12}),
- Schedule Utilization, (C_{13}),
- Adaptability, (C_{14}),
- Precision, (C_{15}),
- Reliability, (C_{16}),
- Safety, (C_{17}),
- Adaptive control machining system, (C_{18}),
- Agility towards production system, (C_{19}),
- Mixed product flexibility, (C_{20}),
- User friendliness, (C_{21}),
- Product Quality, (C_{22}),
- Flaws diagnosing aptitude, (C_{23}),
- Performance under Working progress, (C_{24}),
- Utility ratio per year, (C_{25}),
- Conditional routing, (C_{26}),
- Program generation feature, (C_{27}),
- Readability of changed code, (C_{28}),
- Comprehensiveness of added code, (C_{29}),
- Icon library, (C_{30}),
- Portability, (C_{31}),
- Life cycle maintenance, (C_{32}),
- Output data analysis, (C_{34}),
- Requirement of Service, (C_{35}),
- Programme visualisation, (C_{36}),
- Operator friendly, (C_{37}),

are considered as beneficial criteria, whereas:

- Depreciation, (C_3),
- Requirement of Service, (C_{35}),
- Purchasing Cost, (C_{38}),
- Maintenance cost, (C_{39}) are considered as non-beneficial criteria.

The information pertaining to two non-beneficial criteria such as:

- Purchasing Cost, (C_{38}),
- Maintenance cost, (C_{39}) are defined in Table 2 (see Appendix).

Criteria importance (significance) and ratings against each alternative is expressed in linguistic terms, which are transformed further in triangular interval value fuzzy numbers, as given in Table 3 (see Appendix). Appropriateness ratings (assigned by DMs) for various for respective alternatives A_1, A_2, A_3 and A_4 have been shown in Tables 4, Table 5, Table 6, Table 7 (see Appendix). Similarly, importance weight of criteria for respective alternatives A_1, A_2, A_3 and A_4 are shown in Table 8 (see Appendix). The assigned linguistic ratings transformed into triangular fuzzy number set. Later aggregated into single triangular fuzzy number set by exploration of Equation 5, 6 and 7, shown in Tables 9, Table 10, Table 11 and Table 12 respectively (see Appendix).

Next, aggregated rating is transformed into crisp value in accordance with defuzzification rules, derived by Equation 2, 3 and 4. Then Equations 8 and 9 are applied to normalize the data and also convert non-beneficial criteria into beneficial criteria and then problem became so easy to solve. The weighted normalized matrix is constructed by using Equation 10. Eventually, Equation 11 is applied to procure performance scores and preference orders of various robot MHS alternatives, shown in Table 13 (see Appendix). Hence, alternative sorting is as following.

$A_3 > A_4 > A_2 > A_1$

Proposed general procedural steps, which are valid to evaluate the decision:

Step 1: Form a committee of decision-makers, and then identify the evaluation criteria of defined robot MHS alternatives.
Step 2: Choose the appropriate linguistic variables for importance weight and ratings for electing the best robot MHS alternative.
Step 3: Aggregate the weight of criteria such as weight $\tilde{w}_j$ of criterion C_j, and rating $\tilde{x}_{ij}$ in terms of triangular fuzzy number set against robot MHS alternatives.
Step 4: Defuzzified the ratings and weights of criteria.
Step 5: Construct the weighted normalized matrix.
Step 6: Construct the fuzzy-decision matrix and the normalized fuzzy decision matrix.
Step 7: Construct weighted normalized fuzzy decision matrix.
Step 8: Clubbed all criteria with respect to defined alternatives to compute the performance scores.
Step 9: According to the performance scores, we can understand the status of robot MHS alternatives and can determine the best robot amongst available robots MHS alternatives. Figure 1 depicted the Pie chart, reflected the performance scores against robots
Step 10: According to the performance scores, we can solve the benchmarking problems of robot MHS alternatives and can determine that how much performance score are lost by the close competitor's robots of selected robot.

Figure 1. Pie chart depicts the performance scores against robot MHS

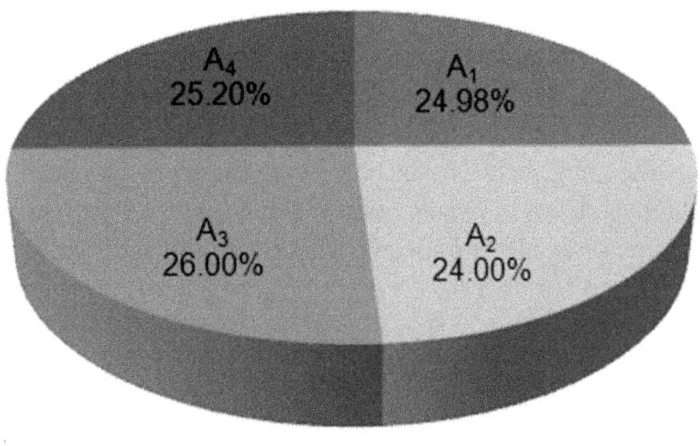

CONCLUSION

The selection of Material Handling Systems (MHS) leads a vital job in achieving the future agendas of firms. Robot MHS is used in handling assembly tasks of production units and referred. The applications of robot MHS is enormously escalated in advanced manufacturing firms, as it is seemed as a terrific dynamic and energetic automation based machine. The MCDM evaluation appraisement (decision making) modules (constituted by criteria and metrics) coupled with MCDM-algorithmic approach is known as Decision Support System (DSS) help to access to potential decision. DSS measures efficiency as well as effectiveness of defined alternative, depending upon DSSs applicants towards tackling problems:

- i.e. Evaluation of CNC machine tool,
- MHS,
- Layout,
- Truck,
- Other alternatives etc.

In the present chapter, a decision support system is built towards undertaking the robot MHS evaluation problems. This paper presented an application of MLMCDM upon a module (consist of thirty-seven contracting criteria) to measure performance of robot MHS. A hypothetical case research is presented. The other outcomes of chapter presented here:

- Emphasized decision making towards tackling advanced minor and major issues of firms via Multi-level multi criterion decision Making approach.
- Recognized the pertinent criteria for solving robot MHS election problem and solved said problem under short period of time via discussion and brainstorming sessions, being conducted in organization.
- Other MCDM approaches such as the AHP, ANP requires more computations and information in comparison to multi-level multi criterion decision making approach.
- The most optimum robot is elected by comparing its performance score with computed performance scores of its close competitors under thirty-seven contradicting criterions.
- The built DSS permit the peoples to share their oral information against defined criteria associated with the mapping of their opinions to a trapezoidal fuzzy number set scale.
- The presented fuzzy based DSS, which could be powered by Microsoft Windows XP or manual computation under fewer time.
- Presented module of DSS comprises power for materializing values of robots under short computation due to its simplicity and the flexibility (can incorporate extensive chain of subjective parameters).
- The constructed module built better by incorporating the more advanced criterion, in order that, elected robot could meet defies of present market and could work well for long period of time.
- The proposed DSS can be availed for gleaning the interest of international firms, as the solution of problem can be carried out under short period of time, where the others software found them so trivial.

- Many problems of society and community of peoples i.e. appraisal the value of bungalow, cottage, community centre, park layout, domestic appliances etc. can be encountered by replacement of defined criterions from other criterion linked to be alternatives.
- The proposed approach possesses the inclusive package for handling the subjective (decision maker's perceptions) or objective (crisp) or subjective or objective information concomitantly.

REFERENCES

Askin, R. G., & Standridge, R. (1993). *Modeling and Analysis of Manufacturing Systems*. New York, NY: Wiley.

Bairagi, B., Dey, B., Sarkar, B., & Sanyal, S. (2012). A novel multiplicative model of multi criteria analysis for robot selection. *International Journal on Soft Computing. Artificial Intelligence and Applications*, *1*(3), 1–9.

Bhangale, P. P., Agrawal, V. P., & Saha, S. K. (2004). Attribute based specification, comparison and selection of a robot. *Mechanism and Machine Theory*, *39*(12), 1345–1366. doi:10.1016/j.mechmachtheory.2004.05.020

Caputo, A. C., & Pelagagge, P. M. (2008). Capacity upgrades criteria of large-intensive material handling and storage systems: A case study. *Journal of Manufacturing Technology Management*, *19*(8), 953–978. doi:10.1108/17410380810911727

Chan, F. T. S., & Lau, H. (2001). Integration of expert system with analytic hierarchy process for the design of material handling equipment selection system. *Journal of Materials Processing Technology*, *116*(2/3), 137–145. doi:10.1016/S0924-0136(01)01038-X

Chu, T. C., & Lin, Y. C. (2003). A fuzzy TOPSIS method for robot selection. *International Journal of Advanced Manufacturing Technology*, *21*(4), 284–290. doi:10.1007/s001700300033

Chu, T. C., & Lin, Y. C. (2003). A fuzzy TOPSIS method for robot selection. *International Journal of Advanced Manufacturing Technology*, *21*(4), 284–290. doi:10.1007/s001700300033

Chu, T. C., & Varma, R. (2012). Evaluating suppliers via a multiple levels multiple criteria decision making method under fuzzy environment. *Computers & Industrial Engineering*, *62*(2), 653–660. doi:10.1016/j.cie.2011.11.036

Ekere, N. N., & Hannam, R. G. (1989). An evaluation of approaches to modeling and simulating manufacturing systems. *International Journal of Production Research*, *27*(4), 599–611. doi:10.1080/00207548908942571

Halgamuge, S. K. (1998). A trainable transparent universal approximator for defuzzification in Mamdani-type neuro-fuzzy controllers. *IEEE Transactions on Fuzzy Systems*, *6*(2), 304–314. doi:10.1109/91.669031

Hassan, M. M. D. (2010). A framework for selection of material handling equipment in manufacturing and logistics facilities. *Journal of Manufacturing Technology Management*, *21*(2), 246–268. doi:10.1108/17410381011014396

Hassan, M. M. D. (2014). An evaluation of input and output of expert systems for selection of material handling equipment. *Journal of Manufacturing Technology Management*, *25*(7), 1049–1067. doi:10.1108/JMTM-08-2012-0077

Hlupic, V., & Paul, R. J. (1999). Guidelines for selection of manufacturing simulation software. *IIE Transactions*, *31*(1), 21–29. doi:10.1080/07408179908969802

Huang, Y., & Oliver, J. (1994). NC milling error assessment and tool path correction.*International Conference on Computer Graphics and Interactive Techniques. Proceedings of the 21st annual conference on computer graphics and interactive techniques*, (pp. 287-294). doi:10.1145/192161.192231

Karsak, E. E. (2008). Robot selection using an integrated approach based on quality function deployment and fuzzy regression. *International Journal of Production Research*, *46*(7), 723–738. doi:10.1080/00207540600919571

Kauffman, A., & Gupta, M. M. (1991). *Introduction to fuzzy arithmetic: theory and application.* Van Nostrand Reinhold New York.

Koulouriotis, D. E., & Ketipi, M. K. (2011). A fuzzy digraph method for robot evaluation and selection. *Expert Systems with Applications*, *38*(9), 11901–11910. doi:10.1016/j.eswa.2011.03.082

Laarhoven, P. J. M. V., & Pedrycz, W. (1983). A fuzzy extension of Satty's priority theory. *Fuzzy Sets and Systems*, *11*, 199–227.

Li, D.-F. (2007a). Compromise ratio method for fuzzy multi-attribute group decision making. *Applied Soft Computing*, *7*(3), 807–817. doi:10.1016/j.asoc.2006.02.003

Li, D.-F. (2007b). A fuzzy closeness approach to fuzzy multi-attribute decision making. *Fuzzy Optimization and Decision Making*, *6*(3), 237–254. doi:10.1007/s10700-007-9010-1

Li, D.-F. (2009). Relative ratio method for multiple attribute decision making problems. *International Journal of Information Technology & Decision Making*, *8*(2), 289–311. doi:10.1142/S0219622009003405

Li, D.-F. (2011). Closeness coefficient based nonlinear programming method for interval-valued intuitionistic fuzzy multiattribute decision making with incomplete preference information. *Applied Soft Computing*, *11*(4), 3402–3418. doi:10.1016/j.asoc.2011.01.011

Li, D.-F. (2013). An effective methodology for solving matrix games with fuzzy payoffs. *IEEE Transactions on Cybernetics*, *43*(2), 610–620. doi:10.1109/TSMCB.2012.2212885 PMID:22987531

Li, D.-F. (n.d.). *Linear Programming Models and Methods of Matrix Games with Payoffs of Triangular Fuzzy Numbers.* Springer.

Li, D.-F., & Cheng, C. T. (2004). Stability on multiobjective dynamic programming problems with fuzzy parameters in the objective functions and in the constraints. *European Journal of Operational Research*, *158*(3), 678–696. doi:10.1016/S0377-2217(03)00374-6

Li, D.-F., & Liu, J. C. (2015). A parameterized non-linear programming approach to solve matrix games with payoffs of I-fuzzy numbers. *IEEE Transactions on Fuzzy Systems*, *23*(4), 885–896. doi:10.1109/TFUZZ.2014.2333065

Li, D.-F., & Nan, J. X. (2011). Extension of the TOPSIS for multi-attribute group decision making under Atanassov IFS environments. *International Journal of Fuzzy System Applications*, *1*(4), 47–61. doi:10.4018/ijfsa.2011100104

Li, D.-F., & Nan, J. X. (2011). An extended weighted average method for MADM using intuitionistic fuzzy sets and sensitivity analysis. *Critical View*, *5*, 5–25.

Li, D.-F., & Wan, S. P. (2014a). A fuzzy inhomogenous multiattribute group decision making approach to solve outsourcing provider selection problems. *Knowledge-Based Systems*, *67*, 71–89. doi:10.1016/j.knosys.2014.06.006

Li, D.-F., & Wan, S. P. (2014b). Fuzzy heterogeneous multiattribute decision making method for outsourcing provider selection. *Expert Systems with Applications*, *41*(6), 3047–3059. doi:10.1016/j.eswa.2013.10.036

Mizumoto, M. (1989). Improvement methods of fuzzy controls. In *Proceedings of Third IFSA Congress* (pp. 60-62).

Ren, H. P., Chen, H. H., Fei, W., & Li, D.-F. (2016). A MAGDM method considering the amount and reliability information of interval-valued intuitionistic fuzzy sets. *International Journal of Fuzzy Systems*, *18*(1), 1–11. doi:10.1109/TFUZZ.2016.2554151

Runkler, T. A., & Glesner, M. (1993). A set of axioms for defuzzification strategies towards a theory of rational defuzzification operators.*Second IEEE International Conference on Fuzzy Systems*, (pp. 1161-1166). doi:10.1109/FUZZY.1993.327350

Sahu, A. K., Datta, S., & Mahapatra, S. S. (2016). Evaluation and selection of suppliers considering green perspectives: Comparative analysis on application of FMLMCDM and Fuzzy-TOPSIS. *Benchmarking: An International Journal*, *23*(6), 1579–1604. doi:10.1108/BIJ-12-2014-0111

Sahu, A. K., Sahu, N. K., & Sahu, A. (2014). Appraisal of CNC machine tool by integrated MULTI MOORA-IGVN circumstances: An empirical study. *International Journal of Grey Systems: Theory and Application*, *4*(1), 104–123.

Sahu, A. K., Sahu, N. K., & Sahu, A. K. (2015b). Benchmarking CNC machine tool using hybrid fuzzy methodology a multi indices decision making approach. *International Journal of Fuzzy System Applications*, *4*(2), 28–46.

Sahu, A. K., Sahu, N. K., & Sahu, A. K. (2016a). Application of Integrated TOPSIS in ASC index: Partners Benchmarking perspective. *Benchmarking: An International Journal*, *23*(3), 540–563. doi:10.1108/BIJ-03-2014-0021

Sahu, A. K., Sahu, N. K., & Sahu, A. K. (2016b). Appraisal of Partner Enterprises under GTFNS Environment in Agile SC. *International Journal of Decision Support System Technology*, *8*(3), 1–19. doi:10.4018/IJDSST.2016070101

Sahu, A. K., Sahu, N. K., & Sahu, A. K. (2016c). Application of Modified MULTI-MOORA for CNC Machine Tool Evaluation in IVGTFNS Environment: An Empirical Study. *International Journal of Computer Aided Engineering and Technology*, *8*(3), 234–259. doi:10.1504/IJCAET.2016.077603

Sahu, N. K., Datta, S., & Mahapatra, S. S. (2012). Establishing green supplier appraisement platform using grey concepts. *International Journal of Grey Systems: Theory and Application*, *2*(3), 395–418.

Sahu, N. K., Datta, S., & Mahapatra, S. S. (2013). Robot Selection based on grey-MULTIMOORA. *Grey Systems: Theory and Applications*, *3*(2), 201–232. doi:10.1108/GS-05-2013-0008

Sahu, N. K., Datta, S., & Mahapatra, S. S. (2013). Decision making for selecting 3PL service provider using three parameter interval grey numbers. *International Journal of Logistic Systems and Management*, *14*(3), 261–297. doi:10.1504/IJLSM.2013.052061

Sahu, N. K., Datta, S., & Mahapatra, S. S. (2014). Green Supplier Appraisement in Fuzzy Environment. *Benchmarking: An International Journal*, *21*(3), 412–429. doi:10.1108/BIJ-06-2012-0042

Sahu, N. K., Datta, S. & Mahapatra, S. S. (2015). Fuzzy Based Appraisement Module for 3PL Evaluation and Selection. *Benchmarking, an International Journal, 22*(3), 354-392.

Sahu, N. K., Sahu A. K., & Sahu, A. K (2015a). Appraisement and Benchmarking of Third Party Logistic Service Provider by Exploration of Risk Based Approach. *Cogent Business and Management, 2*, 1-21.

Samantra, C., Sahu, N. K., Datta, S., & Mahapatra, S. S. (2013). Decision-Making in Selecting Reverse Logistics Alternative using Interval-Valued Fuzzy Sets Combined with VIKOR Approach. *International Journal of Services and Operations Management*, *14*(2), 175–196. doi:10.1504/IJSOM.2013.051828

Semwal, V. B., Chakraborty, P., & Nandi, G. C. (2015). Less computationally intensive fuzzy logic (type-1)-based controller for humanoid push recovery. *Robotics and Autonomous Systems*, *63*, 122–135. doi:10.1016/j.robot.2014.09.001

Sharp, G., Wan, Y. T., McGinnis, L. F., Goetschalckx, M., Bodner, D., Govindaraj, T., & Everette, J. et al. (2001). A structured approach to material handling system selection and ANP for selection of MHS specification for manufacturing.*Proceedings of the 2001 Industrial Engineering Research Conference*, (pp 1-6).

Sujono, S., & Lashkari, R. S. (2005). Modeling the operation allocation and material handling system selection in FMS design: A genetic algorithm-based approach.*Proceedings of the 2005 International Conference on Simulation and Modeling*.

Tansel, Y., Yurdakul, M., & Dengiz, B. (2013). Development of a decision support system for robot selection. *Robotics and Computer-integrated Manufacturing*, *29*(4), 142–157. doi:10.1016/j.rcim.2012.11.008

Tong, R. M. (1978). Synthesis of fuzzy models for industrial processes-some recent results. *International Journal of General Systems*, *4*(3), 143–162. doi:10.1080/03081077808960680

Van, W. L., & Kerre, E. E. (1999). Defuzzification: Criteria and classification. *Fuzzy Sets and Systems*, *108*(2), 159–178. doi:10.1016/S0165-0114(97)00337-0

Webster, D. B., & Reed, R. (1971). A material handling system selection model. *IIE Transactions*, *3*(1), 13–21.

APPENDIX

Table 1. Material Handling Equipment performance measurement module

Goal (C)	Criteria, (C_{ij})	Attitude of Criteria, (C_{ij})
Election of Material Handling Equipment	Effectiveness, (C_1)	(+)
	Efficiency, (C_2)	(+)
	Depreciation, (C_3)	(+)
	Responsiveness towards scheduled production, (C_4)	(-)
	Overall Logistic Capability, (C_5)	(+)
	Quantity flexibility, (C_6)	(+)
	Volume flexibility, (C_7)	(+)
	Reliability, (C_8)	(+)
	Programming flexibility, (C_9)	(+)
	Programming features, (C_{10})	(+)
	Productivity, (C_{11})	(+)
	Coding aspects, (C_{12})	(+)
	Schedule Utilization, (C_{13})	(+)
	Adaptability, (C_{14})	(+)
	Precision, (C_{15})	(+)
	Reliability, (C_{16})	(+)
	Safety, (C_{17})	(+)
	Adaptive control machining system, (C_{18})	(+)
	Agility towards production system, (C_{19})	(+)
	Mixed product flexibility, (C_{20})	(+)
	User friendliness, (C_{21})	(+)
	Product Quality, (C_{22})	(+)
	Flaws diagnosing aptitude, (C_{23})	(+)
	Performance under Working progress, (C_{24})	(+)
	Utility ratio per year, (C_{25})	(+)
	Conditional routing, (C_{26})	(+)
	Program generation feature, (C_{27})	(+)
	Readability of changed code, (C_{28})	(+)
	Comprehensiveness of added code, (C_{29})	(+)
	Icon library, (C_{30})	(+)
	Portability, (C_{31})	(+)
	Life cycle maintenance, (C_{32})	(+)
	Rate of failure of embedded components, (C_{33})	(-)
	Output data analysis, (C_{34})	(+)
	Requirement of Service, (C_{35})	(-)
	Programme visualisation, (C_{36})	(+)
	Operator friendly, (C_{37})	(+)

Table 2. Material Handling Equipment performance measurement module for alternatives A_1, A_2, A3 and A_4

Criteria, (C_{ij})	A_1	A_2	A_3	A_4	Attitude of Criteria, (C_{ij})
Purchasing Cost, (C_{38})	30000$	30100$	30200$	30300$	(-)
Maintenance, (C_{39})	1000$	1020$	1030$	1040$	(-)

Table 3. Set of linguistic variables corresponding to their fuzzy representation

Linguistic Term (Appropriateness Rating)	Corresponding Fuzzy Numbers	Linguistic Term (Priority Weights)	Corresponding Fuzzy Numbers
Unsatisfactory (U)	(0,0,0.25)	Unimportant (UI)	(0,0.1,0.3)
Poor (P)	(0,0.25,0.5)	Slightly Important (SI)	(0,0.2,0.5)
Medium (M)	(0.25,0.5,0.75)	Fairly Important (FI)	(0.3,0.45,0.7)
Satisfactory (S)	(0.5,0.75,1)	Important (I)	(0.5,0.7,0.8)
Excellent (E)	(0.75,1,1)	Very Important (VI)	(0.7,0.9,1)

Table 4. Appropriateness ratings in terms of linguistic variable individual against alternative A_1

Criteria, (C_{ij})	E_1	E_2	E_3	E_4
(C_1)	S	P	E	E
(C_2)	E	P	M	S
(C_3)	E	M	S	S
(C_4)	E	M	S	S
(C_5)	E	M	M	M
(C_6)	S	M	S	S
(C_7)	S	S	S	S
(C_8)	E	E	M	M
(C_9)	S	S	S	S
(C_{10})	E	S	S	S
(C_{11})	E	S	S	S
(C_{12})	E	E	M	M
(C_{13})	E	S	M	S
(C_{14})	E	S	S	S
(C_{15})	E	E	M	M
(C_{16})	E	S	M	S
(C_{17})	M	S	S	S
(C_{18})	M	S	S	S
(C_{19})	M	M	M	M
(C_{20})	E	S	S	S
(C_{21})	E	S	S	S
(C_{22})	M	M	E	S
(C_{23})	P	P	M	M
(C_{24})	P	P	P	P
(C_{25})	M	M	P	P
(C_{26})	P	P	M	M
(C_{27})	P	P	P	P
(C_{28})	M	M	M	M
(C_{29})	M	M	M	M
(C_{30})	M	M	M	M
(C_{31})	S	S	S	S
(C_{32})	S	S	S	S
(C_{33})	S	S	M	S
(C_{34})	S	S	M	S
(C_{35})	S	S	M	S
(C_{36})	E	E	E	S
(C_{37})	E	E	E	S

Table 5. Appropriateness ratings in terms of linguistic variable individual against alternative A_2

Criteria, (C_{ij})	E_1	E_2	E_3	E_4
(C_1)	S	P	E	E
(C_2)	E	P	M	S
(C_3)	E	M	S	S
(C_4)	E	M	S	S
(C_5)	E	M	M	M
(C_6)	S	M	S	S
(C_7)	S	S	S	S
(C_8)	E	E	M	M
(C_9)	S	S	S	S
(C_{10})	E	S	S	S
(C_{11})	E	S	S	S
(C_{12})	E	E	M	M
(C_{13})	E	S	M	S
(C_{14})	E	S	S	S
(C_{15})	E	E	M	M
(C_{16})	E	S	M	S
(C_{17})	M	S	S	S
(C_{18})	M	S	S	S
(C_{19})	M	M	M	M
(C_{20})	E	S	S	S
(C_{21})	E	S	S	S
(C_{22})	M	M	M	E
(C_{23})	P	M	M	E
(C_{24})	M	M	M	E
(C_{25})	M	S	S	S
(C_{26})	M	S	S	S
(C_{27})	P	S	S	S
(C_{28})	M	M	M	M
(C_{29})	M	M	M	M
(C_{30})	M	M	M	M
(C_{31})	S	S	S	S
(C_{32})	S	S	S	S
(C_{33})	S	S	M	S
(C_{34})	S	S	M	S
(C_{35})	S	S	M	S
(C_{36})	E	E	E	S
(C_{37})	E	E	E	S

Table 6. Appropriateness ratings in terms of linguistic variable individual against alternative A_3

Criteria, (C_{ij})	E_1	E_2	E_3	E_4
(C_1)	S	P	E	E
(C_2)	E	P	M	S
(C_3)	E	M	S	S
(C_4)	E	M	S	S
(C_5)	E	M	M	M
(C_6)	S	M	S	S
(C_7)	S	S	S	S
(C_8)	E	E	M	M
(C_9)	S	S	S	S
(C_{10})	E	S	S	S
(C_{11})	E	S	S	S
(C_{12})	E	E	M	M
(C_{13})	E	S	M	S
(C_{14})	E	S	S	S
(C_{15})	E	E	M	M
(C_{16})	E	S	M	S
(C_{17})	M	S	S	S
(C_{18})	M	S	S	S
(C_{19})	M	M	M	M
(C_{20})	E	S	S	S
(C_{21})	E	S	S	S
(C_{22})	E	E	E	P
(C_{23})	E	E	E	P
(C_{24})	E	S	S	P
(C_{25})	E	S	S	P
(C_{26})	E	S	S	P
(C_{27})	E	E	E	P
(C_{28})	M	M	M	M
(C_{29})	M	M	M	M
(C_{30})	M	M	M	M
(C_{31})	S	S	S	S
(C_{32})	S	S	S	S
(C_{33})	S	S	M	S
(C_{34})	S	S	M	S
(C_{35})	S	S	M	S
(C_{36})	E	E	E	S
(C_{37})	E	E	E	S

Table 7. Appropriateness ratings in terms of linguistic variable individual against alternative A_4

Criteria, (C_{ij})	E_1	E_2	E_3	E_4
(C_1)	S	P	E	E
(C_2)	E	P	M	S
(C_3)	E	M	S	S
(C_4)	E	M	S	S
(C_5)	E	M	M	M
(C_6)	S	M	S	S
(C_7)	S	S	S	S
(C_8)	E	E	M	M
(C_9)	S	S	S	S
(C_{10})	E	S	S	S
(C_{11})	E	S	S	S
(C_{12})	E	E	M	M
(C_{13})	E	S	M	S
(C_{14})	E	S	S	S
(C_{15})	E	E	M	M
(C_{16})	E	S	M	S
(C_{17})	M	S	S	S
(C_{18})	M	S	S	S
(C_{19})	M	M	M	M
(C_{20})	E	S	S	S
(C_{21})	E	S	S	S
(C_{22})	P	E	P	P
(C_{23})	P	E	E	E
(C_{24})	P	P	P	E
(C_{25})	M	P	P	E
(C_{26})	M	E	E	P
(C_{27})	M	E	E	P
(C_{28})	M	M	M	M
(C_{29})	M	M	M	M
(C_{30})	M	M	M	M
(C_{31})	S	S	S	S
(C_{32})	S	S	S	S
(C_{33})	S	S	M	S
(C_{34})	S	S	M	S
(C_{35})	S	S	M	S
(C_{36})	E	E	E	S
(C_{37})	E	E	E	S

Table 8. Priority weights in terms of linguistic variable individual against alternatives A_1, A_2, A3 and A_4

Criteria, (C_{ij})	E_1	E_2	E_3	E_4
(C_1)	VI	VI	VI	VI
(C_2)	VI	UI	UI	UI
(C_3)	SI	UI	UI	UI
(C_4)	VI	VI	UI	UI
(C_5)	VI	VI	UI	UI
(C_6)	VI	VI	I	VI
(C_7)	SI	SI	I	VI
(C_8)	SI	SI	I	VI
(C_9)	UI	UI	UI	VI
(C_{10})	UI	UI	UI	VI
(C_{11})	UI	UI	UI	VI
(C_{12})	SI	SI	VI	VI
(C_{13})	VI	VI	VI	VI
(C_{14})	VI	VI	VI	VI
(C_{15})	VI	VI	VI	VI
(C_{16})	VI	VI	VI	VI
(C_{17})	VI	FI	FI	FI
(C_{18})	VI	FI	FI	FI
(C_{19})	VI	FI	FI	FI
(C_{20})	VI	FI	FI	FI
(C_{21})	VI	FI	FI	FI
(C_{22})	VI	FI	FI	FI
(C_{23})	VI	FI	FI	FI
(C_{24})	VI	UI	UI	UI
(C_{25})	VI	UI	UI	UI
(C_{26})	VI	UI	UI	UI
(C_{27})	VI	VI	VI	UI
(C_{28})	VI	VI	VI	UI
(C_{29})	VI	VI	VI	UI
(C_{30})	VI	VI	VI	UI
(C_{31})	VI	UI	UI	UI
(C_{32})	VI	VI	VI	VI
(C_{33})	VI	VI	VI	VI
(C_{34})	VI	VI	VI	VI
(C_{35})	VI	VI	VI	VI
(C_{36})	VI	VI	VI	VI
(C_{37})	VI	VI	VI	VI

Table 9. Computed aggregated fuzzy ratings individual against alternative A_1

Criteria, (C_{ij})	Computed Aggregated Fuzzy-Appropriateness Ratings	Computed Aggregated Fuzzy- Priority Weights
(C_1)	(0.475,0.750,0.875)	(0.700,0.900,1.000)
(C_2)	(0.363,0.625,0.813)	(0.175,0.300,0.475)
(C_3)	(0.488,0.750,0.938)	(0.000,0.125,0.350)
(C_4)	(0.488,0.750,0.938)	(0.350,0.500,0.650)
(C_5)	(0.363,0.625,0.813)	(0.350,0.500,0.650)
(C_6)	(0.438,0.688,0.938)	(0.650,0.850,0.950)
(C_7)	(0.500,0.750,1.000)	(0.300,0.500,0.700)
(C_8)	(0.475,0.750,0.875)	(0.300,0.500,0.700)
(C_9)	(0.500,0.750,1.000)	(0.175,0.300,0.475)
(C_{10})	(0.550,0.813,1.000)	(0.175,0.300,0.475)
(C_{11})	(0.550,0.813,1.000)	(0.175,0.300,0.475)
(C_{12})	(0.475,0.750,0.875)	(0.350,0.550,0.750)
(C_{13})	(0.488,0.750,0.938)	(0.700,0.900,1.000)
(C_{14})	(0.550,0.813,1.000)	(0.700,0.900,1.000)
(C_{15})	(0.475,0.750,0.875)	(0.700,0.900,1.000)
(C_{16})	(0.488,0.750,0.938)	(0.700,0.900,1.000)
(C_{17})	(0.438,0.688,0.938)	(0.400,0.563,0.775)
(C_{18})	(0.438,0.688,0.938)	(0.400,0.563,0.775)
(C_{19})	(0.250,0.500,0.750)	(0.400,0.563,0.775)
(C_{20})	(0.550,0.813,1.000)	(0.400,0.563,0.775)
(C_{21})	(0.550,0.813,1.000)	(0.400,0.563,0.775)
(C_{22})	(0.438,0.688,0.875)	(0.400,0.563,0.775)
(C_{23})	(0.125,0.375,0.625)	(0.400,0.563,0.775)
(C_{24})	(0.000,0.250,0.500)	(0.175,0.300,0.475)
(C_{25})	(0.125,0.375,0.625)	(0.175,0.300,0.475)
(C_{26})	(0.125,0.375,0.625)	(0.175,0.300,0.475)
(C_{27})	(0.000,0.250,0.500)	(0.525,0.700,0.825)
(C_{28})	(0.250,0.500,0.750)	(0.525,0.700,0.825)
(C_{29})	(0.250,0.500,0.750)	(0.525,0.700,0.825)
(C_{30})	(0.250,0.500,0.750)	(0.525,0.700,0.825)
(C_{31})	(0.475,0.750,0.875)	(0.175,0.300,0.475)
(C_{32})	(0.500,0.750,1.000)	(0.700,0.900,1.000)
(C_{33})	(0.438,0.688,0.938)	(0.700,0.900,1.000)
(C_{34})	(0.438,0.688,0.938)	(0.700,0.900,1.000)
(C_{35})	(0.438,0.688,0.938)	(0.700,0.900,1.000)
(C_{36})	(0.650,0.938,1.000)	(0.700,0.900,1.000)
(C_{37})	(0.650,0.938,1.000)	(0.700,0.900,1.000)

Table 10. Computed aggregated fuzzy ratings individual against alternative A_2

Criteria, (C_{ij})	Computed Aggregated Fuzzy-Appropriateness Ratings	Computed Aggregated Fuzzy- Priority Weights
(C_1)	(0.475,0.750,0.875)	(0.700,0.900,1.000)
(C_2)	(0.363,0.625,0.813)	(0.175,0.300,0.475)
(C_3)	(0.488,0.750,0.938)	(0.000,0.125,0.350)
(C_4)	(0.488,0.750,0.938)	(0.350,0.500,0.650)
(C_5)	(0.363,0.625,0.813)	(0.350,0.500,0.650)
(C_6)	(0.438,0.688,0.938)	(0.650,0.850,0.950)
(C_7)	(0.500,0.750,1.000)	(0.300,0.500,0.700)
(C_8)	(0.475,0.750,0.875)	(0.300,0.500,0.700)
(C_9)	(0.500,0.750,1.000)	(0.175,0.300,0.475)
(C_{10})	(0.550,0.813,1.000)	(0.175,0.300,0.475)
(C_{11})	(0.550,0.813,1.000)	(0.175,0.300,0.475)
(C_{12})	(0.475,0.750,0.875)	(0.350,0.550,0.750)
(C_{13})	(0.488,0.750,0.938)	(0.700,0.900,1.000)
(C_{14})	(0.550,0.813,1.000)	(0.700,0.900,1.000)
(C_{15})	(0.475,0.750,0.875)	(0.700,0.900,1.000)
(C_{16})	(0.488,0.750,0.938)	(0.700,0.900,1.000)
(C_{17})	(0.438,0.688,0.938)	(0.400,0.563,0.775)
(C_{18})	(0.438,0.688,0.938)	(0.400,0.563,0.775)
(C_{19})	(0.250,0.500,0.750)	(0.400,0.563,0.775)
(C_{20})	(0.550,0.813,1.000)	(0.400,0.563,0.775)
(C_{21})	(0.550,0.813,1.000)	(0.400,0.563,0.775)
(C_{22})	(0.375,0.625,0.813)	(0.400,0.563,0.775)
(C_{23})	(0.313,0.563,0.750)	(0.400,0.563,0.775)
(C_{24})	(0.375,0.625,0.813)	(0.175,0.300,0.475)
(C_{25})	(0.438,0.688,0.938)	(0.175,0.300,0.475)
(C_{26})	(0.438,0.688,0.938)	(0.175,0.300,0.475)
(C_{27})	(0.375,0.625,0.875)	(0.525,0.700,0.825)
(C_{28})	(0.250,0.500,0.750)	(0.525,0.700,0.825)
(C_{29})	(0.250,0.500,0.750)	(0.525,0.700,0.825)
(C_{30})	(0.250,0.500,0.750)	(0.525,0.700,0.825)
(C_{31})	(0.475,0.750,0.875)	(0.175,0.300,0.475)
(C_{32})	(0.500,0.750,1.000)	(0.700,0.900,1.000)
(C_{33})	(0.438,0.688,0.938)	(0.700,0.900,1.000)
(C_{34})	(0.438,0.688,0.938)	(0.700,0.900,1.000)
(C_{35})	(0.438,0.688,0.938)	(0.700,0.900,1.000)
(C_{36})	(0.650,0.938,1.000)	(0.700,0.900,1.000)
(C_{37})	(0.650,0.938,1.000)	(0.700,0.900,1.000)

Table 11. Computed aggregated fuzzy ratings individual against alternative A_3

Criteria, (C_{ij})	Computed Aggregated Fuzzy-Appropriateness Ratings	Computed Aggregated Fuzzy- Priority Weights
(C_1)	(0.475,0.750,0.875)	(0.700,0.900,1.000)
(C_2)	(0.363,0.625,0.813)	(0.175,0.300,0.475)
(C_3)	(0.488,0.750,0.938)	(0.000,0.125,0.350)
(C_4)	(0.488,0.750,0.938)	(0.350,0.500,0.650)
(C_5)	(0.363,0.625,0.813)	(0.350,0.500,0.650)
(C_6)	(0.438,0.688,0.938)	(0.650,0.850,0.950)
(C_7)	(0.500,0.750,1.000)	(0.300,0.500,0.700)
(C_8)	(0.475,0.750,0.875)	(0.300,0.500,0.700)
(C_9)	(0.500,0.750,1.000)	(0.175,0.300,0.475)
(C_{10})	(0.550,0.813,1.000)	(0.175,0.300,0.475)
(C_{11})	(0.550,0.813,1.000)	(0.175,0.300,0.475)
(C_{12})	(0.475,0.750,0.875)	(0.350,0.550,0.750)
(C_{13})	(0.488,0.750,0.938)	(0.700,0.900,1.000)
(C_{14})	(0.550,0.813,1.000)	(0.700,0.900,1.000)
(C_{15})	(0.475,0.750,0.875)	(0.700,0.900,1.000)
(C_{16})	(0.488,0.750,0.938)	(0.700,0.900,1.000)
(C_{17})	(0.438,0.688,0.938)	(0.400,0.563,0.775)
(C_{18})	(0.438,0.688,0.938)	(0.400,0.563,0.775)
(C_{19})	(0.250,0.500,0.750)	(0.400,0.563,0.775)
(C_{20})	(0.550,0.813,1.000)	(0.400,0.563,0.775)
(C_{21})	(0.550,0.813,1.000)	(0.400,0.563,0.775)
(C_{22})	(0.563,0.813,0.875)	(0.400,0.563,0.775)
(C_{23})	(0.563,0.813,0.875)	(0.400,0.563,0.775)
(C_{24})	(0.438,0.688,0.875)	(0.175,0.300,0.475)
(C_{25})	(0.438,0.688,0.875)	(0.175,0.300,0.475)
(C_{26})	(0.438,0.688,0.875)	(0.175,0.300,0.475)
(C_{27})	(0.563,0.813,0.875)	(0.525,0.700,0.825)
(C_{28})	(0.250,0.500,0.750)	(0.525,0.700,0.825)
(C_{29})	(0.250,0.500,0.750)	(0.525,0.700,0.825)
(C_{30})	(0.250,0.500,0.750)	(0.525,0.700,0.825)
(C_{31})	(0.475,0.750,0.875)	(0.175,0.300,0.475)
(C_{32})	(0.500,0.750,1.000)	(0.700,0.900,1.000)
(C_{33})	(0.438,0.688,0.938)	(0.700,0.900,1.000)
(C_{34})	(0.438,0.688,0.938)	(0.700,0.900,1.000)
(C_{35})	(0.438,0.688,0.938)	(0.700,0.900,1.000)
(C_{36})	(0.650,0.938,1.000)	(0.700,0.900,1.000)
(C_{37})	(0.650,0.938,1.000)	(0.700,0.900,1.000)

Table 12. Computed aggregated fuzzy ratings individual against alternative A_4

Criteria, (C_{ij})	Computed Aggregated Fuzzy-Appropriateness Ratings	Computed Aggregated Fuzzy- Priority Weights
(C_1)	(0.475,0.750,0.875)	(0.700,0.900,1.000)
(C_2)	(0.363,0.625,0.813)	(0.175,0.300,0.475)
(C_3)	(0.488,0.750,0.938)	(0.000,0.125,0.350)
(C_4)	(0.488,0.750,0.938)	(0.350,0.500,0.650)
(C_5)	(0.363,0.625,0.813)	(0.350,0.500,0.650)
(C_6)	(0.438,0.688,0.938)	(0.650,0.850,0.950)
(C_7)	(0.500,0.750,1.000)	(0.300,0.500,0.700)
(C_8)	(0.475,0.750,0.875)	(0.300,0.500,0.700)
(C_9)	(0.500,0.750,1.000)	(0.175,0.300,0.475)
(C_{10})	(0.550,0.813,1.000)	(0.175,0.300,0.475)
(C_{11})	(0.550,0.813,1.000)	(0.175,0.300,0.475)
(C_{12})	(0.475,0.750,0.875)	(0.350,0.550,0.750)
(C_{13})	(0.488,0.750,0.938)	(0.700,0.900,1.000)
(C_{14})	(0.550,0.813,1.000)	(0.700,0.900,1.000)
(C_{15})	(0.475,0.750,0.875)	(0.700,0.900,1.000)
(C_{16})	(0.488,0.750,0.938)	(0.700,0.900,1.000)
(C_{17})	(0.438,0.688,0.938)	(0.400,0.563,0.775)
(C_{18})	(0.438,0.688,0.938)	(0.400,0.563,0.775)
(C_{19})	(0.250,0.500,0.750)	(0.400,0.563,0.775)
(C_{20})	(0.550,0.813,1.000)	(0.400,0.563,0.775)
(C_{21})	(0.550,0.813,1.000)	(0.400,0.563,0.775)
(C_{22})	(0.188,0.438,0.625)	(0.400,0.563,0.775)
(C_{23})	(0.563,0.813,0.875)	(0.400,0.563,0.775)
(C_{24})	(0.188,0.438,0.625)	(0.175,0.300,0.475)
(C_{25})	(0.250,0.500,0.688)	(0.175,0.300,0.475)
(C_{26})	(0.438,0.688,0.813)	(0.175,0.300,0.475)
(C_{27})	(0.438,0.688,0.813)	(0.525,0.700,0.825)
(C_{28})	(0.250,0.500,0.750)	(0.525,0.700,0.825)
(C_{29})	(0.250,0.500,0.750)	(0.525,0.700,0.825)
(C_{30})	(0.250,0.500,0.750)	(0.525,0.700,0.825)
(C_{31})	(0.475,0.750,0.875)	(0.175,0.300,0.475)
(C_{32})	(0.500,0.750,1.000)	(0.700,0.900,1.000)
(C_{33})	(0.438,0.688,0.938)	(0.700,0.900,1.000)
(C_{34})	(0.438,0.688,0.938)	(0.700,0.900,1.000)
(C_{35})	(0.438,0.688,0.938)	(0.700,0.900,1.000)
(C_{36})	(0.650,0.938,1.000)	(0.700,0.900,1.000)
(C_{37})	(0.650,0.938,1.000)	(0.700,0.900,1.000)

Table 13. Evaluated performance scores against robot alternatives

Alternatives	Performances Scores	Preferences Order
A_1	14.294	4
A_2	14.292	3
A_3	15.231	1
A_4	14.821	2

Chapter 12
Benchmarking of Advanced Manufacturing Machines Based on Fuzzy-TOPSIS Method

Anoop Kumar Sahu
J.K.I.E., India

Atul Kumar Sahu
Guru Ghasidas Central Vishwavidyalaya, India

Nitin Kumar Sahu
Guru Ghasidas Central Vishwavidyalaya, India

ABSTRACT

The evaluation of the most appropriate Computer Numerical Control (CNC) machine tool is found foremost issue for sustaining the advanced manufacturing production units at competitive global market place. In last years, many evaluation parameter/ criteria/ issues (i.e. Cost [C], Tool Length [TL], Tool Diameter [TD], Metal Removal Rate [MRR], X-Y-Z-Axis Movement of Tool [AMT], etc.) carried out into account besides productivity, precision, accuracy, reliability etc prior to purchasing/electing CNC machine tools amongst preferred alternatives. This chapter exhibited a hypothetical case research pertaining to election of CNC machine tool under the qualitative parameter/criteria/issues excluding quantitative, where, subjective (qualitative) criteria are considered beside objective (quantitative) criteria and complexity to elect the best CNC machine tool is solved via subjective assessments of expert's panel also called the group decision making. In present research, TOPSIS algorithmic technique conjunction with Trapezoidal Fuzzy Number set (TFNs) is explored for appraising and benchmarking the preferable CNC machine tool options/ alternatives. Finally, a hypothetical case research is carried out to check the validity of proposed algorithmic technique for proposed module and exhibit the path for benchmarking alternative machine tools. The preferences orders are carried out in accordance with descending value of the 'collective index'. Higher value of 'collective index' reflects higher degree of performance.

DOI: 10.4018/978-1-5225-1848-8.ch012

AT A GLANCE

The proposed chapter serves a Fuzzy-TOPSIS (Technique for order Preference by Similarity to Ideal Solution) approach towards learners to solve their real life problems i.e. appraisement, evaluation and benchmarking of Cars, Buses, Trucks, Plant Layouts, Aero Planes, Material Handling Equipments, Robots etc in the context of Multi-Criteria Decision Making. Apart from this, many dilemmas pertaining to advanced and traditional manufacturing realms can be solved with short computing by exploration of proposed Fuzzy-TOPSIS methods in empire of MCDM. The proposed chapter dealt with application of Fuzzy-TOPSIS to solve computer numerical control advanced manufacturing machine tool evaluation and benchmarking problems.

INTRODUCTION: DECISION MAKING

Decision making is actually insightful when considering single criteria problems, since the personnel solely need to elect the alternative/ choices in accordance with highest preference ratings. However, when expert's panel evaluate alternatives under multi criteria, then many problems such as significance (importance) against criteria, preference dependence, and conflicts amongst criteria take an occurs. Then, problems are seemed difficult to be beaten by more complicated techniques.

PREVALENT OVERVIEW ON MULTI-CRITERIA DECISION MAKING

Multi-Criteria Decision Making is the well renowned branch of decision making. It is a division of a common class of Operations Research (OR) models, which deal with decision making problems under chain of multiple criteria. MCDM also deals with economic theory (cost/benefit analysis), applying mathematical modeling in attempts to support decision making involving tradeoffs. MCDM received winning journey, as it deals with a so difficult problems such as evaluation of more economic alternative amongst preferred alternatives under chains of multiple criteria.

Prevalently, MCDM is concerned with structuring and solving decision and planning problems involving multiple criteria. The agenda of MCDM is to support decision makers, facing problems i.e. evaluation of more economic alternative amongst preferred alternatives under chains of multiple criteria. Usually, MCDM does not bring a unique optimal solution for such problems in case of using decision maker's preferences. MCDM bring a unique optimal solution for such problems in case of using experimental / observed data. It could correspond to choose the 'best' alternative from a set of available alternatives.

Decision Making (DM) is divided into two parts: Multi-Objective Decision Making (MODM) and Multi-Criteria Decision Making (MCDM). In MODM, the perusals / learners studies about decision problems in which the decision space is continuous. A classical example is mathematical programming problems with multiple objective functions. While, in MCDM the perusals / learners studies as to decision problems with discrete data. In these problems, the set of decision alternative is predetermined. But, it is indeed necessary in order to understand the meaning of both term i.e. alternatives and criteria, which are the core terms of MCDM.

Alternatives

It represents the many choices of actions available towards the decision makers. Usually, the set of choices is assumed to be finite, ranging from several to hundreds. In MCDM, each alternative is viewed with multiple criteria, which reflects the characteristics of alternatives, which aids the boss of firms to reach to final decision. For example, cars, buses, scooters, trucks etc. are considered as alternatives.

Criteria

The criteria reflect the property, quality or feature of alternatives. Roughly any alternative i.e. an organization, an action plan, a product etc. can be evaluated in accordance with criteria. For example, car is considered as alternative, which stand with multiple criteria such as:

- Fuel consumption,
- Price,
- Maintenance cost,
- Repairment cost,
- Speed,
- Aesthetic look,
- Reliability,
- Efficiency,
- Effectiveness,
- Overall performance etc.

In MCDM, each criterion represents different dimensions / units with alternatives. The following sets are being adapted to undertake the problems pertaining to Multi-Criteria Decision Making (MCDM):

The first step is to recognize the problem, which states about our targeted goals i.e. evaluation of the best alternative amongst preferred alternatives under chain of multiple criteria. Next, it is need to gather the preferences of expert's panel in order to construct the module (consist of chain of multiple criteria). Later, it is need to collect the appropriate information against set of possible alternatives in order to attain the goal (i.e., evaluating the best alternative amongst preferred alternatives). Next footstep is to choose suitable technique to screen, prioritize and eventually assign the preference order against the alternatives as per computed performance scores and elect the optimum alternative.

In MCDM, if the criteria are considered more than a few dozen, then several criteria's are arranged in a hierarchical manner. Some criteria may have chain of its further criteria called major criteria. Each major criteria may be associated with more than a few sub-criteria called interrelated criteria. Similarly, each sub criteria may have further chain of several sub-sub-criteria and so on. The interrelated criteria are used in purpose to broad analysis the activities / actions of preferred alternatives.

To evaluate the best alternative amongst preferred alternatives, a MCDM appraisement module is formed by incorporating chain of multiple criteria, and is set up with defined alternatives. Many times in MCDM, conflict amongst criteria take place, which make MCDM problems so difficult to solve and create stress in brain of problem solvers. For example, a truck must be designed with low fuel consumption and high load carrying capacity. Criteria may have a different unit of measurement, as in the car selection problem, fuel economy is measured by miles per litre and price is expressed by pound etc.

MCDM field works best with objective (quantitative) measures. But, human preferences are complicated to measure objectively. It is stiff to analyze many tradeoffs involved in decision making, especially in times with so many uncertainties presented qualitative information of human preferences. As each decision making problem either dealt with quantitative or non-quantitative (qualitative) criteria. In case of election of car such as fuel consumption, price, speed etc are considered as set of quantitative criteria, while

- Aesthetic look,
- Reliability,
- Efficiency,
- Effectiveness,
- Overall performance of car etc. are considered as set of qualitative criteria.

The quantitative criteria indicated by numerical unit, while non-quantitative criteria indicated in a non-numerical unit. It is possible that several criteria can be measured numerically and other attributes can only be described subjectively. For instance, the price of a car is quantitative and the comfort rating is qualitative. Mixture of deterministic and probabilistic attributes. For example, in the bus selection problem, bus cost is deterministic and fuel economy could be random. Fuel economy changes depending on road and traffic conditions. In present era, the MCDM found its strength for solving many problems under chains of multiple criteria undertaking the quantitative or qualitative or quantitative cum qualitative data.

OBJECTIVES OF MCDM

The core objectives of Multiple-Criteria Decision Making (MCDM) are to circulate MCDM field through academics, researchers, industrialists and professionals, in order that they could solve their professional and own problems. The objectives of MCDM have been articulated below:

- To bring the novel / advanced chains of multiple criteria into decision making appraisement module.
- To enhance the applications of dynamical and statistical algorithmic techniques and approaches in empire of multiple-criteria decision making.
- To solve the several real life problems under the subjective information of expert' panel.
- To enable the forthcoming researchers for solving the alternative appraisement problems under chains of multiple criteria via several algorithmic techniques such as:
    - Analytical hierarchy processes,
    - Decision analysis,
    - Goal programming,
    - Multi-objective evolutionary algorithms,
    - Multi-objective mathematical programming, and
    - Outranking methods,
- To drawn the attention of researchers, academicians and practitioners towards the realm of multiple criteria decision making.

- To aid the managers of firms for solving the performances measurement problems of their firms. Moreover, enabling the managers for solving the benchmarking (preference orders of alternatives under similar criteria) problems via MCDM techniques.
- To enable the managers of firms for diagnostic industrial management dilemmas under the subjective cum objective information.
- To solve the multi-objective optimization problems under non-constrained variables undertaking oral or experimental or oral or experimental information (data).
- To solve the alternative evaluation problems under economic extent, as most of the MCDM problems are being solved by exploration of mutual calculations and excel sheet.
- To facilitate the individuals to solve the supply chain management and so many commercial problems in the context of weight attribute or significance ratio against criteria.
- To aid the community of societies to solve their domestic such as car, bike, other domestic applications evaluation-election problems with short computing under constructed module (consist of multiple criteria).

GROUP DECISION MAKING

Group Decision Making (GDM) is a process, which is conducted in order to undertake the situation faced by personnel cooperatively for choosing an alternative from the alternatives prior to act on them. In GDM, the decision is not taken by single individual, who is a part of the committee. Each personnel, who is member of a constructed committee, deliver its own seminal contribution for making the final consequence against defined problem. The decisions, built by cluster of personnel's (group) are frequently unlike from those made by single individual. Several questions are discussed amongst the individuals to conclude the results.

First question is that what kinds of personnel's are being involved to construct the group. The answer is that it varies from one firm / academy to another and extremely depends upon availability of personnel's /decision makers in such as firm / academy. For example, vice-chancellor, Deputy Vice-chancellor, director, head of department and several departmental faculties would be considered as decision makers in case of academy, while CEO, director, top or middle management employees of cross-functional departments would be considered as decision makers in case of firm, towards solving the individual or cluster of problems. Second question is what kind of ideas are shared by individuals in group decision making. The answer is that a many ideas are created during brainstorming session and mostly 6-12 members are invited to construct the group. In case of larger number of individuals, the groups are being divided into subgroups. Third question is how much time is consumed by group during decision making. The answer is yet not found, as each decision making process has its own process and limitation of time and extremely depends upon the type of decision, as it is diminutive or bigger in nature.

In GDM, there are discussions, which occur amongst the personnel's in regards to make superior decision. How, we can purport that decision is superior or inferior, there is not tool to justify this in case of GDM, but according to the synergy between idea severed by decision makers, decision built cooperatively tends to be more successful rather than decision built by a single individual. Social group behaviors influence the group decisions, for example groups high in cohesion, in combination with other antecedent conditions (e.g. ideological homogeneity and insulation from dissenting opinions) have been noted to have a negative effect on group decision making and hence on group efficiency. The GDM brainstorming process is shown in Figure 1.

Figure 1. The GDM brainstorming process

LINGUISTIC, SUBJECTIVE AND OBJECTIVE ASSESSMENT

Linguistic Information

Linguistic is an ingredient of information, which is concerned in relation to the structure of language. Linguistic is separated into subfields:

- **Phonetics:** It deals with the study of speech sounds in their physical aspects.
- **Phonology:** It deals with the study of speech sounds in their cognitive aspects.
- **Morphology:** It deals with the study of the formation of words.

In the context of MCDM, linguistic variables are explored by decision makers for assigning the appropriateness ratings as well as priority weights against criteria.

The linguistic variables such as {G, VG, MG, PVL, VVP} are found active for assigning the appropriateness ratings against criteria, while linguistic variables such as {VH, H, MH, PHL, VVL} are found valid for assigning the priority weights against criteria. The first linguistic variables set {G, VG, MG, PG, VPG}, where G=good, VG=Very Good, MG= Medium Good, PG= Poor Good and VPG=Very Poor Good, reflects the ratings linguistic scale. Second linguistic variables set {VH, H, MH, PH, VLH}, where VH=Very High, H= High, MH= Medium High, PH= Poor High and VLH=Very Low High reflects the importance linguistic scale. Linguistic variables are found active towards dealing with situations, which are too compound or not well-defined to be reasonably described by traditional quantitative expressions (Zadeh, 1975). Linguistic variables are evaluated by expert's panel for assigning the appropriateness ratings as well as priority weights against criteria to solve industrial problems. Therefore, the fuzzy logic is explored in order to materialize the information provided by expert' panel in the terms of linguistic variables. Momentous of all, linguistic variables under large scale (1-9) aid the decision makers for sharing their verbal information without hesitation, while linguistic variables under small scale (1-4), which occupy hardly four linguistic variables, assist the decision makers for sharing their verbal information with hesitation. The utility of large and small linguistic scale is totally depending upon the perceptions of decision makers against emerging issue during brainstorming time.

Subjective Assessment

The subjectivity is defined as feelings, ideas, and thought, which includes words, meaningfulness, definitions, and undefined concepts. Preceding researchers proposed the critical method as a means to eliminate error in an attempt to regain objective growth of knowledge. The approach recognizes the need to cope with subjective information, while also recognizing its inherent unreliability.

Subjective assessment deals with uncertainty, being covered by the evaluated subjective information (evaluated score) from the group of expert's panel is called qualitative assessment. While decision making procedure requires the assessments (evaluated information) from the group of expert's panel against subjective criteria, then this subjective information is assessed in terms of linguistic variables facilitated by fuzzy representation. A linguistic variable is the explanation / language of the expert's panel, which is facilitated by fuzzy representation. In MCDM framework, linguistic information (subjective information) is evaluated against the subjective (uncertain) measures corresponding to defined alternatives / choices.

Objective Assessment

Quantitative assessment is the values against the evaluated criteria corresponding to defined alternatives in terms of the quantity / crisp or single data / measured data (Objective information). This information does not deal with criteria, which undertake ambiguity, impreciseness and irregularity in information. The decisions are considered in the case of objective information, provides the accurate and precise consequence. For example, Inner and outer diameter of pen, length, thickness of pen etc, require the objective information. Whereas, aesthetic look of pen, overall quality etc, require the subjective information. In case of election of machine tool,

- Cost (C),
- Tool Length (TL),
- Tool Diameter (TD),
- Metal Removal Rate (MRR),
- X-Y-Z-Axis Movement of Tool (AMT),
- Machine Tool Work Space (MTWS),

requirement etc would be considered as quantitative criteria whereas, aesthetic look, effectiveness, durability etc, would be considered as qualitative criteria. The shown counter depicted the hypothetical objective and subjective assessment against quantitative and qualitative criteria respectively (Table 1).

BENCHMARKING PROCESS

Benchmarking is the process for augmenting the performances of companies, public organizations, universities, colleges etc, by recognizing, making out and adapting their efficient practices and processes. Benchmarking is used for improving of any given business process by making use of best practices

Table 1. Qualitative criteria for an advanced machine tool and their subjective and objective assessments

Qualitative Criteria for an Advanced Machine Tool	Subjective Assessment in Terms of Hypothetical Linguistic Variables	Quantitative Criteria for an Advanced Machine Tool	Objective Assessment in Terms of Hypothetical Values
Aesthetic look	G	Number of tools	Max. 3 tools/station
Effectiveness	VG	Max. drilling capability	Ø23mm (29/32 in)
Durability	P	Max. tapping capability	M12XP1.75
Intention for worker	G	Max. depth of hole capability	120mm (4-23/32in)
Complex design	VG	Number of tools	Max. 2 tools/station
Agility to production task	P	Max. drilling capability	Ø10mm (25/64 in)

excluding measuring the best performance of practices, as the best practices are the cause of best performance. Benchmarking is also considered as a process to measure the quality of a firm's policies, products, programs, strategies etc and their comparison with standard measurements or similar measurements of its peers. The aim of benchmarking is to determine what and where improvements are indeed essential.

Performance Measurement (PM) materializes the efficiency and effectiveness of organization under practices, which is the part of benchmarking process. Benchmarking seems as a path for assigning the preferences order against alternatives under the similar criteria module / a set of measurable parameters. Here, the benchmarking process is discussed with two good examples. First example is related to supplier election, a car manufacturing company wants to purchase a gear box from the best supplier. For this purpose, company procured the SC profile of four suppliers and decided to put five criterions such as:

- Reliability,
- Cost of product,
- Delivery time,
- Transportation cost,
- Logistic action of suppliers,

to measure the performances of four suppliers simultaneously under considered five criterions. Then, in accordance with computed performance scores against four suppliers, company can decide that which supplier would be the best supplier for purchasing the gear box.

Benchmarking tool is also valid to solve second example, is pertaining to election of a car. A customer decided to purchase a car from four cars in considering the five common factures of cars such as reliability, cost, aesthetic look, performance, speed. Then, in accordance with computed performance scores against four cars, company can decide that which car would be the best. Therefore, in field of MCDM, performance measurement tool helps the customer to measure the performances of cars under defined criteria and benchmarking assists the customer to elect that car, steady with best performance score. In extant of benchmarking process, the election of alternative depend upon the applications of MCDM methods, few MCDM methods declare the preferences order against alternatives as per maximum scores, while few MCDM methods declare the preferences order against alternatives as per maximum score. Figure 1 shown the MCDM model, is valid to tackle the benchmarking problems of alternatives.

In shown MCDM model, the goal (at 0 Level) reflects the scores against alternative-1-2-3 and election of the best alternative amongst three. Set of four criteria such as:

- Criteria-1,
- Criteria-2,
- Criteria-3,
- Criteria-4,

is defined (at 1st level) having alignment with:

- Alternative-1,
- Alternative-2, and
- Alternative-3 (at 2nd level).

Figure 2 depicted the benchmarking Model, which is valid for solving the problems of industries.

MCDM MODULE

MCDM (Multi-Criteria Decision Making) module is the framework, which consists of the goal, criteria and interrelated criteria (metrics). The core objective is to develop the MCDM module for extremely analysis the process parameters of alternative such as company, public organization, university, college etc. As process parameters helps in order to meet the objectives of firms. The MCDM module construction is easy in case of access of criteria up to 1st level hierarchy and construction become complicated in case of access of criteria up to 2nd, 3rd, 4th so on level hierarchy. The core purpose is to consider the redundant criteria up to 2nd, 3rd, 4th so on level hierarchy for deeply analyzing the processes parameters

Figure 2. The benchmarking model

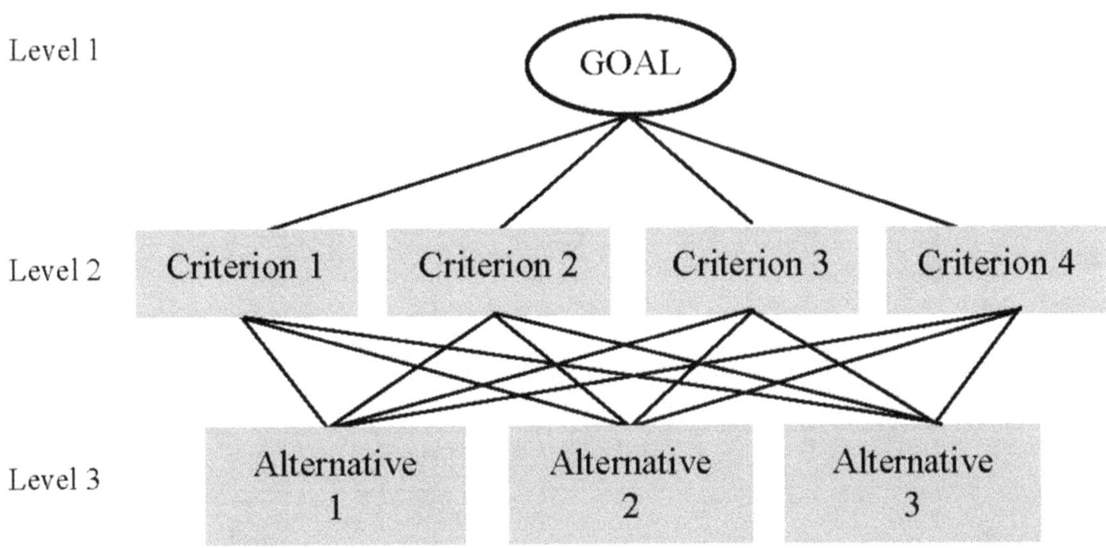

of alternative. The MCDM module consists of several components. The first component is goal, while rest of the components undertakes criteria and further extensive chain of defined criteria (metrics). The Figure 3 shown the goal of module, which is to select the best alternative amongst preferred alternatives, while rest of the levels define criteria at 1st,2nd,3rd,4th level of hierarchy.

An example is represented here for the novel researchers to construct the modules. As if, an academy invited the application form for faculty positions. Academy received ten application forms and short listed three applicants. Then, academy decided to construct the MCDM module to second level hierarchy for checking the quality of three applications. The goal reflects the evaluation of scores against three applications and election of the best applicant amongst three applicants. Then, overall quality will be considered as criteria at 1 level hierarchy, while knowledge, effectiveness, efficiency, personality of applicants will be considered as metrics of (quality) criteria at 2nd level.

Figure 3. Multi level hierarchical module

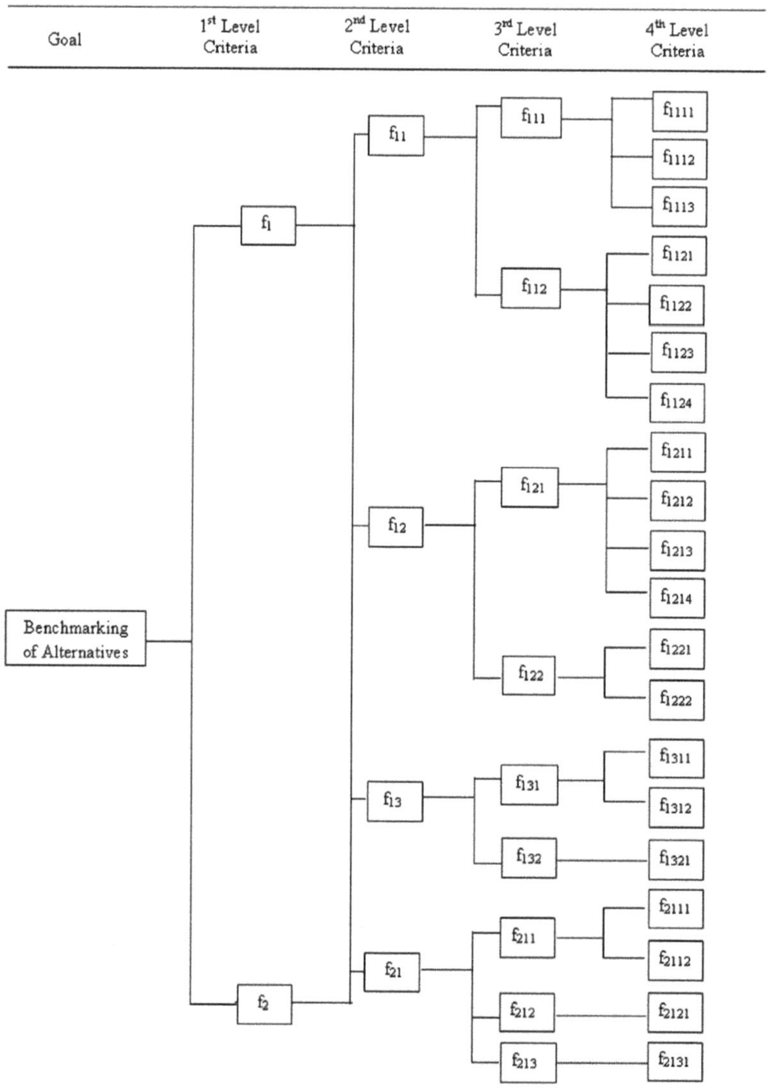

Rating and Weights of Criteria

Rating of Criteria

The rating is the evaluation or assessment of something (alternatives) in terms of quality, quantity or combination of both. It is defined as appraisal of the value of alternatives. The rating varies from one alternative to another alterative, which performing the identical functions. The rating is assigned by individuals in terms of percentage, scrip values etc. When the value of alternative is assigned by individuals or observed via investigations in terms of single or true value is called the crisp rating, when value is assigned by individuals in terms of values, embrace the range of data is called the fuzzy rating. Formed MCDM problem delivers accurate result, in case of assigned crisp rating against criteria, while MCDM problem doesn't deliver accurate result in case of fuzzy rating. The case of crisp and fuzzy rating also does not deliver the accurate results in field of MCDM.

An example is presented here in order to elaborate the rating; a customer goes to market to purchase a car. Customer elected three cars and finally decided to purchase one out of three. Subsequently, customer introduced the four criteria, such as:

- Cost,
- Quality,
- Maintenance cost,
- Repairmen cost for three respective cars,

then considered fist car as alternative-1, second as a alternative and third as alternative-3 and assigned the rating in term of crisp value (Table 2).

Weights of Criteria

It is defined as importance against criteria. It assists individuals to make the feasible, optimum and cost-effective decision. It is also considered as significance against criteria. For example the weight of cost criteria is ever more rather than other criteria. The alternative, which has higher weights and low rating, has the good score as compare to other alternative, which has higher ratings and very low rating. The weight against criteria leads a prominent role in decision making scenario towards making cost effective decision. An example is presented here in order to elaborate the role of weight against criteria: a customer goes to market to purchase a car, where customer introduced four criteria such as:

- Cost,
- Quality,
- Maintenance cost, before purchasing car.

Table 2. Example of rating the criteria

Alternatives	Deprecation of Car	Maintenance Cost	Repairment Cost
Alternative-1	1000$	150$	155$
Alternative-2	1100$	200$	222$
Alternative-3	1200$	125$	155$

Then, cost has a bigger weight than other defined criteria. The assignment of weights of criteria is decided by individuals and worth of criteria. In the field of MCDM, most of the problems are formed and solved under consideration of weights or significance ratio against criteria.

DECISION MAKING IN CNC

Decision making has the successful journey in each field, as it drives the organization with loss or profit. Many decision making tools are developed to solve the decision making problem of organization, where MCDM is found active to solve the problems towards benchmarking and electing the Computer Numerical Control (CNC) machine tool. CNC machine tool is one, where the functions and motions of a machine tool are being controlled by using of prepared programs containing coded alphanumeric information. CNC can control:

- The motions of the metal or tool,
- Input parameters such as feed, depth of cut, speed, and
- The functions such as turning spindle on/off, turning coolant on/off.

CNC is referred as a system that has a local computer to store all required numerical data. Preliminary, CNC machine tool was used to enhance tapes, then tape replaced by other storage media, magnetic tapes and hard disks. The advantages of CNC systems are not limited to the possibility to store and execute a number of large programs, programs and several machining programs can be done. The applications of CNC include both for machine tool as well as non-machine tool areas. In the machine tool area, CNC is widely used for:

- Lathe,
- Drill press,
- Milling machine,
- Grinding unit,
- Laser,
- Sheet-metal press working machine,
- Tube bending machine etc.

Highly automated machine tools such as turning center and machining center is used CNC, where the cutting tools automatically altered under CNC control. In the non-machine tool category, CNC applications include:

- Welding machines (arc and resistance),
- Coordinate measuring machine,
- Electronic assembly,
- Tape laying, and
- Filament winding machines for composites etc.

CNC machine tool is proffered in advanced manufacturing firms due to its several unique features such as:

- High accuracy in manufacturing,
- Short production time,
- Greater manufacturing flexibility,
- Contour machining,
- Reduced human error (Computer Numerical Control, n.d.).

In present scenario, the problem of most feasible Computer Numerical Control (CNC) machine tool selection among available alternatives become a serious issue to:

- Escalate the production capacity,
- Process capacity,
- Provides effective utilization of resources,
- Increase productivity and improves system flexibility.

The selection of a CNC m/c tool amongst the alternatives is a Multi-Criteria Decision Making (MCDM) problem under consideration of qualitative or quantitative criteria or both. In last year's, MCDM field is fruitfully applied to assign the preferences order against alternatives under consideration of objective data such as:

- Cost (C),
- Tool Length (TL),
- Tool Diameter (TD),
- Metal Removal Rate (MRR),
- X-Y-Z-Axis Movement of Tool (AMT) etc,

besides subjective data such as:

- Productivity,
- Precision,
- Accuracy,
- Reliability etc. against criteria corresponding to their alternatives.

So, it became indeed essential to evaluation of CNC machine tool under consideration of subjective criteria beside objective criteria. This chapter deals with the application of Technique for Order Preferences Similarity to Ideal Solution (TOPSIS) algorithmic technique conjunction with Trapezoidal Fuzzy Number set (TFNs) in empire of MCDM for apprising and benchmarking the preferable CNC machine tool options / alternatives. Finally, a hypothetical case research is carried out to check the validity of proposed approach.

STATE OF ART

In the present emerging technology, Advanced Manufacturing Machine (AMM) tools lead crucial role to complete the production task under set time towards achieving the goals of organizations. The capital investment for purchasing machineries and equipments considered as critical factors, which affects the profitability of the facility. Therefore, evaluation and election of a proper machine tool put the manufacturing companies into success, while a wrongly selected machine tool may hamper the performance of operations of manufacturing companies. The authors conducted some literature survey in the context advanced machine tools, which helped the authors to decide their research gaps. Prior move to them, it is indeed necessary to know about CNC machine tool.

Samvedi, Jain, and Chan (2011), stated that Computer Numerical Control (CNC) machine tool is widely used in organizations and its functions are being controlled by the application of digital electronic computers and circuitry. CNC machines are the workhorses of the precision machining industry. CNC stands for Computer Numeric Control. CNC is an industry standard programming language designed specifically for controlling high-precision mills, lathes, cutting and grinding machines. It's the progeny of the marriage between Computer Aided Design (CAD) and Computer Aided Machining (CAM). CNC machine tool uses programs to automatically execute a series of machining operations. It also offers increased productivity and operation flexibility.

Ayag and Ozdemir (2012), proposed modified TOPSIS and the Analytical Network Process (ANP) and presented a performance analysis on machine tool selection problem. The ANP method is used to determine the relative weights of a set of three valuation criteria, as the modified TOPSIS method is utilized to rank competing machine tool alternatives in terms of their overall performance. Abdi (2009), investigated reconfigurable machining system characteristics in order to identify the crucial factors influencing the machine selection and the machine reconfiguration. Duran and Aguilo (2008), proposed an Analytic Hierarchical Process (AHP) based on fuzzy numbers multi-attribute method for the evaluation and justification of an advanced manufacturing system. Lastly, a case study of machine tool selection is used to illustrate and validate the proposed approach. Chu (2009), developed a new fusion method of fuzzy information to manage the information, assessed in different linguistic scales (multi-granularity linguistic term sets) and numerical scales. The flexible manufacturing system adopted in the Taiwanese bicycle industry is employed in this study to demonstrate the computational process of the proposed method. Qi (2010), presented a comprehensive evaluation model for machine tool selection. Then Logarithmic least squares method based on fuzzy pair wise comparison matrix is applied for assessment of uncertain weights of selection criteria, the ways to determine performance value of the alternative with respect to qualitative and quantitative criteria has been discussed respectively. Korena and Shpitalni (2010), defined the core characteristics and design principles of reconfigurable manufacturing systems (RMS) and described the structure recommended for practical RMS with RMS core characteristics. After that, a rigorous mathematical method is introduced for designing RMS with this recommended structure.

Zhang and Chu (2009), presented a group decision making approach incorporating with two optimization models (logarithmic least squares model and weighted least squares model) towards aggregating multi-format and multi-granularity linguistic judgments. Tsai, Wang, Cheng and Kao (2010), showed that how to determine the valuable criteria for selecting machine tools from the vast amount of specifications of CNC machine tools along with consulting the experts with abundant mold manufacturing experiences and also demonstrated the multi-criteria decision making (MCDM) method towards efficient selections of CNC machine tools that would satisfy the needs of an organization. Sahu, Sahu and Sahu

(2014a), utilized MUTI-MOORA approach along with an application of grey number to evaluate the performances score against CNC machine tool alternatives under uncertain information. Sahu, Sahu and Sahu (2015a), integrated a hairy-based approach and applied said approach on novel developed multi hierarchical third-party logistics (3PL) service provider's appraisement module for evaluating the 3PL provider performance score. Sahu, Sahu and Sahu (2015b), employed TOPSIS approach to make the decision in regards to elect the best advanced manufacturing CNC machine tool under subjective information against criterion. Sahu, Sahu and Sahu (2016a), developed interval-valued fuzzy number set and merged this with modified MCDM technique TOPSIS for benchmarking supplier firms under agile supply chain actions. Sahu, Sahu and Sahu (2016b), used a revised ranking approach accompanied with fuzzy performance important index to identify the barriers/ill-strong measures in Agile Supply Chain Management. Sahu, Sahu and Sahu (2016c), explored Ratio system, Reference point, and full Mutification form coupled with aggregated interval valued trapezoidal fuzzy number set to evaluate the CNC machine tool from preferred options. Lin and Yang (1994), explored AHP method for selecting a machine model from a range of machines for the manufacture of a particular part type.

A comparative analysis of VIKOR and TOPSIS is illustrated with a numerical example, showing their similarity and some differences. In order to calculate the results, Generalized Trapezoidal Fuzzy sets (GTFS) embedded with modified TOPSIS (Technique for Order Preference by Similarity to Ideal Solution) methodology is explored towards appraising (and benchmarking) of the preferred candidate alternatives Thorani, Rao & Shankar (2012). Finally, a case study is carried out to display the effectiveness, efficiency and validity of proposed methodology. The raking order of preferred alternative industries has been derived in accordance with ascending value of the 'collective index'. Higher value of 'collective index' reflects higher degree of performance extent. After carrying out the sufficient literature in regards to find the research gaps, it became imperative to know about role of performance measures for measuring the performances of advanced manufacturing machine tools. Performance measures always tied to a goal or an objective (the target). Measures help organizations to identify the weakness, flaws and strength of machine tool. The authors used the TOPSIS with triangular fuzzy set in order to measure the performances of five CNC machine tools under twenty subjective measures, where, all the measures dealt with (+) positive altitude beside power consumption (dealt with (-) negative altitude).

PROBLEM FORMULATION

Decision making modeling usually involves analysis of a set of quantitative as well as qualitative criteria, dealt with appropriateness rating as well as priority weights. The situation becomes uncertain-vague while dealing with subjective criteria due to lack of precise, incomplete and inconsistent data. This invites need of expert opinion in terms of linguistic variable to facilitate such type of decision making. In the present chapter, a Multiple Criteria Decision Making (MCDM) module based on Fuzzy Set theory combined with TOPSIS (Technique by Order Preference by Similarity to Ideal Solution) method (representing a decision support system) is introduced (Mahdavi, Amiri, Heidarzade & Nourifar, 2008; Li, 2010, 2011, 2012, 2014; Beikkhakhian, Javanmardi, Karbasian & Khayambashi, 2015; Zeng, 2016; Zeng, Chen & Li, 2016; Zyoud, Kaufmann, Shaheen, Samhan & Hanusch, 2016; Nan, Wang & An, 2016; Sahu, Datta & Mahapatra, 2012, 2013, 2014, 2015; Samantra, Sahu, Datta & Mahapatra, 2013). The Fuzzy Logic has the capability of dealing with such incomplete, inconsistent and imprecise information by converting decision makers' (DMs) 'linguistic' judgment into appropriate fuzzy numbers. The judgment of the expert panels (linguistic preferences) are transformed into Trapezoidal Fuzzy Numbers (TFNs) and TOPSIS (Technique for Order Preference by Similarity to Ideal Solution) approach is explored towards appraisement (and benchmarking) of the preferred candidate alternatives. Finally, a hypothetical case research is carried out to provide a better understanding of the proposed framework. The raking order of preferred alternative CNC machine tools is derived in accordance with descending value of the 'collective index'. Higher value of 'collective index' reflects higher degree of performance extent.

FUZZY SET THEORY

The fuzzy set theory was first introduced for dealing with problems in which a source of vagueness is present (Zadeh, 1965). It is considered as a modeling language to approximate situations in which fuzzy phenomena and fuzzy criteria exist. In a universe of discourse X, a fuzzy subset A of X is a set defined by a membership function $f_A(x)$ representing a mapping which maps each element x in X to a real number in the closed interval [0, 1]. Here, the value of $f_A(x)$ for the fuzzy set A is called the membership value (or the grade of the membership) of x in X. The membership value represents the degree of x belonging to the fuzzy set A. The greater $f_A(x)$ the stronger is the grade of membership for X in A.

The linguistic value could be used for approximate reasoning within the framework of fuzzy set theory for handling the ambiguity associated with evaluation information and the vague property of linguistic expression (Zadeh, 1975). In order to undertake the vague property of linguistic expression, the normal trapezoidal or triangular fuzzy numbers are used to characterize the fuzzy values of quantitative data. The operations of fuzzy numbers can be understood using the extension principle (Tanaka, 1997). The necessity of applying Fuzzy Logic in data mining is due to the following:

- One is that fuzziness is inherent in many problems of knowledge representation, and the other is that high-level managers or complex decision processes often deal with generalized concepts and linguistic expressions, which are generally fuzzy in nature.
- Fuzziness may prevail in many other association cases in which impression, matching, similarity, implication, partial truth or the like is present.
- The modeling of imprecise and qualitative knowledge, as well as the transmission and handling of uncertainty at various stages are possible through the use of fuzzy sets.
- Fuzzy logic is capable of supporting to a reasonable extent, human type reasoning in natural form.

A fuzzy set $\tilde{A}$ in a universe of discourse X is characterized by a membership function $\mu_{\tilde{A}}\left(x\right)$ which associates with each element x in X a real number in the interval [0, 1]. The function value $\mu_{\tilde{A}}\left(x\right)$ is termed the grade of membership of x in $\tilde{A}$. A trapezoidal fuzzy number can be defined as $\tilde{A} = \left(a_1, a_2, a_3, a_4\right)$, shown in Figure 4. The membership function $\mu_{\tilde{A}}\left(x\right) : R \to \left[0,1\right]$ is defined as follows:

Figure 4. Trapezoidal fuzzy number set $\tilde{A}$

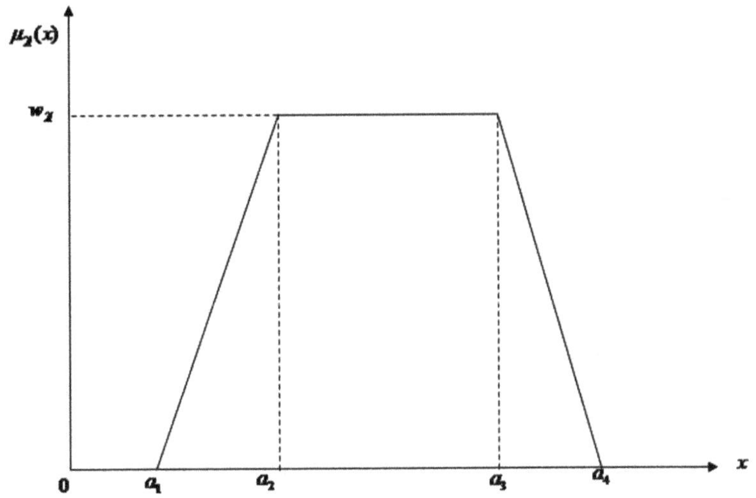

$$\mu_{\tilde{A}}\left(x\right)=\begin{cases} 0, & x < a_1 \\ \dfrac{x-a_1}{a_2-a_1}, & a_1 \le x \le a_2 \\ 1, & a_2 \le x \le a_3 \\ \dfrac{x-a_4}{a_3-a_4}, & a_3 \le x \le a_4 \\ 0, & x > a_4 \end{cases} \tag{1}$$

Here, $a_1 \le a_2 \le a_3 \le a_4$.

The operational rules of the trapezoidal fuzzy numbers $\tilde{a}$ and $\tilde{b}$ are shown as follows (Chen, Lin & Hwang, 2006):

$$\tilde{a} \oplus \tilde{b} = \left(a_1, a_2, a_3, a_4\right) \oplus \left(b_1, b_2, b_3, b_4\right) = \left(a_1 + b_1, a_2 + b_2, a_3 + b_3, a_4 + b_4\right) \tag{2}$$

$$\tilde{a} - \tilde{b} = \left(a_1, a_2, a_3, a_4\right) - \left(b_1, b_2, b_3, b_4\right) = \left(a_1 - b_4, a_2 - b_3, a_3 - b_2, a_4 - b_1\right) \tag{3}$$

$$\tilde{a} \otimes \tilde{b} = \left(a_1, a_2, a_3, a_4\right) \otimes \left(b_1, b_2, b_3, b_4\right) = \left(a_1 \times b_1, a_2 \times b_2, a_3 \times b_3, a_4 \times b_4\right) \tag{4}$$

$$\tilde{a} / \tilde{b} = \frac{\left(a_1, a_2, a_3, a_4\right)}{\left(b_1, b_2, b_3, b_4\right)} = \left(a_1 / b_4, a_2 / b_3, a_3 / b_2, a_4 / b_1\right) \tag{5}$$

Decision Making Problem

Let $E = \{e_1, e_2, ..., e_q\}$ be the set of decision makers in the group decision making process. $A = \{A_1, A_2, ..., A_m\}$ be the set of alternatives, and $C = \{C_1, C_2, ..., C_n\}$ be the set of criteria-attributes.

Suppose that $\tilde{\tilde{a}}_{ijk} = (a_{ijk1}, a_{ijk2}, a_{ijk3}, a_{ijk4})$ is the attribute value given by the decision maker e_k, where $\tilde{\tilde{a}}_{ijk}$ is a trapezoidal fuzzy number for the alternative A_i with respect to the attribute C_j.

Let $\tilde{\tilde{w}}_{kj} = (w_{kj}, w_{kj2}, w_{kj3}, w_{kj4})$ be the attribute weight given by the decision makers e_k, where $\tilde{\tilde{w}}_{kj}$ is also a trapezoidal fuzzy number.

Aggregated Fuzzy Rating of Alternatives

$$\tilde{\tilde{x}}_{ijk} = (a_{ij}, b_{ij}, c_{ij}, d_{ij}) \tag{6}$$

where,

$$a_{ij} = \min_k (a_{ijk})$$

$$b_{ij} = \frac{1}{K} \sum_{k=1}^{K} b_{ijk}$$

$$c_{ij} = \frac{1}{K} \sum_{k=1}^{K} c_{ijk}$$

$$d_{ij} = \max_k (d_{ijk})$$

Then the aggregated fuzzy weight of each criterion can be expressed as:

$$\tilde{\tilde{\varpi}}_j = (\omega_{j1}, \omega_{j2}, \omega_{j3}, \omega_{j4}) \tag{7}$$

where,

$$\omega_{j1} = \min_k (\omega_{kj1})$$

$$\omega_{j2} = \frac{1}{K} \sum_{k=1}^{K} \omega_{kj2}$$

$$\omega_{j3} = \frac{1}{K} \sum_{k=1}^{K} \omega_{kj3}$$

$$\omega_{j4} = \max_k (\omega_{kj4})$$

Normalization of Decision Making Information

In order to eliminate the impact of different physical dimension to the decision making result, the decision making information is to be normalized. Consider that there are generally benefit attributes $\left(I_1\right)$ and cost attributes $\left(I_2\right)$. The normalizing methods are shown as follows:

$$\tilde{\tilde{x}}_{ijk} = \left(\frac{a_{ij}}{m_{jk}}, \frac{b_{ij}}{m_{jk}}, \frac{c_{ij}}{m_{jk}}, \frac{d_{ij}}{m_{jk}}\right) \tag{8}$$

For benefit attributes, where $m_{jk} = \max_i \left(a_{ij}, b_{ij}, c_{ij}, d_{ij}\right)$

$$\tilde{\tilde{x}}_{ijk} = \left(\frac{n_{jk}}{a_{ij}}, \frac{n_{jk}}{b_{ij}}, \frac{n_{jk}}{c_{ij}}, \frac{n_{jk}}{d_{ij}}\right) \tag{9}$$

For cost attributes, where $n_{jk} = \min_i \left(a_{ij}, b_{ij}, c_{ij}, d_{ij}\right)$

The normalization method mentioned above is to preserve the property that the ranges of normalized interval-valued trapezoidal fuzzy numbers belong to [0, 1].

Let, $x_{ij} = \tilde{\tilde{V}} = \left[\tilde{\tilde{v}}_{ij}\right]_{m \times n}$ be the weighted normalized matrix:

$$x_{ij} = \tilde{\tilde{v}}_{ij} = \tilde{\tilde{x}}_{ijk} \otimes \tilde{\tilde{\varpi}}_j \tag{10}$$

$$x_{ij} = \begin{bmatrix} x_{11} & x_{12} & \cdots\cdots & x_{1n} \\ x_{21} & x_{22} & \cdots\cdots & x_{2n} \\ \vdots & \vdots & \vdots & \vdots \\ x_{m1} & x_{m2} & \cdots\cdots & x_{mn} \end{bmatrix}$$

DECISION MAKING BASED ON TOPSIS APPROACH

In recent year, several MCDM-approaches, i.e.,

- Dominance-Based Rough Set Approach (DRSA),
- ELECTRE, Evidential Reasoning (ER),
- Grey Relational Analysis (GRA),
- Aggregated Indices Randomization Method (AIRM),
- Best Worst Method (BWM),
- Characteristic Objects Method (COMET),
- Data Envelopment Analysis (DEA),
- Multi-Attribute Global Inference of Quality (MAGIQ),

- Multi-attribute utility theory (MAUT),
- Multi-attribute value theory (MAVT),
- Nonstructural Fuzzy Decision Support System (NSFDSS),
- PAPRIKA,
- PROMETHEE,
- Stochastic Multi-criteria Acceptability Analysis (SMAA),
- Superiority and inferiority ranking (SIR),
- Weighted product model (WPM),
- Weighted sum model (WSM) TEC,

are proposed by previous researchers to solve the industrial management problem. TOPSIS is also found its great worth amongst entire built MCDM-approaches.

TOPSIS stand for Technique for Order Preferences by Similarity to an Ideal Solution was proposed by Hwang and Yoon (1981). The main idea came from the concept of the compromise solution to choose the best alternative nearest to the positive ideal solution (optimal solution) and farthest from the negative ideal solution (inferior solution). Then, choose the best one of sorting, which will be the best alternative. TOPSIS is used to determine the best alternative based on the concepts of the compromise solution. The compromise solution can be regarded as choosing the solution with the shortest Euclidean distance from the ideal solution and the farthest Euclidean distance from the negative ideal solution. Whereas TOPSIS determine the fuzzy positive ideal solution and the fuzzy negative ideal solution of the computed objects. The fuzzy positive ideal solution and the fuzzy negative ideal solution can be identified as follows:

For benefit attributes:

$$\tilde{\tilde{V}}^{+} = \left[\tilde{\tilde{v}}_j^{+}\right]_{1\times n}, \tilde{\tilde{V}}^{-} = \left[\tilde{\tilde{v}}_j^{-}\right]_{1\times n}$$

and for cost attributes:

$$\tilde{\tilde{V}}^{+} = \left[\tilde{\tilde{v}}_j^{-}\right]_{1\times n}, \tilde{\tilde{V}}^{-} = \left[\tilde{\tilde{v}}_j^{+}\right]_{1xn} \tag{11}$$

$$\tilde{\tilde{v}}_j^{+} = \max_i \left(v_{ij4}\right),\ \tilde{\tilde{v}}_j^{-} = \min_i \left(v_{ij1}\right) \tag{12}$$

Benefit attribute means the larger the rating, the greater the preference. Conversely, cost attribute means, the smaller the rating, the greater the preference.

Measure Separation

The separation of each alternative from the fuzzy positive ideal solution, using the distance measurement between two fuzzy numbers can be currently calculated as:

$$\tilde{d}_i^{+} = \left\{\sum_{j=1}^{n} d\left(\tilde{\tilde{v}}_{ij}, \tilde{\tilde{v}}_j^{+}\right)^2\right\}^{1/2}, i = 1,2,\ldots,m, and\ j = 1,2,\ldots,n \tag{13}$$

Similarly, the separation from the fuzzy negative ideal solution can be calculated as:

$$\tilde{d}_i^- = \left\{ \sum_{j=1}^{n} d\left(\tilde{\tilde{v}}_{ij}, \tilde{\tilde{v}}_j^-\right)^2 \right\}^{1/2}, i = 1,2,...,m, and\ j = 1,2,...,n \tag{14}$$

Closeness Coefficient

A closeness coefficient is defined to determine the ranking order of all alternatives once $\tilde{d}_i^+$ and $\tilde{d}_i^-$ of each alternative A_i has been calculated. The relative closeness to the positive ideal alternatives A_i can be calculated as follows:

$$\tilde{R}_i = \tilde{d}_i^- / \left(\tilde{d}_i^+ + \tilde{d}_i^-\right), i = 1,2,...,m. \tag{15}$$

Performance Score against Alternatives

Rank the alternatives in terms of their relative. Obviously, an alternative A_i is closer to $\tilde{\tilde{V}}^+$ and farther from $\tilde{\tilde{V}}^-$ as $\tilde{R}_i$ approaches to 1. Therefore, according to the closeness coefficient, we can determine the ranking order of all alternatives and select the best one from among a set of feasible alternatives

HYPOTHETICAL CASE RESEARCH

A 3-axis CNC machine tool appraisement module is constructed by incorporating twenty crucial criteria through literature survey (Ayag & Ozdemir, 2012; Qi. 2010; Tsai, Wang, Cheng and Kao, 2010; Abdi, 2009; Zhang & Chu, 2009; Duran & Aguilo, 2008; Korena & Shpitalni, 2010). A hypothetical case research is carried out, where the authors assumed that a committee of six experts' panel, i.e.,

- DM_1,
- DM_2,
- DM_3,
- DM_4,
- DM_5, and
- $DM_{6,}$

elected from quality assurance, production units of host firm. The decision making committee assesses the concerned alternatives based on a structured module (criteria hierarchy), shown in Table 3 for selecting the best CNC machine tool alternative. Structured module involved the twenty criteria in which merely C_{17} is found as a non-beneficial and rests of the criteria are found as beneficial criteria. The definitions of criteria are shown in Table 4.

Table 3. CNC machine tool appraisement module

Goal (C)	Criteria (C_{ij})	Attitude of Criteria
CNC Machine Tool Election	Productivity,(C_1)	(+)
	Coding Aspects,(C_2)	(+)
	Schedule Utilization,(C_3)	(+)
	Adaptability,(C_4)	(+)
	Precision, (C_5)	(+)
	Reliability,(C_6)	(+)
	Safety,(C_7)	(+)
	Machine Tool Compatibility,(C_8)	(+)
	Risk,(C_9)	(+)
	Functionality,(C_{10})	(+)
	Customization,(C_{11})	(+)
	Operative Flexibility,(C_{12})	(+)
	Convenient for Use,(C_{13})	(+)
	Accuracy,(C_{14})	(+)
	Efficiency, (C_{15})	(+)
	Capacity, (C_{16})	(+)
	Resource Consumption,(C_{17})	(-)
	User Friendliness,(C_{18})	(+)
	Product Quality, (C_{19})	(+)
	Working Automation,(C_{20})	(+)

Source: (Ayag & Ozdemir, 2012; Qi. 2010; Tsai, Wang, Cheng and Kao, 2010; Abdi, 2009, Zhang & Chu, 2009; Duran & Aguilo, 2008; Onut, Kara and Efendigil, 2008; Korena & Shpitalni, 2010).

Criteria weights and criteria ratings are expressed in the linguistic terms, which are transformed further into corresponding trapezoidal fuzzy number set, as given in Table 5.

The appropriateness ratings as well as weights of criteria for various alternatives are revealed in Tables 6 and Table 7 respectively.

Equations 6 & 7 and their depended equations 2-5 are utilized to aggregate fuzzy appropriateness ratings and weights, shown in Table 8 and Table 9.

Next, equations 8 & 9 are utilized to normalize the matrix, then, equations 10 is brought to use to construct weighted normalized matrix, shown in Table 10.

To evaluate the performances scores against defined CNC machine tool alternatives under twenty criteria, the equations 11 & 12 are used to compute the positive and negative ideal solution, depicted in Table 11.

Then, equations 13 & 14 are used to compute separation of each alternative from positive and negative ideal solution, shown in Table 12.

Finally, an equation 15 is applied to compute the performances scores of various alternatives, showed in Table 13.

Hence, Alternative sorting is as following proceeding.

Table 4. Definition of criteria

Criteria	Definitions
Productivity, (C_1)	It is defined as a ratio of output to inputs. It states that machine ought to produce max outputs with respect to fewer inputs.
Coding Aspects, (C_2)	Processing of alphanumerical values, which determine the types of operations performed on the CNC machine tool
Schedule Utilization, (C_3)	It may be defined as the action of the CNC machine tool towards finishing the production under allowable time.
Adaptability, (C_4)	It is the aptitude of machine tool to respond the supple market demand.
Precision, (C_5)	Precision is a machining process, where parts are mechanized with a very high tolerance.
Reliability, (C_6)	It is an ability of a machine tool to perform its required functions for a specified period of time under constrained operating parameters.
Safety, (C_7)	It states the protection of operators from operating machine tool.
Machine Tool Compatibility, (C_8)	It helps the operators to produce different kinds of goods by changing defined types of tools.
Risk,(C_9)	It states the losses (an undesirable outcome, however not necessarily so), which are being resulted from actions or inactions of machine tool.
Functionality, (C_{10})	The capacity of application packages of machine tool to perform its useful functions.
Customization, (C_{11})	It is the ability of machine tool to deliver the responses on modified or changed products design as per customer's requirements.
Operative Flexibility, (C_{12})	The machine tools must be utilized in accordance with others except defined operations.
Convenient for Use, (C_{13})	It may be defined as the suitability of machine tool to human comfort, purpose, or needs.
Accuracy, (C_{14})	It is defined as dimensions of machined part must conform to desirable specifications.
Efficiency, (C_{15})	It reflects the ratio of outputs (metal removal rate etc) to the inputs (feed, current etc).
Capacity, (C_{16})	It is the ability to produce units within specified tolerances of quality, based on short-term surveys of continuous production runs.
Resource Consumption, (C_{17})	It may be defined as the acquisition of power to drive the CNC machine tool.
User Friendliness, (C_{18})	It measures the ease of use of a CNC machine tool by minimizing the operator errors.
Product Quality, (C_{19})	It states that the product's ability is made to fulfill the expectations of end users.
Working Automation, (C_{20})	It records the status of machining via Logical signal to control the production take and help to remove the failure in machine tool by systems such as open or closed feedback automation system.

Table 5. The scale for assigning attributes appropriateness ratings and priority weights criteria

Linguistic Terms (Indices Ratings)	Linguistic Terms (Priority Weights)	Generalized Trapezoidal Fuzzy Number Set
Absolutely Poor (AP)	Absolutely Low (AL)	(0, 0, 0, 0)
Very Poor (VP)	Very Low (VL)	(0, 0, 0.02, 0.07)
Poor (P)	Low (L)	(0.04, 0.1, 0.18, 0.23)
Medium Poor (MP)	Medium Low (ML)	(0.17, 0.22, 0.36, 0.42)
Fair (F)	Medium (M)	(0.32, 0.41, 0.58, 0.65)
Medium Good (MG)	Medium High (MH)	(0.58, 0.63, 0.80, 0.86)
Good (G)	High (H)	(0.72, 0.78, 0.92, 0.97)
Very Good (VG)	Very High (VH)	(0.93, 0.98, 1.0, 1.0)
Absolutely Good (AG)	Absolutely High (AH)	(1.0, 1.0, 1.0, 1.0)

Table 6. The appropriateness ratings vs. criteria by DMs

Criteria	Alternatives	DM_1	DM_2	DM_3	DM_4	DM_5	DM_6
Productivity (C_1)	A_1	MP	MG	F	MG	MP	MP
	A_2	MG	MG	MG	MG	MG	MG
	A_3	F	MG	G	G	MG	G
	A_4	F	MG	MG	G	MG	G
	A_5	MP	F	F	G	F	G
Coding Aspects (C_2)	A_1	MG	MG	G	G	MG	G
	A_2	MG	G	G	VG	G	VG
	A_3	F	G	VG	VG	G	VG
	A_4	MP	MG	MG	G	MG	G
	A_5	MP	MG	G	VG	MG	VG
Schedule Utilization (C_3)	A_1	MP	MG	MG	G	MG	G
	A_2	F	G	VG	VG	G	VG
	A_3	F	MG	G	F	MG	F
	A_4	F	MG	VG	G	MG	G
	A_5	F	MG	G	G	MG	G
Adaptability (C_4)	A_1	F	MG	G	G	MG	G
	A_2	MP	G	G	G	G	G
	A_3	F	MG	MG	G	MG	G
	A_4	MP	MG	G	G	MG	G
	A_5	MP	VG	G	G	VG	G
Precision (C_5)	A_1	MP	MG	G	G	MG	G
	A_2	MP	MG	G	G	MG	G
	A_3	F	MG	G	VG	MG	VG
	A_4	MP	MG	G	VG	MG	VG
	A_5	F	MG	G	VG	MG	VG
Reliability (C_6)	A_1	MG	G	AG	AG	G	AG
	A_2	VG	G	G	VG	G	VG
	A_3	AG	MG	AG	G	MG	G
	A_4	VG	VG	G	MG	VG	MG
	A_5	AG	G	MG	AG	G	AG
Safety (C_7)	A_1	MG	VG	AG	G	VG	G
	A_2	AG	VG	G	VG	VG	VG
	A_3	VG	MG	VG	AG	MG	AG
	A_4	VG	VG	AG	MG	VG	MG
	A_5	MG	G	MG	AG	G	AG
Machine Tool Compatibility(C_8)	A_1	AG	VG	G	G	VG	G
	A_2	AG	MG	G	VG	MG	VG
	A_3	VG	G	G	MG	G	MG
	A_4	VG	VG	MG	VG	VG	VG
	A_5	MG	VG	AG	AG	VG	AG

continued on following page

Table 6. Continued

Criteria	Alternatives	DM_1	DM_2	DM_3	DM_4	DM_5	DM_6
Risk (C_9)	A_1	AG	VG	G	G	VG	G
	A_2	VG	MG	VG	AG	MG	AG
	A_3	AG	G	AG	MG	G	MG
	A_4	MG	VG	MG	AG	VG	AG
	A_5	VG	F	VG	G	F	G
Functionality (C_{10})	A_1	MG	MG	G	AG	MG	AG
	A_2	VG	G	AG	MG	G	MG
	A_3	AG	MG	MG	AG	MG	AG
	A_4	AG	VG	AG	MG	VG	MG
	A_5	AG	G	MG	AG	G	AG
Customization (C11)	A_1	MG	G	AG	G	G	G
	A_2	AG	VG	G	G	VG	G
	A_3	VG	MG	VG	AG	MG	AG
	A_4	MG	G	VG	MG	G	MG
	A_5	VG	AG	MG	AG	AG	AG
Operative Flexibility,(C_{12})	A_1	VG	MG	F	AG	MG	AG
	A_2	VG	AG	AG	MG	AG	MG
	A_3	VG	MG	MG	G	MG	G
	A_4	AG	MG	G	G	MG	G
	A_5	VG	MG	G	G	MG	G
Convenient for Use (C_{13})	A_1	MG	G	VG	VG	G	VG
	A_2	F	VG	AG	G	VG	G
	A_3	VG	MG	MG	G	MG	G
	A_4	VG	MG	G	VG	MG	VG
	A_5	AG	MG	MG	G	MG	G
Accuracy (C_{14})	A_1	MG	G	VG	G	G	G
	A_2	VG	VG	AG	G	VG	G
	A_3	VG	MG	G	AG	MG	AG
	A_4	AG	VG	AG	MG	VG	MG
	A_5	AG	G	MG	AG	G	AG
Efficiency (C_{15})	A_1	VG	VG	G	VG	VG	VG
	A_2	VG	G	VG	G	G	G
	A_3	AG	VG	AG	G	VG	G
	A_4	VG	G	VG	VG	G	VG
	A_5	F	AG	G	AG	AG	AG
Capacity (C_{16})	A_1	F	VG	AG	G	VG	G
	A_2	MG	G	G	VG	G	VG
	A_3	F	AG	AG	MG	AG	MG
	A_4	VG	MG	G	G	MG	G
	A_5	VG	G	VG	AG	G	AG

continued on following page

Table 6. Continued

Criteria	Alternatives	DM_1	DM_2	DM_3	DM_4	DM_5	DM_6
Resource Consumption (C_{17})	A_1	MG	MG	MG	AG	MG	AG
	A_2	AG	AG	AG	F	AG	F
	A_3	AG	VG	F	MG	VG	MG
	A_4	F	G	MG	AG	G	AG
	A_5	VG	VG	AG	VG	VG	VG
User Friendliness,(C_{18})	A_1	VG	MG	VG	G	MG	G
	A_2	F	VG	G	MG	VG	MG
	A_3	VG	G	AG	AG	G	AG
	A_4	VG	AG	MG	G	AG	G
	A_5	MG	MG	AG	AG	MG	AG
Product Quality, (C_{19})	A_1	F	G	G	F	G	F
	A_2	AG	AG	AG	MG	AG	MG
	A_3	AG	G	MG	AG	G	AG
	A_4	F	VG	G	G	VG	G
	A_5	MG	MG	AG	AG	MG	AG
Working Automation,(C_{20})	A_1	VG	VG	G	MG	VG	MG
	A_2	F	AG	VG	AG	AG	AG
	A_3	MG	G	MG	MG	G	MG
	A_4	VG	MG	AG	AG	MG	AG
	A_5	MG	AG	G	G	AG	G

$A_3 > A_2 > A_1 > A_4 > A_5$

Proposed fuzzy based CNC machine tool appraisement module: Flow chart representation:

Step 1: Formed a committee of decision makers, and then identify the evaluation criteria of 3-axis CNC machine tools.

Step 2: Evaluated the appropriate linguistic variables for the weight of the criteria and the linguistic ratings for 3-axis CNC machine tools.

Step 3: Aggregated the weight and rating of criteria.

Step 4: Constructed the fuzzy decision matrix and the normalized fuzzy decision matrix.

Step 5: Constructed weighted normalized fuzzy decision matrix.

Step 6: Determine FPIS and FNIS.

Step 7: Calculated the distance of each CNC machine tool alternatives from FPIS and FNIS respectively.

Step 8: Calculated the closeness coefficient of each CNC alternatives.

Step 9: As per the closeness coefficient, it can be made out regarding the status of considered CNC machine tool alternatives under twenty criteria and determined the best CNC machine tool amongst available alternatives.

Table 7. Priority weights assigned by assigned by DMs

Indices/Criteria	Weights					
	DM_1	DM_2	DM_3	DM_4	DM_5	DM_6
Productivity,(C_1)	VH	AH	AH	AH	MH	VH
Coding Aspects,(C_2)	H	VH	H	AH	H	H
Schedule Utilization,(C_3)	VH	VH	AH	H	AH	VH
Adaptability,(C_4)	AH	VH	VH	VH	VH	AH
Precision, (C_5)	MH	MH	ML	M	ML	MH
Reliability,(C_6)	VH	MH	VH	VH	VH	VH
Safety,(C_7)	AH	VH	VH	H	VH	AH
Machine Tool Compatibility,(C_8)	AH	VH	VH	H	VH	AH
Risk,(C_9)	VH	H	VH	VH	VH	VH
Functionality,(C_{10})	VH	MH	AH	VH	AH	VH
Customization,(C_{11})	AH	VH	VH	VH	VH	AH
Operative Flexibility,(C_{12})	AH	VH	VH	VH	VH	AH
Convenient for Use,(C_{13})	MH	MH	H	H	H	MH
Accuracy,(C_{14})	H	AH	VH	AH	VH	H
Efficiency, (C_{15})	MH	MH	VH	MH	VH	MH
Capacity, (C_{16})	H	H	MH	MH	MH	H
Resource Consumption,(C_{17})	VH	H	VH	VH	VH	VH
User Friendliness,(C_{18})	MH	MH	MH	MH	MH	MH
Product Quality, (C_{19})	VH	H	H	AH	H	VH
Working Automation,(C_{20})	MH	MH	MH	MH	MH	MH

Table 8. Aggregated fuzzy appropriateness ratings

Criteria	Alternatives	Aggregated Appropriateness Ratings
Productivity (C_1)	A_1	(0.170, 0.388, 0.543, 0.860)
	A_2	(0.580, 0.630, 0.800, 0.860)
	A_3	(0.320, 0.668, 0.823, 0.970)
	A_4	(0.320, 0.643, 0.803, 0.970)
	A_5	(0.170, 0.502, 0.657, 0.970)
Coding Aspects (C_2)	A_1	(0.580, 0.705, 0.860, 0.970)
	A_2	(0.580, 0.822, 0.927, 1.000)
	A_3	(0.320, 0.818, 0.903, 1.000)
	A_4	(0.170, 0.612, 0.767, 0.970)
	A_5	(0.170, 0.703, 0.813, 1.000)
Schedule Utilization (C_3)	A_1	(0.170, 0.612, 0.767, 0.970)
	A_2	(0.320, 0.818, 0.903, 1.000)
	A_3	(0.320, 0.545, 0.710, 0.970)
	A_4	(0.320, 0.702, 0.837, 1.000)
	A_5	(0.320, 0.668, 0.823, 0.970)

continued on following page

Table 8. Continued

Criteria	Alternatives	Aggregated Appropriateness Ratings
Adaptability (C_4)	A_1	(0.320, 0.668, 0.823, 0.970)
	A_2	(0.170, 0.687, 0.827, 0.970)
	A_3	(0.320, 0.643, 0.803, 0.970)
	A_4	(0.170, 0.637, 0.787, 0.970)
	A_5	(0.170, 0.753, 0.853, 1.000)
Precision (C_5)	A_1	(0.170, 0.637, 0.787, 0.970)
	A_2	(0.170, 0.637, 0.787, 0.970)
	A_3	(0.320, 0.735, 0.850, 1.000)
	A_4	(0.170, 0.703, 0.813, 1.000)
	A_5	(0.320, 0.735, 0.850, 1.000)
Reliability (C_6)	A_1	(0.580, 0.865, 0.940, 1.000)
	A_2	(0.720, 0.880, 0.960, 1.000)
	A_3	(0.580, 0.803, 0.907, 1.000)
	A_4	(0.580, 0.830, 0.920, 1.000)
	A_5	(0.580, 0.865, 0.940, 1.000)
Safety (C_7)	A_1	(0.580, 0.858, 0.940, 1.000)
	A_2	(0.720, 0.950, 0.987, 1.000)
	A_3	(0.580, 0.870, 0.933, 1.000)
	A_4	(0.580, 0.867, 0.933, 1.000)
	A_5	(0.580, 0.803, 0.907, 1.000)
Machine Tool Compatibility(C_8)	A_1	(0.720, 0.883, 0.960, 1.000)
	A_2	(0.580, 0.833, 0.920, 1.000)
	A_3	(0.580, 0.763, 0.893, 1.000)
	A_4	(0.580, 0.922, 0.967, 1.000)
	A_5	(0.580, 0.932, 0.967, 1.000)
Risk (C_9)	A_1	(0.720, 0.883, 0.960, 1.000)
	A_2	(0.580, 0.870, 0.933, 1.000)
	A_3	(0.580, 0.803, 0.907, 1.000)
	A_4	(0.580, 0.870, 0.933, 1.000)
	A_5	(0.320, 0.723, 0.833, 1.000)
Functionality (C_{10})	A_1	(0.580, 0.778, 0.887, 1.000)
	A_2	(0.580, 0.800, 0.907, 1.000)
	A_3	(0.580, 0.815, 0.900, 1.000)
	A_4	(0.580, 0.870, 0.933, 1.000)
	A_5	(0.580, 0.865, 0.940, 1.000)

continued on following page

Table 8. Continued

Criteria	Alternatives	Aggregated Appropriateness Ratings
Customization (C11)	A_1	(0.580, 0.792, 0.913, 1.000)
	A_2	(0.720, 0.883, 0.960, 1.000)
	A_3	(0.580, 0.870, 0.933, 1.000)
	A_4	(0.580, 0.738, 0.873, 1.000)
	A_5	(0.580, 0.935, 0.967, 1.000)
Operative Flexibility,(C_{12})	A_1	(0.320, 0.775, 0.863, 1.000)
	A_2	(0.580, 0.873, 0.933, 1.000)
	A_3	(0.580, 0.738, 0.873, 1.000)
	A_4	(0.580, 0.767, 0.893, 1.000)
	A_5	(0.580, 0.763, 0.893, 1.000)
Convenient for Use (C_{13})	A_1	(0.580, 0.855, 0.940, 1.000)
	A_2	(0.320, 0.822, 0.903, 1.000)
	A_3	(0.580, 0.738, 0.873, 1.000)
	A_4	(0.580, 0.830, 0.920, 1.000)
	A_5	(0.580, 0.742, 0.873, 1.000)
Accuracy (C_{14})	A_1	(0.580, 0.788, 0.913, 1.000)
	A_2	(0.720, 0.917, 0.973, 1.000)
	A_3	(0.580, 0.837, 0.920, 1.000)
	A_4	(0.580, 0.870, 0.933, 1.000)
	A_5	(0.580, 0.865, 0.940, 1.000)
Efficiency (C_{15})	A_1	(0.720, 0.947, 0.987, 1.000)
	A_2	(0.720, 0.847, 0.947, 1.000)
	A_3	(0.720, 0.920, 0.973, 1.000)
	A_4	(0.720, 0.913, 0.973, 1.000)
	A_5	(0.320, 0.865, 0.917, 1.000)
Capacity (C_{16})	A_1	(0.320, 0.822, 0.903, 1.000)
	A_2	(0.580, 0.822, 0.927, 1.000)
	A_3	(0.320, 0.778, 0.863, 1.000)
	A_4	(0.580, 0.763, 0.893, 1.000)
	A_5	(0.720, 0.920, 0.973, 1.000)
Resource Consumption (C_{17})	A_1	(0.580, 0.753, 0.867, 1.000)
	A_2	(0.320, 0.803, 0.860, 1.000)
	A_3	(0.320, 0.772, 0.863, 1.000)
	A_4	(0.320, 0.767, 0.870, 1.000)
	A_5	(0.930, 0.983, 1.000, 1.000)

continued on following page

Table 8. Continued

Criteria	Alternatives	Aggregated Appropriateness Ratings
User Friendliness,(C_{18})	A_1	(0.580, 0.797, 0.907, 1.000)
	A_2	(0.320, 0.735, 0.850, 1.000)
	A_3	(0.720, 0.923, 0.973, 1.000)
	A_4	(0.580, 0.862, 0.940, 1.000)
	A_5	(0.580, 0.815, 0.900, 1.000)
Product Quality, (C_{19})	A_1	(0.320, 0.595, 0.750, 0.970)
	A_2	(0.580, 0.877, 0.933, 1.000)
	A_3	(0.580, 0.865, 0.940, 1.000)
	A_4	(0.320, 0.785, 0.890, 1.000)
	A_5	(0.580, 0.815, 0.900, 1.000)
Working Automation,(C_{20})	A_1	(0.580, 0.830, 0.920, 1.000)
	A_2	(0.320, 0.898, 0.930, 1.000)
	A_3	(0.580, 0.680, 0.840, 0.970)
	A_4	(0.580, 0.873, 0.933, 1.000)
	A_5	(0.580, 0.828, 0.927, 1.000)

Table 9. Aggregated fuzzy priority weights

Criteria	Aggregated Priority Weights
Productivity,(C_1)	(0.580, 0.932, 0.967, 1.000)
Coding Aspects,(C_2)	(0.720, 0.850, 0.947, 1.000)
Schedule Utilization,(C_3)	(0.930, 0.953, 0.987, 1.000)
Adaptability,(C_4)	(0.930, 0.987, 1.000, 1.000)
Precision, (C_5)	(0.170, 0.457, 0.617, 0.420)
Reliability,(C_6)	(0.580, 0.922, 0.967, 1.000)
Safety,(C_7)	(0.720, 0.953, 0.987, 1.000)
Machine Tool Compatibility,(C_8)	(0.720, 0.953, 0.987, 1.000)
Risk,(C_9)	(0.720, 0.947, 0.987, 1.000)
Functionality,(C_{10})	(0.580, 0.928, 0.967, 1.000)
Customization,(C_{11})	(0.930, 0.987, 1.000, 1.000)
Operative Flexibility,(C_{12})	(0.930, 0.987, 1.000, 1.000)
Convenient for Use,(C_{13})	(0.580, 0.705, 0.860, 0.970)
Accuracy,(C_{14})	(0.720, 0.920, 0.973, 1.000)
Efficiency, (C_{15})	(0.580, 0.747, 0.867, 1.000)
Capacity, (C_{16})	(0.580, 0.705, 0.860, 0.970)
Resource Consumption,(C_{17})	(0.720, 0.947, 0.987, 1.000)
User Friendliness,(C_{18})	(0.580, 0.630, 0.800, 0.860)
Product Quality, (C_{19})	(0.720, 0.883, 0.960, 1.000)
Working Automation,(C_{20})	(0.580, 0.630, 0.800, 0.860)

Table 10. Weighted normalized matrix

Criteria	Alternatives	Weighted Normalized Matrix
Productivity (C_1)	A_1	(0.099, 0.362, 0.525, 0.860)
	A_2	(0.336, 0.587, 0.773, 0.860)
	A_3	(0.186, 0.623, 0.796, 0.970)
	A_4	(0.186, 0.599, 0.777, 0.970)
	A_5	(0.099, 0.467, 0.635, 0.970)
Coding Aspects (C_2)	A_1	(0.418, 0.599, 0.814, 0.970)
	A_2	(0.418, 0.698, 0.877, 1.000)
	A_3	(0.230, 0.696, 0.855, 1.000)
	A_4	(0.122, 0.520, 0.726, 0.970)
	A_5	(0.122, 0.598, 0.770, 1.000)
Schedule Utilization (C_3)	A_1	(0.158, 0.583, 0.756, 0.970)
	A_2	(0.298, 0.780, 0.891, 1.000)
	A_3	(0.298, 0.520, 0.701, 0.970)
	A_4	(0.298, 0.669, 0.826, 1.000)
	A_5	(0.298, 0.637, 0.812, 0.970)
Adaptability (C_4)	A_1	(0.298, 0.659, 0.823, 0.970)
	A_2	(0.158, 0.678, 0.827, 0.970)
	A_3	(0.298, 0.635, 0.803, 0.970)
	A_4	(0.158, 0.628, 0.787, 0.970)
	A_5	(0.158, 0.743, 0.853, 1.000)
Precision (C_5)	A_1	(0.029, 0.291, 0.485, 0.407)
	A_2	(0.029, 0.291, 0.485, 0.407)
	A_3	(0.054, 0.336, 0.524, 0.420)
	A_4	(0.029, 0.321, 0.502,0.420)
	A_5	(0.054, 0.336, 0.524, 0.420)
Reliability (C_6)	A_1	(0.336, 0.797, 0.909, 1.000)
	A_2	(0.418, 0.811, 0.928, 1.000)
	A_3	(0.336, 0.740, 0.876, 1.000)
	A_4	(0.336, 0.765, 0.889, 1.000)
	A_5	(0.336, 0.797, 0.909, 1.000)
Safety (C_7)	A_1	(0.418, 0.818, 0.927, 1.000)
	A_2	(0.518, 0.906, 0.974, 1.000)
	A_3	(0.418, 0.829, 0.921, 1.000)
	A_4	(0.418, 0.826, 0.921, 1.000)
	A_5	(0.418, 0.766, 0.895, 1.000)

continued on following page

Table 10. Continued

Criteria	Alternatives	Weighted Normalized Matrix
Machine Tool Compatibility(C_8)	A_1	(0.518, 0.842, 0.947, 1.000)
	A_2	(0.418, 0.794, 0.908, 1.000)
	A_3	(0.418, 0.728, 0.881, 1.000)
	A_4	(0.418, 0.879, 0.954, 1.000)
	A_5	(0.418, 0.888, 0.954, 1.000)
Risk (C_9)	A_1	(0.230, 0.316, 0.357, 0.444)
	A_2	(0.230, 0.325, 0.363, 0.552)
	A_3	(0.230, 0.334, 0.393, 0.552)
	A_4	(0.230, 0.325, 0.363, 0.552)
	A_5	(0.230, 0.364, 0.436, 1.000)
Functionality (C_{10})	A_1	(0.336, 0.723, 0.857, 1.000)
	A_2	(0.336, 0.743, 0.876, 1.000)
	A_3	(0.336, 0.757, 0.870, 1.000)
	A_4	(0.336, 0.808, 0.902, 1.000)
	A_5	(0.336, 0.803, 0.909, 1.000)
Customization (C11)	A_1	(0.539, 0.781, 0.913, 1.000)
	A_2	(0.670, 0.872, 0.960, 1.000)
	A_3	(0.539, 0.858,0.933, 1.000)
	A_4	(0.539, 0.728, 0.873, 1.000)
	A_5	(0.539, 0.923, 0.967, 1.000)
Operative Flexibility,(C_{12})	A_1	(0.298, 0.765, 0.863, 1.000)
	A_2	(0.539, 0.862, 0.933, 1.000)
	A_3	(0.539, 0.728, 0.873, 1.000)
	A_4	(0.539, 0.756, 0.893, 1.000)
	A_5	(0.539, 0.753, 0.893, 1.000)
Convenient for Use (C_{13})	A_1	(0.336, 0.603, 0.808, 0.970)
	A_2	(0.186, 0.579, 0.777, 0.970)
	A_3	(0.336, 0.521, 0.751, 0.970)
	A_4	(0.336, 0.585, 0.791, 0.970)
	A_5	(0.336, 0.523, 0.751, 0.970)
Accuracy (C_{14})	A_1	(0.418, 0.725, 0.889, 1.000)
	A_2	(0.518, 0.843, 0.947, 1.000)
	A_3	(0.418, 0.770, 0.895, 1.000)
	A_4	(0.418, 0.800, 0.908, 1.000)
	A_5	(0.418, 0.796, 0.915, 1.000)

continued on following page

Table 10. Continued

Criteria	Alternatives	Weighted Normalized Matrix
Efficiency (C_{15})	A_1	(0.418, 0.707, 0.855, 1.000)
	A_2	(0.418, 0.632, 0.820, 1.000)
	A_3	(0.418, 0.687, 0.844, 1.000)
	A_4	(0.418, 0.682, 0.844, 1.000)
	A_5	(0.186, 0.646, 0.794, 1.000)
Capacity (C_{16})	A_1	(0.186, 0.579, 0.777, 0.970)
	A_2	(0.336, 0.579, 0.797, 0.970)
	A_3	(0.186, 0.549, 0.742, 0.970)
	A_4	(0.336, 0.538, 0.768, 0.970)
	A_5	(0.418, 0.649, 0.837, 0.970)
Resource Consumption (C_{17})	A_1	(0.418, 0.713, 0.855, 1.000)
	A_2	(0.230, 0.760, 0.849, 1.000)
	A_3	(0.230, 0.731, 0.852, 1.000)
	A_4	(0.230, 0.726, 0.858, 1.000)
	A_5	(0.670, 0.931, 0.987, 1.000)
User Friendliness,(C_{18})	A_1	(0.336, 0.502, 0.725, 0.860)
	A_2	(0.186, 0.463, 0.680, 0.860)
	A_3	(0.418, 0.582, 0.779, 0.860)
	A_4	(0.336, 0.543, 0.752, 0.860)
	A_5	(0.336, 0.513, 0.720, 0.860)
Product Quality, (C_{19})	A_1	(0.230, 0.526, 0.720, 0.970)
	A_2	(0.418, 0.774, 0.896, 1.000)
	A_3	(0.418, 0.764, 0.902, 1.000)
	A_4	(0.230, 0.693, 0.854, 1.000)
	A_5	(0.418, 0.720, 0.864, 1.000)
Working Automation,(C_{20})	A_1	(0.336, 0.523, 0.736, 0.860)
	A_2	(0.186, 0.566, 0.744, 0.860)
	A_3	(0.336, 0.428, 0.672, 0.834)
	A_4	(0.336, 0.550, 0.747, 0.860)
	A_5	(0.336, 0.522, 0.741, 0.860)

Table 11. Computed positive and negative ideal solution

Criteria	A*	A⁻
Productivity,(C_1)	0.970	0.099
Coding Aspects,(C_2)	1.000	0.122
Schedule Utilization,(C_3)	1.000	0.158
Adaptability,(C_4)	1.000	0.158
Precision, (C_5)	0.420	0.029
Reliability,(C_6)	0.928	0.336
Safety,(C_7)	1.000	0.418
Machine Tool Compatibility,(C_8)	1.000	0.418
Risk,(C_9)	1.000	0.230
Functionality,(C_{10})	1.000	0.336
Customization,(C_{11})	1.000	0.539
Operative Flexibility,(C_{12})	1.000	0.298
Convenient for Use,(C_{13})	0.970	0.186
Accuracy,(C_{14})	1.000	0.418
Efficiency, (C_{15})	1.000	0.186
Capacity, (C_{16})	0.970	0.186
Resource Consumption,(C_{17})	1.000	0.230
User Friendliness,(C_{18})	0.860	0.186
Product Quality, (C_{19})	1.000	0.230
Working Automation,(C_{20})	0.860	0.186

Table 12. Computed separation of each alternative from positive and negative ideal solution

Criteria	Alternatives	d_i^*	d_i^-
Productivity (C_1)	A_1	0.5786	0.4558
	A_2	0.3870	0.5766
	A_3	0.4377	0.6180
	A_4	0.4444	0.6077
	A_5	0.5302	0.5438
Coding Aspects (C_2)	A_1	0.3658	0.6147
	A_2	0.3336	0.6631
	A_3	0.4201	0.6417
	A_4	0.5188	0.5569
	A_5	0.4962	0.5949

continued on following page

Table 12. Continued

Criteria	Alternatives	d_i^*	d_i^-
Schedule Utilization (C_3)	A_1	0.4855	0.5472
	A_2	0.3720	0.6428
	A_3	0.4513	0.5252
	A_4	0.3979	0.5989
	A_5	0.4066	0.5780
Adaptability (C_4)	A_1	0.4005	0.5858
	A_2	0.4593	0.5865
	A_3	0.4082	0.5749
	A_4	0.4726	0.5646
	A_5	0.4462	0.6194
Precision (C_5)	A_1	0.2086	0.3240
	A_2	0.2086	0.3240
	A_3	0.1947	0.3511
	A_4	0.2058	0.3398
	A_5	0.1947	0.3511
Reliability (C_6)	A_1	0.3052	0.4950
	A_2	0.2643	0.5055
	A_3	0.3135	0.4731
	A_4	0.3095	0.4821
	A_5	0.3052	0.4950
Safety (C_7)	A_1	0.3072	0.4358
	A_2	0.2457	0.4734
	A_3	0.3060	0.4365
	A_4	0.3065	0.4357
	A_5	0.3183	0.4147
Machine Tool Compatibility(C_8)	A_1	0.2548	0.4500
	A_2	0.3122	0.4247
	A_3	0.3269	0.4033
	A_4	0.2983	0.4581
	A_5	0.2974	0.4605

continued on following page

Table 12. Continued

Criteria	Alternatives	d_i^*	d_i^-
Risk (C_9)	A_1	0.6675	0.1319
	A_2	0.6433	0.1804
	A_3	0.6335	0.1877
	A_4	0.6433	0.1804
	A_5	0.5734	0.4042
Functionality (C_{10})	A_1	0.3667	0.4638
	A_2	0.3612	0.4736
	A_3	0.3593	0.4748
	A_4	0.3489	0.4956
	A_5	0.3491	0.4964
Customization (C11)	A_1	0.2586	0.3203
	A_2	0.1784	0.3593
	A_3	0.2432	0.3425
	A_4	0.2747	0.2998
	A_5	0.2341	0.3679
Operative Flexibility,(C_{12})	A_1	0.3766	0.5078
	A_2	0.2428	0.5644
	A_3	0.2747	0.5170
	A_4	0.2659	0.5285
	A_5	0.2667	0.5278
Convenient for Use (C_{13})	A_1	0.3750	0.5477
	A_2	0.4487	0.5291
	A_3	0.4035	0.5172
	A_4	0.3813	0.5395
	A_5	0.4029	0.5176
Accuracy (C_{14})	A_1	0.3267	0.4050
	A_2	0.2546	0.4504
	A_3	0.3175	0.4158
	A_4	0.3112	0.4262
	A_5	0.3115	0.4271
Efficiency (C_{15})	A_1	0.3340	0.5994
	A_2	0.3559	0.5744
	A_3	0.3397	0.5919
	A_4	0.3409	0.5908
	A_5	0.4558	0.5581

continued on following page

Table 12. Continued

Criteria	Alternatives	d_i^*	d_i^-
Capacity (C_{16})	A_1	0.4487	0.5291
	A_2	0.3821	0.5401
	A_3	0.4595	0.5141
	A_4	0.3964	0.5248
	A_5	0.3264	0.5718
Resource Consumption (C_{17})	A_1	0.3326	0.5592
	A_2	0.4101	0.5602
	A_3	0.4144	0.5542
	A_4	0.4146	0.5550
	A_5	0.1689	0.6797
User Friendliness,(C_{18})	A_1	0.3242	0.4661
	A_2	0.4015	0.4405
	A_3	0.2645	0.5043
	A_4	0.3108	0.4811
	A_5	0.3217	0.4665
Product Quality, (C_{19})	A_1	0.4735	0.4674
	A_2	0.3166	0.5844
	A_3	0.3180	0.5839
	A_4	0.4206	0.5468
	A_5	0.3302	0.5631
Working Automation,(C_{20})	A_1	0.3175	0.4728
	A_2	0.3724	0.4773
	A_3	0.3523	0.4298
	A_4	0.3094	0.4810
	A_5	0.3172	0.4742

Table 13. Computed of d_i^*, d_i^-, and CC_i

Alternatives	Di* (Positive)	Di (Negative)	Di* + Di (Negative)	CC_i	Ranking Order
A_1	7.625	9.204	16.829	0.547	3
A_2	7.642	9.641	17.283	0.558	2
A_3	7.409	9.482	16.891	0.561	1
A_4	8.256	9.207	17.463	0.527	4
A_5	8.243	8.892	17.135	0.519	5

MANAGERIAL IMPLICATIONS

In the present chapter, subjectivity of appropriateness ratings as well as priority weights against CNC machine tool evaluation criteria is assessed by expert's panel, which is finally tackled by trapezoidal Fuzzy Numbers Set (TFNs) theory conjunction with TOPSIS approach. This approach observed to be of quite efficient, simple and flexible of solving such as Multi-Criteria Decision Making (MCDM) problems, which deals with the subjective evaluation information in an efficient manner. The utility of presented research work is for managers of advanced manufacturing firms, as managers can use exposed approach to evaluate the best CNC machine tool amongst preferred machine tools under unified measures. Furthermore, presented DSS can be used to assign the preferences order against alternatives. Additionally, others real life problems can also be resolved under replacement of criteria associated with module. Presented research work has economic impact, as no software is essential to use for carrying out the computation task.

CONCLUSION

In order to the competitive benefit, a lot of companies consider that a well designed MCDM approach play the significant tool. The decision making is found as a key to drive each industry with the profit. In last decade, it is observed that escalation in income of each industry exclusively depends upon:

- Enhancing utilization of industrial assets,
- Providing the best quality of finished goods,
- Augmenting productivity, and
- Diminishing waste (time, high power consumption).

Presently, advanced manufacturing firm such as automobile, automotive etc realized the complexity in regards to evaluate the best advanced manufacturing machine tools prior to establish industry.

The group decision making process in empire of fuzzy Multi-Criteria decision analysis provides an effective framework for benchmarking and evaluating the best and feasible alternatives under objectives of an organization. In the present chapter, a multiple attribute decision making TOPSIS (technique for order positive solution to ideal solution) analytical approach is proposed united with trapezoidal fuzzy number set to undertake the subjective information against criteria. Finally, an hypothetical case research is carried out to provide a better understanding of the proposed framework and the raking order of preferred alternative machine tool has been derived in accordance with descending value of the 'collective index'. Higher value of 'collective index' reflects higher degree of performance extent.

The core contributions of the aforesaid research work are highlighted below.

- Exploration of trapezoidal fuzzy number conjunction with fuzzy TOPSIS method towards selecting the most feasible CNC machine tool alternative amongst all.
- Adaptation of fuzzy logic to tackle subjective evaluation information.
- Multiple criteria decision making module for the selection of CNC turning machine tool.

- A convenient DSS has been proposed to the managers of advanced manufacturing firms for evaluating the best CNC machine tool amongst preferred machine tools under unified measures.
- Presented DSS can be used to assign the preferences order against alternatives.
- Others real life problems can also be resolved under replacement of criteria associated with module.
- Presented research work has economic impact, as no software is essential to use for carrying out the computation task.

REFERENCES

Abdi, M. R. (2009). Fuzzy multi-criteria decision model for evaluating reconfigurable machines. *International Journal of Production Economics*, *117*(1), 1–15. doi:10.1016/j.ijpe.2008.06.009

Ayag, Z., & Ozdemir, R. G. (2012). Evaluating machine tool alternatives through modified TOPSIS and alpha-cut based fuzzy ANP. *International Journal of Production Economics*, *140*(2), 630–636. doi:10.1016/j.ijpe.2012.02.009

Beikkhakhian, Y., Javanmardi, M., Karbasian, M., & Khayambashi, B. (2015). The application of ISM model in evaluating agile suppliers selection criteria and ranking suppliers using fuzzy TOPSIS-AHP methods. *Expert Systems with Applications*, *42*(15-16), 6224–6236. doi:10.1016/j.eswa.2015.02.035

Chen, C. T., Lin, C. T., & Hwang, S. F. (2006). A fuzzy approach for supplier evaluation and selection in supply chain management. *International Journal of Production Economics*, *102*(2), 289–301. doi:10.1016/j.ijpe.2005.03.009

Chu, S. J. (2009). Selecting the advanced manufacturing technology using fuzzy multiple attributes group decision making with multiple fuzzy information. *International Journal of Computers & Industrial Engineering*, *57*(3), 1033–1042. doi:10.1016/j.cie.2009.04.011

Computer Numerical Control. (n.d.) Retrieved from http://wings.buffalo.edu/academic/department/eng/mae/courses/460-564/Course-Notes/CNC%20notes.pdf

Duran, O., & Aguilo, J. (2008). Computer-aided machine-tool selection based on a Fuzzy-AHP approach. *International Journal of Expert Systems with Applications*, *34*(3), 1787–1794. doi:10.1016/j.eswa.2007.01.046

Hwang, C. L., & Yoon, K. (1981). *Multiple Attribute Decision Making-Methods and Applications: A State of the Art Survey*. Springer-Verlag Berlin Heidelberg. doi:10.1007/978-3-642-48318-9

Korena, Y., & Shpitalni, M. (2010). Design of reconfigurable manufacturing systems. *International Journal of Manufacturing Systems*, *29*(4), 130–141. doi:10.1016/j.jmsy.2011.01.001

Li, D.-F. (2010). TOPSIS-based nonlinear-programming methodology for multiattribute decision making with interval-valued intuitionistic fuzzy sets. *IEEE Transactions on Fuzzy Systems*, *18*(2), 299–311.

Li, D.-F. (2011). Closeness coefficient based nonlinear programming method for interval-valued intuitionistic fuzzy multiattribute decision making with incomplete preference information. *Applied Soft Computing*, *11*(4), 3402–3418. doi:10.1016/j.asoc.2011.01.011

Li, D.-F. (2012). TOPSIS for multiattribute group decision making using IFSs. *Advances in Fuzzy Sets and Systems*, *11*(2), 105–129.

Li, D.-F. (2014). *Decision and Game Theory in Management with Intuitionistic Fuzzy Sets*. Heidelberg, Germany: Springer. doi:10.1007/978-3-642-40712-3

Li, D.-F., & Nan, J. X. (2011). Extension of the TOPSIS for multi-attribute group decision making under Atanassov IFS environments. *International Journal of Fuzzy System Applications*, *1*(4), 47–61. doi:10.4018/ijfsa.2011100104

Lin, Z. C., & Yang, C. B. (1994). Evaluation of machine selection by the AHP method. *Journal of Materials Processing Technology*, *57*(3/4), 253–258.

Mahdavi, I., Amiri, N. M., Heidarzade, A., & Nourifar, R. (2008). Designing a model of fuzzy TOPSIS in multiple criteria decision making. *Applied Mathematics and Computation*, *206*(2), 607–617. doi:10.1016/j.amc.2008.05.047

Nan, J., Wang, T., & An, J. (2016). Intuitionistic Fuzzy Distance Based TOPSIS Method and Application to MADM. *International Journal of Fuzzy System Applications*, *5*(1), 43–56. doi:10.4018/IJFSA.2016010103

Onut, S., Kara, S. S., & Efendigil, T. (2008). A hybrid fuzzy MCDM approach to machine tool selection. *Journal of Intelligent Manufacturing*, *19*(4), 443–453. doi:10.1007/s10845-008-0095-3

Qi, J. (2010). Machine tool selection model based on fuzzy MCDM approach.*International Conference on Intelligent Control and Information Processing*, (pp. 282-285). doi:10.1109/ICICIP.2010.5564228

Sahu, A. K., Sahu, N. K., & Sahu, A. (2014). Appraisal of CNC machine tool by integrated MULTI MOORA-IGVN circumstances: An empirical study. *International Journal of Grey Systems: Theory and Application*, *4*(1), 104–123.

Sahu, A. K., Sahu, N. K., & Sahu, A. K. (2015b). Benchmarking CNC machine tool using hybrid fuzzy methodology a multi indices decision making approach. *International Journal of Fuzzy System Applications*, *4*(2), 28–46.

Sahu, A. K., Sahu, N. K., & Sahu, A. K. (2016a). Application of Integrated TOPSIS in ASC index: Partners Benchmarking perspective. *Benchmarking: An International Journal*, *23*(3), 540–563. doi:10.1108/BIJ-03-2014-0021

Sahu, A. K., Sahu, N. K., & Sahu, A. K. (2016b). Appraisal of Partner Enterprises under GTFNS Environment in Agile SC.[IJDSST]. *International Journal of Decision Support System Technology*, *8*(3), 1–19. doi:10.4018/IJDSST.2016070101

Sahu, A. K., Sahu, N. K., & Sahu, A. K. (2016c). Application of Modified MULTI-MOORA for CNC Machine Tool Evaluation in IVGTFNS Environment: An Empirical Study. *International Journal of Computer Aided Engineering and Technology*, *8*(3), 234–259. doi:10.1504/IJCAET.2016.077603

Sahu, N. K., Datta, S., & Mahapatra, S. S. (2012). Establishing green supplier appraisement platform using grey concepts. *International Journal of Grey Systems: Theory and Application*, *2*(3), 395–418.

Sahu, N. K., Datta, S., & Mahapatra, S. S. (2013). Decision making for selecting 3PL service provider using three parameter interval grey numbers. *International Journal of Logistic Systems and Management*, *14*(3), 261–297. doi:10.1504/IJLSM.2013.052061

Sahu, N. K., Datta, S., & Mahapatra, S. S. (2014). Green Supplier Appraisement in Fuzzy Environment. *Benchmarking: An International Journal*, *21*(3), 412–429. doi:10.1108/BIJ-06-2012-0042

Sahu, N. K., Datta, S. & Mahapatra, S. S. (2015). Fuzzy Based Appraisement Module for 3PL Evaluation and Selection. *Benchmarking, an International Journal, 22*(3), 354-392.

Sahu, N. K., Sahu A. K., & Sahu, A. K (2015a). Appraisement and Benchmarking of Third Party Logistic Service Provider by Exploration of Risk Based Approach. *Cogent Business and Management, 2*, 1-21.

Samantra, C., Sahu, N. K., Datta, S., & Mahapatra, S. S. (2013). Decision-Making in Selecting Reverse Logistics Alternative using Interval-Valued Fuzzy Sets Combined with VIKOR Approach. *International Journal of Services and Operations Management*, *14*(2), 175–196. doi:10.1504/IJSOM.2013.051828

Samvedi, V., Jain, V., & Chan, F. T. S. (2011). An integrated approach for machine tool selection using fuzzy analytical hierarchy process and grey relational analysis. *International Journal of Production Research*, *50*(12), 3211–3221. doi:10.1080/00207543.2011.560906

Tanaka, K. (1997). *An Introduction to Fuzzy Logic for practical Applications*. Springer-Verlag New York.

Thorani, Y. L. P., Rao, P. B., & Shankar, N. R. (2012). Ordering generalized trapezoidal fuzzy numbers using orthocenter of centroids. *International Journal of Algebra*, *6*(22), 1069–1085.

Tsai, J. P., Wang, S. Y., Cheng, H. Y., & Kao, Y. C. (2010). Multi-criteria decision making method for selection of machine tool.*International Symposium on Computer, Communication, Control and Automation*, (pp. 49-52). doi:10.1109/3CA.2010.5533376

Wan, S. P., & Li, D.-F. (2013). Fuzzy LINMAP approach to heterogeneous MADM considering comparisons of alternatives with hesitation degrees. *Omega: The International Journal of management. Science*, *41*(6), 925–940.

Zadeh, L. A. (1965). Fuzzy sets. *Information and Control*, *8*(3), 338–353. doi:10.1016/S0019-9958(65)90241-X

Zadeh, L. A. (1975). The concept of a linguistic variable and its application to approximate reasoning-II. *Information Sciences*, *8*(4), 301–357. doi:10.1016/0020-0255(75)90046-8

Zeng, S., Chen, J., & Li, X. (2016). A Hybrid Method for Pythagorean Fuzzy Multiple-Criteria Decision Making. *International Journal of Information Technology & Decision Making*, *15*(2), 403–422. doi:10.1142/S0219622016500012

Zeng, S., & Xiao, Y. (2016). TOPSIS method for intuitionistic fuzzy multiple-criteria decision making and its application to investment selection. *Kybernetes*, *45*(2), 282–296. doi:10.1108/K-04-2015-0093

Zhang, Z., & Chu, X. (2009). Fuzzy group decision-making for multi-format and multi-granularity linguistic judgments in quality function deployment. *International Journal of Expert Systems with Applications, 36*(5), 9150-9158.

Zyoud, S. H., Kaufmann, L. G., Shaheen, H., Samhan, S., & Hanusch, D. F. (2016). A framework for water loss management in developing countries under fuzzy environment: Integration of Fuzzy AHP with Fuzzy TOPSIS. *Expert Systems with Applications*, *61*, 86–105. doi:10.1016/j.eswa.2016.05.016

Chapter 13
Prediction of Solar and Wind Energies by Fuzzy Logic Control

Sanaa Faquir
University Sidi Mohamed Ben Abdallah, Morocco

Ali Yahyaouy
University Sidi Mohamed Ben Abdallah, Morocco

Hamid Tairi
University Sidi Mohamed Ben Abdallah, Morocco

Jalal Sabor
Ecole Nationale Superieure d'Arts et Metiers (ENSAM), Morocco

ABSTRACT

Nowadays, the use of renewable energy has become increasingly significant, and cost-effective. Between all existing sources of energy, solar and wind energies are the highly exploited. However, solar and wind energies are not available all the time and their performance is affected by unpredictable weather changes, therefore, it is not always feasible to obtain an accurate mathematical model of the controlled system. Various mathematical modeling methods were used to predict wind and solar powers using natural parameters but considering multiple parameters in equations makes the solution more complex. In addition to complexity, some coefficients are uncertain and based on probability. Fuzzy logic is a perfect tool to model any kind of uncertainty related to vagueness. This chapter presents a computer algorithm based on fuzzy logic control (FLC) to estimate the wind and solar energies using natural factors. As input parameters, the wind speed was used to predict the wind power and the temperature and the lightening were used to estimate the solar power.

INTRODUCTION

Recently, governments and industries all around the world are progressively trying to find alternatives to overcome the significant increase in the prices of conventional energy sources and to minimize the greenhouse gas emissions (GHC) and the resulting global warming. For these reasons, the implementation and diffusion of renewable energies as well as the establishment of a low-carbon economy have been subject of research in many regional and international initiatives, institutions and financial mechanisms

DOI: 10.4018/978-1-5225-1848-8.ch013

(Fritzsche, Zejli, & Tanzler, 2011). In contrast to fossil energy, renewable energy comes from resources which are continually refilled such as:

- Sunlight,
- Wind,
- Rain,
- Tides,
- Waves, and
- Geothermal Heat, and it must be harvested when it is available.

From all existing renewable energy sources, solar energy especially photovoltaic panels (PV) and wind turbines are the best shared resources of energies and therefore those that lend themselves best to the decentralized production of electricity. However, solar and wind energies are not available all the time and their performance is affected by unpredictable weather changes. They also rely on natural factors which require the use of many variables to simulate the operation of the renewable sources as accurate as possible. It is also difficult to predict the power as it is not always feasible to obtain an accurate mathematical model of the simulated system.

Over the past three decades, several models have been developed in order to generate the solar irradiance data based on stochastic models such as:

- Autoregressive (AR),
- Autoregressive moving average (ARMA),
- Autoregressive integrated moving average (ARIMA), and
- Markov chain (Aguiar & Collares-Pereira, 1992; Mora-Lopez & Sidrach De Cardona, 1998; Santos, Pinazo, & Canada, 2003; Maafi & Adnane, 1989).

Nevertheless, these models, based on the probability estimation, do not always give accurate results as they require the precise definition of problem domains as well as the identification of mathematical functions. This is why most stochastic models were found with relatively big errors and sometimes difficult to be adopted widely. Fuzzy modeling was also used with neural networks in (Chen, Gooi, & Wang, 2013) to forecast solar radiation at different weather conditions for a grid-connected photovoltaic system. It was also used in (Rizwan, Majid, Kirmani, & Kothari, 2014) to model and estimate the global solar energy in smart-grid applications using various parameters that require precise measurement. These parameters are:

- The mean duration sunshine per hour,
- Temperature,
- Latitude,
- Longitude,
- Altitude, and
- Months of the year.

The method presented a flexible, tolerant of imprecise data, model for nonlinear functions of arbitrary complexity, blended with conventional control techniques, based on natural language but uses many parameters that must be measured and with measurements errors can happen.

On the other hand, energy output estimation for WTs of different power ranges has been the subject of a number of papers as discussed in (Jafarian & Ranjbar, 2010). The wind speed pattern of a region was used to predict the energy output of the turbine in most of these studies. Under many conditions, crisp data are insufficient to represent mathematically real-life decision problems. Indeed, human judgments including preference information are vague or fuzzy in nature and may not be appropriate to represent them by accurate numerical values. The choice of a mathematical model to determine the output power of the renewable sources of energy depends on different parameters. Combining more parameters and considering different weather conditions makes the mathematical modeling too complex to be implemented. In general, the mathematical modeling aims to specify parameters that deliver optimal powers under the best weather conditions while in reality, this is not possible given the different climatic conditions.

The difficulty of management environments and decision problems themselves can lead decision makers to give their rating or judgment to some certain degree which can cause some hesitancy degree. This last represents a vital factor to be considered for the construction of decision problems models expressed with intuitionistic fuzzy sets in comparison with exact numerical values. (Deng-Feng, 2014) (Shu-Ping Wan, 2013). In the same vein, complications inherent in systems that combine cooling, heating, and power make it difficult to attain exact numerical data for quantitative criteria. Fuzzy theory can, therefore, play an essential role in the decision-making process whereby fuzzy evaluation can quantitatively represent the qualitative evaluation from decision makers (Jing, Bai, & Wang, 2012).

This chapter presents a study of PV panels and wind turbines and proposes a simulator based on fuzzy logic control (FLC) to estimate the wind and solar powers under any weather conditions without dealing with complex mathematical formulas. The production of electricity by a photovoltaic generator is related to many parameters, two of which will be used in this controller:

- Temperature and
- Lighting.

For the wind turbine, the amount of energy depends on many characteristics as well. The wind speed will be the one used as input parameter to the Fuzzy controller.

Different climatic conditions were combined to handle all cases of weather changes. PV power and WT power simulations performed in previous research along with the expert knowledge were used to implement the four phases of the FLC simulator: fuzzification, knowledge base, Inference rules and deffuzification.

HISTORY

Numerous developments were conducted recently in renewable energies. These trends reflected the increasing significance of renewable energy in energy markets, investments and industries compared to conventional energy sources such as gas, oil, coal, and nuclear. Renewable energy attained a clear tipping point in the context of global energy supply by the end of 2010. In a number of countries, renewable energy distribution had known a rapid increase in supplying total energy—including heat and transport.

Figure 1. Average annual growth rates of renewable energy capacity, end of 2004 to 2009

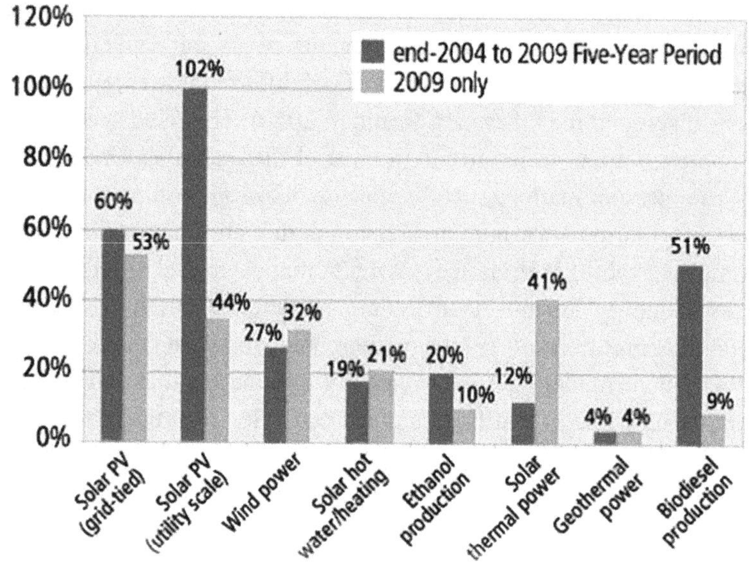

In 2008 and 2009, over half of total global investment in new power generation was issued to the new renewable power capacity (Renewables 2010 Global Status Report, 2010).

Since 2000, Grid-connected solar photovoltaic (PV) technologies have known an important growth by an average of 60 percent every year. From 2005 to 2009, all other types of renewable technologies marked virtually a constant high growth year-after-year. Wind power capacity was growing annually by an average of 27 percent, solar hot water was growing annually by 19 percent annually, and ethanol production was increasing annually by 20 percent. The Biomass and geothermal energies have also known an important growth during these years (Renewables 2010 Global Status Report, 2010) especially for home and business heating.

For all the benefits it presents, renewable energy substitutes conventional fuels in four distinct markets: transport fuels, power generation, hot water and space heating, and off-grid energy services. For these reasons, the global renewable energy capacity knew an important growth at rates of 10–60 percent annually for many technologies during the five year period from the end of 2004 through 2009 as shown in Figure 1.

ADVANTAGES AND DISADVANTAGES OF RENEWABLE ENERGIES

Renewable energies are inexhaustible sources of energy related to their site and can be used independently (used in the same place where they are produced). They don't require transformers, underground channels and distribution networks. They are constantly replenished and will never run out in terms of human lifetimes. They resist to the different weather conditions such as:

- Rain,
- Snow, and
- Wind.

Compared to fossil fuels, the use of renewable energy can reduce the green house gas (GHG) emissions depending heavily on the specific geographical circumstances. In addition to that, renewable energies do not contaminate the nature and destroy the landscape with towers, stations and power lines. For these reasons, new techniques are being developed to improve their performance and accurately predict their powers (Nema, Nema, & Rangnerak, 2009).

Renewable energy technologies require units of equipment or appliances that can capture a renewable energy resource and convert a significant portion of it into useful heat, electricity, or mobility. A wide range of these technologies exists but developments are still continuing to develop new units and improve their designs to offer lower capital costs, better reliability and improved conversion efficiencies (Sims, 2014).

SOLAR AND WIND ENERGIES

Solar Energy

Almost every source of energy that isn't a fossil fuel is a form of renewable energy. This type of energy comes either directly or indirectly from the sun (Sims, 2014). The power from the sun intercepted by the earth is about 1.8×10^{11} MW, which is many times larger than the present rate of all the energy consumption (Bhubaneswari, Iniyan, & Goicc, 2011). Sunlight or solar energy (or photovoltaic energy) is the direct conversion of sunlight into electricity without interference of any heat engine. This phenomenon of photovoltaic effect was highlighted in 1839 by the French scholar Alexandre Becquerel (Bhubaneswari, Iniyan, & Goicc, 2011).

Solar energy is inexhaustible with adequate financial support since it provides energy as long as the sun shines (approximately 4.5 billion years) it requires no toxic fuel and relatively little maintenance, and is capable of becoming directly competitive with conventional technologies in many fields. Despite the benefits that the solar energy presents, there are two limitations of this energy:

- The amount of solar radiation reaching the earth's surface is not constant. The amount of sunlight varies depending on the site location where the solar system is implemented, the time of the day, the time of the year, and also the variable weather conditions.
- Since the sun doesn't deliver the required energy in any place at any time, a large system implementation surface area is required to collect the energy at an important rate.

Solar energy can be converted into electricity in two ways:

- Thermal energy and
- photovoltaic energy, which will be considered in next sections.

Photovoltaic Energy

The simplest method for the production of solar energy is the photovoltaic conversion, which consist on transforming the solar energy into electrical energy by solar collectors. Panels and arrays of panels are sets of Individual PV cells. These panels can be used in a variety of applications ranging from single small cells that charge calculator and watch batteries, to systems that power single houses, to large power plants covering many areas. Photovoltaic (PV) Panels uses photovoltaic cells to collect the energy delivered by the sun which is then converted into electrical energy in the form of continuous current. This energy can be stored in accumulators such as batteries.

The power supplied by the PV cells depends on the intensity of the lighting and the voltage at the borders of the cell. An association in series of several cells (to increase the voltage) gives a module and an association in series or parallel of several modules (to increase the current) makes a solar panel (Gergaud, 2002). The series-parallel wiring is therefore used to make a PV generator (or PV Panel) as shown in Figure 2.

A photovoltaic power generation systems are rated in peak kilowatts (KWp) which represents the maximum amount of electrical power a system is expected to deliver under the best Test Conditions (TTC) (Bhubaneswari, Iniyan, & Goicc, 2011).

Photovoltaic Panels (PV) Composition

Solar panel electricity system, also known as solar photovoltaic (PV), captures the sun's energy using photovoltaic cells. The cells are composed of two silicon layers, a P-type silicon ("p" for positive) and an N-type silicon ("n" for negative).

Lighting photons are absorbed by the semiconductor. They transmit their energy to the atoms of the PN junction by freeing the electrons which creates electrons (N loads) and holes (P loads). This phenomena creates a difference between the n- and p- regions that can be measured between the positive and negatives ends of the cell producing a voltage which is in turn used to drive a current though a circuit (Bernard, Sebrao– Oliveira, Laval, & Vaudouer, 2009) as shown in Figure 3 (Natsheh, 2013).

There are many types of cells as explained in (Pandeya, Tyagi, Selvaraj, Rahim, & Tyagi, 2016) but the mostly used type of is the silicon based solar cells. Panels composed of these cells live the longest and provides an efficiency rate between 15 and 20%.

Figure 2. Solar panels

Figure 3. Solar cells composition

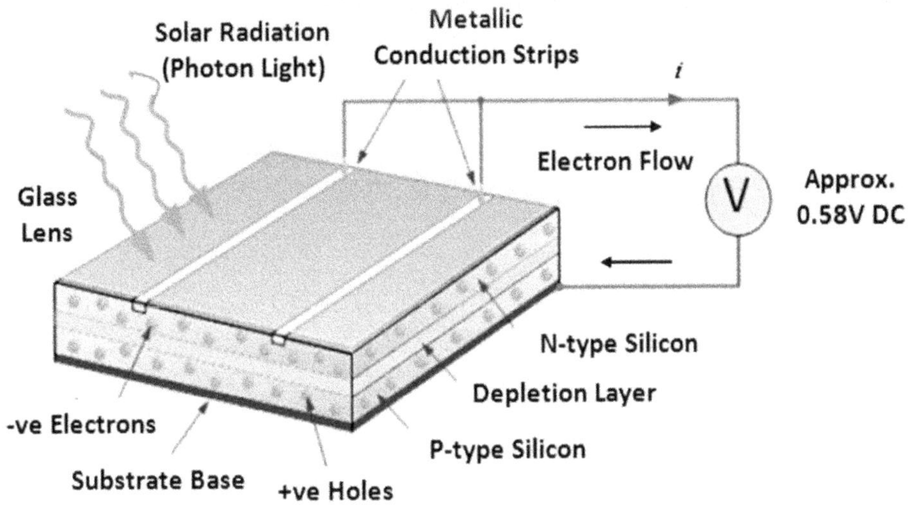

An association in series of several cells (to increase the voltage) gives a module and an association in series or parallel of several modules (to increase the current) makes a solar panel (Gergaud, 2002). The series-parallel wiring is therefore used to get a PV generator (or PV Panel) as Figure 4 shows.

Wind Energy

The sun's heat also drives the winds, whose energy, is captured with wind turbines (Figure 5). A wind turbine extracts energy from the wind and transforms it at the same time to available energy. The wind energy is the primary source of energy that has been exploited by humans for at least the past 3000 years. By the 20th century wind power was used to generate a mechanical power that can pump water or

Figure 4. Cells, modules, and photovoltaic panel

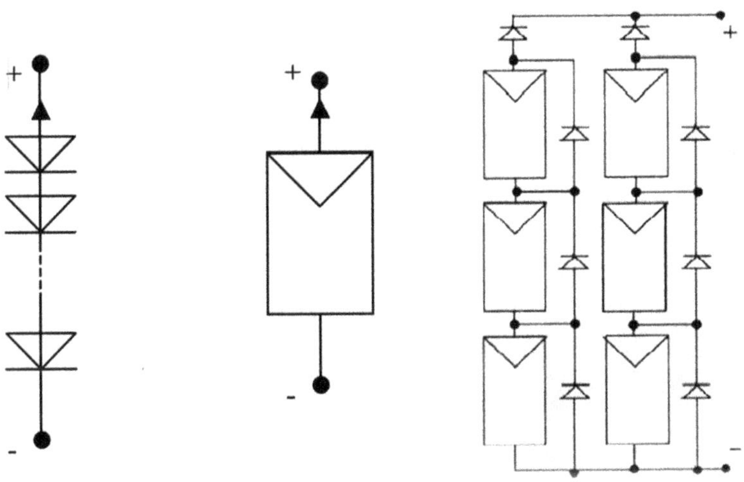

grind grain. To produce a more reliable source of power, wind energy was replaced by fossil fuel fired engines and electrical grids at the beginning of modern industrialization. At the beginning of the 1970s, and the raise of oil price, the interest to use the wind power appeared again with the focus on providing electrical energy rather than mechanical energy. By the end of the 1990s, wind energy has become one of the most essential sustainable energy resources (Ackermann & Lennart, 2000).

Wind energy is exploited to propellers installed on top of pylons and running under the action of the wind. The rotation of the propeller operates a system generating electricity. The whole system is called a wind turbine.

Figure 5. A wind generator

Wind Turbine Composition

A wind turbine is basically built of three main components as shown in Figure 6., a mechanical turbine (left) which includes a low speed rotor and blades, a gearbox (middle) and an electric generator (right).

The use of asynchronous generators with double power supply and a segmental command presents excellent performance for the conversion of energy in wind turbines, since they offer the possibility to operate at variable wind speeds, which eventually allows to significantly reduce efforts on the mechanical part of the wind turbine (WT), and control active and reactive power of the machine, in order to extract the maximum power from the wind (Hafiane, Sabor, & Gualous, 2015). The wind power extracted can be determined by a mathematical formula as described in next section.

Hybrid Renewable Energy Systems (HRES)

A hybrid renewable energy system (HRES) usually consists of two or more renewable energy sources combined together in one electrical system. With this type of combination, it will be possible to overcome the reliability problem of each source, increase the system's total energy efficiency and provide a greater balance in energy supply. With the advances in renewable energies technology, HRES are becoming a useful source of energy and a better substitution of fossil fuels in future energy scenarios (Pérez-Navarro, et al., 2015).

The most common type of these HRES is the one combining solar and wind energy systems (PV-Wind). Solar and wind energies have the advantage of complementing each other (Courtecuisse, et al., 2012):

- Solar radiation provides the energy during cleared days (usually days with little wind), while during cloudy and windy days; it is the wind that provides sufficient energy to operate the turbine. Also, wind turbines can't operate in high or low wind speeds.

Figure 6. Wind turbine model composition [74]

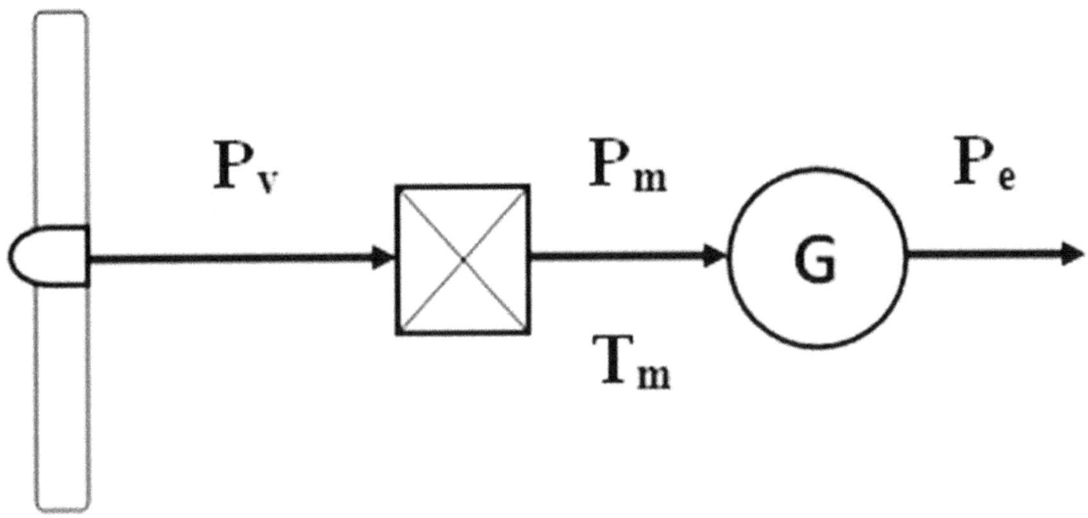

HRESs composed of Photovoltaic (PV) panels, a wind turbine (WT), and one or two energy storage systems (ESS) have a long lifetime and normally low maintenance cost. They also offer many advantages over other generation systems:

- Low pollution,
- High efficiency,
- Reusability of exhaust heat, and
- On-site installation (Atia, Fahmey, Ninet, & Dorrah, 2012).

For all the benefits they present, these systems are becoming more popular for remote area power generation applications due to advances in renewable energy technologies and subsequent rise in prices of petroleum products (Ounnar, Naamane, Pujo, & M'Sirdi, 2012). They also present the most cost-effective and reliable way to produce power since they generate a little environmental disturbance through the use of clean sources of energy. Advances in WT and PV energy technologies have increased their use in these systems as mentioned in (Mermoud, 2010) and (El-Shatter, Eskander, & El Hagry, 2006). Figure 7 shows an example of a hybrid renewable energy system.

In addition to the advantages they present, PV-WT HRES also became the most cost-effective and consistent way to produce power. HRESs composed of PV panels and wind turbines along with storage units such as batteries are the most commonly used systems due to the benefits they represent. Many studies have been completed to simulate and optimize their design as indicated in Pedrazzi, Zini, & Tartarini (2010), Tao & Bruno (2009), and Kuo-Hao & Grace (2015). These systems are becoming popular for remote area power generation applications due to advances in renewable energy technologies and subsequent rise in prices of petroleum products (Ounnar, Naamane, Pujo, & M'Sirdi, 2012). To manage the flow of energy between the input sources and the output sources of energy while satisfying the load, an energy management system (EMS) is required. EMS requires the load demand of the proposed site where the HRES will be implemented and the powers of the renewable sources of energy. To overcome the problems of using complex mathematical formulas, a fuzzy modeling is proposed to predict the PV power and the wind powers.

Figure 7. Example of a hybrid renewable energy system

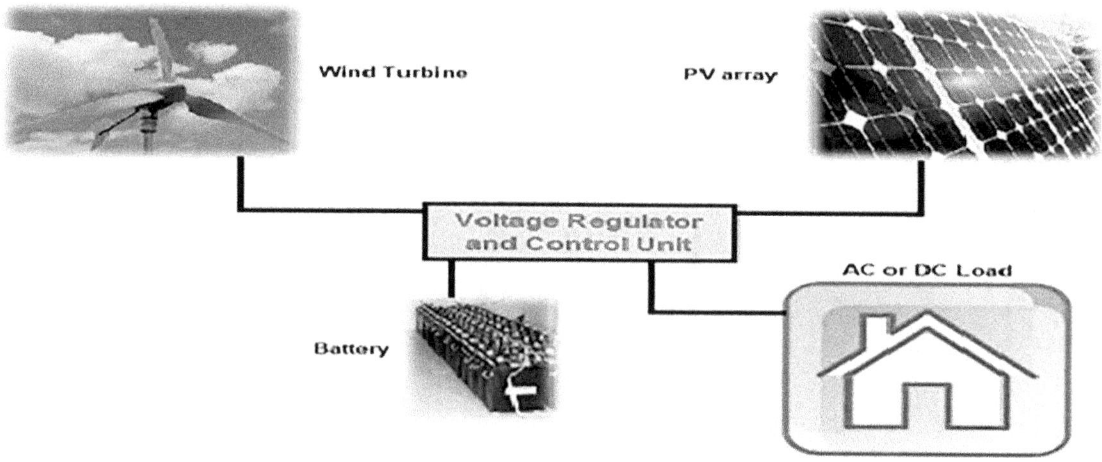

PV AND WT POWERS MATHEMATICAL MODELING

PV Power Mathematical Modeling

The photovoltaic effect converts directly the sunlight into electrical energy. The photovoltaic energy consists on emerging the difference of the potential at the ends of a semiconductor material's structure produced by the absorption of light. The power supplied by the PV cell depends on the intensity I of the lighting and the voltage V at the borders of the cell.

$$P_{pv} = V \times I \tag{1}$$

The equations to determine the power of the PV panels depend on the factors used. In (Aziz, 2006), the power depends on the number of cells connected. For a PV panel composed of M cells connected in series and N columns of cells connected in parallel, the output power P_{pv} is calculated by the equation bellow:

$$P_{pv} = M \times N \times V_{cel} \times I_{cel} \tag{2}$$

where V_{cel} is the voltage at the borders of the cell and I_{cel} is the current generated by the cell.

The equation can be determined in a different way if other factors are considered. The PV power can be determined by the following formula as described in (Garcıa, Torreglosa, Fernandez, & Jurado, 2013):

$$P_{pv} = I \times \eta_{pv} \times \eta_{red} \times S_{pv} \tag{3}$$

where I is the sun irradiation (W/m2), η_{pv} is the efficiency of the panel, η_{red} is the efficiency reduction factor due to the shades and orientation, and S_{pv} represents the total area of the installed panels (m^2).

More geographical features such as latitude, longitude, meteorological parameters like mean duration of the sunshine per hour and temperature were used in (Rizwan, Majid, Kirmani, & Kothari, 2014) to predict the solar energy power as follows:

$$P_{\max} = I_s \times V_{oc} \times Fill\ Factor \tag{4}$$

where I_s represents the short circuit current, V_{oc} is the open circuit voltage and the fill factor parameter is calculated as follows:

$$Fill\ factor = \frac{P_{\max}\left(rated\right)}{I_s\left(at\ 100W / m^2\right) \times V_{oc}\left(rated\right)} \tag{5}$$

More mathematical modeling formulas of the solar power were used in Chen, Gooi, and Wang (2013), and Alam (2009) where other parameters were considered.

Considering different factors to estimate the solar power will make the mathematical formulas more complex.

WT Power Mathematical Modeling

The amount of energy depends on many characteristics, such as the speed ratio of the rotor blade (in degrees), the wind speed, the blade pitch angle (in degrees), the rotation of the blades and the wind speed (Mengi & Hakkı, 2012; Laid, 2010). The wind carries kinetic energy that is transformed by the energy generator into electrical energy. The aerodynamic power at the level of the turbine rotor is given by the following equation:

$$P_{turbine} = \frac{1}{2}\rho\pi R^2 C_p\left(\lambda,\beta\right)V^3 \tag{6}$$

where ρ is the air density, its typical value is 1.225 kg/m^3, R is the radius of the turbine blades which depends on the turbine chosen, Cp is the performance coefficient of the turbine. It is a variable that depends on λ the tip speed ratio of the rotor blade (in degrees), and β the blade pitch angle (in degrees) (Gergaud, 2002; Smaïl & Benatiallah, 2011; Kendouli, Nabti, Abed, & Benalla, 2011), and V is the wind speed which varies with weather conditions in (m/s).

In some cases and to simplify the equation, the speed ratio of the blades and the blades pitch angle are ignored which means that the Cp coefficient is constant (also called the Betz limit). The Betz' law states that a turbine cannot capture more than 16/27 (59.3%) of the kinetic energy in wind (Gijis & Kuik, 2007). The performance coefficient called Betz's coefficient is a constant factor equals to 16/27 (0.593). Therefore, the wind energy per unit of area (W/m^2) depends only on the wind speed.

In other cases, the speed ratio of the blades and the blades pitch angle are calculated and their values are used to determine the value of the Cp coefficient as shown by the equation below (Hafiane, Sabor, & Gualous, 2015):

$$C_p\left(\lambda,\beta\right) = \left(0.44 - 0.0167\beta\right)\sin\left(\frac{\pi\left(\lambda-3\right)}{15-0.3\beta}\right)_0 - 0.00184\left(\lambda-3\right)\beta \tag{7}$$

where λ is calculated with the equation in function of the wind speed V, the radius of the blades R and the turbine speed Ω as shown by the next equation:

$$\lambda = \frac{\Omega t R}{V} \tag{8}$$

The evolution of the performance coefficient Cp is specific to each turbine. For these reasons, the coefficient can be determined by different equations as described in Bhattacharjee and Roy, (2016); Smaïl and Benatiallah (2011), and Kendouli, Nabti, Abed, & Benalla (2011).

Predictions of how much energy will be produced by a turbine are important. This means that accurate calculations of the energy are very important to balancing the energy. By including more parameters into the equation, results will be more accurate but will make the formula more complicated and therefore difficult to implement.

FUZZY LOGIC CONTROL (FLC)

Definition

Fuzzy control refers to the incorporation of the theory of Fuzzy Logic in the application of control. It was developed to improve the capabilities of industrial automation, and is suitable for control level tasks generally performed in Programmable Controllers (PC) (IEC, 1997).

Fuzzy Control is based on linguistic rules that applies knowledge rather than analytical (either empirical or theoretical) models. Fuzzy Control can be integrated in situations where available knowledge can be used to improve processes and execute a variety of tasks. It is used when the available process model is not precise or when the analytical model is too difficult to evaluate (e.g. multiple input multiple output systems) or when the model is too complicated to evaluate it in real time (IEC, 1997). Human experience can be integrated directly in fuzzy control which represents another advantage of it.

Fuzzy logic controllers, compared with other methods, are well adapted tools to manage complex hybrid systems where we cannot obtain or use accurate models. Their approach is used to represent the uncertainty in inputs or the behavior of the HRES by the fuzzification of uncertain variables and assigning membership functions based on preference or experience (Elhadidy & Shaahid, 2000).

Although other intelligent methods such as genetic algorithms and neural networks can perform just as well as fuzzy logic in many applications, fuzzy logic has the advantage that the solution to the problem can be cast in terms that human operators can understand, so that their experience can be used in the design of the controller. This limits the automation of tasks that are already successfully completed by humans.

Phases of FLC

Many control tasks are apparently simple for humans, but they create a continuous challenge for machines. Examples of such systems include:

- Walking through a cluttered environment,
- Lifting fragile objects, or
- Parking a car.

Normally humans are able to deal with vague and imprecise data. To replicate the control actions of a human operator, a model must be built so that it is capable to deal with vague information.

In the conventional control, the amount of control is determined at a number of data entries using a set of equations to express the control of the whole process. Expressing the human experience in the form of a mathematical formula is a very difficult task, if not impossible. Fuzzy logic provides a simple tool to interpret this experience in reality (Dernoncourt, 2011). As mentioned in the section before, a fuzzy system is a repository of fuzzy expert knowledge that can reason about data in vague terms instead of precise Boolean logic. Expert knowledge is represented by a collection of fuzzy membership functions and a set of fuzzy rules, known as the rule-base, of the form:

IF (conditions are fulfilled) THEN (consequences are inferred)

The FLC structure resembles a knowledge base controller except that FLC uses the principles of the theory of fuzzy sets in its representation of data and its logic. A typical fuzzy system is composed of four main parts, a fuzzifier, a knowledge base, an inference engine, and a defuzzifier. The basic configuration of a fuzzy system is shown in Figure 8.

Fuzzification Phase

This is the first step of the FLC implementation. The crisp control variables are measured and converted into their corresponding linguistic variables. The degree of membership of each input for each state variable is defined using the membership function chosen in the system.

Knowledge Base

The knowledge base contains knowledge about input and output fuzzy partitions. This step includes the definitions of the membership functions of for each variable to be controlled as well as the necessary fuzzy rules to specify the control objectives along with the predefined linguistic variables.

Inference Engine

This is the stage where the fuzzy rules previously defined are applied on the variables and the membership value of each is thus obtained. Unfortunately, the handling of unspecified rules can generate a large number of errors. It should be able to simulate human decision making and influence the actions of control based on fuzzy logic.

Figure 8. Schematic diagram of the FLC building blocks

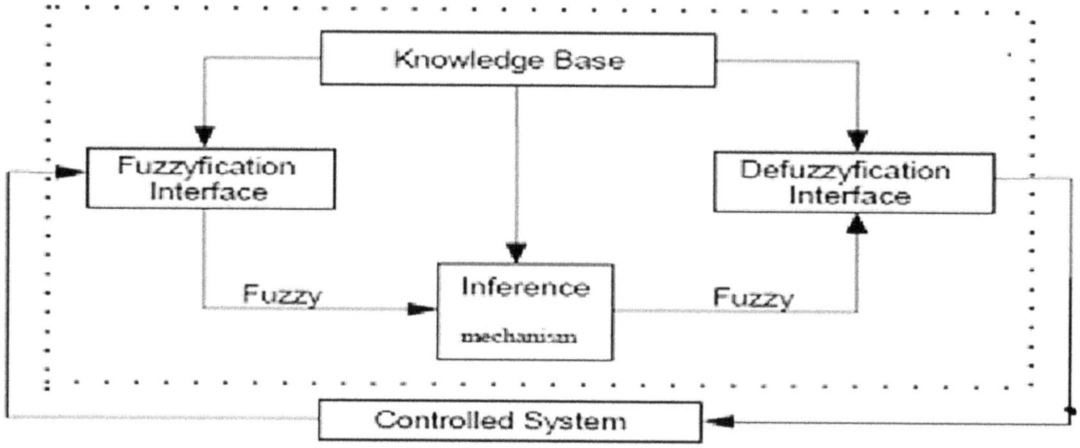

Defuzzification Phase

As an output, the fuzzy system gives several fuzzy output values according to the membership functions that were operated by the system. Since every control system requires a crisp output value, the system must convert in this step the fuzzy output value of the linguistic variables generated from the inference rules into numerical values and therefore, presenting more precise values as a result (Balmisse, 2002) .

The prediction of the PV power and the WT power was done by a computer algorithm based of FLC as explained in the next section.

PREDICTION BASED ON FUZZY LOGIC CONTROL

To predict the WT and PV powers under any weather conditions without dealing with complex mathematical formulas, a modeling algorithm based on Fuzzy logic control (FLC) was developed. The algorithm was implemented by an open source Java library called jFuzzyLogic. Fuzzy control language was incorporated into the library to define the linguistic variables, implements the fuzzy sets and inputs the inference rules. jFuzzyLogic package proposes a fully functional and complete implementation of fuzzy inference systems (FIS), and provides a programming interface (API) and an Eclipse plug-in in order to make it easier to design and test FCL code.

To presume that for any particular input there is only one dominant fuzzy subset, the Trapezoid shape was chosen to represent the membership functions for each input and output linguistic variable. The composition operation by which the controlled output was generated is the Max–Min method. The output membership function of each rule is given by the Minimum. As a system usually requires a non-fuzzy value of control, the defuzzification method applied for this system was the center of gravity (COG) method.

PV Power Prediction

The four phases of the fuzzy controller were implemented to predict the PV power as shown in Figure 9.

Figure 9. Four Phases of the FLC applied to PV power (PPV) prediction

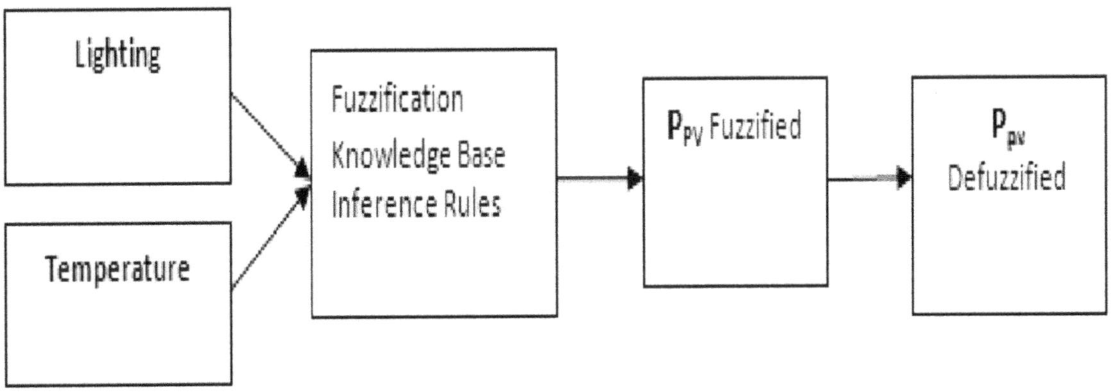

Data Acquisition for One Panel

In a photovoltaic system using solar cells, the available energy depends on the lighting conditions and the temperature (Ounnar, Naamane, Pujo, & M'Sirdi, 2012). Figure 10 shows a simulation of the variation of the optimal photovoltaic power with the lighting and the temperature for a solar panel. The lighting varies between 0 and 300W/m^2 and the temperature values vary between 0 and 50 °C.

The graph represents for each point the optimal PV power (the maximum power it can reach under different weather conditions (Kassmi, Hamdaoui, & Olivié, 2007)).

Fuzzification Phase

The FLC phases were encapsulated in one Function Block (Function Block PV). The Function Block instances contain the specific data of the Fuzzy Control such as the input and output parameters and the Fuzzy Control specific rules and declarations.

To implement the function block for the PV power, linguistic variables and values were chosen as follows:

- **Input Linguistic Variables:**
    - Lighting,
    - Temperature.
- **Output Linguistic Variable:** Photovoltaic Power (P_{PV})
- **Membership Functions Fuzzy Sets:**
    - **Lighting:** Null, VeryLow, Low, Medium, High, VeryHigh,
    - **Temperature:** VeryLow, Low, Medium, High, VeryHigh,
    - **P_{PV}:** Null, VeryLow, Low, Medium, High, VeryHigh.

Figure 10. Optimal Power (PPVOpt) in function of the lighting and the temperature for one cell
For a more accurate representation of this figure, please see the electronic version.

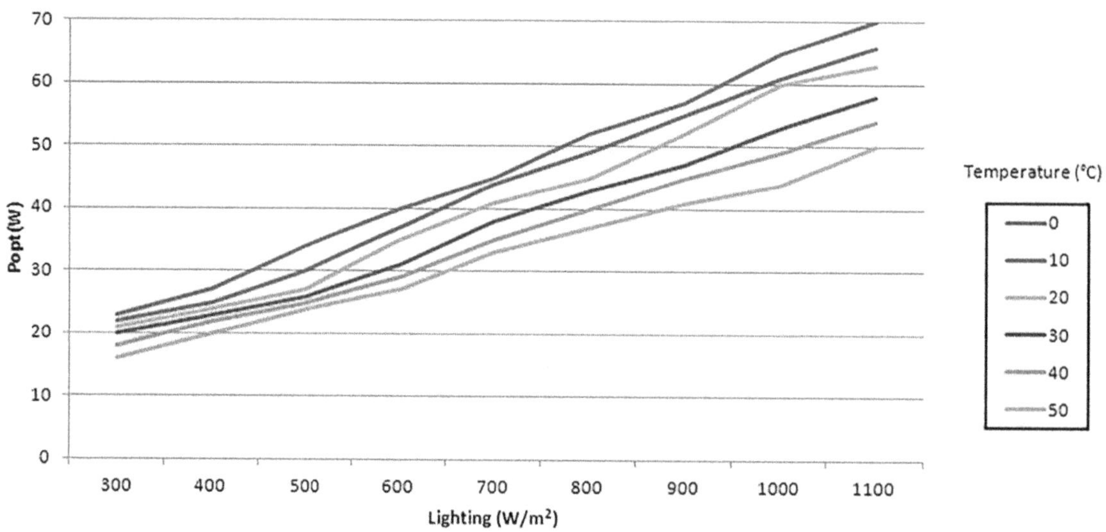

Figure 11. Fuzzy control language building block for modeling PV power (PPV)

```
FUNCTION_BLOCK PV
//fuzzification for one panel of a maximum of 72W
// Define input variables
VAR_INPUT
    Lighting : REAL;
    Temperature: REAL;

END_VAR

// Define output variable
VAR_OUTPUT
    Ppv : REAL;
END_VAR

// Fuzzify input variable 'Lighting'
FUZZIFY Lighting
    TERM Null:=(0, 0);
    TERM VeryLow:=(0, 0) (180, 1) (200,1)(220,0);
    TERM Low := (180, 0) (220, 1) (380, 1) (420,0);
    TERM Medium := (380, 0) (420, 1) (580, 1) (620,0);
    TERM High := (580, 0) (620, 1) (780, 1) (820,0);
    TERM VeryHigh := (780, 0) (820, 1) (1100, 1);

END_FUZZIFY

// Fuzzify input variable 'Temperature'
FUZZIFY Temperature
    TERM VeryLow:=(0, 1) (8, 1) (12,0);
    TERM Low:= (8, 0) (12, 1) (18, 1) (22,0);
    TERM Medium := (18,0) (22,1) (28,1)(32,0);
    TERM High:= (28, 0) (32,1) (38,1) (42,0);
    TERM VeryHigh := (38, 0) (42, 1) (55, 1) ;
END_FUZZIFY
```

The function block implementation for the PV power is shown Figure 11.

Inference Rules

Fuzzy inference rules are the assembly of different rules combining the fuzzy input variables of a system to fuzzy output variables of that system. They are represented by the following form:

If condition1 and/or condidtion2 (and/or) then action output

To determine the inference rules that can predict all different fuzzy values of the PV power, the data taken from the simulation chart above was taken into account. The simulation shows that the optimal power increases when the lighting increases and decreases when the temperature increases. To predict the PV power, each possible weather condition combining input variables (temperature and lighting) with one output value (PV power) was implemented with a fuzzy rule (See Figure 12).

26 inference rules were generated to determine the fuzzy value of the PV for each temperature and lighting. The inference rules were implemented in the function block for the PV power with the FLC programming.

The inference rules were summarized in Table 1.

A fuzzy set is completely characterized by its membership function (MF). It represents the degree of truth of each value in the fuzzy set defined. Due to their simple formulas and computational efficiency, both triangular MFs and trapezoidal MFs have been used extensively. The fuzzy sets related to the Lighting (L) and the Temperature (T) are represented by trapezoidal membership functions as shown in the charts in Figure 13 and Figure 14.

Figure 12. Fuzzy inference rules to generate the PV power implemented the PV Function Block

```
PV.fcl
RULE 1 : IF Lighting IS Null THEN Ppv IS Null;
RULE 2 : IF Lighting IS VeryLow and Temperature is VeryLow THEN Ppv IS Low;
RULE 3 : IF Lighting IS VeryLow and Temperature is Low THEN Ppv IS VeryLow;
RULE 4 : IF Lighting IS VeryLow and Temperature is Medium THEN Ppv IS VeryLow;
RULE 5 : IF Lighting IS VeryLow and Temperature is High THEN Ppv IS VeryLow;
RULE 6 : IF Lighting IS VeryLow and Temperature is VeryHigh THEN Ppv IS VeryLow;
RULE 7 : IF Lighting IS Low and Temperature is VeryLow THEN Ppv IS Medium;
RULE 8 : IF Lighting IS Low and Temperature is Low THEN Ppv IS Low;
RULE 9 : IF Lighting IS Low and Temperature is Medium THEN Ppv IS Low;
RULE 10 : IF Lighting IS Low and Temperature is High THEN Ppv IS Low;
RULE 11 : IF Lighting IS Low and Temperature is VeryHigh THEN Ppv IS Low;
RULE 12 : IF Lighting IS Medium and Temperature is VeryLow THEN Ppv IS High;
RULE 13 : IF Lighting IS Medium and Temperature is Low THEN Ppv IS Medium;
RULE 14 : IF Lighting IS Medium and Temperature is Medium THEN Ppv IS Medium;
RULE 15 : IF Lighting IS Medium and Temperature is High THEN Ppv IS Medium;
RULE 16 : IF Lighting IS Medium and Temperature is VeryHigh THEN Ppv IS Medium;
RULE 17 : IF Lighting IS High and Temperature is VeryLow THEN Ppv IS High;
RULE 18 : IF Lighting IS High and Temperature is Low THEN Ppv IS High;
RULE 19 : IF Lighting IS High and Temperature is Medium THEN Ppv IS High;
RULE 20 : IF Lighting IS High and Temperature is High THEN Ppv IS Medium;
RULE 21 : IF Lighting IS High and Temperature is VeryHigh THEN Ppv IS Medium;
RULE 22 : IF Lighting IS VeryHigh and Temperature is VeryLow THEN Ppv IS VeryHigh;
RULE 23 : IF Lighting IS VeryHigh and Temperature is Low THEN Ppv IS VeryHigh;
RULE 24 : IF Lighting IS VeryHigh and Temperature is Medium THEN Ppv IS VeryHigh;
RULE 25 : IF Lighting IS VeryHigh and Temperature is High THEN Ppv IS VeryHigh;
RULE 26 : IF Lighting IS VeryHigh and Temperature is VeryHigh THEN Ppv IS High;
```

Table 1. Inference rules to generate the fuzzy PV power in function of the lighting and the temperature values

Lighting	**Temperature**				
	VeryLow	**Low**	**Medium**	**High**	**VeryHigh**
Null	Null	Null	Null	Null	Null
VeryLow	Low	VeryLow	VeryLow	VeryLow	VeryLow
Low	Medium	Low	Low	Low	Low
Medium	High	Medium	Medium	Medium	Medium
High	High	High	High	Medium	Medium
VeryHigh	VeryHigh	VeryHigh	VeryHigh	VeryHigh	High

Figure 13. Graphical representation of the fuzzy sets for the Lighting
For a more accurate representation of this figure, please see the electronic version.

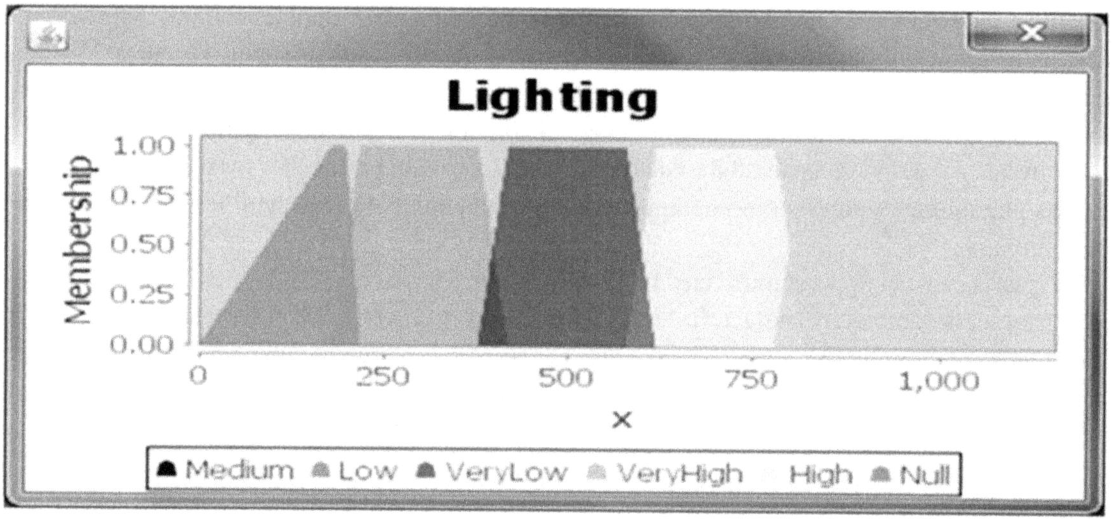

Figure 14. Graphical representation of the fuzzy sets for the Temperature
For a more accurate representation of this figure, please see the electronic version.

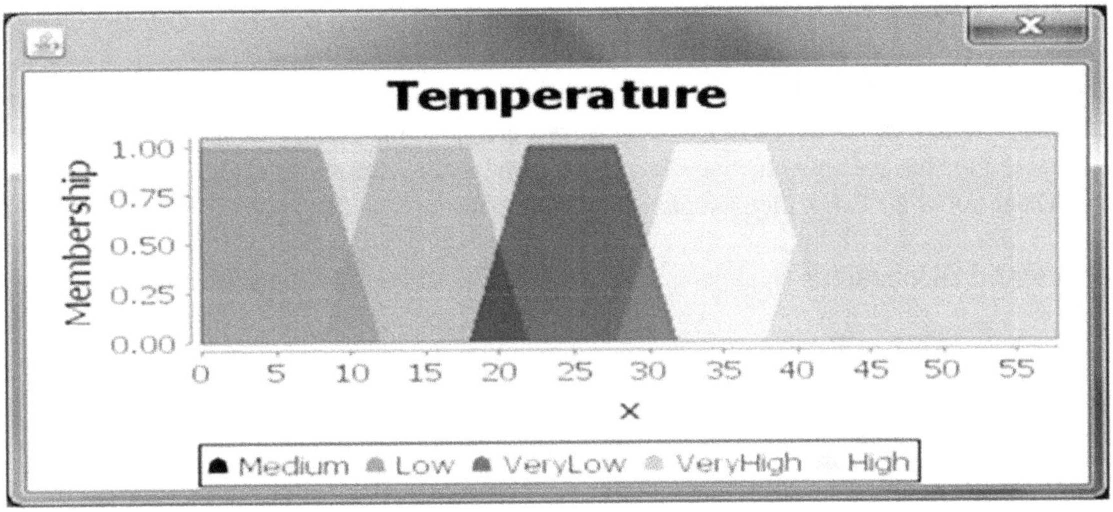

Given the data from the simulation above, a text file was generated containing all possible weather combinations of temperature and lighting. By applying the inference rules generated, the PV power fuzzy values were predicted. The chart in Figure 15 represents the fuzzy sets for each linguistic variable (Null, Very Low, Low, Medium, High and very high) of the PV power (P_{pv}).

DE morgan's Law (pair of transformation rules relating the set operators "union" and "intersection" in terms of each other by means of negation) was applied to use the MIN function for activation method (union) and the MAX function as an accumulation method (Intersection).

As a system usually requires a non-fuzzy value of control, a defuzzification stage is needed.

Figure 15. Graphical representation of the fuzzy sets for the PV power (P_{pv})
For a more accurate representation of this figure, please see the electronic version.

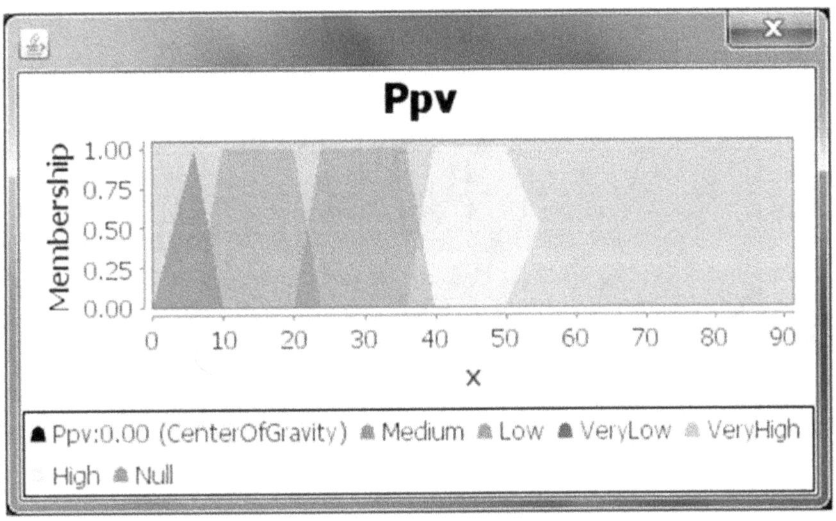

Defuzzification of the Fuzzy PV Power (P_{PV})

Inference rules provide as a fuzzy result. For this result to be directly processed by a control element it must be converted into a numerical value that should better represent the information contained in the fuzzy set. Among all deffuzification methods used, COG (Center of Gravity) is the most commonly used. The fuzzy PV values were calculated by corresponding to the abscissa of the center of gravity of the surface of the membership function characterizing the fuzzy set from the aggregation of the conclusions. The result of the deffuzification phase is shown in the next section.

Results and Discussions

After the deffuzification steps, the PV power generated by the fuzzy modeling ($P_{pv\_simulated}$) was compared to the optimal PV values ($P_{pv\_Optimal}$) generated by the simulation above. Results are shown in the comparaison chart in Figure 16.

As shown from the chart, the predicted PV power values are less than the optimal P_v power values as those last are taken under best test conditions the two curves converge as shown from the figure.

The PV power predicted will be used as an input power for the proposed HERS.

WT Power Prediction

The four phases of the fuzzy logic controller were implemented to predict the WT power as shown in the Figure 17.

To implement the FLC phases, a WT power simulation is required.

Figure 16. Comparison chart of the optimal PV power (Ppv_Optimal) taken from the simulation and the deffuzified PV power (Ppv_simulated) predicted by the fuzzy modeling

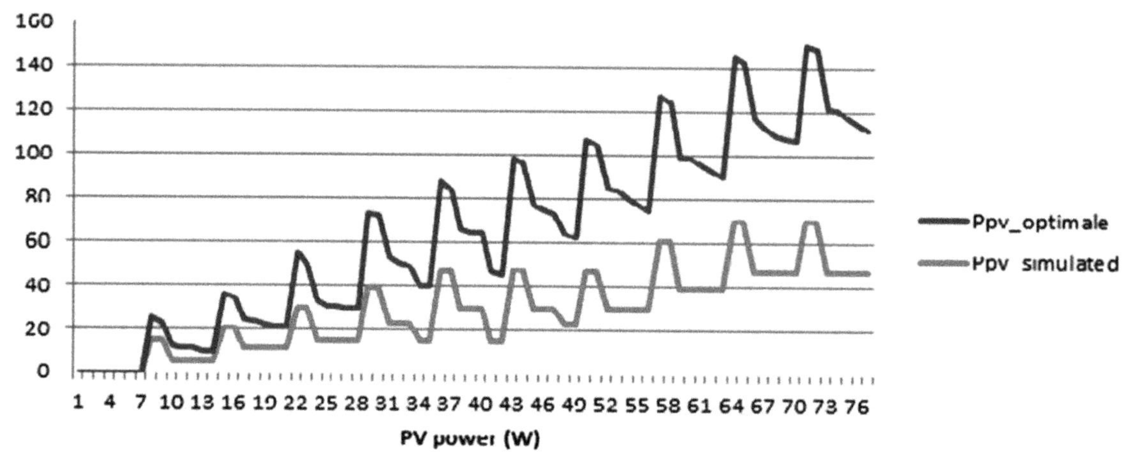

Figure 17. Four phases of the FLC applied to WT power (Pw) prediction

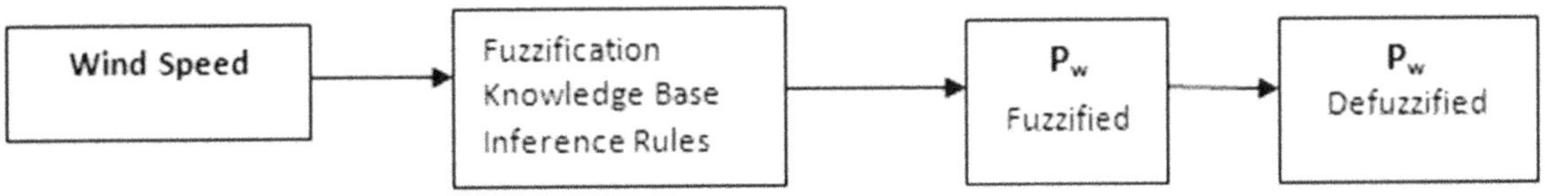

Figure 18. WT power generation [84]

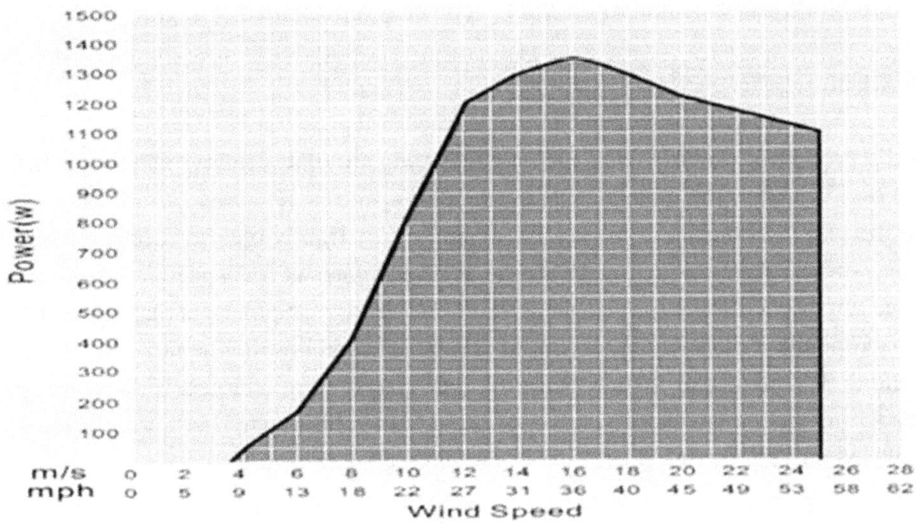

Data Acquisition for the Generic WT

To assign the membership functions for each cluster, appropriate weather data are required. The power generation curve for the generic wind turbine used with a maximum power rate of 1KW is shown in Figure 18.

Fuzzification Phase

The four phases of the fuzzy controller are encapsulated in one Function Block (Function Block Pw). The Function Block instances contain the specific data of the Fuzzy Control such as the input and output parameters and the Fuzzy Control specific rules and declarations.

To implement the function block for the Pw power, fuzzy variables and linguistic values for the Wind Turbine input and output were specified as follows:

- **Input Linguistic Variable:** Wind speed (Speed)
- **Output Linguistic Variable:** Wind Power (P_w).
- **Membership Functions Fuzzy Sets:**
  - **Speed:**
    - VVeryLow,

    - Verylow,
    - Low,
    - Medium,
    - VeryMedium,
    - High,
    - Very High.
- **Pw:** Null, Very low, Low, Medium, High.

The function block implementation for the P_W power is represented in Figure 19.

Inference Rules Generated

The graphical representation of the wind speed/wind power simulation above shows that the wind power status can be divided into three phases: when the wind speed is very low (wind speed below 3.5m/s) the blades of the wind turbine are exposed to wind loads very lightly but they don't produce any energy. When the wind speed is between 3.5 and 16m/s, the rotor rotates at low speed and begins to extract energy. When the wind speed is between 16m/s and 25 m/s the rotor reaches its maximum intended speed. It is the optimal operation of the wind turbine. It is kept constant despite the increase of the speed of the cross wind due to the use of a speed box and a regulation system of charges. If the wind speed is above 25m/s the blades are placed in the off position, parallel to the cross wind. In this way, they don't produce any energy. Based on that, the inference rules were generated in Figure 20.

Figure 19. Fuzzy control language building block for modeling wind power (P_w)

```
wind.fcl
// Block definition (there may be more than one block per file)
FUNCTION_BLOCK power

// Define input variables
VAR_INPUT
    Speed : REAL;

END_VAR

// Define output variable
VAR_OUTPUT
   Pw : REAL;
END_VAR

// Fuzzify input variable 'Wind Speed: speed'
FUZZIFY Speed
    TERM VVeryLow := (0, 1) (2.5, 1) (3.5, 0);
    TERM VeryLow := (2.5, 0) (3.5, 1) (5, 1) (7,0);
    TERM Low := (5, 0) (7, 1) (9, 1) (11,0);
    TERM Medium := (9, 0) (11, 1) (13,1) (14,0);
    TERM VeryMedium := (13, 0) (14, 1) (17,1)(19, 0);
    TERM High := (17,0) (19,1)(24,1)(26,0);
    TERM VeryHigh := (24,0) (26,1)(30,1);
END_FUZZIFY

// Defzzzify output variable 'Wind Power: Pw'
DEFUZZIFY Pw
    TERM Null := (0,0);
    TERM VeryLow := (0, 1) (50, 1)(250,0);
    TERM Low := (50, 0) (250, 1) (900,1)(1100,0);
    TERM Medium := (900, 0)(1100, 1) (1200,1) (1350,0);
    TERM High := (1200, 0)(1350,1)(1400,1);

    // Use 'Center Of Gravity' defuzzification method
    METHOD : COG;
```

Figure 20. Fuzzy inference rules to generate the Pw power implemented the Pw Function Block

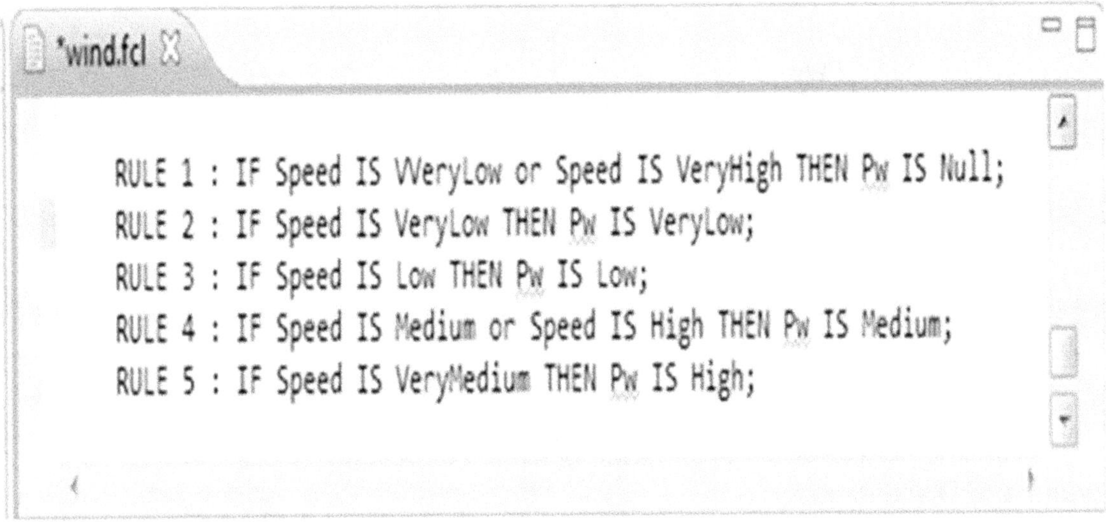

To simulate the human decision making and influence the control actions based on fuzzy logic, the max-min method of fuzzification is used to set the fuzzy rules of the controller. 5 inference rules were generated to determine the fuzzy value of the Pw for each wind speed. The inference rules were summarized in Table 2.

The membership functions related to the Wind Speed and the Wind Power are represented in the charts in Figure 21.

Defuzzification of the Fuzzy WT Power (P_W)

For the WT power, the COG (Center of Gravity) deffuzification method was also applied to convert the fuzzy Pw values to their corresponding crisp values. The result of the deffuzification phase is shown in the next section.

Table 2. Inference rules to generate the fuzzy wind power in function of the wind speed

Speed	Pw
VVeryLow or VeryHigh	Null
VeryLow	VeryLow
Low	Low
Medium or High	Medium
VeryMedium	High

Figure 21. Graphical representation of the fuzzy sets for the wind speed and the WT power (Pw)

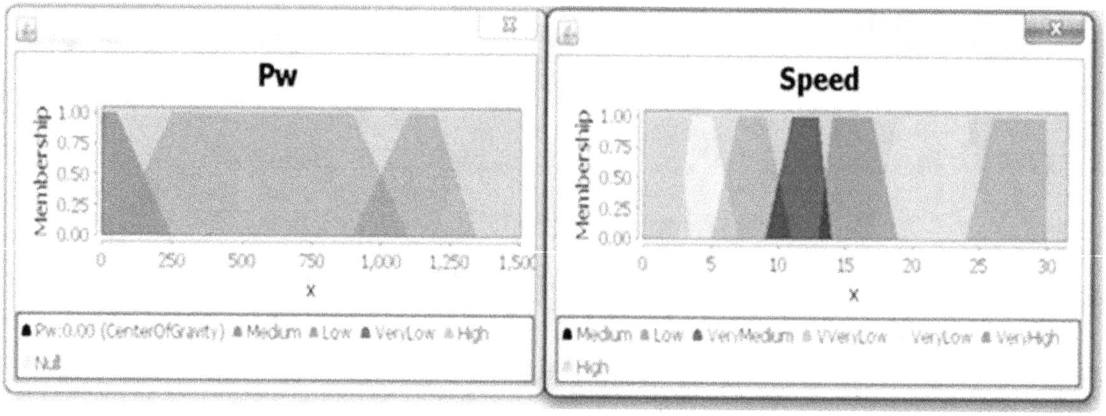

Results and Discussions

In both graphs, it is shown that the two curves converge and that the difference error is very small. This is explained by the choice of the intervals representing each fuzzy set, the choice of the membership functions and the inference rules generated. In all cases, the wind power generated by the fuzzy modeling is less than the optimal wind power taken under the best weather conditions. A comparison between the PV power curve generated by a previous simulation and the power curve generated by the fuzzy controller was also established and showed in Figure 22. Another comparison between the WT power curve given by the manufacturer and the power curve generated by the fuzzy controller was also presented. Results found showed a convergence between the curves. Since they are completing each other, hybrid renewable systems composed of PV panels and wind turbines along with storage units such as batteries are the most commonly used systems.

Figure 22. Comparison between the WT power generated (P_w_simulated) by the fuzzy modeling and the optimal wind

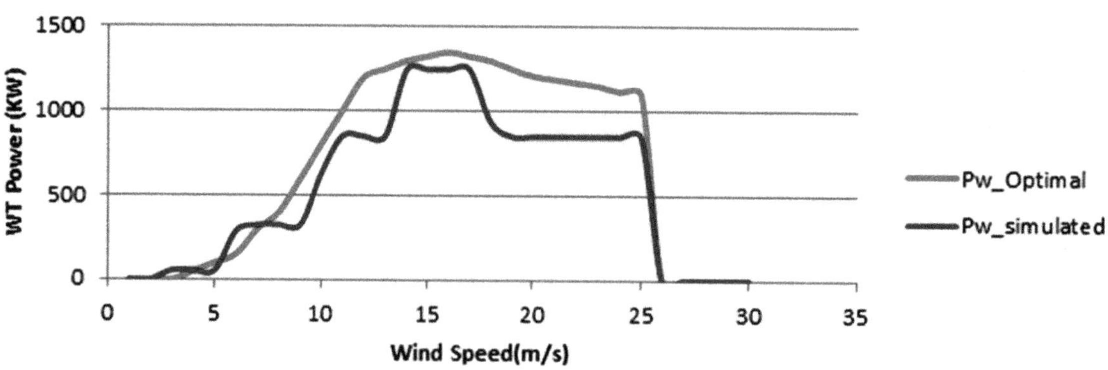

Advantages of the Fuzzy Modeling

The production of renewable energy is to a large extent unpredictable and difficult to control because it is not always feasible to obtain an accurate mathematical model of the controlled system. It is also difficult to predict the behavior of the wind or the sun as this depends on many natural factors. In general, the mathematical modeling aims to specify parameters that deliver optimal powers under the best weather conditions while in reality, this is not possible given the different climatic conditions. The fuzzy modeling adopted in this research presents a way dealing with a real world problem. It is implementing linguistic, non-formal expressed control laws derived from expert knowledge to predict the power of the PV panels and the wind turbine. The PV power fuzzy controller used data taken from a previous simulation based on two natural factors:

- Temperature and
- Solar lighting,

to determine the power of the PV panel for each possible weather condition. The WT fuzzy controller used data taken from a previous simulation based on the wind speed, to determine the power of the WT for each possible weather condition. The fuzzy controllers were implemented by the fuzzy control language (FCL) integrated into an open source Java package. The fuzzy modeling represents a simple way that doesn't require the use and the measurement of many parameters and therefore errors can be minimized and the implementation of more complex formulas can be avoided.

CONCLUSION

The two energy sources discussed in this chapter are very useful renewable energies that can be combined together in one hybrid system to increase system's efficiency. The choice of a mathematical model to determine the output power of each source depends on different parameters. Combining more parameters and considering different weather conditions makes the mathematical modeling too complex to be implemented. Fuzzy modeling, as a solution to these problems, was proposed in this research to predict the PV and WT powers from natural factors, temperature, lighting and wind speed, and under any weather conditions. Results obtained from previous work was compiled and used to determine the values of the fuzzy variables' intervals. The PV and WT power values predicted from the fuzzy modeling were compared to the data obtained from the simulation and the difference error was very small. The fuzzy modeling values obtained can be used as input values for a hybrid renewable system composed of PV panels, a wind turbine and storage units such as batteries that can be implemented to supply a load demand for a site. As perspectives, since complicated systems may require several iterations to find a set of rules resulting in a stable system, combining Neural Networks with fuzzy logic reduces time to establish rules by analyzing clusters of data. Other models and techniques can help in finding an optimal action-selection policy for any given Markov decision process (MDP). These models can be functions such as parametric functions or algorithms such as Q-learning. This last represents an efficient learning technique for generating rules and has the advantage to compare the expected utility of the available actions without requiring a model of the environment.

REFERENCES

Ackermann, T., & Lennart, S. (2000). Wind energy technology and current status: A review. *Renewable & Sustainable Energy Reviews*, *4*(4), 315–374. doi:10.1016/S1364-0321(00)00004-6

Aguiar, R., & Collares-Pereira, M. T. (1992). A time-dependent, autoregressive, Gaussian model for generating synthetic hourly radiation. *Solar Energy*, *49*(3), 167–174. doi:10.1016/0038-092X(92)90068-L

Alam, M. S. (2009). *Real-time maximum power tracking and robust load matching of a stand-alone photovoltaic system*. UMI.

Atia, D. A., Fahmey, F. H., Ninet, M. A., & Dorrah, A. T. (2012). Modeling and Control PV-Wind Hybrid System Based On Fuzzy Logic Control Technique. *TELKOMNIKA*, *10*(3), 431–441. doi:10.12928/telkomnika.v10i3.821

Aziz, A. (2006). *Propriétés électriques des composants électroniques minéraux et organiques, Conception et modélisation d'une chaîne photovoltaïque pour une meilleure exploitation de l'énergie solaire*. Toulouse: Thesis. Retrieved from https://tel.archives-ouvertes.fr/tel-00143296

Balmisse, G. (2002, Septembre). *La logique floue*. Retrieved from GB_Fuzzy.pdf: https://sites.google.com/site/cours1900/logique

Bernard, C., Sebrao-Oliveira, C., Laval, B., & Vaudouer, C. (2009). *Panneau photovoltaïque et algorithme MPPT à base de logique floue*. Academic Press.

Bhattacharjee, C., & Roy, B. K. (2016). Advanced fuzzy power extraction control of wind energy conversion system for power quality improvement in a grid tied hybrid generation system. *IET Generation, Transmission and Distribution, 10*(5), 1179-1189.

Bhubaneswari, P., Iniyan, S., & Goicc, R. (2011). A review of solar photovoltaic technologies. *Renewable & Sustainable Energy Reviews*, *15*(3), 1625–1636. doi:10.1016/j.rser.2010.11.032

Chen, S. X., Gooi, H. B., & Wang, M. Q. (2013). Solar radiation forecast based on fuzzy logic and neural networks. *Renewable Energy*, *60*, 195–201. doi:10.1016/j.renene.2013.05.011

Courtecuisse, V., Sprooten, J., Robyns, B., Petit, M., Francois, B., & Deuse, J. (2012). A methodology to design a fuzzy logic based supervision of Hybrid Renewable Energy Systems. *Mathematics and Computers in Simulation*, *81*(2), 208–224. doi:10.1016/j.matcom.2010.03.003

D.-F., Li. (2014). *Decision and Game Theory in Management with Intuitionistic Fuzzy Sets*. Heidelberg, Germany: Springer.

Dernoncourt, F. (2011, April 18). *Introduction a la logique floue*. Retrieved 04 06, 2016, from http://franck-dernoncourt.developpez.com/tutoriels/algo/introduction-logique-floue/

El-Shatter, T. F., Eskander, M. N., & El Hagry, M. T. (2006). Energy flow and management of a hybrid wind/PV/fuel cell generation system. *International Journal of Sustainable Energy*, *47*(9-10), 91–106. doi:10.1080/14786450600631483

Elhadidy, M., & Shaahid, S. (2000). Parametric study of hybrid (wind plus solar plus diesel). *Renewable Energy, 21*, 129–139. doi:10.1016/S0960-1481(00)00040-9

Fritzsche, K., Zejli, D., & Tanzler, D. (2011). The relevance of global energy governance for Arab countries: The case of Morocco. *Energy Policy, 39*(8), 4497–4506. doi:10.1016/j.enpol.2010.11.042

Garcıa, P., Torreglosa, J. P., Fernandez, L. M., & Jurado, F. (2013). Optimal energy management system for standalone wind turbine/photovoltaic/hydrogen/ battery hybrid system with supervisory control based on fuzzy logic, international journal of hydrogen energy. *International Journal of Hydrogen Energy, 38*(33), 14146–14158. doi:10.1016/j.ijhydene.2013.08.106

Gergaud, O. (2002). *Modélisation énergétique et optimisation économique d'un système de production éolien et photovoltaïque couplé au réseau et associé à un accumulateur.* Retrieved 4 26, 2014, from http://www.enrdd.com/documents/documents/Construction-Ecoconstruction/Developpement%20Durable/TH%C3%88SE%20de%20DOCTORAT%20O.GERGAUD.pdf

Gijis, A. M., & Kuik, V. (2007). The Lanchester-Betz-Joukowsky limit. *Wind Energy (Chichester, England), 10*(3), 289–291. doi:10.1002/we.218

Hafiane, M., Sabor, J., & Gualous, H. (2015). Gestion d'énergie d'une Eolienne par une approche mode glissant Application sur le site Cap Sim à Essaouira (Maroc). *Congrès international de génie industriel - EducationCM (CIGIMS).* FEZ.

IEC. (1997). Programmable Controllers-Part 7: Fuzzy Control Programming. International Electrotechnical Commission (IEC).

Jafarian, M., & Ranjbar, A. (2010). Fuzzy modeling techniques and artificial neural networks to estimate annual energy output of a wind turbine. *Renewable Energy, 35*(9), 2008–1014. doi:10.1016/j.renene.2010.02.001

Jing, Y.-Y., Bai, H., & Wang, J.-J. (2012). A fuzzy multi-criteria decision-making model for CCHP systems driven by different energy sources. *Energy Policy, 42*, 286–296. doi:10.1016/j.enpol.2011.11.085

Kassmi, K., Hamdaoui, M., & Olivié, F. (2007). Conception et modélisation d'un système photovoltaïque adapté par une commande MPPT analogique. *Revue des Energies Renouvelables, 10*(4), 451–462.

Kendouli, F., Nabti, K., Abed, K., & Benalla, H. (2011). Modélisation, simulation et contrôle d'une turbine éolienne à vitesse variable basée sur la génératrice asynchrone à double alimentation. *Revue des Energies Renouvelables, 14*(1), 109–120.

Kuo-Hao, C., & Grace, L. (2015). Optimal design of hybrid renewable energy systems using simulation optimization. *Simulation Modelling Practice and Theory, 52*, 40–51. doi:10.1016/j.simpat.2014.12.002

Laid, Z. (2010). *Etude technique d'un système d'énergie hybride photovoltaïque-éolien hors réseau.* Retrieved from http://www.umc.edu.dz/theses/electrotec/ZAR5745.pdf

Maafi, A., & Adnane, A. (1989). A two-state markovian model of global irradiation suitable for photovoltaic conversion. *Solar & Wind Technology, 6*(3), 247–252. doi:10.1016/0741-983X(89)90076-3

Mengi, O., & Hakkı, I. (2012). Fuzzy logic control for a wind/battery renewable energy production system. *Turkish Journal of Elecrical Engineering & Computer Sciences*, *20*(2), 187–206.

Mermoud, A. (2010). *Le potentiel Photovoltaïque*. Retrieved from http://etat.geneve.ch/dt/SilverpeasWebFileServer/Potentiel_PV_ScanE_250510.pdf?ComponentId=kmelia197&SourceFile=1275571643860.pdf&MimeType=application/pdf&Directory=Attachment/Images/

Mora-Lopez, L., & Sidrach De Cardona, M. (1998). Multiplicative arma models to generate hourly series of global irradiation. *Solar Energy*, *63*(5), 283–291. doi:10.1016/S0038-092X(98)00078-4

Natsheh, E. M. (2013). *Hybrid Power Systems Energy Management Based on Artificial Intelligence* (Thesis). Manchester Metropolitan University, School of Engineering. Manchester, UK.

Nema, P., Nema, R. K., & Rangnerak, S. (2009). A current and future state of art development of hybrid energy system using wind and PV-solar:A Review. *Renewable & Sustainable Energy Reviews*, *13*(8), 2096–2103. doi:10.1016/j.rser.2008.10.006

Ounnar, F., Naamane, A., Pujo, P., & M'Sirdi, N. (2012). Pilotage multicritère d'un système énergétique multi-sources et multiutilisateurs. 1er Congrès International en Génie Industriel et Management des Systèmes. Marseille-France.

Pandeya, A. K., Tyagi, V. V., Selvaraj, J., Rahim, N. A., & Tyagi, S. K. (2016). Recent advances in solar photovoltaic systems for emerging trends and advanced applications. *Renewable & Sustainable Energy Reviews*, *53*, 859–884. doi:10.1016/j.rser.2015.09.043

Pedrazzi, S., Zini, G., & Tartarini, P. (2010). Complete modeling and software implementation of a virtual solar hydrogen hybrid system. *Energy Conversion and Management*, *51*(1), 122–129. doi:10.1016/j.enconman.2009.08.036

Pérez-Navarro, A., Alfonso, D., Ariza, H. E., Cárcel, J., Correcher, A., Escrivá-Escrivá, G., & Vargas, C. et al. (2015). Experimental verification of hybrid renewable systems as feasible energy source. *Renewable Energy*, *86*, 384–391. doi:10.1016/j.renene.2015.08.030

Piette, F. (2011). *La logique floue, intérêts et limites*. Retrieved 4 6, 2016, from http://www.ferdinand-piette.com/blog/2011/08/la-logique-floue-interets-et-limites(2010)

Rizwan, M., Majid, J. I., Kirmani, S., & Kothari, D. (2014). Fuzzy logic based modeling and estimation of global solar energy using meteorological parameters. *Energy*, *70*, 685–691. doi:10.1016/j.energy.2014.04.057

Ross, T. J. (2009). *Fuzzy Logic with Engineering applications*. John Wiley & Sons.

Santos, J. M., Pinazo, J., & Canada, J. (2003). Methodology for generating daily clearness index values kt starting from the monthly average daily value Kt determining the daily sequence using stochastic models. *Renewable Energy*, *28*(10), 1523–1544. doi:10.1016/S0960-1481(02)00217-3

Shu-Ping, W., & Deng-Feng, L. (2013). Fuzzy LINMAP approach to heterogeneous MADM considering comparisons of alternatives with hesitation degrees. *Omega: The International Journal of management. Science*, *41*(6), 925–940.

Sims, R. E. (2014). *Renewable Energy Potential*. Retrieved from http://seat.massey.ac.nz/research/centres/energy/documents/Renewable%20Energy%20Potential.pdf

Smaïl, M., & Benatiallah, A. (2011). Modélisation d'une chaîne de conversion éolienne équipée D'une machine asynchrone double alimentation. *Revue des Energies Renouvelables ICES*, *11*, 233–238.

Tao, Z., & Bruno, F. (2009). Modeling and control design of hydrogen production process for an active hydrogen/wind hybrid power system. *International Journal of Hydrogen Energy*, *34*(1), 21–30. doi:10.1016/j.ijhydene.2008.10.030

Compilation of References

Abbasbandy, S., & Hajjari, T. (2009). A new approach for ranking of trapezoidal fuzzy numbers. *Computers & Mathematics with Applications (Oxford, England)*, *57*(3), 413–419. doi:10.1016/j.camwa.2008.10.090

Abbasgholizadeh Rahimi, S., & Jamshidi, A. (2014a). *Prioritization of Organ Transplant Patients using Analytic Network Process.* Montreal, Canada: Academic Press.

Abbasgholizadeh Rahimi, S., Jamshidi, A., Ruiz, A. & Ait-Kadi, D. (2014b). *Applied Methods in Prioritization of Patients in Surgery Waiting Lists.* Montreal, Canada: Academic Press.

Abbasgholizadeh Rahimi, S., Jamshidi, A., Ruiz, A., & Ait Kadi, D. (2016). A new dynamic integrated framework for surgical patients prioritization considering risks and uncertainties. *Decision Support Systems*, *88*, 112–120. doi:10.1016/j.dss.2016.06.003

Abdi, M. R. (2009). Fuzzy multi-criteria decision model for evaluating reconfigurable machines. *International Journal of Production Economics*, *117*(1), 1–15. doi:10.1016/j.ijpe.2008.06.009

Abdurazzag, A. A., & Vladimir, M. (2008). Fuzzy Logic Based Algorithm for Uniprocessor Scheduling.*Proceedings of the International Conference on Computer and Communication Engineering* (pp. 499-504). IEEE.

Ackermann, T., & Lennart, S. (2000). Wind energy technology and current status: A review. *Renewable & Sustainable Energy Reviews*, *4*(4), 315–374. doi:10.1016/S1364-0321(00)00004-6

Addis, B., Carello, G., Grosso, A., & Tanfani, E. (2015). *Operating room scheduling and rescheduling: a rolling horizon approach.* Flex Serv Manuf J.

Aguiar, R., & Collares-Pereira, M. T. (1992). A time-dependent, autoregressive, Gaussian model for generating synthetic hourly radiation. *Solar Energy*, *49*(3), 167–174. doi:10.1016/0038-092X(92)90068-L

Ahlert, M., & Kliemt, H. (2013). Problems of priority change in kidney allocation and beyond. *The European Journal of Health Economics*, *14*(3), 383–390. doi:10.1007/s10198-012-0382-y PMID:22358456

Akkucuk, U. (2015). *Handbook of Research on Waste Management Techniques for Sustainability*. Hershey, PA: IGI Global.

Aktas, H., & Cagman, N. (2007). Soft sets and soft groups. *Inf Sci*, *177*(13), 2726–2735. doi:10.1016/j.ins.2006.12.008

Alam, B., Doja, M. N., & Biswas, R. (2009). Improving the performance of fair share scheduling algorithm using fuzzy logic.*International Conference on Advances in Computing, Communication and Control* (pp. 567-570). doi:10.1145/1523103.1523216

Alam, B., Doja, M. N., Biswas, R., & Alam, M. (2011). Fuzzy Priority CPU Scheduling Algorithm. *International Journal of Computer Science and Issues*, *8*(6), 386–390.

Alam, M. S. (2009). *Real-time maximum power tracking and robust load matching of a stand-alone photovoltaic system.* UMI.

Al-Ebbinia, L., Oztekin, A., & Chen, Y. (2016). FLAS: Fuzzy lung allocation system for US-based transplantations. *European Journal of Operational Research*, *248*(3), 1051–1065. doi:10.1016/j.ejor.2015.08.001

Ali, M., Feng, F., Liu, X., Min, W. K., & Shabir, M. (2009). On some new operations in soft set theory. *Computers & Mathematics with Applications (Oxford, England)*, *57*(9), 1547–1553. doi:10.1016/j.camwa.2008.11.009

Ali, M., & Shabir, M. (2010). Comments on De Morgan's law in fuzzy soft sets. *J. Fuzzy Math.*, *18*(3), 679–686.

Allahviranloo, T., & Firozja, M. A. (2010). Ranking of fuzzy numbers by a new metric. *Soft Computing*, *14*(7), 773–782. doi:10.1007/s00500-009-0464-7

Allahviranloo, T., & Saneifard, R. (2012). Defuzzification method for ranking fuzzy numbers based on center of gravity. *Iranian Journal of Fuzzy Systems*, *9*(6), 57–67.

Alonso, J. A., & Lamata, M. T. (2006). Consistency in the Analytic Hierarchy Process: A new approach. *International Journal of Uncertainty, Fuzziness and Knowledge-based Systems*, *14*(4), 445–459. doi:10.1142/S0218488506004114

An, L., & Wilfred, N. (2005). Vague Sets or Intuitionistic Fuzzy Sets for Handling Vague Data: Which One Is Better? Lecture Notes in Computer Science, 376, 401–416.

Andersson, A., Omberg, M., & Svedlund, M. (2006). Triage in the emergency department—a qualitative study of the factors which nurses consider when making decisions. *Nursing in Critical Care*, *11*(3), 136–145. doi:10.1111/j.1362-1017.2006.00162.x PMID:16719019

Arbos, L. C. (2002). Design of a rapid response and high efficiency service by lean production principles: Methodology and evaluation of variability of performance. *International Journal of Production Economics*, *80*(2), 169–183. doi:10.1016/S0925-5273(02)00316-X

Argon, N., & Ziya, S. (2009). Priority Assignment Under Imperfect Information on Customer Type Identities. *Manuf Serv Oper Manag*, *11*, 674–693.

Asady, B., & Zendehman, A. (2007). Ranking of fuzzy numbers by distance minimization. *Applied Mathematical Modelling*, *31*(11), 2589–2598. doi:10.1016/j.apm.2006.10.018

Ashour, O. & Okudan, G. (2010). *Patient sorting through emergency severity index and descriptive variables' utility.* Academic Press.

Ashour, O., & Kremer, G. (2016). Dynamic patient grouping and prioritization: A new approach to emergency department flow improvement. *Health Care Management Science*, *19*(2), 192–205. doi:10.1007/s10729-014-9311-1 PMID:25487711

Askin, R. G., & Standridge, R. (1993). *Modeling and Analysis of Manufacturing Systems.* New York, NY: Wiley.

Atanassov, K. (1999). Intuitionistic Fuzzy Sets: Theory and Applications. Physica-Verlag.

Atanassov, K. (2000). Intuitionistic Fuzzy Sets: Theory and Applications. In Studies in Fuzziness and Soft Computing. Physica-Verlag.

Atanassov, K. T. (1983). *Intuitionistic fuzzy sets.* In Seventh Scientific Session of ITKR, Sofia.

Atanassov, K. (1986). Intuitionistic fuzzy sets. *Fuzzy Sets and Systems*, *20*(1), 87–96. doi:10.1016/S0165-0114(86)80034-3

Atanassov, K. (1994). Operators over interval-valued intuitionistic fuzzy sets. *Fuzzy Sets and Systems*, *64*(2), 159–174. doi:10.1016/0165-0114(94)90331-X

Atanassov, K. T. (1999). *Intuitionistic fuzzy sets*. Heidelberg, Germany: Springer-Verlag. doi:10.1007/978-3-7908-1870-3

Atanassov, K. T., & Gargov, G. (1989). Interval valued intuitionistic fuzzy sets. *Fuzzy Sets and Systems*, *31*(3), 343–349. doi:10.1016/0165-0114(89)90205-4

Atia, D. A., Fahmey, F. H., Ninet, M. A., & Dorrah, A. T. (2012). Modeling and Control PV-Wind Hybrid System Based On Fuzzy Logic Control Technique. *TELKOMNIKA*, *10*(3), 431–441. doi:10.12928/telkomnika.v10i3.821

Ayag, Z., & Ozdemir, R. G. (2012). Evaluating machine tool alternatives through modified TOPSIS and alpha-cut based fuzzy ANP. *International Journal of Production Economics*, *140*(2), 630–636. doi:10.1016/j.ijpe.2012.02.009

Azar, A. T. (2010). Adaptive Neuro-Fuzzy Systems. In A. T. Azar (Ed.), *Fuzzy Systems*. Vienna, Austria: IN-TECH. doi:10.5772/7220

Azar, A. T. (2010). *Fuzzy Systems*. Vienna, Austria: IN-TECH.

Azar, A. T. (2012). Overview of Type-2 Fuzzy logic systems. *International Journal of Fuzzy System Applications*, *2*(4), 1–28. doi:10.4018/ijfsa.2012100101

Aziz, A. (2006). *Propriétés électriques des composants électroniques minéraux et organiques, Conception et modélisation d'une chaîne photovoltaïque pour une meilleure exploitation de l'énergie solaire*. Toulouse: Thesis. Retrieved from https://tel.archives-ouvertes.fr/tel-00143296

Bairagi, B., Dey, B., Sarkar, B., & Sanyal, S. (2012). A novel multiplicative model of multi criteria analysis for robot selection. *International Journal on Soft Computing. Artificial Intelligence and Applications*, *1*(3), 1–9.

Balmisse, G. (2002, Septembre). *La logique floue*. Retrieved from GB_Fuzzy.pdf: https://sites.google.com/site/cours1900/logique

Barua, B. (2015). *Waiting Your Turn: Wait Times for Health Care in Canada, 2015 Report*. Fraser institute.

Barua, B., & Esmail, N. (2013). *Why are we waiting so long? Health care wait times nearly double over the past two decades to 18.2 weeks*. Fraser Institute.

Bayazit, O. (2012). An Application of The AHP In Selection of a Third Party Logistics Provider.*Orlando International Academic Conference*, (pp. 315-317).

Beamon, B. M. (1999). Designing the green supply chain. *Logist. Inform. Manage*, *12*(4), 332–342. doi:10.1108/09576059910284159

Beikkhakhian, Y., Javanmardi, M., Karbasian, M., & Khayambashi, B. (2015). The application of ISM model in evaluating agile suppliers selection criteria and ranking suppliers using fuzzy TOPSIS-AHP methods. *Expert Systems with Applications*, *42*(15-16), 6224–6236. doi:10.1016/j.eswa.2015.02.035

Bernard, C., Sebrao-Oliveira, C., Laval, B., & Vaudouer, C. (2009). *Panneau photovoltaïque et algorithme MPPT à base de logique floue*. Academic Press.

Bhangale, P. P., Agrawal, V. P., & Saha, S. K. (2004). Attribute based specification, comparison and selection of a robot. *Mechanism and Machine Theory*, *39*(12), 1345–1366. doi:10.1016/j.mechmachtheory.2004.05.020

Bhattacharjee, C., & Roy, B. K. (2016). Advanced fuzzy power extraction control of wind energy conversion system for power quality improvement in a grid tied hybrid generation system. *IET Generation, Transmission and Distribution*, *10*(5), 1179-1189.

Bhatti, R. S., Kumar, P., & Kumar, D. (2010). Analytical modeling of third party service provider selection in lead Logistics provider environments. *Journal of Modelling in Management*, *5*(3), 275–286. doi:10.1108/17465661011092641

Bhubaneswari, P., Iniyan, S., & Goicc, R. (2011). A review of solar photovoltaic technologies. *Renewable & Sustainable Energy Reviews*, *15*(3), 1625–1636. doi:10.1016/j.rser.2010.11.032

Bo, Y., & Klir, G. J. (1995). *Fuzzy Sets and Fuzzy Logic Theory and Applications*. PHI Learning Private Limited.

Boren, F. E., Gene, S., Kurt, M., & Akay, D. (2009). A multi-criteria intuitionistic fuzzy group decision making for supplier selsection with TOPSIS method. *Expert Systems with Applications*, *36*(8), 11363–11368. doi:10.1016/j.eswa.2009.03.039

Brans, J. P., & Source, Ph. V. (1985). A Preference Ranking Organisation Method: The PROMETHEE Method for Multiple Criteria Decision-Making. *Management Science*, *31*(6), 647–656. doi:10.1287/mnsc.31.6.647

Brauers, W. K. M., & Zavadskas, E. K. (2006). The MOORA method and its application to privatization in transition economy. *Control and Cybernetics*, *35*(2), 443–468.

Brauers, W. K. M., & Zavadskas, E. K. (2010). Project management by MULTI-MOORA as an instrument for transition economies. *Technological and Economic Development of Economy*, *16*(1), 5–24. doi:10.3846/tede.2010.01

Brinch Hansen, R. (1973). *Operating System Principles*. ACM Digital Library, Prentice Hall.

Bui, H., Emeneker, W., Apon, A., Hoffman, D., & Dowdy, L. (2010). Fair share Scheduling – a case study. In *Proceedings of 11th LCI Conference* (1-10).

Burillo, P., Bustince, H., & Mohedano, V. (1994). Some definitions of intuitionistic fuzzy number. Fuzzy based expert systems, fuzzy Bulgarian enthusiasts, Sofia, Bulgaria.

Bustince, H., & Burillo, P. (1995). Correlation of interval-valued intuitionistic fuzzy sets. *Fuzzy Sets and Systems*, *74*(2), 237–244. doi:10.1016/0165-0114(94)00343-6

Bustince, H., & Burillo, P. (1996). Vague sets are intuitionistic fuzzy sets. *Fuzzy Sets and Systems*, *79*(3), 403–405. doi:10.1016/0165-0114(95)00154-9

Butterfield, W. (1968). *Priorities in medicine*. London: The Nuffield Provincial Hospitals Trust.

Cabral, I., Grilo, A., & Machado, V. C. (2012). A decision-making model for lean, agile, resilient and green supply chain management. *International Journal of Production Research*, *50*(17), 4830–4845. doi:10.1080/00207543.2012.657970

Cabrerizo, F. J., Perez, I. J., & Herrera Viedma, E. (2010). Managing the consensus in group decision making in an unbalanced fuzzy linguistic context with incomplete information. *Knowledge-Based Systems*, *23*(2), 169–181. doi:10.1016/j.knosys.2009.11.019

Cai, X. S., & Han, L. G. (2014). Some induced Einstein aggregation operators based on the data mining with interval-valued intuitionistic fuzzy information and their application to multiple attribute decision making. *Journal of Intelligent & Fuzzy Systems*, *27*, 331–338.

Cakir, E., Tozan, H., & Vayvay, O. (2009). A Method For Selecting Third Party Logistic Service Provider Using Fuzzy AHP. *Journal of Naval Science and Engineering*, *5*(3), 38–54.

Caprita, B., Nieh, J., & Chan, W. C. (2005). Group Round-Robin: Improving the Fairness and Complexity of Packet Scheduling.*Proceedings of the 2005 ACM symposium on Architecture for networking and communications systems* (pp. 29-40). doi:10.1145/1095890.1095896

Caputo, A. C., & Pelagagge, P. M. (2008). Capacity upgrades criteria of large-intensive material handling and storage systems: A case study. *Journal of Manufacturing Technology Management*, *19*(8), 953–978. doi:10.1108/17410380810911727

Celik, Y., & Yamak, S. (2013). Fuzzy soft set theory applied to medical diagnosis using fuzzy arithmetic operations. *Journal of Inequalities and Applications*, *82*, 1–9.

Chakraborty, C., & Chakraborty, D. (2006). A theoretical development on a fuzzy distance measure for fuzzy numbers. *Mathematical and Computer Modelling*, *43*(3-4), 254–261. doi:10.1016/j.mcm.2005.09.025

Chan, F. T. S., & Lau, H. (2001). Integration of expert system with analytic hierarchy process for the design of material handling equipment selection system. *Journal of Materials Processing Technology*, *116*(2/3), 137–145. doi:10.1016/S0924-0136(01)01038-X

Chan, F. T. S., Qi, H. J., Chan, H. K., Lau, H. C. W., & Ip, R. W. L. (2003). A conceptual model of performance measurement for supply chains. *Management Decision*, *41*(7), 635–642. doi:10.1108/00251740310495568

Chang, D. Y. (1996). Application of the extent analysis method on fuzzy AHP. *European Journal of Operational Research*, *65*(3), 649–655. doi:10.1016/0377-2217(95)00300-2

Chen, C. T. (2000). Extensions of the TOPSIS for group decision-making under fuzzy environment. *Fuzzy Sets and Systems*, *114*(1), 1–9. doi:10.1016/S0165-0114(97)00377-1

Chen, C. T., Lin, C. T., & Hwang, S. F. (2006). A fuzzy approach for supplier evaluation and selection in supply chain management. *International Journal of Production Economics*, *102*(2), 289–301. doi:10.1016/j.ijpe.2005.03.009

Chen, D. F., Zhang, L., & Jiao, J. (2010). Triagular Fuzzy Number Intuitionistic Fuzzy Aggregation Operators and Their Group Decision Making. *LNAI*, *6320*, 350–357.

Cheng, C. H. (1998). A new approach for ranking fuzzy numbers by distance method. *Fuzzy Sets and Systems*, *95*(3), 307–317. doi:10.1016/S0165-0114(96)00272-2

Cheng, J. H., Chen, S. S., & Chuang, Y. W. (2008). Using Fuzzy Multiple Criterion Methods for Fourth Party Logistics Criteria Selection, *7th WSEAS Int. Conf. on Artificial Intelligence, Knowledge Engineering and Data Bases (AIKED'08)*, (pp. 282-288).

Chen, K. Y., & Wu, W. T. (2011). Applying Analytic Network Process in Logistics Service Provider selection– A Case Study of the industry investing in Southeast Asia. *International Journal of Electronic Business Management*, *9*(1), 24–36.

Chen, S. H., & Hsieh, C. H. (1999). Ranking generalized fuzzy number with graded mean integration representation. In *Proceedings of the Eighth International Conference of Fuzzy Sets and Systems Association World Congress.*

Chen, S. J. (1985). Operations on fuzzy numbers with function principal. *Tamkang Journal of Management Sciences*, *6*(1), 13–25.

Chen, S. J., & Chen, S. M. (2007). Fuzzy risk analysis based on the ranking of generalized trapezoidal fuzzy numbers. *Applied Intelligence*, *26*(1), 1–11. doi:10.1007/s10489-006-0003-5

Chen, S. J., & Hwang, C. L. (1992). *Fuzzy multiple attribute decision making*. Berlin: Springer-Verlag. doi:10.1007/978-3-642-46768-4

Chen, S. M., & Chiou, C. H. (2015). Multiattribute decision making based on interval-valued intuitionistic fuzzy sets, PSO techniques and evidential reasoning methodology. *IEEE Transactions on Fuzzy Systems*, *23*(6), 1905–1912. doi:10.1109/TFUZZ.2014.2370675

Chen, S. M., & Horng, Y. J. (1996). Finding inheritance hierarchies in interval-valued fuzzy concept networks. *Fuzzy Sets and Systems*, *84*(1), 75–83. doi:10.1016/0165-0114(95)00315-0

Chen, S. M., Lee, L. W., Liu, H. C., & Yang, S. W. (2012). Multi-attribute decision making based on interval valued intuitionistic fuzzy values. *Expert Systems with Applications*, *39*(12), 10343–10351. doi:10.1016/j.eswa.2012.01.027

Chen, S. M., & Li, T. S. (2013). Evaluating students answer scripts based on interval-valued intuitionistic fuzzy sets. *Information Sciences*, *235*, 308–322. doi:10.1016/j.ins.2012.12.031

Chen, S. M., & Tan, J. M. (1994). Handling multi-criteria fuzzy decision making problems based on vague sets. *Fuzzy Sets and Systems*, *67*(2), 163–172. doi:10.1016/0165-0114(94)90084-1

Chen, S. X., Gooi, H. B., & Wang, M. Q. (2013). Solar radiation forecast based on fuzzy logic and neural networks. *Renewable Energy*, *60*, 195–201. doi:10.1016/j.renene.2013.05.011

Chen, T. Y. (2015). The inclusion-based TOPSIS method with interval-valued intuitionistic fuzzy sets for multiple criteria group decision making. *Applied Soft Computing*, *26*, 57–73. doi:10.1016/j.asoc.2014.09.015

Chetia, B., & Das, P. (2010). An application of interval valued fuzzy soft set in medical diagnosis. *Int. J. Contemp. Math. Sci.*, *5*(38), 1887–1894.

Chiang, D. A., & Lin, N. P. (1999). Correlation of fuzzy sets. *Fuzzy Sets and Systems*, *102*(2), 221–226. doi:10.1016/S0165-0114(97)00127-9

Chiclana, F., Herrera, F., Herrera-Viedma, E., & Alonso, S. (2004). Induced ordered weighted geometric operators and their use in the aggregation of multiplicative preference relations. *International Journal of Intelligent Systems*, *19*(3), 233–255. doi:10.1002/int.10172

Cholongitas, E., & Burroughs, A. (2012). The evolution in the prioritization for liver transplantation. *Annals of Gastroenterology*, *25*(1). PMID:24713804

Cholongitas, E., Germani, G., & Burroughs, A. K. (2010). Prioritization for liver transplantation. *Gastroenterology & Hepatology Nature Reviews*, *7*(12), 659–668. doi:10.1038/nrgastro.2010.169 PMID:21045793

Christopher, M. (1998). Logistics and supply chain management: Strategies for reducing cost and improving service. *International Journal of Logistics Research and Applications: A Leading Journal of Supply Chain Management*, *2*(1), 103-104.

Chung, S., Lee, A., & Pearn, W. (2005). Analytic network process (ANP) approach for product mix Planning in semiconductor fabricator. *International Journal of Production Economics*, *96*(1), 15–36. doi:10.1016/j.ijpe.2004.02.006

Chung, T. A., & Chung, T. (2002). Ranking fuzzy numbers with an area between the centroid point and the original point. *An International Journal of Computer and Mathematics with Application*, *43*(1-2), 111–117. doi:10.1016/S0898-1221(01)00277-2

Chu, S. J. (2009). Selecting the advanced manufacturing technology using fuzzy multiple attributes group decision making with multiple fuzzy information. *International Journal of Computers & Industrial Engineering*, *57*(3), 1033–1042. doi:10.1016/j.cie.2009.04.011

Chu, T. C., & Lin, Y. C. (2003). A fuzzy TOPSIS method for robot selection. *International Journal of Advanced Manufacturing Technology*, *21*(4), 284–290. doi:10.1007/s001700300033

Chu, T. C., & Varma, R. (2012). Evaluating suppliers via a multiple levels multiple criteria decision making method under fuzzy environment. *Computers & Industrial Engineering*, *62*(2), 653–660. doi:10.1016/j.cie.2011.11.036

Clare, P. (1973). *Waiting Lists: Priority Formulae*. Birmingham, UK: Notes taken at a meeting of the Midlands Health Services Discussion Group.

Claudio, D., & Okudan, G. (2010). Utility function based patient prioritization in the emergency department. *European Journal of Industrial Engineering*, *4*(1), 59–77. doi:10.1504/EJIE.2010.029570

Comas, M., Castells, X., Hoffmeister, L., Román, R., Cots, F., Mar, J., & Espallargues, M. et al. (2008). Discrete-Event Simulation Applied to Analysis of Waiting Lists.Evaluation of a Prioritization System for Cataract Surgery. *Value in Health*, *11*(7), 1203–1213. doi:10.1111/j.1524-4733.2008.00322.x PMID:18494754

Computer Numerical Control. (n.d.) Retrieved from http://wings.buffalo.edu/academic/department/eng/mae/courses/460-564/Course-Notes/CNC%20notes.pdf

Courtecuisse, V., Sprooten, J., Robyns, B., Petit, M., Francois, B., & Deuse, J. (2012). A methodology to design a fuzzy logic based supervision of Hybrid Renewable Energy Systems. *Mathematics and Computers in Simulation*, *81*(2), 208–224. doi:10.1016/j.matcom.2010.03.003

Culyer, A., & Cullis, J. (1976). Some economics of hospital waiting lists in the NHS. *Journal of Social Policy*, *5*(3), 239–264. doi:10.1017/S0047279400004748

Davis, B., & Johnson, S. (1998). Real-time priority scoring system must be used for prioritisation on waiting lists. *British Medical Journal*. PMID:10373189

Day, B. (2013). *Reducing Wait Times for Health Care*. Fraser Institute.

Dennett, E., & Parry, B. (1998). Generic surgical priority criteria scoring system: The clinical reality. *The New Zealand Medical Journal*, *111*, 163–166. PMID:9612483

Dernoncourt, F. (2011, April 18). *Introduction a la logique floue*. Retrieved 04 06, 2016, from http://franck-dernoncourt.developpez.com/tutoriels/algo/introduction-logique-floue/

De, S., Biswas, R., & Roy, A. (2001). An application of intuitionistic fuzzy sets in medical diagnosis. *Fuzzy Sets and Systems*, *117*(2), 209–213. doi:10.1016/S0165-0114(98)00235-8

Deshmukh, S. C. (2013). Preference Ranking Organization Method of Enrichment Evaluation. *International Journal of Engineering Science Invention*, *2*(11), 28–34.

Detcharat, S., & Pongpun, A. (2013). Using DEMATEL Method to Analyze the Causal Relations on Technological Innovation Capability Evaluation Factors in Thai Technology-Based Firms. *International Transaction Journal of Engineering, Management, &. Applied Sciences & Technologies*, *4*(2), 81–103.

Dhamdhere, D. M. (2006). *Operating Systems A concept Based Approach* (2nd ed.). Tata: McGraw-Hill.

Ding, J. F., & Chou, C. C. (2011). Middle managers selection for third-party Logistics service providers. *International Journal of the Physical Sciences*, *6*(3), 610–619.

Dolan, P., & Cookson, R. (2000). A qualitative study of the extent to which health gain matters when choosing between groups of patients. *Health Policy (Amsterdam)*, *51*(1), 19–30. doi:10.1016/S0168-8510(99)00079-2 PMID:11010223

Domènech, M., Adam, P., Tebé, C., & Espallargues, M. (2013). Developing a universal tool for the prioritization of patients waiting for elective surgery. *Journal of Health Policy*, *113*(1-2), 118–126. doi:10.1016/j.healthpol.2013.07.006 PMID:23932414

Dowseya, M., Gunn, J., & Choong, P. (2014). Selecting those to refer for joint replacement: Who will likely benefit and who will not? *Best Practice & Research. Clinical Rheumatology*, *28*(1), 157–171. doi:10.1016/j.berh.2014.01.005 PMID:24792950

Dubois, D., & Prade, H. (1980). *Fuzzy Sets and Systems Theory and Applications*. New York, NY, USA.

Duran, O., & Aguilo, J. (2008). Computer-aided machine-tool selection based on a Fuzzy-AHP approach. *International Journal of Expert Systems with Applications*, *34*(3), 1787–1794. doi:10.1016/j.eswa.2007.01.046

Ekere, N. N., & Hannam, R. G. (1989). An evaluation of approaches to modeling and simulating manufacturing systems. *International Journal of Production Research*, *27*(4), 599–611. doi:10.1080/00207548908942571

Elhadidy, M., & Shaahid, S. (2000). Parametric study of hybrid (wind plus solar plus diesel). *Renewable Energy*, *21*, 129–139. doi:10.1016/S0960-1481(00)00040-9

El-Shatter, T. F., Eskander, M. N., & El Hagry, M. T. (2006). Energy flow and management of a hybrid wind/PV/fuel cell generation system. *International Journal of Sustainable Energy*, *47*(9-10), 91–106. doi:10.1080/14786450600631483

Eltringham, D., & Clare, P. (1973). *Waiting List Management by Computer*. OR Unit Regional Hospital Board.

Feng, F., & Li, Y. (2013). Soft subsets and soft product operations. *Inf Sci*, *232*, 44–57. doi:10.1016/j.ins.2013.01.001

Fieldsa, E., Okudana, G., & Ashour, O. (2013). Rank aggregation methods comparison: A case for triage prioritization. *Expert Systems with Applications*, *40*(4), 1305–1311. doi:10.1016/j.eswa.2012.08.060

Folinas, D. (2012). *Outsourcing Management for Supply Chain Operations and Logistics Service*. Hershey, PA: IGI Global.

Fordyce, A., & Phillips, R. (1970). Waiting list management by computer. *The Hospital*, *66*(9), 303–305.

Fritzsche, K., Zejli, D., & Tanzler, D. (2011). The relevance of global energy governance for Arab countries: The case of Morocco. *Energy Policy*, *39*(8), 4497–4506. doi:10.1016/j.enpol.2010.11.042

Galvin, S., & Gagne. (2009). *Operating systems concepts* (8th ed.). Wiley.

Garcıa, P., Torreglosa, J. P., Fernandez, L. M., & Jurado, F. (2013). Optimal energy management system for standalone wind turbine/photovoltaic/hydrogen/ battery hybrid system with supervisory control based on fuzzy logic, international journal of hydrogen energy. *International Journal of Hydrogen Energy*, *38*(33), 14146–14158. doi:10.1016/j.ijhydene.2013.08.106

Garvin, A. O. (1993). The 12 commandments of environmental compliance,'. *Industrial Engineering (American Institute of Industrial Engineers)*, *25*(9), 18–22.

Gau, W. L., & Buehrer, D. J. (1993). Vague Sets. *IEEE Transactions on Systems, Man, and Cybernetics*, *23*(2), 610–614. doi:10.1109/21.229476

Gergaud, O. (2002). *Modélisation énergétique et optimisation économique d'un système de production éolien et photovoltaïque couplé au réseau et associé à un accumulateur*. Retrieved 4 26, 2014, from http://www.enrdd.com/documents/documents/Construction-Ecoconstruction/Developpement%20Durable/TH%C3%88SE%20de%20DOCTORAT%20O.GERGAUD.pdf

Gerstenkorn, T., & Manko, J. (1991). Correlation of intuitionistic fuzzy sets. *Fuzzy Sets and Systems*, *44*(1), 39–43. doi:10.1016/0165-0114(91)90031-K

Gijis, A. M., & Kuik, V. (2007). The Lanchester-Betz-Joukowsky limit. *Wind Energy (Chichester, England)*, *10*(3), 289–291. doi:10.1002/we.218

Godber, G. (1970). Priorities in medicine. *The Journal of the Royal College of General Practitioners, 20*, 313–322. PMID:5500451

Gokhan, Y. M., & Gorener, A. (2016). Decision Making for Company Acquisition by ELECTRE Method. *Int. J Sup. Chain. Mgt. Exceling Tech Pub, 5*(1), 75–83.

Gol, H., & Catay, B. (2007). Third-party Logistics provider selection: Insights from a Turkish automotive company. *Supply Chain Management: An International Journal, 12*(6), 379–384. doi:10.1108/13598540710826290

Gorzalczany, M. B. (1987). A method of inference in approximate reasoning based on Interval-valued fuzzy sets. *Fuzzy Sets and Systems, 21*(1), 1–17. doi:10.1016/0165-0114(87)90148-5

Govindan, K. (2009). Fuzzy approach for the selection of third party reverse Logistics provider. *Asia Pacific Journal of Marketing and Logistics, 21*(3), 397–416. doi:10.1108/13555850910973865

Govindan, K., & Jepsen, M. B. (2016b). ELECTRE: A comprehensive literature review on methodologies and applications. *European Journal of Operational Research, 250*(1), 1–29. doi:10.1016/j.ejor.2015.07.019

Govindan, K., Khodaverdi, R., & Vafadarnikjoo, A. (2016a). A grey DEMATEL approach to develop third-party logistics provider selection criteria. *Industrial Management & Data Systems, 116*(4), 690–722. doi:10.1108/IMDS-05-2015-0180

Govindan, K., & Murugesan, P. (2011). Selection of third-party reverse logistics provider using fuzzy extent analysis. *Benchmarking: An International Journal, 18*(1), 149–167. doi:10.1108/14635771111109869

Green, F. B., Turner, W., Roberts, S., Nagendra, A., & Wininger, E. (2008). A Practitioner's Perspective on the role of a Third-Party Logistics Provider. *Journal of Business & Economics Research, 6*(6), 9–14.

Green, K., Morton, B., & New, S. (1998). Green purchasing and supply policies: Do they improve companies environmental performance. *Supply Chain Management, 3*(2), 89–95. doi:10.1108/13598549810215405

Grzegorzewski, P. (2003). Distances and orderings in a family of intuitionistic fuzzy numbers. Conference of the European Society for Fuzzy Logic and Technology3rd (EUSFLAT'03), Zittau, Germany.

Grzgorzewski, P. (2004). Distance between intuitionistic fuzzy sets and/or interval fuzzy sets based on Hausdorff metric. *Fuzzy Sets and Systems, 148*(2), 319–328. doi:10.1016/j.fss.2003.08.005

Guha, D., & Chakraborty, D. (2010). A new approach to fuzzy distance measure and similarity measure between two generalized fuzzy numbers. *Applied Soft Computing, 10*(1), 90–99. doi:10.1016/j.asoc.2009.06.009

Guha, D., & Chakraborty, D. (2010). A Theoretical development of distance measure for intuitionistic fuzzy numbers. *International Journal of Mathematics and Mathematical Sciences, 2010*, 1–25. doi:10.1155/2010/949143

Gunasekaran, A. (1999). Agile manufacturing: A framework for research and development. *International Journal of Production Economics, 62*(1-2), 87–105. doi:10.1016/S0925-5273(98)00222-9

Gupta, R., Sachdeva, A., & Bhardwaj, A. (2010). Selection of 3pl Service Provider using Integrated Fuzzy Delphi and Fuzzy TOPSIS.*Proceedings of the World Congress on Engineering and Computer Science 2010 (WCECS 2010)*, (pp. 1092-1097).

Gupta, R., Sachdeva, A., & Bhardwaj, A. (2011). Criteria of Selecting 3pl Provider: A Literature Review. *World Academy of Science. Engineering and Technology, 59*, 2020–2024.

Hadorn, D. (2000). Setting priorities for waiting lists: Defining our terms. *Canadian Medical Association Journal, 163*(7), 857–860. PMID:11033717

Hadorn, D., & Holmes, A. (1997). The New Zealand priority criteria project Part 2: Coronary artery bypass graft surgery. *British Medical Journal, 314*(7074), 135–138. doi:10.1136/bmj.314.7074.135 PMID:9006478

Hafiane, M., Sabor, J., & Gualous, H. (2015). Gestion d'énergie d'une Eolienne par une approche mode glissant Application sur le site Cap Sim à Essaouira (Maroc). *Congrès international de génie industriel - EducationCM (CIGIMS)*. FEZ.

Halgamuge, S. K. (1998). A trainable transparent universal approximator for defuzzification in Mamdani-type neuro-fuzzy controllers. *IEEE Transactions on Fuzzy Systems, 6*(2), 304–314. doi:10.1109/91.669031

Halim Zaim, A. (2003). Design of A Scheduler: Comparison of Different Scheduling Algorithms. *Journal of Electronic and Electrical Engineering, 3*(2), 859–877.

Hamdan, A., & Rogers, K. J. (2008). Evaluating the efficiency of 3PL Logistics operations. *International Journal of Production Economics, 113*, 235–244. doi:10.1016/j.ijpe.2007.05.019

Hassan, M. M. D. (2010). A framework for selection of material handling equipment in manufacturing and logistics facilities. *Journal of Manufacturing Technology Management, 21*(2), 246–268. doi:10.1108/17410381011014396

Hassan, M. M. D. (2014). An evaluation of input and output of expert systems for selection of material handling equipment. *Journal of Manufacturing Technology Management, 25*(7), 1049–1067. doi:10.1108/JMTM-08-2012-0077

Hellendoorn, H., & Thomas, C. (1995). On Quality Defuzzification. In Fuzzy Logic & its Application to Engineering. Information Sciences & Intelligent Systems. doi:10.1007/978-94-009-0125-4_16

Herrera, F., & Herrera, V. E. (2000). Linguistic decision analysis: Steps for solving decision problems under linguistic information. *Fuzzy Sets and Systems, 115*(1), 67–82. doi:10.1016/S0165-0114(99)00024-X

Herrera, F., Herrera-Viedma, E., & Chiclana, F. (2003). A study of the origin and uses of the ordered weighted geometric operator in multicriteria decision making. *International Journal of Intelligent Systems, 18*(6), 689–707. doi:10.1002/int.10106

Herrera, F., Martinez, L., & Sanchez, P. J. (1999). Managing non-homogenous information in group decision making. *European Journal of Operational Research, 116*, 115–132.

Hlupic, V., & Paul, R. J. (1999). Guidelines for selection of manufacturing simulation software. *IIE Transactions, 31*(1), 21–29. doi:10.1080/07408179908969802

Hong, D. H., & Hwang, S. W. (1995). Correlation of intuitionistic fuzzy sets in probability spaces. *Fuzzy Sets and Systems, 75*(1), 77–81. doi:10.1016/0165-0114(94)00330-A

Hsiao, H. I., Kemp, R. G. M., Vorst, V. J. G. A. J., & Omta, S. W. F. (2010). A classification of Logistics outsourcing levels and their impact on service performance: Evidence from the food processing industry. *International Journal of Production Economics, 124*, 75–86. doi:10.1016/j.ijpe.2009.09.010

Hsu, C. W., & Hu, A. H. (2008). Green supply chain management in the electronic industry'. *International Journal of Science and Technology, 5*(2), 205–216.

Hsu, P. F. (2012). Selection model based on ANP and GRA for independent media agencies. *Quality & Quantity, 46*(1), 1–17. doi:10.1007/s11135-010-9323-y

Huang, B., Zhuang, Y. L., & Li, H. X. (2013). Information granulation and uncertainty measures in interval-valued intuitionistic fuzzy information systems. *European Journal of Operational Research, 23*(1), 162–170. doi:10.1016/j.ejor.2013.05.006

Huang, Y., & Oliver, J. (1994). NC milling error assessment and tool path correction.*International Conference on Computer Graphics and Interactive Techniques. Proceedings of the 21st annual conference on computer graphics and interactive techniques*, (pp. 287-294). doi:10.1145/192161.192231

Huiyu, C., & Weiwei, W. (2010). *Green supply chain management for a Chinese auto manufacturer.* Department of Technology and Built Environment, University of GAVLE.

Hult, G. T. M., Ketchen, D. J. K., & Arrelti, M. (2007). Strategic supply chain management: Improving performance through a culture of competitiveness and knowledge development. *Strategic Management Journal, 28*(10), 1035–1052. doi:10.1002/smj.627

Hung, K. C., & Wang, P. K. (2014). An integrated intuitionistic fuzzy similarity measures for medical problems. *International Journal of Computational Intelligent Systems, 7*(2), 327–343. doi:10.1080/18756891.2013.856259

Hwang, C. L., & Yoon, K. (1981). *Multiple Attribute Decision Making Methods and Applications.* Berlin: Springer. doi:10.1007/978-3-642-48318-9

IEC. (1997). Programmable Controllers-Part 7: Fuzzy Control Programming. International Electrotechnical Commission (IEC).

Isiklar, G., Alptekin, E., & Buyukozkan, G. (2007). Application of a hybrid intelligent decision support model in Logistics outsourcing. *Computers & Operations Research, 34*(12), 3701–3714. doi:10.1016/j.cor.2006.01.011

Jafarian, M., & Ranjbar, A. (2010). Fuzzy modeling techniques and artificial neural networks to estimate annual energy output of a wind turbine. *Renewable Energy, 35*(9), 2008–1014. doi:10.1016/j.renene.2010.02.001

Jing, Y.-Y., Bai, H., & Wang, J.-J. (2012). A fuzzy multi-criteria decision-making model for CCHP systems driven by different energy sources. *Energy Policy, 42*, 286–296. doi:10.1016/j.enpol.2011.11.085

Kadhim, S. J., & Al-Aubidy, K. M. (2010). Design and Evaluation of a Fuzzy-Based CPU Scheduling Algorithm. *Information Processing and Management. Communications in Computer and Information Science, 70*, 45–52. doi:10.1007/978-3-642-12214-9_9

Kahraman, C. (2008). Multi-Criteria Decision Making-Theory and Applications with Recent Develoments. Springer Optimization and its Applications, 16.

Kahraman, C., Cebeci, U., & Ulukan, Z. (2003). Multi-criteria supplier selection using fuzzy AHP. *Logistics Information Management, 16*(6), 382–394. doi:10.1108/09576050310503367

Kainuma, Y., & Tawara, N. (2006). A multiple attribute utility theory approach to lean and green supply chain management. *International Journal of Production Economics, 101*(1), 99–108. doi:10.1016/j.ijpe.2005.05.010

Karagul, H., & Albayrakoglu, M. M. (2007). *Selecting a Third-Party Logistics Provider for an Automotive Company: An Analytic Hieararchy Process Model.* Vina Del Mar, Chile: ISAHP.

Karayiannis, N. (2000). Soft learning vector quantization and clustering algorithms based on ordered weighted aggregation operators. *IEEE Transactions on Neural Networks, 11*(5), 1093–1105. doi:10.1109/72.870042 PMID:18249837

Karsak, E. E. (2008). Robot selection using an integrated approach based on quality function deployment and fuzzy regression. *International Journal of Production Research, 46*(7), 723–738. doi:10.1080/00207540600919571

Kassmi, K., Hamdaoui, M., & Olivié, F. (2007). Conception et modélisation d'un système photovoltaïque adapté par une commande MPPT analogique. *Revue des Energies Renouvelables, 10*(4), 451–462.

Kauffman, A., & Gupta, M. M. (1991). *Introduction to fuzzy arithmetic: theory and application.* Van Nostrand Reinhold New York.

Kaufmann, A., & Gupta, M. (1991). Introduction to Fuzzy Arithmetic Theory & Applications. Van Nostrand.

Kay, J., & Lauder, P. (1988). A Fair Share Scheduler. *Communications of the ACM*, *31*(1), 44–55. doi:10.1145/35043.35047

Kendouli, F., Nabti, K., Abed, K., & Benalla, H. (2011). Modélisation, simulation et contrôle d'une turbine éolienne à vitesse variable basée sur la génératrice asynchrone à double alimentation. *Revue des Energies Renouvelables*, *14*(1), 109–120.

Knadel, A. (1986). *Fuzzy mathematical techniques with applications*. Boston, MA: Addison-Wesley Longman Publishing Co.

Koh, S. C. L., & Tan, Z. (2005). Using e-commerce to gain a competitive advantage in 3PL enterprises in China. *International Journal of Logistics Systems and Management*, *1*(2/3), 87–210. doi:10.1504/IJLSM.2005.005970

Korena, Y., & Shpitalni, M. (2010). Design of reconfigurable manufacturing systems. *International Journal of Manufacturing Systems*, *29*(4), 130–141. doi:10.1016/j.jmsy.2011.01.001

Koulouriotis, D. E., & Ketipi, M. K. (2011). A fuzzy digraph method for robot evaluation and selection. *Expert Systems with Applications*, *38*(9), 11901–11910. doi:10.1016/j.eswa.2011.03.082

Kuo-Hao, C., & Grace, L. (2015). Optimal design of hybrid renewable energy systems using simulation optimization. *Simulation Modelling Practice and Theory*, *52*, 40–51. doi:10.1016/j.simpat.2014.12.002

Kuo, M. S. (2011). A novel interval-valued fuzzy MCDM method for improving airlines service quality in Chinese cross-strait airlines. *Transportation Research Part E, Logistics and Transportation Review*, *47*(6), 1177–1193. doi:10.1016/j.tre.2011.05.007

Kurien, G. P., & Qureshi, M. N. (2011). Study of performance measurement practices in supply chain management. *International Journal of Business. Management and Social Sciences*, *2*(4), 19–34.

Laarhoven, P. J. M. V., & Pedrycz, W. (1983). A fuzzy extension of Satty's priority theory. *Fuzzy Sets and Systems*, *11*, 199–227.

Lack, A., Edwards, R., & Boland, A. (2000). Weights for waits: Lessons from Salisbury. *Journal of Health Services Research & Policy*, *5*(2), 83–88. PMID:10947552

Laid, Z. (2010). *Etude technique d'un système d'énergie hybride photovoltaïque-éolien hors réseau.* Retrieved from http://www.umc.edu.dz/theses/electrotec/ZAR5745.pdf

Lai, Y. J., Liu, T. Y., & Hwang, C. L. (1994). TOPSIS for MODM. *European Journal of Operational Research*, *76*(3), 486–500. doi:10.1016/0377-2217(94)90282-8

Lakshmana, V., Nayagam, G., Muralikrishnan, G., & Sivaraman, G. (2011). Multi-criteria decision-making method based on intervalvalued intuitionistic fuzzy sets. *Expert Systems with Applications*, *38*(3), 1464–1467. doi:10.1016/j.eswa.2010.07.055

Lambert, D. M. (2008). *Supply Chain Management: Processes, Partnerships, Performance*. Sarasota, FL: Supply Chain Management Institute.

Langham, S., & Thorogood, M. (1996). Waiting lists, Bypassing the time. *The Health Service Journal*, *106*(33). PMID:10156829

Lavee, J., & Brock, D. (2012). Prioritizing registered donors in organ allocation: An ethical appraisal of the Israeli organ transplant law. *Current Opinion in Critical Care*, *18*(6), 707–711. doi:10.1097/MCC.0b013e328357a2e2 PMID:22914426

Lee, C. C. (1990). Fuzzy Logic in Control Systems: Fuzzy Logic Controller. *Part II. IEEE Transactions on Systems, Man, and Cybernetics*, *20*(2), 419–435. doi:10.1109/21.52552

Leekwijck, V., & Kerre, E. E. (1999). Defuzzification: Criteria and classification. *Journal of Fuzzy Sets and Systems*, *108*(2), 159–178. doi:10.1016/S0165-0114(97)00337-0

Li D.-, F. (2011c). The GOWA operator based approach to multi attribute decision making using intuitionistic fuzzy sets. *Mathematical and Computer Modelling*, *53*(5-6), 1182–1196. doi:10.1016/j.mcm.2010.11.088

Li, D.-, F. (2011a). Linear programming approach to solve interval-valued matrix games. OMEGA. *International Journal of Management Sciences*, *39*(6), 655–666.

Li, D.-, F. (2011b). Closeness coefficient based nonlinear programming method for interval-valued intuitionistic fuzzy multiattribute decision making with incomplete preference information. *Applied Soft Computing*, *11*(4), 3402–3418. doi:10.1016/j.asoc.2011.01.011

Li, D.-, F. (2012). A fast approach to compute fuzzy values of matrix games with payoffs of triangular fuzzy numbers. *European Journal of Operational Research*, *223*(2), 421–429. doi:10.1016/j.ejor.2012.06.020

Li, D.-, F. (2013). An effective methodology for solving matrix games with fuzzy payoffs. *IEEE Transactions on Cybernetics*, *43*(2), 610–621. doi:10.1109/TSMCB.2012.2212885 PMID:22987531

Li, D.-, F. (2014). *Decision and Game Theory in Management with Intuitionistic Fuzzy Sets*. Heidelberg, Germany: Springer.

Li, D.-F. (2014). Decision and Game Theory in Management with Intuitionistic fuzzy sets. Springer. doi:10.1007/978-3-642-40712-3

Li, D.-, F. (2016a). *Linear Programming Models and Methods of Matrix Games with Payoffs of Triangular Fuzzy Numbers*. Berlin: Springer.

Li, D.-, F. (2016b). *Models and Methods of Interval-Valued Cooperative Games in Economic Management*. Cham, Switzerland: Springer.

Li, D.-F. (n.d.). *Linear Programming Models and Methods of Matrix Games with Payoffs of Triangular Fuzzy Numbers*. Springer.

Li, D.-, F.,, & Cheng, C. T. (2002). Fuzzy multi objective programming methods for fuzzy constrained matrix games with fuzzy numbers. *International Journal of Uncertainty, Fuzziness and Knowledge-based Systems*, *10*(4), 385–400. doi:10.1142/S0218488502001545

Li, D.-F. (2005). Multi attribute decision making models and methods using intuitionistic fuzzy sets. *Journal of Computer and System Sciences*, *70*(1), 73–85. doi:10.1016/j.jcss.2004.06.002

Li, D.-F. (2007a). Compromise ratio method for fuzzy multi-attribute group decision making. *Applied Soft Computing*, *7*(3), 807–817. doi:10.1016/j.asoc.2006.02.003

Li, D.-F. (2007b). A fuzzy closeness approach to fuzzy multi-attribute decision making. *Fuzzy Optimization and Decision Making*, *6*(3), 237–254. doi:10.1007/s10700-007-9010-1

Li, D.-F. (2008). Extension of the LINMAP for multiattribute decision making under Atanassovs intuitionistic fuzzy environment. *Fuzzy Optimization and Decision Making*, *7*(1), 17–34. doi:10.1007/s10700-007-9022-x

Li, D.-F. (2009). Multiattribute group decision making method using extended linguistic variables. *International Journal of Uncertainty, Fuzziness and Knowledge-based Systems*, *17*(6), 793–806. doi:10.1142/S0218488509006273

Li, D.-F. (2009). Relative ratio method for multiple attribute decision making problems. *International Journal of Information Technology & Decision Making*, *8*(2), 289–311. doi:10.1142/S0219622009003405

Li, D.-F. (2010). A new methodology for fuzzy multi-attribute group decision making with multi-granularity and non-homogeneous information. *Fuzzy Optimization and Decision Making*, *9*(1), 83–103. doi:10.1007/s10700-010-9071-4

Li, D.-F. (2010). Linear programming method for MADM with interval valued intuitionistic fuzzy sets. *Expert Systems with Applications*, *37*(8), 5939–5945. doi:10.1016/j.eswa.2010.02.011

Li, D.-F. (2010). Mathematical-programming approach to matrix games with payoffs represented by Atanassovs interval-valued intuitionistic fuzzy sets. *IEEE Transactions on Fuzzy Systems*, *18*(6), 1112–1128. doi:10.1109/TFUZZ.2010.2065812

Li, D.-F. (2010). TOPSIS-based nonlinear-programming methodology for multiattribute decision making with interval-valued intuitionistic fuzzy sets. *IEEE Transactions on Fuzzy Systems*, *18*, 299–311.

Li, D.-F. (2010a). TOPSIS-based nonlinear-programming methodology for multi-attribute decision making with interval-valued intuitionistic fuzzy sets. *IEEE Transactions on Fuzzy Systems*, *18*(2), 299–311.

Li, D.-F. (2010b). A ratio ranking method of triangular intuitionistic fuzzy numbers and its application to MADM problems. *Computers & Mathematics with Applications (Oxford, England)*, *60*(6), 1557–1570. doi:10.1016/j.camwa.2010.06.039

Li, D.-F. (2010c). Multiattribute decision making method based on generalized OWA operators with intuitionistic fuzzy sets. *Expert Systems with Applications*, *37*(12), 8673–8678. doi:10.1016/j.eswa.2010.06.062

Li, D.-F. (2012). TOPSIS for multiattribute group decision making using IFSs. *Advances in Fuzzy Sets and Systems*, *11*(2), 105–129.

Li, D.-F. (2012). TOPSIS for multi-attribute group decision making using IFSs. *Advances in Fuzzy Sets and Systems*, *11*(2), 105–129.

Li, D.-F., & Cheng, C. T. (2004). Stability on multiobjective dynamic programming problems with fuzzy parameters in the objective functions and in the constraints. *European Journal of Operational Research*, *158*(3), 678–696. doi:10.1016/S0377-2217(03)00374-6

Li, D.-F., & Hong, F.-X. (2013). Alfa-cut based linear programming methodology for constrained matrix games with payoffs of trapezoidal fuzzy numbers. *Fuzzy Optimization and Decision Making*, *12*(2), 191–213. doi:10.1007/s10700-012-9148-3

Li, D.-F., & Liu, J. C. (2015). A parameterized non-linear programming approach to solve matrix games with payoffs of I-fuzzy numbers. *IEEE Transactions on Fuzzy Systems*, *23*(4), 885–896. doi:10.1109/TFUZZ.2014.2333065

Li, D.-F., & Nan, J. X. (2011). An extended weighted average method for MADM using intuitionistic fuzzy sets and sensitivity analysis. *Critical View*, *5*, 5–25.

Li, D.-F., & Nan, J. X. (2011). Extension of the TOPSIS for multi-attribute group decision making under Atanassov IFS environments. *International Journal of Fuzzy System Applications*, *1*(4), 47–61. doi:10.4018/ijfsa.2011100104

Li, D.-F., Nan, J. X., & Zhang, M. J. (2010). A ranking method of triangular intuitionistic fuzzy numbers and application to decision making. *International Journal of Computational Intelligence Systems*, *3*(5), 522–530. doi:10.1080/18756891.2010.9727719

Li, D.-F., & Nan, J.-X. (2014). An interval-valued programming approach to matrix games with payoffs of triangular intuitionistic fuzzy numbers. *Iranian Journal of Fuzzy Systems*, *11*(2), 45–57.

Li, D.-F., & Ren, H. P. (2015). Multi-attribute decision making method considering the amount and reliability of intuitionistic fuzzy information. *Journal of Intelligent & Fuzzy Systems*, *28*, 1877–1883.

Li, D.-F., & Sun, T. (2007a). Fuzzy linear programming approach to multiattribute decision making with linguistic variables and incomplete information. *Advances in Complex Systems*, *10*(4), 505–525. doi:10.1142/S0219525907001306

Li, D.-F., & Sun, T. (2007b). Fuzzy LINMAP method for multiattribute group decision making with linguistic variables and incomplete information. *International Journal of Uncertainty, Fuzziness and Knowledge-based Systems*, *15*(2), 153–173. doi:10.1142/S0218488507004509

Li, D.-F., Wang, Y., Liu, S., & Shan, F. (2009). Fractional programming methodology for multi-attribute group decision-making using IFS. *Applied Soft Computing*, *9*(1), 219–225. doi:10.1016/j.asoc.2008.04.006

Li, D.-F., & Wan, S. P. (2014). A fuzzy inhomogenous multiattribute group decision making approach to solve outsourcing provider selection problems. *Knowledge-Based Systems*, *67*, 71–89. doi:10.1016/j.knosys.2014.06.006

Li, D.-F., & Wan, S. P. (2014). Fuzzy heterogeneous multiattribute decision making method for outsourcing provider selection. *Expert Systems with Applications*, *41*(6), 3047–3059. doi:10.1016/j.eswa.2013.10.036

Li, D.-F., & Yang, J. (2015). A difference-index based ranking method of trapezoidal intuitionistic fuzzy numbers and application to multiattribute decision making. *Mathematical and Computational Applications*, *20*(1), 25–38. doi:10.19029/mca-2015-003

Lim, S., & Cho, S. (2007). Intelligent OS Process Scheduling Using Fuzzy Inference with User Models. *Lecture Notes in Computer Science*, *45*, 725–734. doi:10.1007/978-3-540-73325-6_72

Lin, C., Chiu, T. H., & Tseng, Y. H. (2006). Agility evaluation using fuzzy logic. *International Journal of Production Economics*, *101*(2), 353–368. doi:10.1016/j.ijpe.2005.01.011

Lin, C., & Harris, S. (2012). A Unified Framework for the Prioritization of Organ Transplant Patients: Analytic Hierarchy Process, Sensitivity and Multifactor Robustness Study. *Journal of Multi-Criteria Decision Analysis*, *20*(3-4), 157–172. doi:10.1002/mcda.1480

Lin, F., Li, L., Jin, C., Wanga, R., Wange, H., & Yang, L. (2012). A 3PL supplier selection model based on fuzzy sets. *Computers & Operations Research*, *39*(8), 1879–1884. doi:10.1016/j.cor.2011.06.022

Lin, Z. C., & Yang, C. B. (1994). Evaluation of machine selection by the AHP method. *Journal of Materials Processing Technology*, *57*(3/4), 253–258.

Liu, B. (2008). Uncertainty Theory (3rd ed.). Retrieved from http://orsc.edu.cn/liu/ut.pdf

Liu, F., & Yuan, X.H. (2007). Fuzzy number intuitionistic fuzzy set. *Fuzzy System and Mathematics*, *21*, 88-91.

Liu, B., & Liu, Y. K. (2002). Expected value of fuzzy variable and fuzzy expected models. *IEEE Transactions on Fuzzy Systems*, *4*(10), 445–450.

Liu, H. T., & Wang, W. K. (2009). An integrated fuzzy approach for provider evaluation and selection in third-party Logistics. *Expert Systems with Applications*, *36*(3), 4387–4398. doi:10.1016/j.eswa.2008.05.030

Liu, P. D. (2013). Some generalized dependent aggregation operators with intuitionistic linguistic numbers and their application to group decision making. *Journal of Computer and System Sciences*, *79*(1), 131–143. doi:10.1016/j.jcss.2012.07.001

Liu, P. D. (2013). Some geometric aggregation operators based on interval intuitionistic uncertain linguistic variables and their application to group decision making. *Applied Mathematical Modelling*, *37*(4), 2430–2444. doi:10.1016/j.apm.2012.05.032

Liu, P. D. (2014). Some hamacher aggregation operators based on the interval valued intuitionistic fuzzy numbers and their application to group decision making. *IEEE Transactions on Fuzzy Systems*, *22*(1), 83–97. doi:10.1109/TFUZZ.2013.2248736

Liu, P. D., He, L., & Yu, X. C. (2016). Generalized Hybrid Aggregation Operators Based on the 2-Dimension Uncertain Linguistic Information for Multiple Attribute Group Decision Making. *Group Decision and Negotiation*, *25*(1), 103–126. doi:10.1007/s10726-015-9434-x

Liu, P. D., & Jin, F. (2012). A multi-attribute group decision-making method based on weighted geometric aggregation operators of interval-valued trapezoidal fuzzy numbers. *Applied Mathematical Modelling*, *36*(6), 2498–2509. doi:10.1016/j.apm.2011.09.006

Liu, P. D., Jin, F., Zhang, X., Su, Y., & Wang, M. (2011). Minghe Wang, Research on the multi-attribute decision-making under risk with interval probability based on prospect theory and the uncertain linguistic variables. *Knowledge-Based Systems*, *24*(4), 554–561. doi:10.1016/j.knosys.2011.01.010

Liu, P. D., Liu, Z. M., & Zhang, X. (2014). Some intuitionistic uncertain linguistic heronian mean operators and their application to group decision making. *Applied Mathematics and Computation*, *230*, 570–586. doi:10.1016/j.amc.2013.12.133

Liu, P. D., & Shi, L. L. (2015). The Generalized Hybrid Weighted Average Operator Based on Interval Neutrosophic Hesitant Set and Its Application to Multiple Attribute Decision Making. *Neural Computing & Applications*, *26*(2), 457–471. doi:10.1007/s00521-014-1736-4

Liu, P. D., & Wang, Y. W. (2014). Multiple attribute group decision making methods based on intuitionistic linguistic power generalized aggregation operators. *Applied Soft Computing*, *7*, 90–104. doi:10.1016/j.asoc.2013.12.010

Liu, P. D., & Yu, X. C. (2014). 2-dimension uncertain linguistic power generalized weighted aggregation operator and its application for multiple attribute group decision making. *Knowledge-Based Systems*, *57*(1), 69–80. doi:10.1016/j.knosys.2013.12.009

Liu, P. D., Zhang, X., & Jin, F. (2012). A multi-attribute group decisionmaking method based on interval valued trapezoidal fuzzy numbers hybrid harmonic averaging operators. *Journal of Intelligent & Fuzzy Systems*, *23*, 159–168.

Liu, P. D., Zhang, X., & Jin, F. (2012). A multi-attribute group decision-making method based on interval-valued trapezoidal fuzzy numbers hybrid harmonic averaging operators. *Journal of Intelligent & Fuzzy Systems*, *23*(5), 159–168.

Liu, S., & Fang, Z. (2009). Study on the Relationship between Customer Satisfaction and 3PL Costs. *International Journal of Business and Management*, *4*(6), 23–28. doi:10.5539/ijbm.v4n6p23

Luckman, J., Mackenzie, M., & Stringer, J. (1969). *Management Policies for Large Ward Units*. Coventry, UK: Institute for Operational Research.

Maafi, A., & Adnane, A. (1989). A two-state markovian model of global irradiation suitable for photovoltaic conversion. *Solar & Wind Technology*, *6*(3), 247–252. doi:10.1016/0741-983X(89)90076-3

MacCormick, A., Collecutt, W., & Parry, B. (2003). Prioritizing patients for elective surgery: A systematic review. *ANZ Journal of Surgery*, *73*(8), 633–642. doi:10.1046/j.1445-2197.2003.02605.x PMID:12887536

Mahdavi, I., Amiri, N. M., Heidarzade, A., & Nourifar, R. (2008). Designing a model of fuzzy TOPSIS in multiple criteria decision making. *Applied Mathematics and Computation*, *206*(2), 607–617. doi:10.1016/j.amc.2008.05.047

Maji, P., Biswas, R., & Roy, A. (2001). Fuzzy soft set. *J. Fuzzy Math.*, *9*(3), 677–692.

Maji, P., Biswas, R., & Roy, A. (2003). *Soft set theory*. Comp.&Math.Appl.

Mariotti, G., Siciliani, L., Rebba, V., Fellini, R., Gentilini, M., Benea, G., & Liva, C. et al. (2014). Waiting time prioritisation for specialist services in Italy: The homogeneous waiting time groups approach. *Health Policy (Amsterdam)*, *117*(1), 54–63. doi:10.1016/j.healthpol.2014.01.018 PMID:24576498

Mengi, O., & Hakkı, I. (2012). Fuzzy logic control for a wind/battery renewable energy production system. *Turkish Journal of Elecrical Engineering & Computer Sciences*, *20*(2), 187–206.

Meredith, S., & Francis, D. (2000). Journey towards agility: The agile wheel explored. *The TQM Magazine*, *12*(2), 137–143. doi:10.1108/09544780010318398

Merigo, J. M., & Casanovas, M. (2008). Generalized aggregation operators and fuzzy numbers in a unified Model between the weighted average and the OWA operator. *International Journal of Intelligent Systems*, *24*, 934–954.

Merigo, J. M., & Casanovas, M. (2009). Induced aggregation operators in decision making with Dempster-Shafer belief structure. *International Journal of Intelligent Systems*, *24*(8), 934–954. doi:10.1002/int.20368

Merigo, J. M., & Gil-Lafuente, A. M. (2008a). On the use of the OWA operator in the Euclidean distance. *International Journal on Computer Science and Engineering*, *2*, 170–176.

Merigo, J. M., & Gil-Lafuente, A. M. (2009a). OWA operators in generalized distances. *International Journal of Applied Mathematics and Computer Science*, *5*, 11–18.

Mermoud, A. (2010). *Le potentiel Photovoltaïque*. Retrieved from http://etat.geneve.ch/dt/SilverpeasWebFileServer/Potentiel_PV_ScanE_250510.pdf?ComponentId=kmelia197&SourceFile=1275571643860.pdf&MimeType=application/pdf&Directory=Attachment/Images/

Meyer, T., & Raspe, H. (2012). Priority setting: Priorisierung: Was ist das und wie geht das? *Rehabilitation*, *51*(2), 73–80. doi:10.1055/s-0032-1306288 PMID:22570153

Min, D., & Yih, Y. (2010). An elective surgery scheduling problem considering patient priority. *Computers & Operations Research*, *37*(6), 1091–1099. doi:10.1016/j.cor.2009.09.016

Min, H., & Galle, W. P. (1997). Green purchasing strategies: Trends and implications. *International Journal of Purchasing and Materials Management*, *33*(2), 10–17. doi:10.1111/j.1745-493X.1997.tb00026.x

Mizumoto, M. (1989). Improvement methods of fuzzy controls. In *Proceedings of Third IFSA Congress* (pp. 60-62).

Molodtsov, D. (1999). Soft set theory first results. *Computers & Mathematics with Applications (Oxford, England)*, *37*(4-5), 19–31. doi:10.1016/S0898-1221(99)00056-5

Mondal, K. T., & Samanta, S. K. (2001). Topology of interval-valued intuitionistic fuzzy sets. *Fuzzy Sets and Systems*, *119*(3), 483–494. doi:10.1016/S0165-0114(98)00436-9

Montoya, S., González, M., & López, S. et al.. (2014). Study to Develop a Waiting List Prioritization Score for Varicose Vein Surgery. *Annals of Vascular Surgery*, *28*(2), 306–312. doi:10.1016/j.avsg.2012.11.017 PMID:24084264

Mora-Lopez, L., & Sidrach De Cardona, M. (1998). Multiplicative arma models to generate hourly series of global irradiation. *Solar Energy*, *63*(5), 283–291. doi:10.1016/S0038-092X(98)00078-4

Mousavi, S. M., Vahdani, B., Moghaddam, R. T., Ebrahimnejad, S., & Amiri, M. (2012). A multistage decision making process for multiple attributes analysis under an interval-valued fuzzy environment. *International Journal of Advanced Manufacturing Technology*, *64*(9-12), 1263–1273. doi:10.1007/s00170-012-4084-5

Mula, P., Roy, S. K., & Li, D.-, F. (2015). Birough programming approach for solving bi-matrix games with birough payoff elements. *Journal of Intelligent & Fuzzy Systems*, *29*(2), 863–875. doi:10.3233/IFS-151616

Mullen, P. (2003). Prioritizing waiting lists: How and why? *European Journal of Operational Research*, *150*(1), 32–45. doi:10.1016/S0377-2217(02)00779-8

Nagel, E., & Lauerer, M. (2016). *Prioritization in Medicine An International Dialogue*. Springer International. doi:10.1007/978-3-319-21112-1

Nambirajan, T., & Ganeshkumar, C. (2010). Measurement of performance of supply chains and their impact on the competitiveness of manufacturing industries in union territory of Pondicherry (India).*International Research Conference on Business and Information*. University of Kelaniya.

Nan, J., Wang, T., & An, J. (2016). Intuitionistic Fuzzy Distance Based TOPSIS Method and Application to MADM. *International Journal of Fuzzy System Applications*, *5*(1), 43–56. doi:10.4018/IJFSA.2016010103

Nan, J.-X., Li, D.-F., & Zhang, M.-J. (2010). A lexicographic method for matrix games with payoffs of triangular intuitionistic fuzzy numbers. *International Journal of Computational Intelligence Systems*, *3*(3), 280–289. doi:10.1080/18756891.2010.9727699

Nan, J.-X., Zhang, M.-J., & Li, D.-F. (2014a). Intuitionistic fuzzy programming models for matrix games with payoffs of trapezoidal intuitionistic fuzzy numbers. *International Journal of Fuzzy Systems*, *16*(4), 444–456.

Nan, J.-X., Zhang, M.-J., & Li, D.-F. (2014b). A methodology for matrix games with payoffs of triangular intuitionistic fuzzy number. *Journal of Intelligent & Fuzzy Systems*, *26*(6), 2899–2912.

Natsheh, E. M. (2013). *Hybrid Power Systems Energy Management Based on Artificial Intelligence* (Thesis). Manchester Metropolitan University, School of Engineering. Manchester, UK.

Naylor, C. (1993). Waiting for coronary revascularization in Toronto: 2 years' experience with a regional referral office. *Canadian Medical Association Journal*, *149*(7), 955–962. PMID:8402424

Neely, A. (2005). The evolution of performance measurement research. *International Journal of Operations & Production Management*, *1*(1), 1264–1277. doi:10.1108/01443570510633648

Nema, P., Nema, R. K., & Rangnerak, S. (2009). A current and future state of art development of hybrid energy system using wind and PV-solar:A Review. *Renewable & Sustainable Energy Reviews*, *13*(8), 2096–2103. doi:10.1016/j.rser.2008.10.006

Nie, B., Du, J., Guoliang, X., Liu, H., Yu, R., & Wen, Q. (2011). A New Operating System Scheduling Algorithm. *Advanced Research on Electronic Commerce, Web Application, and Communication, Springer*, *143*, 92–96. doi:10.1007/978-3-642-20367-1_15

Noci, G., & Toletti, G. (2000). Selecting quality-based programmes in small firms: A comparison between the fuzzy linguistic approach and the analytic hierarchy process. *International Journal of Production Economics*, *67*(2), 113–133. doi:10.1016/S0925-5273(99)00131-0

Noseworthy, T., McGurran, J., & Hadorn, D. C. (2003). Waiting for scheduled services in Canada: Development of priority-setting scoring systems. *Journal of Evaluation in Clinical Practice*, *9*(1), 23–31. doi:10.1046/j.1365-2753.2003.00377.x PMID:12558699

Olugu, E. U., & Wong, K. Y. (2009). Supply Chain Performance Evaluation: Trends and Challenges. *American J. of Engineering and Applied Sciences*, *2*(1), 202–211. doi:10.3844/ajeas.2009.202.211

Olugu, E. U., & Wong, K. Y. (2011). Fuzzy logic evaluation of reverse Logistics performance in the automotive industry. *Scientific Research and Essays*, *6*(7), 1639–1649.

Onut, S., Kara, S. S., & Efendigil, T. (2008). A hybrid fuzzy MCDM approach to machine tool selection. *Journal of Intelligent Manufacturing*, *19*(4), 443–453. doi:10.1007/s10845-008-0095-3

Ounnar, F., Naamane, A., Pujo, P., & M'Sirdi, N. (2012). Pilotage multicritère d'un système énergétique multi-sources et multiutilisateurs. 1er Congrès International en Génie Industriel et Management des Systèmes. Marseille-France.

Pandeya, A. K., Tyagi, V. V., Selvaraj, J., Rahim, N. A., & Tyagi, S. K. (2016). Recent advances in solar photovoltaic systems for emerging trends and advanced applications. *Renewable & Sustainable Energy Reviews*, *53*, 859–884. doi:10.1016/j.rser.2015.09.043

Panizzolo, R. (1998). Applying the lessons learned from 27 lean manufacturers: The relevance of relationships management. *International Journal of Production Economics*, *55*(3), 223–240. doi:10.1016/S0925-5273(98)00066-8

Park, D. G., Kwun, Y. C., Park, J. H., & Park, I. Y. (2009). Correlation coefficient of interval-valued intuitionistic fuzzy sets and its application to multiple attribute group decision making problems. *Mathematical and Computer Modelling*, *50*(9-10), 1279–1293. doi:10.1016/j.mcm.2009.06.010

Park, M., Jin, Y. H., Chae, J., & Kim, C. (2008). Quantum-Based Fixed Priority Scheduling. *International Journal of Computer and Information Sciences*, *11*(5), 341–356.

Pavnaskar, S. J., Gershenson, J. K., & Jambekar, A. B. (2003). Classification scheme for lean manufacturing tools. *International Journal of Production Research*, *41*(13), 3075–3090. doi:10.1080/0020754021000049817

Peacock, S., et al. (2009). Overcoming barriers to priority setting using interdisciplinary methods. *Health Policy, 92*(2-3), 124-32. doi: 10.1016/j.healthpol.2009.02.006

Pedrazzi, S., Zini, G., & Tartarini, P. (2010). Complete modeling and software implementation of a virtual solar hydrogen hybrid system. *Energy Conversion and Management*, *51*(1), 122–129. doi:10.1016/j.enconman.2009.08.036

Percin, S. (2009). 'Evaluation of third-party Logistics (3PL) providers by using a two-phase AHP and TOPSIS methodology. *Benchmarking: An International Journal*, *16*(5), 588–604. doi:10.1108/14635770910987823

Pérez-Navarro, A., Alfonso, D., Ariza, H. E., Cárcel, J., Correcher, A., Escrivá-Escrivá, G., & Vargas, C. et al. (2015). Experimental verification of hybrid renewable systems as feasible energy source. *Renewable Energy*, *86*, 384–391. doi:10.1016/j.renene.2015.08.030

Peter, C. B., & Thomas, W. S. (2000). Using the balanced scorecard to measure supply chain performance. *Journal of Business Logistics*, *21*(1), 75.

Peter, F. C. (1967). Letter to the Editor-Additive Utilities with Incomplete Product Sets: Application to Priorities and Assignments. *Operations Research*, *15*(3), 537–542. doi:10.1287/opre.15.3.537

Petrovic, D., Roy, R., & Petrovic, R. (1999). Supply chain modeling using fuzzy sets. *International Journal of Production Economics*, *59*(1), 443–453. doi:10.1016/S0925-5273(98)00109-1

Phoenix, C. J. (1972). Waiting list management and admission scheduling. Butterworth.

Piette, F. (2011). *La logique floue, intérêts et limites.* Retrieved 4 6, 2016, from http://www.ferdinandpiette.com/blog/2011/08/la-logique-floue-interets-et-limites(2010)

Pornak, S., & Raspe, H. (2015). Die dänische Debatte über Priorisierung und Posteriorisierung in der Medizin – Was gibt es Neues? *Gesundheitswesen (Bundesverband der Arzte des Offentlichen Gesundheitsdienstes (Germany)), 77*, 1–7.

Power, D. J. (2013). *Engineering effective decision support technologies: New models and applications*. Hershey, PA: IGI Global. doi:10.4018/978-1-4666-4002-3

Pradhan, R., & Madhumangal, P. (2014). The generalized inverse of Atanassovs intuitionistic fuzzy matrices. *International Journal of Computational Intelligence Systems, 7*(6), 1083–1095. doi:10.1080/18756891.2014.963976

Prentice, J., & Pizer, S. (2007). Delayed Access to Health Care and Mortality. *Health Services Research, 42*(2), 644–662. doi:10.1111/j.1475-6773.2006.00626.x PMID:17362211

Qi, J. (2010). Machine tool selection model based on fuzzy MCDM approach.*International Conference on Intelligent Control and Information Processing*, (pp. 282-285). doi:10.1109/ICICIP.2010.5564228

Qureshi, M. N., Kumar, P., & Kumar, D. (2009a). Selection of 3PL service providers: A combined approach of AHP and Graph theory. *International Journal of Services Technology and Management, 12*(1), 35–60. doi:10.1504/IJSTM.2009.025035

Qureshi, M. N., Kumar, P., & Kumar, D. (2009b). Framework for benchmarking Logistics performance using fuzzy AHP. [IJBPSCM]. *International Journal of Business Performance and Supply Chain Modelling, 1*(1), 82–98. doi:10.1504/IJBPSCM.2009.026267

Raheja, S., Dhadich, R., & Rajpal, S. (2013). Vague Oriented Highest Response Ratio Next Scheduling Algorithm. *IEEE-International Conference on Fuzzy Systems* (pp. 1-7).

Raheja, S., & Dhadich, R. (2013). Many Valued Logics for Modeling Vagueness. *International Journal of Computers and Applications, 61*(7), 35–39. doi:10.5120/9942-4579

Raheja, S., Dhadich, R., & Rajpal, S. (2012). An Optimum Time Quantum using Linguistic Synthesis for Round Robin Scheduling Algorithm. *International Journal of Soft Computing, 3*(1), 57–66. doi:10.5121/ijsc.2012.3105

Raheja, S., Dhadich, R., & Rajpal, S. (2014). Designing of 2-Stage CPU Scheduler using Vague Logic. *Advances in Fuzzy Systems, 2014*, 1–10. doi:10.1155/2014/841976

Raheja, S., Dhadich, R., & Rajpal, S. (2015). Designing of Vague Logic Based Fair-Share CPU Scheduler. *International Journal of Fuzzy System Applications, IGI Global, 4*, 25–49.

Rajesh, R., Pugazhendhi, S., & Ganesh, K. (2010). A survey of literature on selection of third party Logistics service provider.[IJBPSCM]. *International Journal of Business Performance and Supply Chain Modelling, 2*(2), 146–170. doi:10.1504/IJBPSCM.2010.036167

Ramik, J., & Korviny, P. (2010). Inconsistency of pair wise comparison matrix with fuzzy elements based on geometric mean. *Fuzzy Sets and Systems, 161*(11), 1604–1613. doi:10.1016/j.fss.2009.10.011

Raspe, H. (2016). Prioritisation: (At Least) Two Normative. In *Prioritization in Medicine: An International Dialogue* (pp. 85–99). Springer International Publishing. doi:10.1007/978-3-319-21112-1_7

Ren, H. P., Chen, H. H., Fei, W., & Li, D.-F. (2016). A MAGDM method considering the amount and reliability information of interval-valued intuitionistic fuzzy sets. *International Journal of Fuzzy Systems, 18*(1), 1–11. doi:10.1109/TFUZZ.2016.2554151

Riet, C., & Demeulemeester, E. (2015). Trade-offs in operating room planning for electives and emergencies: A review. *Operations Research for Health Care, 7*, 52–69. doi:10.1016/j.orhc.2015.05.005

Rizwan, M., Majid, J. I., Kirmani, S., & Kothari, D. (2014). Fuzzy logic based modeling and estimation of global solar energy using meteorological parameters. *Energy*, *70*, 685–691. doi:10.1016/j.energy.2014.04.057

Robinson, J.P., & Amirtharaj, E.C.H. (2012b). A Search for the Correlation coefficient of Triangular and Trapezoidal intuitionistic Fuzzy sets for Multiple Attribute Group Decision Making. *Communications in Computer and Information Science 283,* 333-342.

Robinson, J. P., & Amirtharaj, E. C. H. (2011a). A short primer on the Correlation coefficient of Vague sets. *International Journal of Fuzzy System Applications*, *1*(2), 55–69. doi:10.4018/ijfsa.2011040105

Robinson, J. P., & Amirtharaj, E. C. H. (2011b). Extended TOPSIS with correlation coefficient of Triangular Intuitionistic fuzzy sets for Multiple Attribute Group Decision Making. *International Journal of Decision Support System Technology*, *3*(3), 15–40. doi:10.4018/jdsst.2011070102

Robinson, J. P., & Amirtharaj, E. C. H. (2012a). Vague Correlation coefficient of Interval Vague sets. *International Journal of Fuzzy System Applications*, *2*(1), 18–34. doi:10.4018/ijfsa.2012010102

Robinson, J. P., & Amirtharaj, E. C. H. (2014a). Efficient Multiple Attribute Group Decision Making models with Correlation coefficient of Vague sets. *International Journal of Operations Research and Information Systems*, *5*(3), 27–49. doi:10.4018/ijoris.2014070102

Robinson, J. P., & Amirtharaj, E. C. H. (2014b). MAGDM-MINER: A new algorithm for mining trapezoidal intuitionistic fuzzy correlation rules. *International Journal of Decision Support System Technology*, *6*(1), 34–59. doi:10.4018/ijdsst.2014010103

Ross, T. J. (2009). *Fuzzy Logic with Engineering applications*. John Wiley & Sons.

Rouhparvar, H., & Panahi, A. (2015). A new definition for defuzzification of generalized fuzzy numbers and its application. *Applied Soft Computing*, *30*, 577–584. doi:10.1016/j.asoc.2015.01.053

Runkler, T. A., & Glesner, M. (1993). A set of axioms for defuzzification strategies towards a theory of rational defuzzification operators.*Second IEEE International Conference on Fuzzy Systems*, (pp. 1161-1166). doi:10.1109/FUZZY.1993.327350

Russell, C., Roberts, M., Williamson, T. G., McKercher, J., Jolly, S. E., & McNeil, J. (2003). Clinical categorization for elective surgery in Victoria. *ANZ Journal of Surgery*, *73*(10), 839–842. doi:10.1046/j.1445-2197.2003.02797.x PMID:14525579

Saaty, T. L. (1980). *The Analytic Hierarchy Process*. New York: McGraw-Hill.

Saaty, T. L., & Vargas, L. G. (2006). *Decision Making with the Analytic Network Process: Economic, Political, Social and Technological Applications with Benefits, Opportunities, Costs and Risks*. New York: Springer.

Saaty, T., & Özdemir, M. (2005). *The Encyclion: A Dictionary of Decisions with Dependence and Feedback based on the ANP*. RWS Publications.

Sabin, G., Kochhar, G., & Sadayappan, P. (2004). Job fairness in non-preemptive job scheduling.*4th International Conference on Parallel processing* (pp. 186-194).

Sadeghi, M., & Hosseini, H. M. (2013). Evaluation of Fuzzy Linear Programming Application in Energy Models. *International Journal of Energy Optimization and Engineering*, *2*(1), 50–59. doi:10.4018/ijeoe.2013010104

Sadegh-Zadeh, K. (2012). *Handbook of analytic philosophy of medicine*. New York: Springer Dordrecht Heidelberg. doi:10.1007/978-94-007-2260-6

Saen, R. F. (2010a). A New Model for Ranking 3PL Providers. *Australian Journal of Basic and Applied Sciences*, *4*(8), 3762–3769.

Saen, R. F. (2010b). A new model for selecting third-party reverse Logistics providers in the presence of multiple dual-role factors. *International Journal of Advanced Manufacturing Technology*, *46*(1-4), 405–410. doi:10.1007/s00170-009-2092-x

Sahu, N. K., Datta, S. & Mahapatra, S. S. (2015). Fuzzy Based Appraisement Module for 3PL Evaluation and Selection. *Benchmarking, an International Journal, 22*(3), 354-392.

Sahu, N. K., Sahu A. K., & Sahu, A. K (2015a). Appraisement and Benchmarking of Third Party Logistic Service Provider by Exploration of Risk Based Approach. *Cogent Business and Management, 2*, 1-21.

Sahu, N. K., Sahu, A. K., & Sahu, A. K. (2015b). Appraisement and Benchmarking of Third Party Logistic Service Provider by Exploration of Risk Based Approach. Cogent business and management, 2, 1-21.

Sahu, A. K., Datta, S., & Mahapatra, S. S. (2016). Evaluation and selection of suppliers considering green perspectives: Comparative analysis on application of FMLMCDM and Fuzzy-TOPSIS. *Benchmarking: An International Journal*, *23*(6), 1579–1604. doi:10.1108/BIJ-12-2014-0111

Sahu, A. K., Sahu, N. K., & Sahu, A. (2014). Appraisal of CNC machine tool by integrated MULTI MOORA-IGVN circumstances: An empirical study. *International Journal of Grey Systems: Theory and Application*, *4*(1), 104–123.

Sahu, A. K., Sahu, N. K., & Sahu, A. K. (2015a). Benchmarking CNC machine tool using hybrid fuzzy methodology a multi indices decision making approach. *International Journal of Fuzzy System Applications*, *4*(2), 28-46..

Sahu, A. K., Sahu, N. K., & Sahu, A. K. (2016a). Application of Integrated TOPSIS in ASC index: Partners Benchmarking perspective. *Benchmarking: An International Journal*, *23*(3), 540–563. doi:10.1108/BIJ-03-2014-0021

Sahu, A. K., Sahu, N. K., & Sahu, A. K. (2016b). Appraisal of Partner Enterprises under GTFNS Environment in Agile SC. *International Journal of Decision Support System Technology*, *8*(3), 1–19. doi:10.4018/IJDSST.2016070101

Sahu, A. K., Sahu, N. K., & Sahu, A. K. (2016c). Application of Modified MULTI-MOORA for CNC Machine Tool Evaluation in IVGTFNS Environment: An Empirical Study. *International Journal of Computer Aided Engineering and Technology*, *8*(3), 234–259. doi:10.1504/IJCAET.2016.077603

Sahu, N. K., Datta, S., & Mahapatra, S. S. (2012). Establishing green supplier appraisement platform using grey concepts. *Grey Systems: Theory and Application*, *2*(3), 395–418. doi:10.1108/20439371211273276

Sahu, N. K., Datta, S., & Mahapatra, S. S. (2012). Establishing green supplier appraisement platform using grey concepts. *International Journal of Grey Systems: Theory and Application*, *2*(3), 395–418.

Sahu, N. K., Datta, S., & Mahapatra, S. S. (2013). Decision making for selecting 3PL service provider using three parameter interval grey numbers. *International Journal of Logistic Systems and Management*, *14*(3), 261–297. doi:10.1504/IJLSM.2013.052061

Sahu, N. K., Datta, S., & Mahapatra, S. S. (2013). Robot Selection based on grey-MULTIMOORA. *Grey Systems: Theory and Applications*, *3*(2), 201–232. doi:10.1108/GS-05-2013-0008

Sahu, N. K., Datta, S., & Mahapatra, S. S. (2014). Green Supplier Appraisement in Fuzzy Environment. *Benchmarking: An International Journal*, *21*(3), 412–429. doi:10.1108/BIJ-06-2012-0042

Samantra, C., Sahu, N. K., Datta, S., & Mahapatra, S. S. (2013). Decision-Making in Selecting Reverse Logistics Alternative using Interval-Valued Fuzzy Sets Combined with VIKOR Approach. *International Journal of Services and Operations Management*, *14*(2), 175–196. doi:10.1504/IJSOM.2013.051828

Sampath, S., & Ramya, B. (2015). Credibility Hypothesis Testing of Fuzzy Triangular Distributions. *Journal of Uncertain Systems, 9*(1), 62-74.

Sampath, S. (2009). Hybrid Single Sampling Plan. *World Applied Sciences Journal, 12*(6), 1685–1690.

Sampath, S., & Deepa, S. P. (2013). Determination of Optimal Chance Double Sampling Plan using Genetic Algorithm. *Model Assisted Statistics and Applications*, *8*(4), 265–273.

Sampath, S., & Kalaivani, R. (2010). Clustering of fuzzy data using credibilistic critical values.*Proceedings of the IEEE Xplore, ICSIP*, (pp. 227 – 232). doi:10.1109/ICSIP.2010.5697474

Sampath, S., & Ramya, B. (2013). Distance-based test for uncertainty hypothesis testing. *Journal of Uncertainty Analysis and Applications*, *1*(1), 4. doi:10.1186/2195-5468-1-4

Sampath, S., & Ramya, B. (2014). Credibility Hypothesis Testing of Expectation of Fuzzy Normal Distribution. *Journal of Intelligent and Fuzzy Systems*, *27*(4), 1837–1847.

Sampath, S., & Ramya, B. (2015a). Credibility hypothesis testing of fuzzy Exponential distributions. *International Journal of Fuzzy Computation and Modelling*, *1*(3), 287–311. doi:10.1504/IJFCM.2015.069954

Sampath, S., & Senthilkumar, R. (2014). clustering of fuzzy data using credibilistic expected and critical values.*Proceedings of the IEEE Xplore, ICCCS*. doi:10.1109/ICCCS.2014.7068189

Sampath, S., & Vaidyanathan, V. S. (2009). *Credibilistic ISODATA algorithm. In Mathematical and Computational models recent trends*. New Delhi, India: Narosha Publishing House.

Samvedi, V., Jain, V., & Chan, F. T. S. (2011). An integrated approach for machine tool selection using fuzzy analytical hierarchy process and grey relational analysis. *International Journal of Production Research*, *50*(12), 3211–3221. doi: 10.1080/00207543.2011.560906

Sanchez, E. (1979). Inverse of fuzzy relations, application to possibility distributions and medical diagnosis. *Fuzzy Sets and Systems*, *2*(1), 75–86. doi:10.1016/0165-0114(79)90017-4

Santos, J. M., Pinazo, J., & Canada, J. (2003). Methodology for generating daily clearness index values kt starting from the monthly average daily value Kt determining the daily sequence using stochastic models. *Renewable Energy*, *28*(10), 1523–1544. doi:10.1016/S0960-1481(02)00217-3

Seddon, M., French, J. K., Amos, D. J., Ramanathan, K., McLaughlin, S. C., & White, H. D. (1999). Waiting times and prioritisation for coronary artery bypass survey in New Zealand. *Heart (British Cardiac Society)*, *81*(6), 586–592. doi:10.1136/hrt.81.6.586 PMID:10336915

Selviaridis, K., & Spring, M. (2007). Third party Logistics: A literature review and research agenda. *International Journal of Logistics Management*, *18*(1), 125–150. doi:10.1108/09574090710748207

Semwal, V. B., Chakraborty, P., & Nandi, G. C. (2015). Less computationally intensive fuzzy logic (type-1)-based controller for humanoid push recovery. *Robotics and Autonomous Systems*, *63*, 122–135. doi:10.1016/j.robot.2014.09.001

Shah, R., & Ward, P. T. (2003). Lean manufacturing: Context, practice bundles, and performance. *Journal of Operations Management*, *21*(2), 129–149. doi:10.1016/S0272-6963(02)00108-0

Shan, L. (2012). Research on Green Logistics Service Providers Selection based on Intuitionistic Language Fuzzy Entropy. *Journal of Computers*, *7*(2), 540–546. doi:10.4304/jcp.7.2.540-546

Sharifi, H., Colquhoun, G., Barclay, I., & Dann, Z. (2001). Agile manufacturing: A management and operational framework. *Proceedings of the Institution of Mechanical Engineers. Part B, Journal of Engineering Manufacture*, *215*(6), 857–869. doi:10.1243/0954405011518647

Sharifi, H., & Zhang, Z. (2001). Agile manufacturing in practice: Application of a methodology. *International Journal of Operations & Production Management*, *21*(5/6), 772–794. doi:10.1108/01443570110390462

Sharifi, H., & Zhang, Z. A. (1999). Methodology for achieving agility in manufacturing organizations: An introduction. *International Journal of Production Economics*, *62*(1-2), 7–22. doi:10.1016/S0925-5273(98)00217-5

Sharp, G., Wan, Y. T., McGinnis, L. F., Goetschalckx, M., Bodner, D., Govindaraj, T., & Everette, J. et al. (2001). A structured approach to material handling system selection and ANP for selection of MHS specification for manufacturing.*Proceedings of the 2001 Industrial Engineering Research Conference*, (pp 1-6).

Sharp, J. M., Irani, Z., & Desai, S. (1999). Working towards agile manufacturing in the UK industry. *International Journal of Production Economics*, *62*(1), 155–169. doi:10.1016/S0925-5273(98)00228-X

Simpson, D., & Samson, D. (2008). *Developing strategies for green supply chain management. In Decision Line* (pp. 12–15). University of Melbourne.

Sims, R. E. (2014). *Renewable Energy Potential.* Retrieved from http://seat.massey.ac.nz/research/centres/energy/documents/Renewable%20Energy%20Potential.pdf

Smaïl, M., & Benatiallah, A. (2011). Modélisation d'une chaîne de conversion éolienne équipée D'une machine asynchrone double alimentation. *Revue des Energies Renouvelables ICES*, *11*, 233–238.

Soh, S. (2010). A decision model for evaluating third-party Logistics providers using fuzzy analytic hierarchy process. *African Journal of Business Management*, *4*(3), 339–349.

Soljak, M. (1997). *The Use of Scoring Systems in the NHS: Cardiac Surgery in London. From Waiting Lists to Booking Systems HSMC*. University of Birmingham.

So, S., Kim, J., Cheong, K., & Cho, G. (2006). Evaluating The Service Quality of Third party Logistics Service Providers using the Analytic Hierarchy Process. *Journal of Information Systems and Technology Management*, *3*(3), 261–270. doi:10.4301/S1807-17752006000300001

Srivastava, S. K. (2007). Green supply-chain management: A state-of-the-art literature review. *International Journal of Management Reviews*, *9*(1), 53–80. doi:10.1111/j.1468-2370.2007.00202.x

Stallings, W. (2014). Operating Systems Internal and Design Principles (8th ed.). Academic Press.

Sugeno, M. (1985). An introductory survey of fuzzy control. *Information Sciences*, *36*(1-2), 59–83. doi:10.1016/0020-0255(85)90026-X

Sujono, S., & Lashkari, R. S. (2005). Modeling the operation allocation and material handling system selection in FMS design: A genetic algorithm-based approach.*Proceedings of the 2005 International Conference on Simulation and Modeling*.

Szmidt, E., & Kacprzyk, J. (2000). Distance between intuitionistic fuzzy sets. *Fuzzy Sets and Systems*, *114*(3), 505–518. doi:10.1016/S0165-0114(98)00244-9

Szmidt, E., & Kacprzyk, J. (2001). Entropy for intuitionistic fuzzy sets. *Fuzzy Sets and Systems*, *118*(3), 467–477. doi:10.1016/S0165-0114(98)00402-3

Szmidt, E., & Kacprzyk, J. (2002). Using intuitionistic fuzzy sets in group decision making. *Control and Cybernetics*, *31*, 1037–1053.

Szmidt, E., & Kacprzyk, J. (2003). A consensus-reaching process under intuitionistic fuzzy preference relations. *International Journal of Intelligent Systems*, *18*(7), 837–852. doi:10.1002/int.10119

Tanabe, P., Travers, D., Gilboy, N., Rosenau, A., Sierzega, G., Rupp, V., & Adams, J. G. et al. (2005). Refining emergency severity index (ESI) triage criteria. *Academic Emergency Medicine*, *12*(6), 497–501. doi:10.1111/j.1553-2712.2005.tb00888.x PMID:15930399

Tanaka, K. (1997). *An Introduction to Fuzzy Logic for Practical Applications*. New York: Springer-Verlag.

Tanaka, K. (1997). *An Introduction to Fuzzy Logic for practical Applications*. Springer-Verlag New York.

Tanenbaum, S. A., & Woodfhull, S. A. (2008). Operating Systems Design and Implementation (2nd ed.). Academic Press.

Tansel, Y., Yurdakul, M., & Dengiz, B. (2013). Development of a decision support system for robot selection. *Robotics and Computer-integrated Manufacturing*, *29*(4), 142–157. doi:10.1016/j.rcim.2012.11.008

Tao, Z., & Bruno, F. (2009). Modeling and control design of hydrogen production process for an active hydrogen/wind hybrid power system. *International Journal of Hydrogen Energy*, *34*(1), 21–30. doi:10.1016/j.ijhydene.2008.10.030

Testi, A., Tanfani, E., Valente, R., Ansaldo, G. L., & Torre, G. C. (2008). Prioritizing surgical waiting lists. *Journal of Evaluation in Clinical Practice*, *14*(1), 59–64. doi:10.1111/j.1365-2753.2007.00794.x PMID:18211645

Thorani, Y. L. P., Rao, P. B., & Shankar, N. R. (2012). Ordering generalized trapezoidal fuzzy numbers using orthocenter of centroids. *International Journal of Algebra*, *6*(22), 1069–1085.

Thorani, Y. L. P., Rao, P. P. B., & Shankar, N. R. (2012). Ordering generalized trapezoidal fuzzy numbers. *International Journal of Contemporary Mathematical Sciences*, *7*(12), 555–573.

Timothy, J. R. (2010). *Fuzzy Logic with Engineering Applications* (3rd ed.). John Wiley & Sons.

Toke, L. K., Gupta, R. C., & Dandekar, M. (2010). Green supply chain management; critical research and practices. *Proceedings of the 2010 International Conference on Industrial Engineering and Operations Management*, (pp. 9-10).

Tong, R. M. (1978). Synthesis of fuzzy models for industrial processes-some recent results. *International Journal of General Systems*, *4*(3), 143–162. doi:10.1080/03081077808960680

Tran, L., & Duckstein, L. (2002). Comparison of fuzzy numbers using a fuzzy distance measure. *Fuzzy Sets and Systems*, *130*(3), 331–341. doi:10.1016/S0165-0114(01)00195-6

Tsai, J. P., Wang, S. Y., Cheng, H. Y., & Kao, Y. C. (2010). Multi-criteria decision making method for selection of machine tool.*International Symposium on Computer, Communication, Control and Automation*, (pp. 49-52). doi:10.1109/3CA.2010.5533376

Tzeng, G. H., Chiang, C. H., & Li, C. W. (2007). Evaluating intertwined effects in e-learning programs: A novel hybrid MCDM model based on factor analysis and DEMATEL. *Expert Systems with Applications*, *32*(4), 1028–1044. doi:10.1016/j.eswa.2006.02.004

Urea, M. R., Chiclana, F., Morente Molinera, J. A., & Herrera Viedma, E. (2015). Managing incomplete preference relations in decision making: A review and future trends. *Information Sciences*, *302*, 14–32. doi:10.1016/j.ins.2014.12.061

Vahdani, B., Mousavi, S. M., & Tavakkoli, M. R. (2011). Group decision making based on novel fuzzy modified TOPSIS method. *Applied Mathematical Modelling*, *35*(9), 4257–4269. doi:10.1016/j.apm.2011.02.040

Valente, R., Testi, A., Tanfani, E., Fato, M., Porro, I., Santo, M., & Ansaldo, G. et al. (2009). A model to prioritize access to elective surgery on the basis of clinical urgency and waiting time. *BMC Health Services Research*, *9*(1), 1. doi:10.1186/1472-6963-9-1 PMID:19118494

Vinay, V. P., Govindan, K., & Sasikumar, P. (2009). Conceptual study on 3PL/4PL/ new trends for service industry. *International Journal of Services Technology and Management*, *12*(1), 3–22. doi:10.1504/IJSTM.2009.025033

Voxman, W. (1998). Some remarks on distances between fuzzy numbers. *Fuzzy Sets and Systems*, *100*(1-3), 353–365. doi:10.1016/S0165-0114(97)00090-0

Wang, X., & Peng, Z. (2012). *Method of moments for estimating uncertainty distributions*. Retrieved from http://www.orsc.edu.cn/online/ 100408.pdf

Wang, G., & Li, X. (1999). Correlation of information energy of interval valued fuzzy numbers. *Fuzzy Sets and Systems*, *103*(1), 169–175. doi:10.1016/S0165-0114(97)00303-5

Wang, J. Q., Nie, R. R., Zhang, H. Y., & Chen, X. (2013). Intuitionistic fuzzy multi-criteria decision-making method based on evidential reasoning. *Applied Soft Computing*, *13*(4), 1823–1831. doi:10.1016/j.asoc.2012.12.019

Wang, Q., Zantow, K., Lai, F., & Wang, X. (2006). Strategic postures of third party Logistics providers in Mainland China. *International Journal of Physical Distribution & Logistics Management*, *36*(10), 793–819. doi:10.1108/09600030610714607

Wang, W., & Xin, X. (2005). Distance measure between intuitionistic fuzzy sets. *Pattern Recognition Letters*, *26*(13), 2063–2069. doi:10.1016/j.patrec.2005.03.018

Wang, X. F. (2008). Fuzzy number intuitionistic fuzzy geometric aggregation operators and their application to decision making. *Control and Decision*, *23*, 68–73.

Wang, X., Gao, Z., & Guo, H. (2012). Delphi method for estimating uncertainty distributions. *An International Interdisciplinary Journal*, *12*(2), 449–460.

Wang, X., Gao, Z., & Guo, H. (2012a). Uncertain hypothesis testing for two experts empirical data. *Mathematical and Computer Modelling*, *55*(3-4), 1478–1482. doi:10.1016/j.mcm.2011.10.039

Wan, S. P., & Li, D.-F. (2013). Fuzzy LINMAP approach to heterogeneous MADM considering comparisons of alternatives with hesitation degrees. *Omega: The International Journal of management. Science*, *41*(6), 925–940.

Wan, S. P., & Li, D.-F. (2013). Possibility mean and variance based method for multi-attribute decision making with triangular intuitionistic fuzzy numbers. *Journal of Intelligent and Fuzzy Systems*, *24*(4), 743–754.

Wan, S. P., & Li, D.-F. (2014). Atanassovs intuitionistic fuzzy programming method for heterogeneous multiattribute group decision making with Atanassovs intuitionistic fuzzy truth degrees. *IEEE Transactions on Fuzzy Systems*, *22*(2), 300–312. doi:10.1109/TFUZZ.2013.2253107

Wan, S. P., & Li, D.-F. (2015). Fuzzy mathematical programming approach to heterogeneous multiattribute decision-making with interval-valued intuitionistic fuzzy truth degrees. *Information Sciences*, *325*, 484–503. doi:10.1016/j.ins.2015.07.014

Wan, S. P., Li, D.-F., & Rui, Z. F. (2013). Possibility mean, variance and covariance of triangular intuitionistic fuzzy numbers. *Journal of Intelligent and Fuzzy Systems*, *24*(4), 847–858.

Webster, D. B., & Reed, R. (1971). A material handling system selection model. *IIE Transactions*, *3*(1), 13–21.

Wei, C. P., Wang, P., & Zhang, Y. Z. (2011). Entropy, similarity measure of interval-valued intuitionistic fuzzy sets and their applications. *Information Sciences*, *181*(19), 4273–4286. doi:10.1016/j.ins.2011.06.001

Wei, G. (2010). Some arithmetic aggregation operators with intuitionistic trapezoidal fuzzy numbers and their application to group decision making. *Journal of Computers*, *5*(3), 345–351. doi:10.4304/jcp.5.3.345-351

Wei, G. W., Wang, H. J., & Lin, R. (2011). Application of correlation coefficient to interval-valued intuitionistic fuzzy multiple attribute decision-making with incomplete weight information. *Knowledge and Information Systems*, *26*(2), 337–349. doi:10.1007/s10115-009-0276-1

Williams, I., Robinson, S., & Dickinson, H. (2012). *Rationing in health care: the theory and practice of priority setting*. Bristol: The Policy Press.

Wong, J. T. (2012). DSS for 3PL provider selection in global supply chain: Combining the multi-objective optimization model with experts opinions. *Journal of Intelligent Manufacturing*, *23*(3), 599–614. doi:10.1007/s10845-010-0398-z

Xia & Feng. (2005). Fuzzy logic based feedback scheduler for embedded control systems. In *Advances in Intelligent Computing*. Springer Berlin Heidelberg.

Xu, R., & Li, C. (2001). Multidimensional least-aquares fitting with a fuzzy model. *Fuzzy Sets and Systems*, *119*(2), 215–223. doi:10.1016/S0165-0114(98)00350-9

Xu, Z. S. (2005). An Overview of Methods for Determining OWA Weights. *International Journal of Intelligent Systems*, *20*(8), 843–865. doi:10.1002/int.20097

Xu, Z. S. (2006). Induced uncertain linguistic OWA operators applied to group decision making. *Information Fusion*, *7*(2), 231–238. doi:10.1016/j.inffus.2004.06.005

Xu, Z. S. (2007). Methods for aggregating interval-valued intuitionistic fuzzy information and their application to decision making. *Control and Decision*, *2*, 215–219.

Xu, Z. S. (2007a). Models for multiple attribute decision-making with intuitionistic fuzzy information. *International Journal of Uncertainty, Fuzziness and Knowledge-based Systems*, *15*(3), 285–297. doi:10.1142/S0218488507004686

Xu, Z. S. (2007b). Intuitionistic fuzzy aggregation operators. *IEEE Transactions on Fuzzy Systems*, *15*(6), 1179–1187. doi:10.1109/TFUZZ.2006.890678

Xu, Z. S. (2007c). Multi-person multi-attribute decision making models under intuitionistic fuzzy environment. *Fuzzy Optimization and Decision Making*, *6*(3), 221–236. doi:10.1007/s10700-007-9009-7

Xu, Z. S. (2007d). Intuitionistic preference relations and their application in group decision making. *Information Science*, *177*(11), 2363–2379. doi:10.1016/j.ins.2006.12.019

Xu, Z. S. (2010). A method based on distance measure for intervalvalued intuitionistic fuzzy group decision making. *Information Sciences*, *180*(1), 181–190. doi:10.1016/j.ins.2009.09.005

Xu, Z. S., & Chen, J. (2007). On geometric aggregation over interval-valued intuitionistic fuzzy information. In *Fourth International Conference on Fuzzy Systems and Knowledge Discovery (FSKD 2007)*, (vol. 2, pp. 466–471). doi:10.1109/FSKD.2007.427

Xu, Z. S., & Da, Q. L. (2003). An overview of operators for aggregating the information. *International Journal of Intelligent Systems*, *18*(9), 953–969. doi:10.1002/int.10127

Xu, Z. S., & Yager, R. R. (2006). Some geometric aggregation operators based on Intuitionistic Fuzzy sets. *International Journal of General Systems*, *35*(4), 417–433. doi:10.1080/03081070600574353

Yager, R. R., & Filev, D. P. (1995). Defuzzification with Constraints. In Fuzzy Logic & its Application to Engineering. Information Sciences & Intelligent Systems. doi:10.1007/978-94-009-0125-4_15

Yager, R. R. (1981). A procedure for ordering fuzzy subsets of the unit interval. *Information Sciences*, *24*(2), 143–161. doi:10.1016/0020-0255(81)90017-7

Yager, R. R. (1988). On ordered weighted averaging aggregation operators in multi-criteria decision making. *IEEE Transactions on Systems, Man, and Cybernetics*, *18*(1), 183–190. doi:10.1109/21.87068

Yager, R. R. (1993). Families of OWA operators. *Fuzzy Sets and Systems*, *59*(2), 125–148. doi:10.1016/0165-0114(93)90194-M

Yager, R. R. (2004). Generalized OWA aggregation operators. *Fuzzy Optimization and Decision Making*, *3*(1), 93–107. doi:10.1023/B:FODM.0000013074.68765.97

Yager, R. R. (2007). Centered OWA operators. *Soft Computing*, *11*(7), 631–639. doi:10.1007/s00500-006-0125-z

Yager, R. R., & Filev, D. P. (1999). Induced ordered weighted averaging operators. *IEEE Transactions on Systems, Man, and Cybernetics. Part B, Cybernetics*, *29*(2), 141–150. doi:10.1109/3477.752789 PMID:18252288

Yager, R. R., & Filev, D. P. (2006). On the issue of defuzzification and selection based on a fuzzy set. *Fuzzy Sets and Systems*, *55*(3), 255–272. doi:10.1016/0165-0114(93)90252-D

Yang, J., Fei, W., & Li, D.-F. (2016b). Nonlinear programming approach to solve bi-matrix games with payoffs represented by I-fuzzy numbers. *International Journal of Fuzzy Systems*, *18*(3), 492–503. doi:10.1007/s40815-015-0052-1

Yang, J., Li, D.-F., & Lai, L.-B. (2016a). Parameterized bilinear programming methodology for solving triangular intuitionistic fuzzy number bi-matrix games. *International Journal of Fuzzy Systems*, *31*(1), 115–125. doi:10.3233/IFS-162125

Yang, M., Fry, M., & Raikhelkar, J. (2013). A Model to Create an Efficient and Equitable Admission Policy for Patients Arriving to the Cardiothoracic ICU. *The Clinical Investigator*, *41*(2), 414–422. PMID:23263573

Yan, J., Chaudhry, P. E., & Chaudhry, S. S. (2003). A model of a decision support system based on case-based reasoning for third-party Logistics evaluation. *Expert Systems: International Journal of Knowledge Engineering and Neural Networks*, *20*(4), 196–207. doi:10.1111/1468-0394.00244

Ye, J. (2009). Multicriteria fuzzy decision-making method based on a novel accuracy function under interval-valued intuitionistic fuzzy environment. *Expert Systems with Application, 36*, 6899-6902.

Yu, D. J. (2013). Decision making based on generalized geometric operator under interval-valued intuitionistic fuzzy environment. *Journal of Intelligent & Fuzzy Systems*, *25*, 471–480.

Yu, D., Li, D.-F. , & Merigo, J. M. (2015). Dual hesitant fuzzy group decision making method and its application to supplier selection. *International Journal of Machine Learning and Cybernetics,* 1-13.

Yu, C. (1993). Correlation of fuzzy numbers. *Fuzzy Sets and Systems*, *55*(3), 303–307. doi:10.1016/0165-0114(93)90256-H

Yu, D.-J., & Li, D.-F. (2014). Dual hesitant fuzzy multi-criteria decision making and its application to teaching quality assessment. *Journal of Intelligent and Fuzzy Systems*, *27*(4), 1679–1688.

Yue, Z. L., & Jia, Y. Y. (2013). An application of soft computing technique in group decision making under interval-valued intuitionistic fuzzy environment. *Applied Soft Computing*, *13*, 308–322.

Yu, P. J., & Zhang, J. (2014). An algorithmic method to extend TOPSIS for multiple attribute decision making under intuitionistic fuzzy setting. *Journal of Intelligent and Fuzzy Systems*, *26*(5), 2315–2322.

Yu, P. L. (1973). A Class of Solutions for Group Decision Problems. *Management Science*, *19*(8), 936–946. doi:10.1287/mnsc.19.8.936

Yusuf, Y. Y., Sarhadi, M., & Gunasekaran, A. (1999). Agile manufacturing: The drivers, concepts and attributes. *International Journal of Production Economics*, *62*(1-2), 33–43. doi:10.1016/S0925-5273(98)00219-9

Zadeh, L. A. (1965). Fuzzy Sets. *Information and Control*, *8*(3), 338–353. doi:10.1016/S0019-9958(65)90241-X

Zadeh, L. A. (1975). The concept of a linguistic variable and its application to approximate reasoning- II. *Information Sciences*, *8*(4), 301–357. doi:10.1016/0020-0255(75)90046-8

Zadeh, L. A. (1975). The Concept of a Linguistic Variable and its Application to Approximate Reasoning. *Information Sciences*, *8*(3), 199–249. doi:10.1016/0020-0255(75)90036-5

Zadeh, L. A. (1978). Fuzzy sets as a basis for a theory of possibility. *Fuzzy Sets and Systems*, *1*(1), 3–28. doi:10.1016/0165-0114(78)90029-5

Zadeh, L. A. (1984). Making computers think like people. *IEEE Spectrum*, *8*(8), 26–32. doi:10.1109/MSPEC.1984.6370431

Zadeh, L. A. (2008). Is there a need for Fuzzy Logic. *Information Science*, *178*(13), 2751–2779. doi:10.1016/j.ins.2008.02.012

Zaim, S., Sevkli, M., & Tarim, M. (2003). Fuzzy Analytic Hierarchy Based Approach for Supplier Selection. *Journal of Euromarketing*, *12*(3/4), 147–176. doi:10.1300/J037v12n03_09

Zeng, S., Chen, J., & Li, X. (2016). A Hybrid Method for Pythagorean Fuzzy Multiple-Criteria Decision Making. *International Journal of Information Technology & Decision Making*, *15*(2), 403–422. doi:10.1142/S0219622016500012

Zeng, S., & Xiao, Y. (2016). TOPSIS method for intuitionistic fuzzy multiple-criteria decision making and its application to investment selection. *Kybernetes*, *45*(2), 282–296. doi:10.1108/K-04-2015-0093

Zeng, W., & Li, H. (2007). Correlation Coefficient of Intuitionistic Fuzzy sets. *Journal of Industrial Engineering International*, *3*, 33–40.

Zhang, Z., & Chu, X. (2009). Fuzzy group decision-making for multi-format and multi-granularity linguistic judgments in quality function deployment. *International Journal of Expert Systems with Applications*, *36*(5), 9150-9158.

Zhang, H., Li, X., Liu, W., Li, B., & Zhang, Z. (2004). An Application of the AHP in 3PL Vendor Selection of a 4PL System. *IEEE International Conference on Systems, Man and Cybernetics*, (pp. 1255-1260). doi:10.1109/ICSMC.2004.1399797

Zhang, Q. S., Jiang, S. Y., Jia, B. G., & Luo, S. H. (2010). Some information measures for interval-valued intuitionistic fuzzy sets. *Information Sciences*, *180*(24), 5130–5145. doi:10.1016/j.ins.2010.08.038

Zhang, X. L., & Xu, Z. S. (2015). Soft computing based on maximizing consensus and fuzzy TOPSIS approach to interval-valued intuitionistic fuzzy group decision making. *Applied Soft Computing*, *26*, 42–56. doi:10.1016/j.asoc.2014.08.073

Zhang, X., & Liu, P. D. (2010). Method for Aggregating Triangular Fuzzy Intuitionistic Fuzzy Information and its Application to Decision making. *Technological and Economic Development of Economy. Baltic Journal of Sustainability*, *16*(2), 280–290.

Zhou, G., Min, H., Xu, C., & Cao, Z. (2008). Evaluating the comparative efficiency of Chinese third-party Logistics providers using data envelopment analysis. *International Journal of Physical Distribution & Logistics Management*, *38*(4), 262–279. doi:10.1108/09600030810875373

Zimmerman, J. (2001). *Fuzzy set theory And Its Application*. Norwell, MA: Kluwer Academic Publishers. doi:10.1007/978-94-010-0646-0

Zimmermann, H. J. (2010). Fuzzy set theory. John Wiley & Sons Inc.

Zyoud, S. H., Kaufmann, L. G., Shaheen, H., Samhan, S., & Hanusch, D. F. (2016). A framework for water loss management in developing countries under fuzzy environment: Integration of Fuzzy AHP with Fuzzy TOPSIS. *Expert Systems with Applications*, *61*, 86–105. doi:10.1016/j.eswa.2016.05.016

About the Contributors

Deng-Feng Li was born in 1965. He received the B.Sc. and M.Sc. degrees in applied mathematics from the National University of Defense Technology, Changsha, China, in 1987 and 1990, respectively, and the Ph.D. degree in system science and optimization from the Dalian University of Technology, Dalian, China, in 1995. From 2003 to 2004, he was a Visiting Scholar with the School of Management, University of Manchester Institute of Science and Technology, Manchester, U. K. He is currently a "Changjiang Scholar" Professor with the School of Economics and Management, Fuzhou University, Fuzhou, China. He has authored or coauthored more than 300 journal papers and seven monographs. He has coedited one proceeding of the international conference and won 23 academic achievements and awards such as Chinese State Natural Science Award and 2013 IEEE Computational Intelligence Society IEEE Transactions on Fuzzy Systems Outstanding paper award. His current research interests include game theory, fuzzy decision analysis, group decision making, fuzzy game theory, supply chain, fuzzy sets and system analysis, fuzzy optimization, and differential game. He is Editor-in-chief (or Associate Editors) and Editors of several international journals.

* * *

Samira Abbasgholizadeh Rahimi is completing her postdoctoral fellowship at Department of Family and Emergency Medicine, Laval University, Canada working on development and assessment of an interdisciplinary shared decision making continuous professional training program for health care providers. She achieved her BEng. in 2011, and her PhD in 2016 at Laval University. Her PhD studies had been supported by graduate funding from Natural Sciences and Engineering Research Council of Canada (NSERC), and doctorate scholarship from IRDPQ Quebec Institute for Physical Rehabilitation. Dr. Rahimi is founder and president of Québec Women in Engineering association. She received numerous awards including the 2015 Canada IEEE Women in Engineering Award. She has developed and/or implemented decision support tools for delivery of health services in orthopaedic surgery, general surgery, organ transplantation, and in rehabilitation, along with tools for diagnosis of arthritis. She is currently leading different international projects in medical decision making with top researchers and medical experts in the USA, Europe, the Middle East and Canada. Her research interests include: Medical decision making, Patient Centered Care, Decision support tools, Quality Improvement of clinical decision making, Implementation and evaluation of patient-centered eHealth Systems.

E. C. Amirtharaj works as an Associate Professor of Mathematics in Bishop Heber College. Has teaching experience of more than two decades. Research interest is Fuzzy Optimization.

Harendra Kumar is an Assistant Professor in the Department of Mathematics & Statistics, Gurukula Kangari University Haridwar-Uttarakhand (India). He obtained his Master of Science in 2003 from I.I.T Roorkee and Ph.D. from Gurukula Kangari University in 2008. He has published several research papers in National and International Journals in the field of Operation Research & Distributed Real Time System. He has more than 8 years teaching experience and has written two books for undergraduate students. He has qualified NET (UGCJRF)- Dec.2003, NET(CSIR-JRF)-June 2005 and GATE-2004.

Supriya Raheja is currently Senior Assistant Professor in Department of CSE & IT, School of Engineering & Technology, NCU. She has 8+ years of extensive teaching experience. She is PhD in Computer Science & Engg. Her areas of interest include Object Oriented Programming, Rapid Application Development, Operating System, Networks, Fuzzy Logic and Vague Logic. She had guided 10 M.Tech Projects and more than 30 B.Tech Projects. She has Published 20 Papers in International Journals with good indexing and in IEEE International Conferences. She has one book chapter, She is the reviewer and assistant editor of various international journals. She is the Faculty Coordinator for the FOSS.

B. Ramya is a part-time research scholar working for her Ph.D. degree in the Department of Statistics, University of Madras, Chennai, India since 2012. She got her Masters degree from the University of Madras and worked as a Junior Research Fellow in a Department of Science and Technology funded Project. Her research interest includes Fuzzy, Rough and Uncertainty theories.

John Robinson P. works as an Assistant professor, Department of Mathematics, Bishop Heber College, India. Completed Ph.D. in Mathematics. Have 7 years of teaching experience. Interested in the area of fuzzy set theory and decision making.

Anoop Kumar Sahu is an assistant professor in J.K.I.E, Bilaspur, C.G, India. He earned his BE Degree from Central University, Bilaspur, C.G. He earned his Master Degree in Manufacturing Engineering from N.I.F.F.T, Ranchi, India. His PhD work is on supply chain performance appraisement and benchmarking for manufacturing industries from NIT, Rourkela, India.His research realm deals with Multi-Criteria Decision Making (MCDM) applications in production and operation management dilemmas, Supply Chain Performance Measurement (SCPM), Reverse Logistic (RL), Multi-Objective Optimization Approaches (MOOA), Industrial Inspection, Labour and Workforce Utilization in Industries.

Atul Kumar Sahu is faculty in Department of Industrial and Production Engineering, Guru Ghasidas (Central) Vishwavidyalaya, Bilaspur, C.G, India. His research realm deal with Multi-Criteria Decision Making (MCDM) applications in production and operation management dilemmas, Supply Chain Performance Measurement (SCPM), Reverse Logistic (RL), Multi Objective Optimization Approaches (MOOA), Industrial Inspection, Labour and Workforce Utilization in Industries. He has published various peer reviewed journals and presented various National cum International conferences.

Nitin Kumar Sahu, faculty in Department of Industrial and Production Engineering, Guru Ghasidas (Central) Vishwavidyalaya, Bilaspur, C.G, India. His research realm deal with Multi-Criteria Decision Making (MCDM) applications in production and operation management dilemmas, Supply Chain Performance Measurement (SCPM), Reverse Logistic (RL), Multi Objective Optimization Approaches (MOOA), Industrial Inspection, Labour and Workforce Utilization in Industries. He has published various peer reviewed journals and presented various National cum International conferences.

S. Sampath is currently holding the position of Professor of Statistics in the University of Madras, Chennai,India. He got his Ph.D. degree from the University of Madras. He has served in Loyola College, Chennai and Birla Institute of Science and Technology, Dubai, UAE before joining University of Madras in 2008. He has 34 years experience in teaching and research. He has guided many scholars for their research degrees and currently four are working for their Ph.D. degree under his guidance. His main areas of research are Inference for Finite Populations, Credibility Theory, Classical Statistical Inference, and Data Mining. He has to his credit nearly 50 research articles in highly rated research journals. His book titled "Sampling Theory and Methods" published by CRC Press and Narosha Publishing house is a popular book in the academic circles.

Gao-Feng Yu was born in 1986. He received the M.Sc. degree in system theory from the Kunming University of Science and Technology, in 2012. He has authored or coauthored more than 20 journal papers. He has been leading/participate in more than six research projects. His current research interests include fuzzy decision analysis, group decision making and fuzzy game theory.

Index

A

B

C

D

E

F

Encyclopedia of Information Science and Technology, Third Edition (10 Vols.)

Mehdi Khosrow-Pour, D.B.A. (Information Resources Management Association, USA)
ISBN: 978-1-4666-5888-2; **EISBN:** 978-1-4666-5889-9; © 2015; 10,384 pages.

The **Encyclopedia of Information Science and Technology, Third Edition** is a 10-volume compilation of authoritative, previously unpublished research-based articles contributed by thousands of researchers and experts from all over the world. This discipline-defining encyclopedia will serve research needs in numerous fields that are affected by the rapid pace and substantial impact of technological change. With an emphasis on modern issues and the presentation of potential opportunities, prospective solutions, and future directions in the field, it is a relevant and essential addition to any academic library's reference collection.

Take An Extra 30% Off[1]

[1] 30% discount offer cannot be combined with any other discount and is only valid on purchases made directly through IGI Global's Online Bookstore (www.igi-global.com/books), not intended for use by distributors or wholesalers. Offer expires December 31, 2016.

Free Lifetime E-Access with Print Purchase

Take 30% Off Retail Price:

Hardcover with Free E-Access:[2] $2,765
~~List Price: $3,950~~

E-Access with Free Hardcover:[2] $2,765
~~List Price: $3,950~~

Recommend this Title to Your Institution's Library: www.igi-global.com/books

[2] IGI Global now offers the exclusive opportunity to receive free lifetime e-access with the purchase of the publication in print, or purchase any e-access publication and receive a free print copy of the publication. You choose the format that best suits your needs. This offer is only valid on purchases made directly through IGI Global's Online Bookstore and not intended for use by book distributors or wholesalers. Shipping fees will be applied for hardcover purchases during checkout if this option is selected.

The lifetime of a publication refers to its status as the current edition. Should a new edition of any given publication become available, access will not be extended on the new edition and will only be available for the purchased publication. If a new edition becomes available, you will not lose access, but you would no longer receive new content for that publication (i.e. updates). Free Lifetime E-Access is only available to single institutions that purchase printed publications through IGI Global. Sharing the Free Lifetime E-Access is prohibited and will result in the termination of e-access.

www.ingramcontent.com/pod-product-compliance
Lightning Source LLC
LaVergne TN
LVHW060400300926
849162LV00006B/1315